Developments in Geotechnical Engineering, 56

Subsidence

Occurrence, Prediction and Control

Further titles in this series:
Volumes 2, 3, 5, 6, 7, 9, 10, 13, 16 and 26 are out of print

1. G. SANGLERAT — THE PENETROMETER AND SOIL EXPLORATION
4. R. SILVESTER — COASTAL ENGINEERING, 1 and 2
8. L.N. PERSEN — ROCK DYNAMICS AND GEOPHYSICAL EXPLORATION
   Introduction to Stress Waves in Rocks
11. H.K. GUPTA AND B.K. RASTOGI — DAMS AND EARTHQUAKES
12. F.H. CHEN — FOUNDATIONS ON EXPANSIVE SOILS
14. B. VOIGHT (Editor) — ROCKSLIDES AND AVALANCHES, 1 and 2
15. C. LOMNITZ AND E. ROSENBLUETH (Editors) — SEISMIC RISK AND ENGINEERING DECISIONS
17. A.P.S. SELVADURAI — ELASTIC ANALYSIS OF SOIL-FOUNDATION INTERACTION
18. J. FEDA — STRESS IN SUBSOIL AND METHODS OF FINAL SETTLEMENT CALCULATION
19. Á. KÉZDI — STABILIZED EARTH ROADS
20. E.W. BRAND AND R.P. BRENNER (Editors) — SOFT-CLAY ENGINEERING
21. A. MYSLIVE AND Z. KYSELA — THE BEARING CAPACITY OF BUILDING FOUNDATIONS
22. R.N. CHOWDHURY — SLOPE ANALYSIS
23. P. BRUUN — STABILITY OF TIDAL INLETS
   Theory and Engineering
24. Z. BAŽANT — METHODS OF FOUNDATION ENGINEERING
25. Á. KÉZDI — SOIL PHYSICS
   Selected Topics
27. D. STEPHENSON — ROCKFILL IN HYDRAULIC ENGINEERING
28. P.E. FRIVIK, N. JANBU, R. SAETERSDAL AND L.I. FINBORUD (Editors) — GROUND FREEZING 1980
29. P. PETER — CANAL AND RIVER LEVÉES
30. J. FEDA — MECHANICS OF PARTICULATE MATERIALS
   The Principles
31. Q. ZÁRUBA AND V. MENCL — LANDSLIDES AND THEIR CONTROL
   Second completely revised edition
32. I.W. FARMER (Editor) — STRATA MECHANICS
33. L. HOBST AND J. ZAJÍC — ANCHORING IN ROCK AND SOIL
   Second completely revised edition
34. G. SANGLERAT, G. OLIVARI AND B. CAMBOU — PRACTICAL PROBLEMS IN SOIL MECHANICS AND FOUNDATION ENGINEERING, 1 and 2
35. L. RÉTHÁTI — GROUNDWATER IN CIVIL ENGINEERING
36. S.S. VYALOV — RHEOLOGICAL FUNDAMENTALS OF SOIL MECHANICS
37. P. BRUUN (Editor) — DESIGN AND CONSTRUCTION OF MOUNDS FOR BREAKWATERS AND COASTAL PROTECTION
38. W.F. CHEN AND G.Y. BALADI — SOIL PLASTICITY
   Theory and Implementation
39. E.T. HANRAHAN — THE GEOTECTONICS OF REAL MATERIALS: THE ϵ_g, ϵ_k METHOD
40. J. ALDORF AND K. EXNER — MINE OPENINGS
   Stability and Support
41. J.E. GILLOTT — CLAY IN ENGINEERING GEOLOGY
42. A.S. CAKMAK (Editor) — SOIL DYNAMICS AND LIQUEFACTION
42. A.S. CAKMAK (Editor) — SOIL-STRUCTURE INTERACTION
44. A.S. CAKMAK (Editor) — GROUND MOTION AND ENGINEERING SEISMOLOGY
45. A.S. CAKMAK (Editor) — STRUCTURES, UNDERGROUND STRUCTURES, DAMS, AND STOCHASTIC METHODS
46. L. RÉTHÁTI — PROBABILISTIC SOLUTIONS IN GEOTECTONICS
47. B.M. DAS — THEORETICAL FOUNDATION ENGINEERING
48. W. DERSKI, R. IZBICKI, I. KISIEL AND Z. MRÓZ — ROCK AND SOIL MECHANICS
49. T. ARIMAN, M. HAMADA, A.C. SINGHAL, M.A. HAROUN AND A.S. CAKMAK (Editors) — RECENT ADVANCES IN LIFELINE EARTHQUAKE ENGINEERING
50. B.M. DAS — EARTH ANCHORS
51. K. THIEL — ROCK MECHANICS IN HYDROENGINEERING
52. W.F. CHEN AND X.L. LIU — LIMIT ANALYSIS IN SOIL MECHANICS
53. W.F. CHEN AND E. MIZUNO — NONLINEAR ANALYSIS IN SOIL MECHANICS
54. F.H. CHEN — FOUNDATIONS ON EXPANSIVE SOILS
55. J. VERFEL — ROCK GROUTING AND DIAPHRAGM WALL CONSTRUCTION

Developments in Geotechnical Engineering, 56

Subsidence

Occurrence, Prediction and Control

Barry N. Whittaker and **David J. Reddish**

Department of Mining Engineering, The University of Nottingham, University Park, Nottingham NG7 2RD (U.K.)

ELSEVIER

Amsterdam — Oxford — New York — Tokyo 1989

ELSEVIER SCIENCE PUBLISHERS B.V.
Sara Burgerhartstraat 25
P.O. Box 211, 1000 AE Amsterdam, The Netherlands

Distributors for the United States and Canada:

ELSEVIER SCIENCE PUBLISHING COMPANY INC.
655, Avenue of the Americas
New York, NY 10010, U.S.A.

ISBN 0-444-87274-4 (Vol. 56)
ISBN 0-444-41662-5 (Series)

Printed in The Netherlands

PREFACE

Surface subsidence is recognised as a problem in most countries, particularly those with significant mining and other underground resource extraction industries. This book addresses the problems relating to subsidence whether caused naturally, or arising from mining or other forms of underground extractive activity.

The character of subsidence can range from general lowering of the surface thereby often resulting in a saucer or trough shaped depression to that of a distinct hole with vertical sides and of appreciable depth and width. In many cases the occurrence of surface subsidence from mining operations can be predicted but in some instances there is a degree of uncertainty with the potential for some form of subsidence to occur later. The role played by time, although generally recognised as a factor, has not been always thoroughly appreciated in terms of the development or likely occurrence of subsidence at the surface.

Subsidence is usually associated with underground mining operations even though in several cases subsequent surface subsidence may not be of any significance. There are also many areas possessing a history of past mining, and the stability of the surface in respect of existing and future structures and particular land uses is frequently questioned. There is a pressing need for an improved understanding of the mechanics of mining subsidence particularly regarding predicting its occurrence and characteristics. The mining subsidence problem is compounded by the fact that its surface manifestation can take different forms and have far reaching effects on different types of structure at or near the surface. There is a world-wide wealth of knowledge on surface subsidence although such information is scattered widely and not necessarily readily accessible or expressed in a form allowing its ease of application to and comparison with related problems.

The main purpose of this book has been to bring together subsidence knowledge, experiences and research findings in many countries and rationalise such information especially in respect of its particular field of application. Emphasis has been given to collating field data on subsidence from different countries in order to make direct comparisons. Prediction of subsidence, particularly its occurrence and general characteristics has been seen as an important area where the book can contribute significantly in terms of reviewing available knowledge, methods, scope of application and orders of accuracy achieved. The book also examines methods of controlling subsidence and discusses the response of surface structures to and protection against subsidence.

The book refers extensively to research into mining subsidence carried out at the Department of Mining Engineering, University of Nottingham, and the following persons have made special contributions: Dr. T. R. C. Aston, Professor T. Atkinson, Dr. D. S. Berry, Dr. C. D. Breeds, Dr. D. J. Forrester, Dr. R. Firman, Dr. D. J. Fitzpatrick, P. Gaskell, the late Emeritus Professor H. J. King, J. C. Malcolm, J. N. van der Merwe, Dr. A. G. Pasamehmetoglu, Dr. J. H. Pye, G. Ren, Dr. C. H. Shadbolt, S. F. Smith, A. Szeki, and K. Wardell. Thanks are expressed to these and many other persons who have given valuable help to the authors during preparation of this book.

The authors in their review of current knowledge pertaining to mining subsidence have been greatly assisted by various publications and particularly from the British Coal Corporation, Elsevier Science Publications, Pergamon Press Ltd., and the transactions of the Institution of Mining Engineers and the Institution of Mining and Metallurgy to whom the authors express their gratitude.

The authors hope the book will create greater interest in achieving a better understanding of the mechanics of mining subsidence and its associated aspects and provide a basis for making improved assessments and predictions on subsidence related problems.

Barry N. Whittaker and David J. Reddish
Department of Mining Engineering
University of Nottingham
Nottingham.

CONTENTS

CHAPTER 1

NATURAL SUBSIDENCE AND INFLUENCE OF GEOLOGICAL PROCESSES

The occurrence of subsidence in nature is well-known. It can take the form of regional sinking of major parts of the earth's crust in one extreme whilst the localised occurrence of small surface depressions due to formation of solution cavities in limestone represents perhaps another extreme form of natural subsidence, although with widely differing time scales.

The geological cycle involving transition from earth movement to denudation followed by deposition, and thereafter further earth movement initiating the next cycle, is responsible for the occurrence of significant natural subsidence. The effect of the geological cycle in allowing selective and localised sedimentation to occur, is to produce disturbances in the loading of the earth's crust and thus lead to earth movements involving combinations of subsidence and uplift.

The meaning of subsidence

The term subsidence as applied to the earth's surface normally refers to a surface point sinking to a lower level, and can include a structure settling into the ground or the ground itself lowering and carrying the structure with it, or even a surface layer collapsing into an underground cavity. Subsidence usually refers to vertical displacement of a point, but also implies a measure of horizontal movement of adjacent points by virtue of the lateral shift of ground generated by the accompanying downward movement.

Soil properties influence their settlement (which is a form of subsidence) behaviour, particularly their consolidation characteristics in relation to the conditions and magnitude of loading. Additionally soils exhibit shrinkage characteristics following loss of moisture, and the obverse occurs on gaining moisture. Consequently, a certain degree of short-term settlement is to be expected with many new structures depending upon the type of foundations and the soils on which such foundations are built. Some structures experience long-term settlement effects owing to fluctuating moisture changes in the foundation soils, especially where the local water-table is highly sensitive to climatic conditions.

Subsidence of the surface commonly refers to *en masse* lowering of the ground rather than the localised effect of consolidation or shrinkage of soils. Such sinking of the surface may arise from regional geological reasons including tectonic or volcanic activities, or from removal of material from below the surface as with tunnelling or mining operations, or from localised natural causes such as occurs with swallow-holes or sink-holes in limestone country.

Some examples of natural subsidence involving relatively localised (as opposed to regional) effects are illustrated in Figure 1; movement of land masses in the form of gravitational slides is well-known particularly in mountainous regions, whilst earthquakes can trigger additional and sudden lowering of unconsolidated deposits; collapse of cavities created by groundwater solution of salt domes has resulted in subsidence of the surface similar to collapse of large natural voids in lava flows.

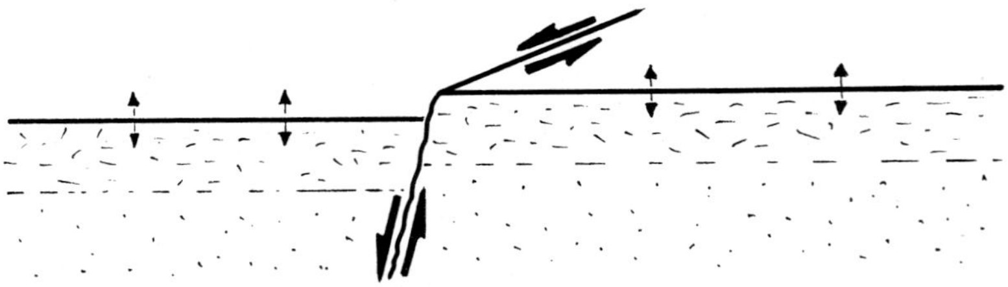

(a) Earthquake induced movement along fault planes

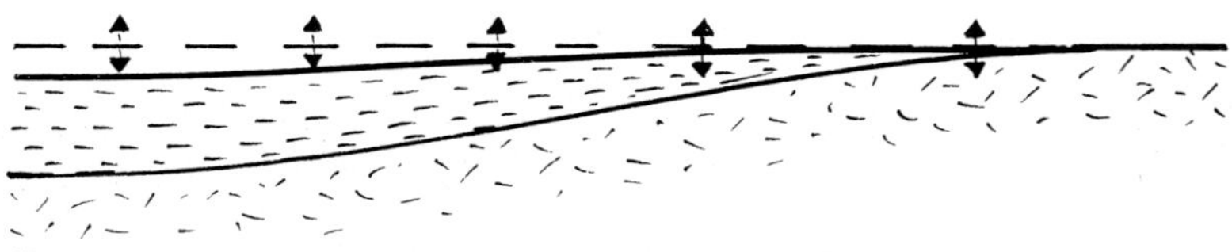

(b) Earthquake induced additional and sudden subsidence of deltaic and other unconsolidated deposits

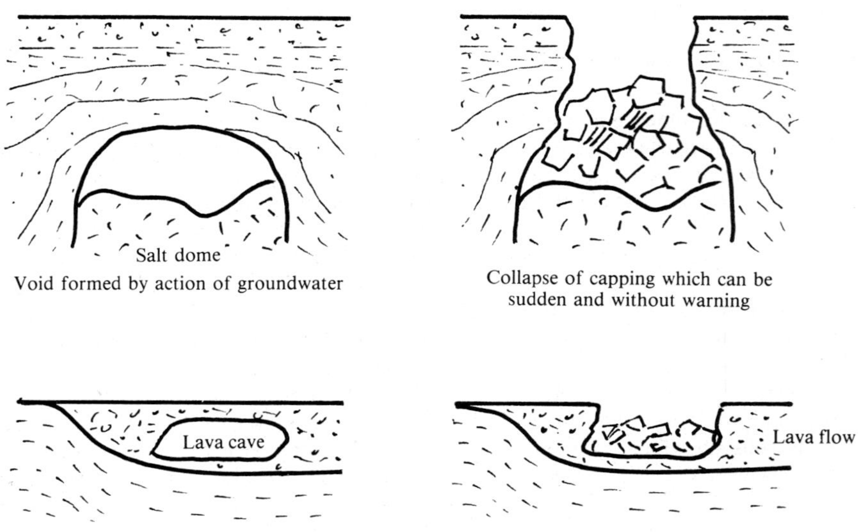

(c) Collapse of natural voids associated with salt domes and lava flows

(d) Natural instability of land masses resulting in gravitational slides which may creep or be triggered by other natural phenomena

Figure 1 Examples of natural subsidence

Geological cycle in relation to natural subsidence

Legget (1962) draws attention to the role played by the geological cycle in producing the structural features of the earth's crust as are evident today. The relevant processes involved with the cycle have a major bearing on natural subsidence. There is ample evidence around the world of major subsidence occurrences in the past. Natural subsidence continues to occur in various parts of the world either as a fairly continuous or intermittent phenomenon or as sudden manifestations triggered by earthquakes or collapse of capping over an underground cavity.

Subsidence of the earth's surface is associated with the geological cycle, namely those processes responsible for the structure of the earth's crust. The basic elements of the geological cycle are (1) denudation, (2) deposition, and (3) earth movement, which repeats itself.

The denudation, or weathering, element of the cycle includes the action of water resulting in decomposition of rocks by erosion either mechanically or chemically, temperature changes resulting in comminution of exposed rock surfaces, the effects of wind action especially in arid regions, the action and effect of glaciation, and that of wave action of the sea. The depositional element of the geological cycle mainly includes the action of water, and to a significantly less extent that of wind, in the creation of accumulated deposits. The effect of progressive deposition, accompanied by denudation, leads to disturbing the state of equilibrium of rock pressures in parts of the earth's crust and can induce local and regional vertical and consequential horizontal movements of the surface.

The earth movement element of the cycle is spasmodic in relation to the other two. The effects of major earth movement occurrences in the past are strongly evident world-wide. Volcanic activity occasionally triggers significant natural subsidence, although not on the same scale or frequency as earthquakes.

Legget (1962) when discussing ground subsidence refers to substantial vertical displacements triggered by earthquakes in New Zealand and the USA; in the first example at Karamea the town was situated on deltaic deposits and sank 60cm, whilst in the second example, this occurred following the 1934 earthquake where subsidence of 38cm was observed near to Kosmo, Utah.

Sedimentary basin subsidence

The mechanisms of sedimentary basin subsidence have been reviewed by Bott (1976) and he points out that such basins occur in various tectonic settings in relation to the proximity of plate boundaries. The two main types of basin subsidence encountered within plate interiors are described by Bott as broad regional subsidence dissociated from obvious faulting and narrow graben-type basins. The causation of subsidence on such major scales has been attributed to three main factors:

1. gravity loading by sediments and water producing flexure of the earth's crust;
2. thermal events inducing raising the temperature of crustal rocks producing consequential uplift by way of thermal expansion and later followed by erosion and subsidence due to subsequent cooling; and
3. deformational behaviour of the continental crust in relation to mainly tensional stresses.

The factors listed here are largely hypothetical and are still the subject of considerable debate amongst geologists. Subsidence of sedimentary basins is discussed by Bott (1976) and for further information the reader is advised to consult Tectonophysics, 36 (1976).

Figure 2 illustrates the principal features of subsidence arising from sedimentary basin, rift valley (graben) fault structures and plate boundary interaction mechanisms.

Dolines, swallow-holes, sink-holes and crown-holes

The occurrence of depressions at the surface overlying limestone country are referred to by four common terms. Dolines refer to simple closed depressions in karst (rough limestone country with water drainage through fractures, fissures, joints and other cavities giving rise to potential widening by solution of limestone due to chemical weathering) which have also manifested themselves as a natural surface subsidence phenomenon. The term 'doline' as is used by geomorphologists is all embracing and includes swallow-holes, sink-holes* and crown-holes.

A swallow-hole refers to a fissure system in limestone which intercepts the surface, in some cases as a well-defined open void and in others as a bottle-necked opening or simply as a widened fissure, and where such holes have free ingress and egress for running water.

A sink-hole in connection with natural subsidence in limestone country is another term for swallow-hole, namely a subsidence hole down which water drains away naturally. Swallow-holes usually refer to those holes where water continuously, or intermittently, flows into and through such structures; sink-holes commonly include those structures which now appear dry but have drainage potential.

Crown-holes refer to the natural phenomenon of a localised subsidence depression caused by progressive upward collapse from a subterranean cavity.

Holmes (1965) in his book 'Principles of Physical Geology' p.422 associates swallow-holes with sink-holes in connection with solution of limestone by the action of water giving rise to funnel-shaped holes at the surface; the process loosens locally the jointed block structure producing enlargement of such voids or drainage paths. He draws attention to the well-known swallow-hole of Ingleborough namely Gaping Ghyll whose shaft has a depth of 111m which connects with a chamber of 146m length and 33m height; the water traverses an underground labyrinth of flow paths to emerge several kilometres from its surface entry point. The potential for creating large caverns by solution of limestone is demostrated by the Carlsbad Cavern of New Mexico where the 'Big Cavern' is reported by Holmes to be almost 1220m long, with a height of 90m and width up to 180m.

Subsidence arising from development of subterranean voids

The occurrence of sink-holes and crown-holes in limestone country is probably the most common form of natural subsidence. This can be a frequent occurrence where carbonate rocks, rock salt and gypsum deposits lie relatively close to the surface. The formation of

*The term sink-hole has a broader meaning in connection with mining subsidence, and refers to the appearance of a sharply defined subsidence funnel-shaped depression or shaft-like hole at the surface which may or may not have free drainage potential owing to possible plugging of its collapse chimney by clay puddling action, and consequently embraces the meaning of crown-holes also.

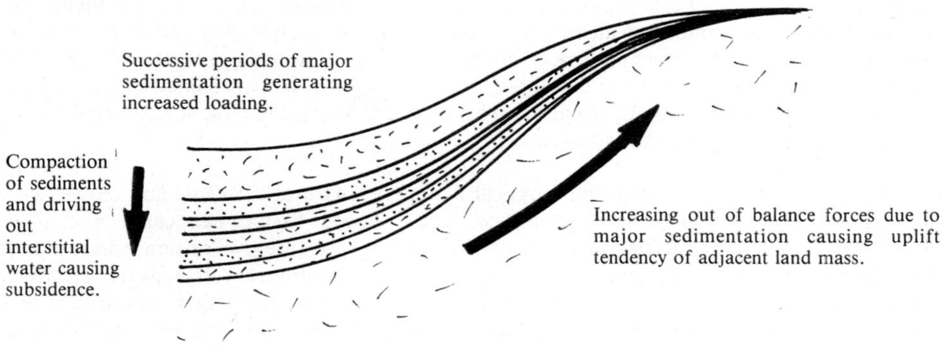

(a) Subsidence of sedimentation basin

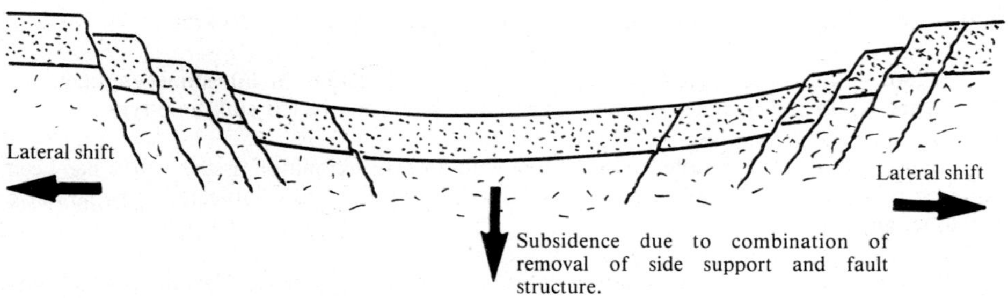

(b) Faulting associated with rift valley (graben) and lateral shift earth movement producing subsidence.

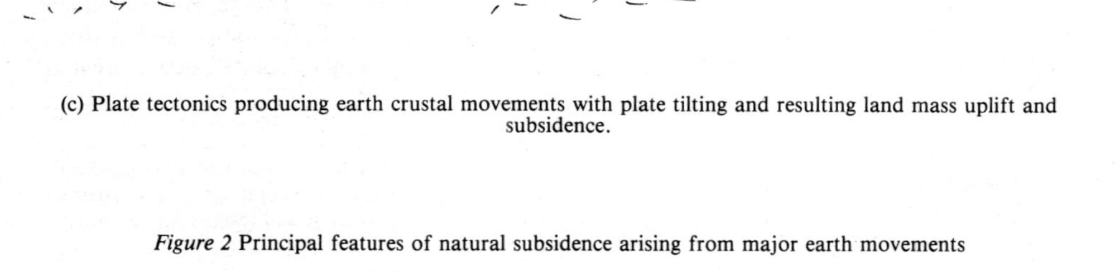

(c) Plate tectonics producing earth crustal movements with plate tilting and resulting land mass uplift and subsidence.

Figure 2 Principal features of natural subsidence arising from major earth movements subterranean voids in such rocks by solution action* creates the conditions favouring subsequent collapse and/or widening by washing down of overlying materials thereby resulting in sink-hole or crown-hole depressions occurring at the surface. Subterranean voids require significant time periods to develop to a size in order for major collapse of the surface to occur. However, much smaller fissures on being widened by solution action can give rise to minor subsidence effects at the surface.

Figure 3 illustrates the principal physical features of natural sink-hole and crown-hole development in limestone rock formations. The widened fissure can cause a conical depression to occur in superficial deposits, but if such deposits are thin or non-existent then the widened fissure can become exposed. Where a major subterranean network of voids allows considerable inflow of surface material then large conical depressions can arise at the surface as demonstrated. Conversely, a large underground cavity can form whereby its width exceeds the critical dimensions which the rocks can withstand by natural spanning taking account of the strength of the overlying rocks; progressive upward collapse is initiated as shown in Figure 3. The appearance of such crown-holes at the surface is in the form of a subsidence hole with overhanging or vertical sides which usually weather and decrease the severity of their overhanging character; however a number of holes retain almost vertical sides and a few still possess overhanging sides many years after they first appeared at the surface.

Figure 4 illustrates typical natural subsidence depressions in limestone country; the subsidence features shown are funnel-shaped and correspond to the sink-hole type.

Figure 5 shows a sink-hole/swallow-hole with an exposed major fissure which has been widened by the action of water; the sides of the sink-hole are being subjected to progressive erosional agencies.

Figure 6 illustrates a major natural subsidence feature in limestone country which has experienced significant chemical weathering, direct erosional action of water and later collapse. The feature shown probably began as a relatively small sink-hole and was enlarged to its existing size by subsequent action of water and progressive collapse.

Herak and Stringfield (1972) in their book entitled 'Karst' give detailed discussions on important regions of the Northern Hemisphere and highlight hydrogeological and geomorphological factors. They illustrate typical sink-hole characteristics (page 476) common in central Florida and these appear to conform with Figure 3 (2); Herak and Stringfield do indicate, however, the potential for such sink-holes to interconnect.

Legget (1962) draws attention to the relatively large size which some sink-holes and crown-holes attain, particularly Culpepper's Dish, England of about 60m diameter and depth around 45m, whilst the Cockpit Country of white limestone, in Jamaica is renowned for its sink-holes and crown-holes of various sizes with some attaining a depth of as much as 150m. He gives an interesting example of triggering the occurrence of small sink-holes/crown-holes at a sewage farm at Whitlingham, Norwich, England where the land on which discharge was taking place soon became covered in holes from 1 to 1·5m diameter. The explanation given for this localised subsidence was the presence of sand galls in a chalk formation and as these became washed out they allowed surface material to sink or collapse into the newly created voids.

---

*The water causing solution of limestone carries dissolved oxygen and carbon-dioxide and other chemicals and associated products picked up whilst percolating through surface deposits and gives it chemical weathering properties. The acidity of the water arises mainly from the dissolved carbon dioxide and reacts with the calcium carbonate of limestone dissolving it progressively to a bicarbonate.

1. Partial or complete infilling of hidden fissure in limestone.
2. Usually gives rise to small conical subsidence depression where significant superficial deposits exist.

1. If superficial deposits are thin (or non-existent) the fissure is likely to be exposed.
2. Such features can become re-exposed during wet seasons.

(1) Fissure in limestone widened by action of water

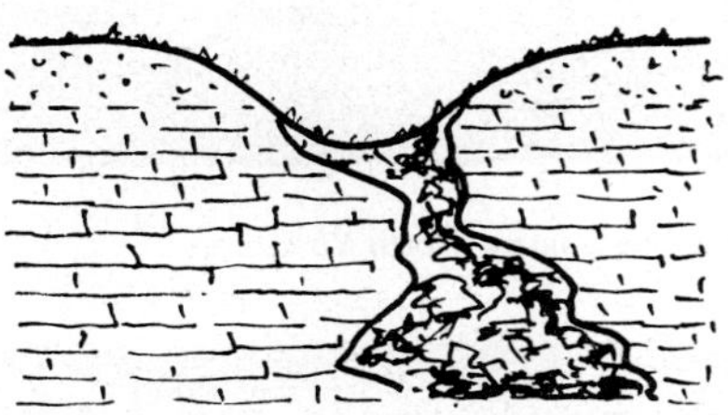

1. Major conical depressions with erosion of bedrock are common in limestone country and usually exhibit good drainage characteristics.
2. Surface material has previously sunk and washed into a solution cavity network.

(2) Large conical depression at surface

1 2 3 4

(3) Progressive upward collapse to form crown-hole

Stages

1. Underground cavity formed by solution of limestone.
2. Progressive collapse is related to weakening the stability of limestone structure and loss of natural spanning ability of the opening; internal weathering occurs.
3. Cavity can "day-light" suddenly and size of initial surface hole is related to cavity dimensions and nature of near surface bedrock and deposits.
4. Extent and duration of side weathering of subsidence hole is mainly related to thickness of superficial deposits and condition of bedrock.

Figure 3 Principal features of development of natural subsidence in limestone rock formations

Figure 4 Natural subsidence depressions in limestone country (North Yorkshire)

Figure 5 Sink-hole with exposed major fissure (North Yorkshire)

Figure 6 Major natural subsidence feature in limestone country exhibiting maturity (North Yorkshire)

North (1952) points out that the downwards flow of water in limestones makes its way through joints and bedding planes thereby giving rise to chemical weathering of the rock and widening the channels; eventually a network of fissures extending from the surface becomes established with possible connection to deep-seated caverns. The present authors have observed that fault planes in limestone are particularly susceptible to widening by solution of limestone rock in such flow channels and readily give rise to the opportunity for major and extensive subterranean voids to form. North explains that the Carboniferous Limestone is the most prone of all the limestones in the UK to development of caverns and notes that they occur in hundreds on the Millstone Grit and Carboniferous Limestone outcrops that run parallel to the South Wales Coalfield North Crop. A relatively thin layer of Millstone Grit covering cavernous Carboniferous Limestone is highly prone to sink-hole development. North (1952) examined the relationship between sink-hole occurrences and rainfall at Bridgend but his results did not appear to suggest any direct correlation when considering the local yearly rainfall; he does point out, however, that taking only the yearly rainfall into account may tend to mask wet and dry spells within particular years since some of his sink-hole occurrences were observed in relatively dry years. The authors' personal observations have been that superficial deposits are in their weakest state during wet spells and are more likely to sink or flow into fissured limestone bedrock than in dry seasons; the same observations of the authors also apply to sink-hole development.

North (1952) refers to an example involving a building on an outcrop of Carboniferous Limestone in Pembrokeshire, South Wales, where crater-like and funnel-shaped holes appeared owing to surface material collapsing into unsuspected cavities. The local limestone outcrop did not give any indication of an extensive cave system. Subsequent site investigation work indicated that superficial material was sinking into underlying hollows in the limestone and were spread over a limited area associated with the outcrop of a certain part of the Carboniferous Limestone. There was gradual washing of the fill material in the limestone bedrock to lower levels thereby creating the conditions for localised subsidences to occur.

Waltham (1978) in his Figure 45 shows a sink-hole which appeared during foundation digging operations for a warehouse. He points out that such excavation work was sufficient to disturb the ground to cause the sink-hole's occurrence at the Birmingham, Alabama site; the triggering mechanism for its development was the washing down of surface sediments into underlying cavernous limestone.

Collapse features of natural subsidence in limestone country

Thomas (1952) carried out an investigation of natural subsidence arising from solution of underlying calcareous beds in South Wales particularly in connection with the Millstone Grit and Carboniferous Limestone, and he reported that the collapse features as observed at the surface were generally swallow-holes or shallow depressions. The following general observations were made by him:

1. Where cavernous Carboniferous Limestone underlies other near surface rocks there is a risk of natural subsidence holes developing at the surface. Millstone Grit surface rocks were found to experience the larger swallow-holes, some attaining diameters of 90m and depths of 18 to 21m particularly where that formation had a cover thickness up to 75m above the Limestone.

2. In eastern Carmarthenshire, South Wales, a few shallow hollows can be observed filled with water forming lakes (indicating sealing of the bases by drift material) where the Millstone Grit cover thickness is up to 120m; these shallow hollows are up to 370m diameter. Thomas suggests that bearing in mind the thickness of cover of 120m, it reflects the effects of subterranean solution of the underlying Limestone and the ability of solution cavity effects to penetrate through such thicknesses of cover rocks. Where shales predominantly exist in the cover rocks Thomas reports that the effective upper limit of penetration by natural subsidence from the Limestone below appears to be around 60 to 76m.

3. Along Carboniferous Limestone outcrops swallow-hole distribution generally reflected the hydrological conditions both within the beds and as a result of lithological and structural influences. Linear development of concentrated swallow-holes can be observed as being associated with major fault lines. Similarly the character of lithological contacts, whether massive or well-jointed limestones abutting against oolitic, dolomitic or argillaceous contacts influences the degree of concentration of swallow-holes at the surface. The presence of a thin but irregular cover of superficial drift material was associated with major concentrations of swallow-holes.

4. An area was surveyed for swallow-hole development involving Carboniferous Limestone and Millstone Grit outcrops north of Tredegar. The sample areas covered 100 acre blocks of country. Within the Limestone area some 200 to 300 small swallow-holes (averaging 4·5 to 9m diameter) were observed and represented three per cent of the subsidence area; the junction area contained 70 to 100 medium-sized swallow-holes (averaging 14 to 18m diameter) again representing about three per cent of the affected area; the Millstone Grit outcrop block contained 15 to 20 large swallow-holes (averaging 37 to 46m diameter, but up to 90m diameter in some cases) together with a similar number of medium-sized holes and the subsidence holes covered some seven per cent of the affected area.

Kerridge (1952) has reported the effects of cavernous limestones underlying coal seams in the Cherapunji Coalfield of Assam. Swallow-holes (probably crown-holes) were observed on the surface indicating that such subsidence holes penetrated through the coal seams. Special problems occurred in assessing the reserves of coal and also in safely working the coal seams in view of the presence of a risk of possible collapse into cavities beneath the coal seams. Such underlying cavities in the limestone were made more hazardous due to the general area being within an earthquake belt, and there were records of accidents involving subsidence of underground workings into such cavities during periods of earth tremors.

Natural subsidence associated with Permian gypsum beds

Thick gypsum beds within the Edlington and Roxby Formations of the Zechstein sequence in North Yorkshire exhibit natural subsidence over a 3 by 100 km belt, particularly in the Ripon and Bedale areas according to Cooper (1986). Over the past 150 years, this area has experienced around 40 episodes of subsidence; the existence of several hundred other subsidence depressions is indicative of prolonged activity since the later stages of the Devensian. Cooper (1986) draws attention to the linear and grid-like character of these subsidence depressions and suggests that they appear to coincide with formation of cavities in a joint-controlled cave system. He also points out that it is the transition from deeper anhydrite to near-surface secondary gypsum that gives rise to this subsidence active belt. The subsidence activity appears to be mainly confined to the area where the gypsum is within 100m of the surface.

Surface subsidence appears to result from the creation of solution cavities within the gypsum which subsequently collapse and progressively propagate upwards. The form of surface subsidence is generally that of an approximately circular depression which can have clearly defined vertical, or near-vertical sides or be funnel-shaped even if only occurring as a slight hollow. Cooper (1986) presents a table of sizes of subsidence hollows in the Ripon-Bedale area since around 1870 and shows that the maximum diameter is around 30m whilst the majority are less than 15m diameter; the maximum depth of such subsidence holes is given as 10m whilst most have a depth of not more than 2-3m. Where competent rock is present at the surface and is underlain by the gypsum beds the subsidence hole tends to exhibit a shaft-like character whilst where appreciable drift deposits exist subsidence takes the form of a conical depression. Tute (1868 and 1870) reported a major cylindrical subsidence hole of about 14m diameter and 15m depth in sandstone near Ripon Railway Station which appeared in July 1834. Similar subsidence hole dimensions are discussed later in this book when considering the occurrence of sink-holes above room and pillar mine workings.

Cooper (1986) concludes that the orientations of natural structural weaknesses within the near-surface rocks, and which appear to influence the occurrence of subsidence holes triggered by solution cavities in underlying gypsum, should offer some guidance to assessing likely places where such holes might appear. Additionally he suggests that these subsidence depressions generally occur in groups, and that further subsidence in the Ripon area is predicted on the basis of the history of such events.

Conclusion

A number of factors contribute to natural subsidence, and several aspects have been discussed in this chapter. A summary of the principal factors and comments on their significance in relation to subsidence is presented in Table 1 which demonstrates the importance of a thorough site investigation where structures are planned and there is a possibility of natural subsidence affecting the site.

Table 1. Summary of main forms of natural subsidence

Principal factor contributing to natural subsidence	General comments on aspects pertaining to its significance
1. Soil compaction	1. Magnitude of settlement is a function of loading intensity and drainage conditions of the loaded material. 2. Settlement is normally greatest during the early stages and gradually decreases with time. 3. Foundation design has a major bearing on the character of the ensuing soil compaction. 4. Uniformity of foundation loading has a marked influence on reducing differential settlement tendencies. 5. Settlement problems can arise when superficial deposits are thin and irregular bed rock conditions exist. 6. Initial settlement during construction loading is due to elastic deformation of the soil mass and vertical displacement is accompanied by lateral deformation.
2. Soil shrinkage	1. Since volume change of soil is a function of moisture content, it follows that drainage conditions influence soil shrinkage behaviour and consequently the potential for settlement changes to occur. 2. Removal of moisture from soils occurs as a result of climatic factors and in the presence of significant tree roots which can lead to appreciable soil shrinkage and accompanying subsidence effects.
3. Subsidence due to lowering of water table	1. As water drains from soil pore spaces, this allows particles to settle into voids vacated by the water. 2. The rate of subsidence is influenced by the permeability of the soils. 3. The subsidence process is not reversible on restoration of the water table to its original position. 4. Fluctuating water tables can weaken soil structures and induce structural collapse of soils resulting in subsidence.
4. Subsidence due to development of subterranean voids by solution of host rocks.	1. Formation of subterranean voids in rocks susceptible to solution by groundwater can result in superficial unconsolidated material subsiding into such cavities. 2. Limestones are the common rock types where sink-holes and crown-holes are encountered as natural subsidence, although such subsidence holes have occurred above salt and gypsum deposits also. 3. Sink-hole developments at the surface appear as shallow saucer-like or funnel-shaped depressions, and on occasions exhibit steep or overhanging sides depending upon their age and the nature and thickness of superficial deposits.
5. Subsidence due to tectonic and volcanic activities	1. Earthquakes can result in sudden sinking of relatively large surface areas; deltaic deposits are particularly prone to sinking due to earthquake effects. 2. Areas adjacent to major faults can experience the effects of subsidence during earthquakes. 3. Volcanic deposits can exhibit surface subsidence owing to collapse of void structures.

Notes:

1. The table considers those aspects of natural subsidence of consequence to surface structures and excludes sedimentary basin subsidence theory and plate tectonics theories and their relevance to regional subsidence behaviour.
2. Material used in this table has been taken from North (1952) and Legget (1962) and supplemented with the authors' own personal observations.

Important conclusions drawn from the chapter are listed as follows:

1. Natural subsidence continues to occur in various parts of the world either as a fairly continuous or intermittent phenomenon or as sudden manifestations triggered by earthquakes or collapse of capping over an underground cavity.

2. The occurrence of sink-holes and crown-holes in limestone country is probably the most common form of natural subsidence. Where cavernous Carboniferous Limestone underlies other near surface rocks there is a risk of natural subsidence holes developing at the surface. Sink-holes from cavernous limestone can penetrate a Millstone Grit cover thickness of 120m, although an upper limit for shales has been indicated to be around 60 - 76m.

3. Linear development of concentrated swallow-holes/sink-holes has been observed as being associated with major fault lines; hydrological conditions both within the beds and as a result of lithological and structural influences have a marked effect on the occurrence of subsidence holes at the surface.

CHAPTER 2

DEVELOPMENT OF CONCEPTS AND THEORIES OF MINING SUBSIDENCE

Principal Developments up to 1900

A comprehensive review of early developments in mining subsidence concepts and theories is given by Young and Stoek (1916), whilst Bulman (1906) presents an interesting summary of important Nineteenth Century observations and views in the discussion on the paper by Halbaum (1905). From these publications it appears that Belgian engineers played a leading role in the scientific study of mining subsidence, and in 1825 a commission examined damage to the surface at Liege overlying mine workings; that commission concluded that surface damage was principally caused by mining operations within a depth range of 90m below the surface. A second Belgian commission in 1839 also considered the same problems and reported similar findings to the 1825 commission that depth played a decisive role in respect of surface damage and it was virtually confined to mining depths of 90m and less.

The early concepts of mining subsidence appear to be of Belgian and French origin according to Young and Stoek (1916) and they are based on the extraction of coal seams, both flat and inclined, and involved relatively large extraction areas in relation to the depth below surface. Mining subsidence associated with narrow working such as room and pillar operation does not appear to have attracted the attention of these early investigators. The mode and mechanism of transmission of movements from the mining extraction to the surface appears to have been the starting point in the development of mining subsidence concepts, and especially the influence of seam inclination. The concept that mining subsidence was within the vertical confines of the extracted area must have been confounded by the little known influence of dip and angle of draw, and thus resulted in the idea that movements were transmitted normal to the seam, Gonot (1858). Further refinements to the early concepts of subsidence devised by Gonot were advanced by Callon (1874) and Dumont (1875) who argued that bedding plane effects would modify the development of subsidence at the surface.

Some interesting observations were made in the Austrian coal mining region of Ostrau-Karwin and published by Jicinsky (1884) and Rziha (1882), with much of this early work being directed towards protecting railways. The idea emerged of a paraboloidal region overlying the coal extraction which was referred to as a "zone of falling", see Young and Stoek (1916). The dome theory of Rziha was an important concept and provided a foundation for future developments in the principles of mining subsidence.

Mining subsidence studies in Germany appear to have been launched by the work of Schulz (1867). He considered the projection of angles of fracture towards the surface and the pillar sizes to afford protection to the surface against mining subsidence. Von Dechen (1869) drew particular attention to the part played by significant thicknesses of marl overlying the Coal Measures around Essen and which formed the foundation for surface structures; he suggested that the surface subsidence would be extended over an appreciably greater area by the presence of such marls. Hausse (1885) put forward his theory of the occurrence of "break angles" of differing significance, and the progressive development of subsidence from initial collapse of roof beds to give the propagation of "main and after breaks".

Fayol (1885) made a significant contribution to developing a scientific understanding of the mechanism of mining subsidence, and identified many important aspects included amongst which were:

(i) movement is transmitted to the surface from all depths of working,
(ii) surface effects become minimal, below a certain depth,
(iii) movements become increasingly feeble as they progress upwards from the working level,
(iv) subsidence always occurs vertically above mine workings,
(v) subsidence occurs within limit lines drawn from the edge of the mine workings,
(vi) packing (stowing) of mined out areas reduces the effect of subsidence and affords protection to the surface.

Fayol carried out several experiments and lengthy investigations into mining subsidence, and he subsequently concluded that the caving process around a working face gave rise to an active zone of limited extent which was responsible for the manner in which subsidence extended laterally as the face progressed.

Important subsidence observations were made by early British investigators notably Dixon (1885) and Dickinson (1898), the former reporting the existence of a leading and following wave of disturbance which exhibited regularity with respect to each other. Dickinson recognised the existence of draw and recommended a margin of one-tenth of the depth be taken to provide surface security against subsidence damage in fairly flat seams of 1·88m (6ft) thickness and less in Coal Measures strata of average hardness similar to those encountered in Lancashire; where softer strata exist he recommends increasing the margin of draw to one-sixth or even one-quarter of the depth of working. On the other hand, however, Dickinson suggests that in stronger than average strata the draw margin can be less than one-tenth of the depth. Dickinson was one of the pioneers in rational analysis of mining subsidence and its interpretation in respect of surface behaviour.

Trompeter (1899) drew attention to the three-dimensional effect of mining subsidence and particularly the role played by the rock movements in the immediate vicinity of the extraction. A well-known expression developed from his work, namely that of "Trompeter's Zone", although he appears to have been the first investigator to recognise the significance of horizontal displacement of surface points within the zone of subsidence.

Figures 7 and 8 illustrate the principal early developments in concepts and theories of mining subsidence.

The Dortmund Board of Mines (1897) issued a report on mining subsidence in the Westphalian Coalfield, and drew particular attention to the angle of draw being greater where a significant thickness of marl overlies the Coal Measures. The report contributed to knowledge on angle of draw in the case of level and dipping seams and Table 2 gives a summary of the main findings. The type of goaf treatment involving stowing did not appreciably influence the limits of mining subsidence.

Principal Developments 1900 to 1950

Halbaum (1903, 1905) gave increased clarity to the understanding of mining subsidence and endeavoured to standardise the terminology; he considered the subsiding strata as a cantilever acting under its own weight with a neutral axis a relatively short distance above the coalface. Halbaum's concept visualised the bulk of the strata between the extraction and the

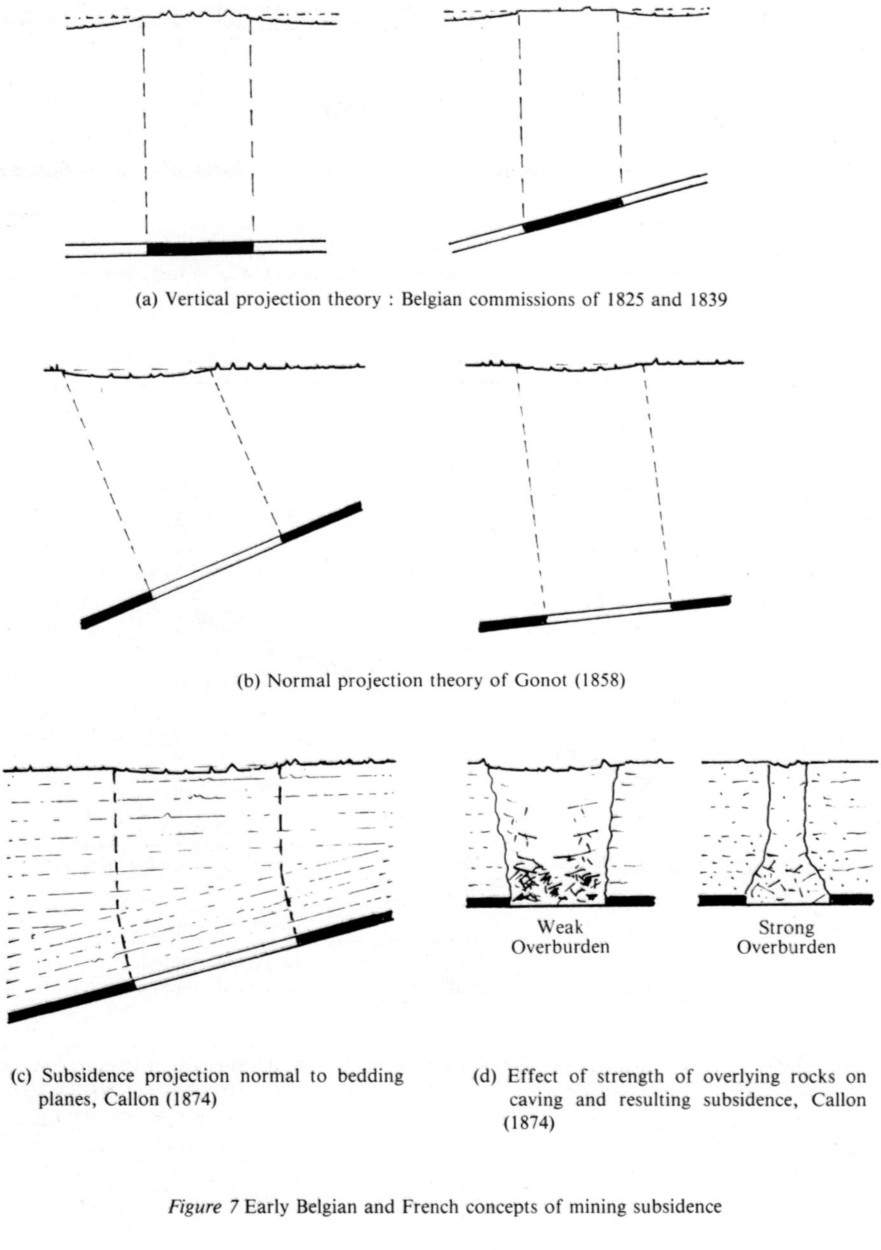

(a) Vertical projection theory : Belgian commissions of 1825 and 1839

(b) Normal projection theory of Gonot (1858)

(c) Subsidence projection normal to bedding planes, Callon (1874)

(d) Effect of strength of overlying rocks on caving and resulting subsidence, Callon (1874)

Figure 7 Early Belgian and French concepts of mining subsidence

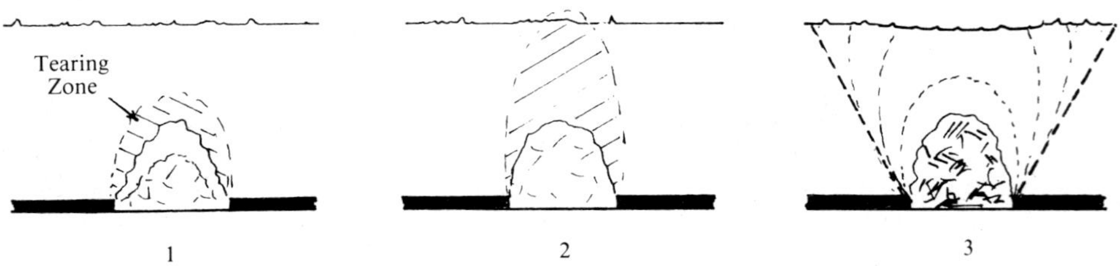

(a) Illustrating progressive development of mining subsidence according to Rziha (1882)

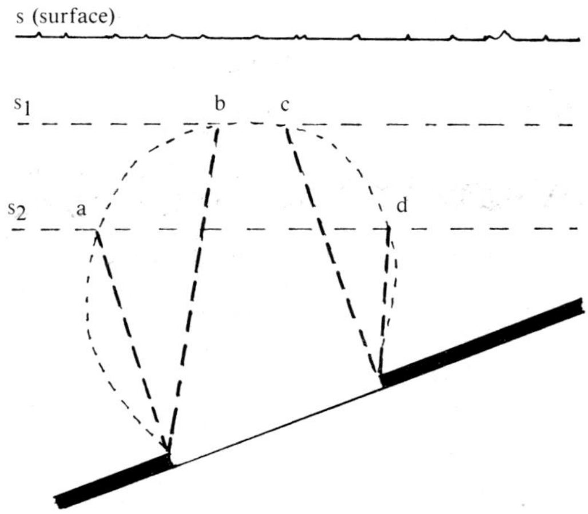

(b) Illustrating concept of mining subsidence according to Fayol (1885) as established from model experiments

Notes 1. For surface at s, no subsidence is indicated
2. For surface at s_1, subsidence would occur between the limits b-c
3. For surface at s_2, subsidence would occur between the limits a-d

Figure 8 Early dome theories and concepts and their extrapolation to the development of surface subsidence

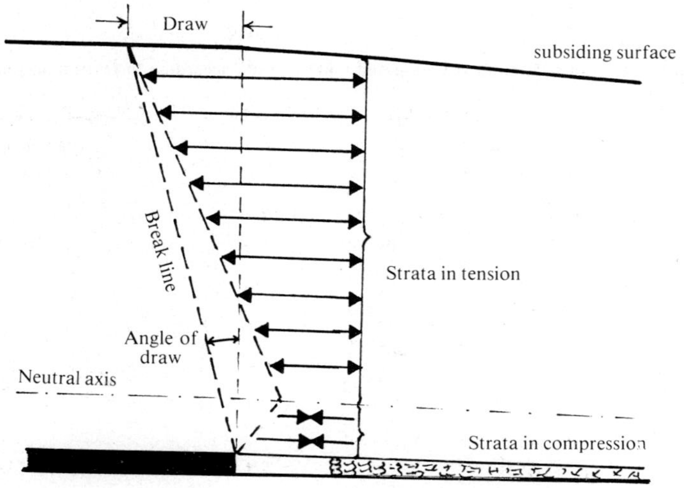

(a) Representation of strata as composite cantilever

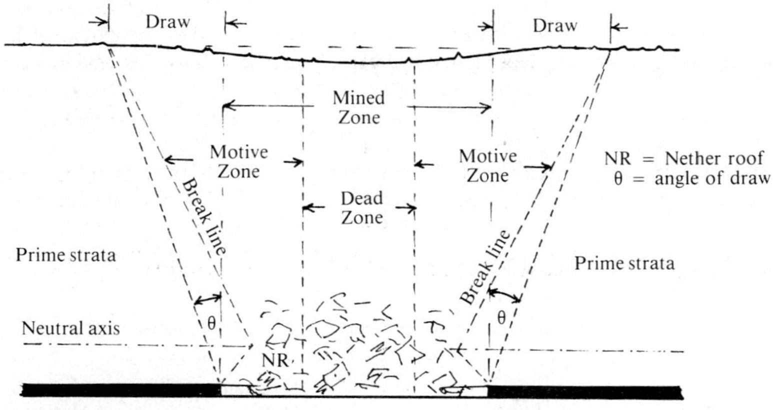

(b) Halbaum's terminology

Figure 9 Mining subsidence representation and terminology according to Halbaum (1903) surface to be in a state of tension during the subsidence phase and this was largely responsible for his postulation that a 'line of break' existed ahead of the face-line and this could be extrapolated towards the surface.

Table 2. Angle of draw in relation to seam dip and type of strata according to Dortmund Board (1897)

Range of seam dip	Angle of draw in Coal Measures (from vertical)		Angle of draw in surface marls (from vertical)	
	dip side	rise side	dip side	rise side
0 - 15°	15°	15°	20°	20°
15 - 35°	dip of seam ie 15 - 35°	15°	20°	20°
> 35°	35°	15°	20°	20°

Figure 9 illustrates Halbaum's concepts of mining subsidence, particularly his representation of the strata as a cantilever together with the terminology he suggested.

The publication by O'Donahue (1907) extends the ideas of Dickinson (1898), to a wider range of situations giving special regard to factors influencing break lines and extent of subsidence by considering the relative strength of the strata, seam inclination and thickness, and the influence of surface deposits. He draws attention to the false presumption which was widely held at the turn of the century that at mining depths of around 600m no subsidence would manifest itself at the surface. O'Donahue recognised that the extent of underground extraction in relation to the depth had a bearing on surface subsidence, and indeed if all of the mine was worked out subsidence on the surface would occur at all workable depths. He supports his statement with field observations and these are presented in Table 3.

Hausse (1907) presented the findings of his 20 years of study of mining subsidence in German coalfields, especially those in Saxony and he put forward the concept of a 'main break' (over the extraction) being responsible for a subsequent 'after break' associated with an 'angle of break' which haded over the solid sides and in effect governed the limits of mining subsidence.

Table 3. Observed subsidence data according to O'Donahue (1907)

Depth below surface (m)	Extraction thickness (m)	Maximum observed subsidence (m)	Maximum subsidence (%)
180	1·2	0·9	75
730	1·2	0·3 to 0·45	25 to 37

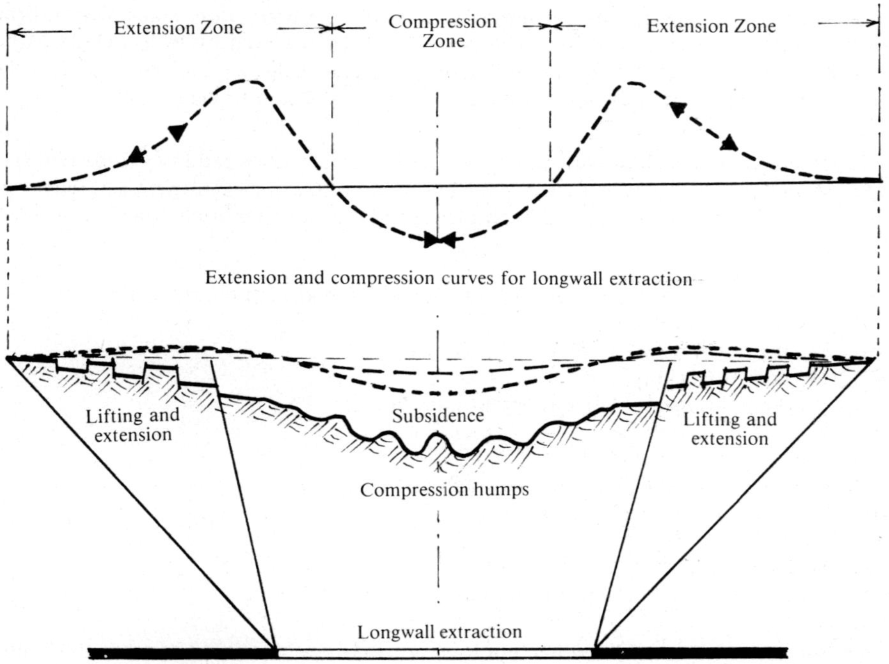

Figure 10 Representation of surface ground extension and compression in relation to mining subsidence and character of surface effects according to Lehmann (1919). Note : surface profile exaggerated for purpose of demonstrating Lehmann's concept

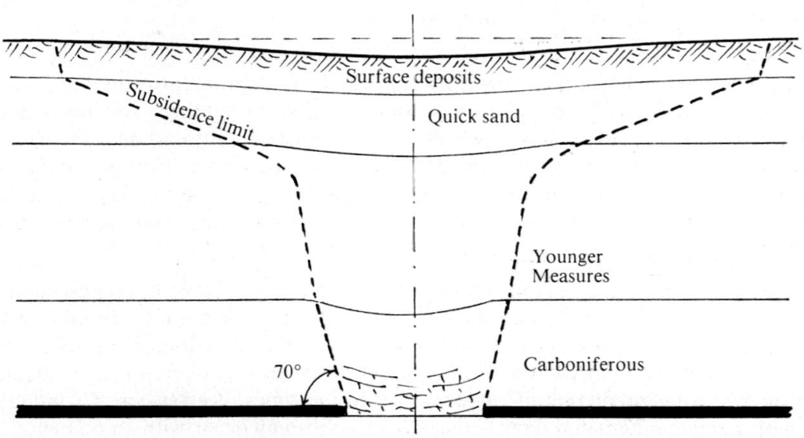

Figure 11 Subsidence region above longwall extraction according to Groothoff (1922)

Goldreich (1913) published a major works on mining subsidence in which he discussed various theories of that period and included several results of observations made in the Ostrau Karwin Coalfield of Czechoslovakia. This work has been reviewed by Briggs (1929), who considered Goldreich's data made a useful addition to knowledge particularly regarding the extent of subsidence at the surface and its significance in relation to depth.

The interaction of geology with mining was given special consideration by Knox (1913) in a paper on mining subsidence presented to the International Geological Congress of that year. He listed the principal factors affecting subsidence and the more important ones are included as follows:

(i) Extent of unmined area ie degree of permanent support left to protect the surface.
(ii) Extracted seam thickness.
(iii) Mining depth below surface.
(iv) Method of working.
(v) Working direction relative to strata jointing.
(vi) Rate of extraction.
(vii) Nature of the overburden.
(viii) Presence of major discontinuities such as faults.
(ix) Permeability of overlying rocks.
(x) Strata inclination.
(xi) Surface contour.
(xii) Nature of stress field prior to mining.

Some of the factors listed by Knox had not been previously appreciated and many have since been shown to play important roles in influencing mining subsidence.

The extensive study of mining subsidence carried out by Young and Stoek (1916), was prompted by increasing concern in North America over its effects at the surface and they describe several examples of such effects in the case of fairly shallow mining. They do point out, however, that actual subsidence data in the United States was very sparse and their report resorts extensively to field data from European sources.

Empirical formulae have been put forward which attempt to give an approximate assessment of the magnitude of surface subsidence due to mining, see for example O'Donahue (1907), Louis (1922) and Briggs (1929), but none take into account the width of extraction and as Grond (1953) points out they are patently defective. These formulae were obviously developed for a particular set of conditions and form of mining layout, but they were very restricted even in local use and were unsatisfactory for universal application.

The representation of a surface ground strain profile showing zones of compression and tension across a longwall extraction appears to be first due to Lehmann (1919) in Germany who referred to it as his 'trough theory', and this form of representation has been in general use ever since. Lehmann's representation of subsidence strains is shown in Figure 10 and of particular interest is his prediction of surface uplift in the zones of extension. Grond (1953) disagrees with Lehmann here and reports that whilst uplift may occur with an odd case, as a general rule field observations in German and Dutch (South Limburg) coalfields show no such phenomenon and that pseudoplastic movement appears to be a major factor. Compression humps were indicated by Lehmann in the subsidence trough compression zone whilst differential movement in the tension zones was indicated to occur with an uplift tendency.

The line drawn from the edge of the longwall extraction to the limit of mining subsidence at the surface is generally shown to be straight, but Groothoff (1922) in considering Dutch experiences especially the nature of near-surface geology suggested that the limit angle would be much shallower on approaching the surface. Groothoff's concept of the region of subsidence movement between the longwall extraction and the surface is illustrated in Figure 11. The overlying saturated sands (quick sand) allow the limits of mining subsidence to be appreciably increased according to Groothoff and most probably explains why angles of draw of up to 45° have been observed in the South Limburg coalfield.

Two text books on mining subsidence appeared at the end of the 1920's, namely Briggs (1929) and Lane and Roberts (1929). Both books give fairly comprehensive reviews of the state-of-the-art of mining subsidence as it was at that time. They provide useful sources of data and experiences on mining subsidence in addition to aspects relating to the law of support.

Emergence of Subsidence Calculation Methods

The first major step in the development of subsidence calculation methods appears to have been taken by Keinhorst (1928). He produced a zone-calculation of subsidence which is demonstrated in Figure 12, and this shows the total zone of influence to be limited by a circle since the seam is assumed to be flat. The angle of draw determines the radius of the circle. The critical extracted area is divided into two zones referred to as the middle and the edge zones and these occupy areas defined by a concentric circle, the magnitude of which depends on the angle of break. Keinhorst used field observations to arrive at the deduction that extraction of the middle zone exerted twice the influence of the edge zone. He also established that the subsidence factor was 0·6. (This corresponded to the maximum subsidence being 0·6 of the extraction thickness for a critical area of extraction.) The example given by Keinhorst in Figure 12 is a situation where the middle zone is one-quarter worked (surface area A_m) and the edge zone is one-tenth worked (surface area A_e); consequently the anticipated subsidence (s) at point P is as given by equation (1).

$$s = 0{\cdot}6\,M\left[\frac{2A_m}{3Z_m} + \frac{A_e}{3Z_e}\right] \qquad \ldots (1)$$

where s = predicted subsidence at point P
M = extracted seam height
Z_m = surface area of middle zone
Z_e = surface area of edge zone
A_m = surface area worked of middle zone
A_e = surface area worked of edge zone.

In Keinhorst's example, the data he gives when inserted in equation (1) gives the following result:

$$s = 0{\cdot}6\,M\,[(2/3)(1/4) + (1/3)(1/10)]$$

$$= 0{\cdot}12\,M$$

Bals (1931/32) made a significant contribution to the calculation of subsidence in horizontal strata and added considerable refinement to the early development of Keinhorst (1928). Bals divided the influencing area into different zones to achieve greater accuracy. He employed the basic principle of Newton's Law of Gravitation, ie force inversely proportional

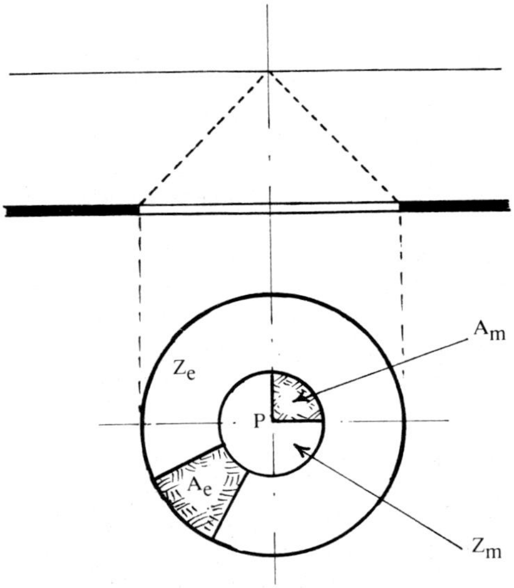

Figure 12 Zone-calculation of subsidence of point P according to Keinhorst (1928)

to the square of the distance, in order to find ratios of the attraction caused by the different zones associated with the extraction. He integrated a simple function and arrived at an expression with allowed the limits of zones of equal influence to be determined. Although Bals' analysis has been criticised for some of the assumptions and simplifications he makes, it has, however, been used extensively as a graphical integration method in subsidence prediction assessments. Bals pointed the way for other investigators in the application of integral calculus to performing calculation of zones of influence.

Schleier (1937) extended the work of Bals to inclined seams, and arrived at the conclusion that the action of the goaf is inversely proportional to the distance and not the square as used in Bals' hypothesis. Schleier proposed some minor modification to Bals' original expression.

Fläschenträger (1938) considered Bals' and Schleier's calculations of zones in a study of the distribution of costs resulting from mining subsidence damage. He drew attention to comparing the spatial movements as a practical basis for assessing the effects of subsidence.

Recognition of the important role played by the dynamic phase of mining subsidence was made by Perz (1948) during his attempt to include time factors into subsidence formulae. Perz suggests that the movement of points is towards the momentary centre of gravity of the extraction. A similar view is expressed by Niemczyk (1949). Perz gave considerable attention to seeking a mathematical relationship between subsidence and displacement but of special interest was his representation of the time related development of subsidence which is shown in Figure 13. The point P is subsiding due to an extraction travelling across the entire range of influence which is expressed as 2R and this is illustrated in Figure 13(a). In Figure 13(b) Perz has not considered the speed of extraction as he does in Figure 13(c), and where he indicates that subsidence will have ceased within 3 to 4 years for an infinite rate of advance. Perz has embodied rate of advance with residual mining subsidence effects in these graphs. Perz, however, assumes that subsidence will follow the course indicated by Figure 13(b), the details of which are given in Table 4.

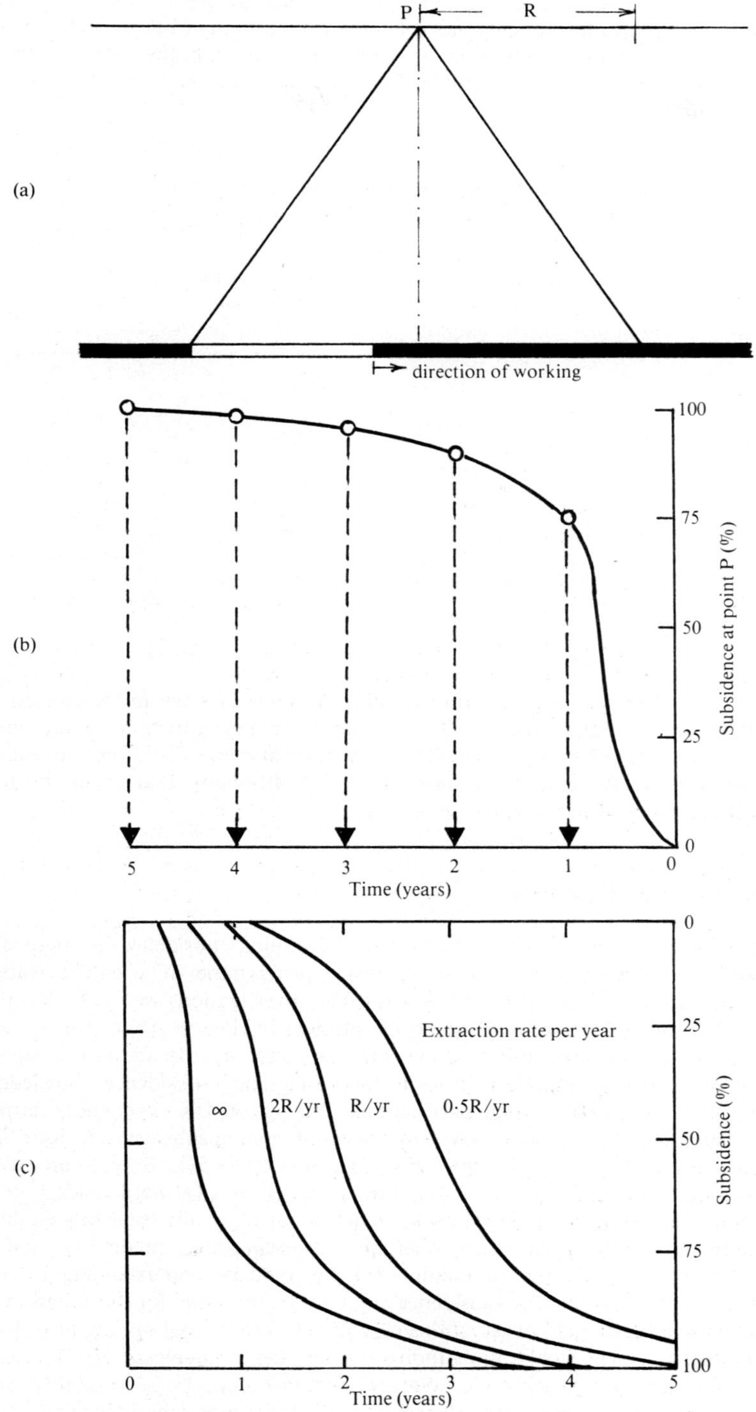

Figure 13 Time related development of subsidence according to Perz (1949)

The course of subsidence development put forward by Perz is based on mining operations in deep European coalfields, notably Germany, although Baron (1951) expressed some reservations about those situations in very deep conditions with slow moving extractions since their movements may still be discernible after five years if they had not moved outside of the range of influence (R).

Table 4. Course of subsidence developments after Perz (1948).

Time duration since mining began	Amount of subsidence occurring (of total)
After the 1st year	75%
After the 2nd year	90%
After the 3rd year	95%
After the 4th year	98%
After the 5th year	100%

The presence of major tectonic strains within a coalfield have been referred to by Weissner (1932) in the Ruhr Mining District and he draws attention to an observed northwards directed force being present. Other investigators have also noted its existence. Shafts are particularly susceptible to such effects. Additionally faults can disturb the symmetry of the picture of movement.

Principal Developments Since 1950

Special interests in the prediction and effects of mining subsidence in UK coalfields developed following the launching of an extensive programme of scientific study and investigation by the National Coal Board. Several field investigations were set up in the UK whereby detailed measurements were made of subsidence, lateral displacement, ground strains and tilt in many different mining and geological conditions. In addition comparisons were made between British and German experiences of mining subsidence knowledge and observations, Wardell (1950). Several publications emerged on UK experiences during the early 1950's, and some of the more noteworthy papers on mining subsidence include Sinclair (1950-51), Wardell (1952, 1953-54, and 1954), Webster (1951-52), Beevers and Wardell (1954-55), and Orchard (1954 and 1956-57). Both Wardell and Orchard made impressive contributions to knowledge on mining subsidence behaviour especially regarding establishing ground movement mechanisms, interpretation, representation, prediction and field monitoring. Their work laid the foundation for the modern understanding of ground movement behaviour due to mining subsidence and they set the scene for the collation of the most comprehensive set of field observations ever undertaken in coal mining history which ultimately resulted in the UK Coal Mining Industry's Subsidence Engineers' Handbook which is respected and employed world-wide. Wardell demonstrated the repeatability of field observations, consistency of results and their potential for forming a basis for prediction of

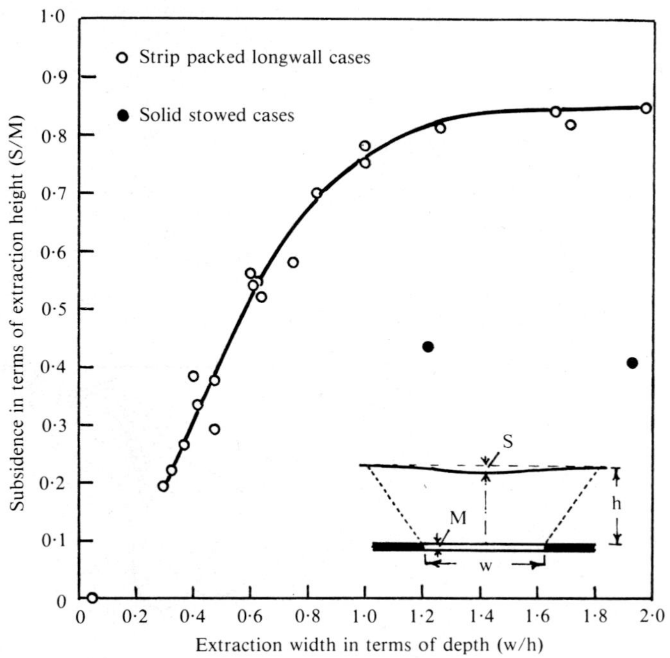

Figure 14 Partial subsidence from longwall extraction in relation to width and depth of working, after Orchard (1956-57)

ground movements. Orchard's interpretation of subsidence data allowed comparisons to be made between results from widely differing conditions; he put forward new forms for the representation of results and significantly advanced the rationalisation of subsidence observations which turned what was largely regarded as an art into an established science. An important example of the latter was Orchard's partial subsidence graph which is shown here as Figure 14 and demonstrated the inter-relationship between subsidence, extraction height, width of extraction and mining depth.

The 1957 European Congress on Ground Movement held at the University of Leeds brought together the leading experts in mining subsidence and allowed the state of knowledge to be presented and discussed and thereafter compiled in the Congress Proceedings (1957). An important contribution by Wardell (1957) was his demonstration of the consistency of the travelling ground strain profile, shown here as Figure 15. This he employed to good effect to highlight how harmonious mining layouts can interact to partly compensate for their respective strains and thus reduce the level of strains imparted to surface structures. The mechanics of mining subsidence were effectively demonstrated by King and Whetton (1957), especially the interpretation of displacements. Of particular interest was the paper of Orchard (1957) who presented an analysis of strain data which formed the basis of the NCB Subsidence Engineers' Handbook method of calculating surface ground strains due to mining subsidence. Drent (1957) presented observations made in the South Limburg Coalfield and showed that heavily water-bearing near-surface deposits can substantially increase the angle of draw. Some important contributions were made to the Congress regarding subsidence and ground movement behaviour in Germany, Austria, The Netherlands and France especially by

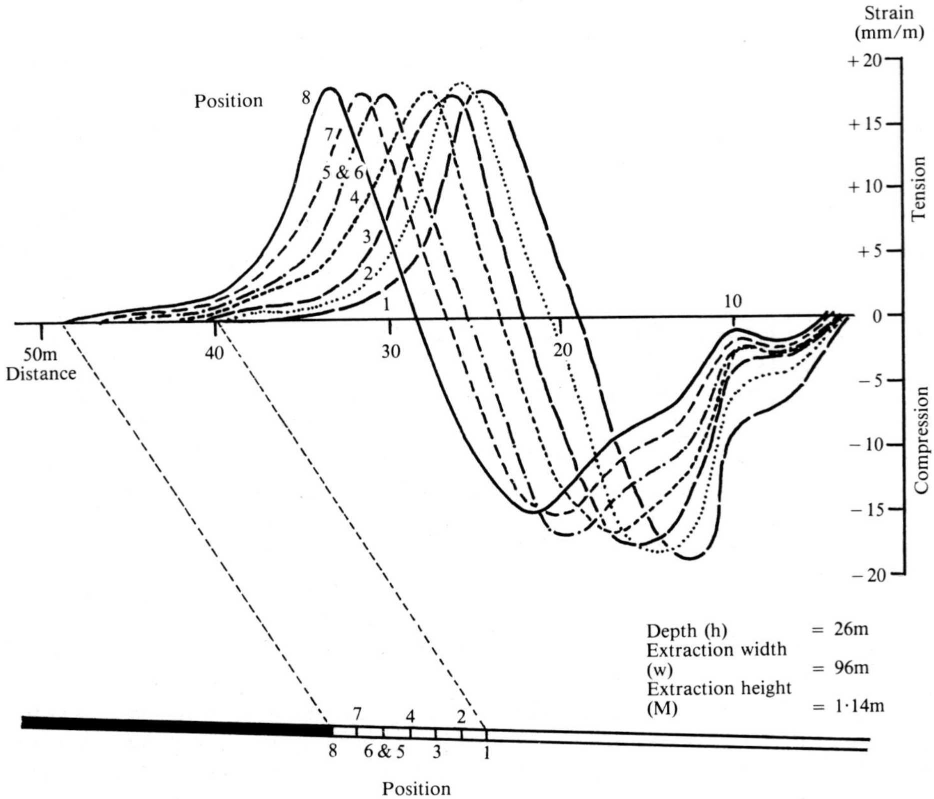

Figure 15 Demonstrating consistent pattern of the travelling strain profile above a longwall extraction, after Wardell (1957)

Notes 1. Observations taken on successive days 1 - 8.

2. Strain profile moves forward at more or less the same rate as the advancing face.

3. Travelling strains measured in central area of extraction i.e. in critical area of uniform strain change.

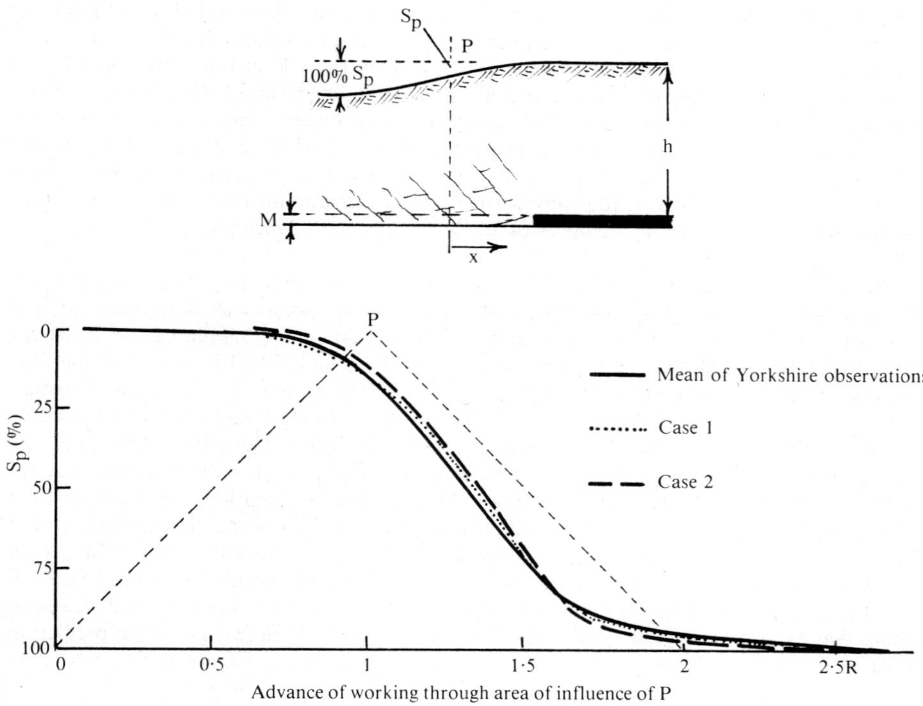

Figure 16(a) Comparison of subsidence development curves, after Wardell and Webster (1957)

Data:

	h (m)	M (m)	100% S_p (m)	Full dip	Rate of advance (m/yr)	Overburden
Case 1	25	1·14	0·646	1 in 15	400	Coal Measures
Case 2	80	1·32	0·726	1 in 13	180	Coal Measures

Note:

(a) Angle of draw : Case 1 : 34°, Case 2 : 36° (to vertical)

Fläschenträger (1957), Rellensmann and Wagner (1957), Perz (1957), Lenge (1957), Grond (1957) and Schwartz and Dubois (1957).

Consideration was given to surface observations and their relationship to movement underground by Wardell and Webster (1957). They examined the development of surface subsidence and underground convergence in relation to depth and produced Figures 16 (a) and (b). This form of representation was widely used in the UK in the 1950's, and it was mainly introduced by Wardell following his detailed assessment of German subsidence knowledge. Figure 16(a) suggests that the subsidence development curve is not influenced by depth of working providing the results are normalised as carried out by these authors. Figure 16(b) shows underground convergence behind the face line in relation to the surface subsidence development curve; the convergence behind the face line occurs over a very short distance and rapidly settles by comparison to the surface subsidence phase.

Much attention has been devoted to the pre-calculation and theoretical modelling of mining subsidence since the late 1950's and a measure of success has been achieved in a number of countries, see for example Cherny (1966) and Akimov (1958) in the USSR; Berry and Sales (1960-62), Orchard and Allen (1965) and King and Whetton (1957) in the UK; Litwiniszyn (1957, 1964) and Knothe (1957) in Poland ; Martos (1958) in Hungary; Salamon (1963-64) in South Africa; Hiramatsu and Oka (1968) in Japan; Dahl (1972) in the USA; Kratzsch (1967, 1975, 1983) and Bräuner (1973) in West Germany. The principal developments of a number of these publications are discussed in later chapters. Several theoretical models have been devised which satisfy particular conditions but have not met universal approval owing to their fairly limited applications. Empirical models have found wider acceptance, especially that of the NCB (1965, 1975) Subsidence Engineers' Handbook in view of their ability to substantially take geological variations into account. The NCB empirical model is based on several hundred field observations carried out in a wide range of mining depths and geological conditions; it is the most widely used subsidence prediction model.

Computer programs have been written to perform multiple calculations of mining subsidence and associated surface ground strains. A program that uses data from the NCB Subsidence Engineers' Handbook as the basis for its calculations has been described by Whittaker and Reddish (1984), and Whittaker, Reddish and Fitzpatrick (1985) whereby the principle of superposition is applied to extend single-panel data into multi-panel data. Such computer programs in addition to introducing speed of calculation, they add the capability of refinement and allow for a measure of correction of data where this is needed. Such programs are proving to be vital planning tools in assessing the effects of mining subsidence in new and existing projects.

Conclusion

The most significant progress achieved in the development of knowledge relating to mining subsidence phenomena has undoubtedly been the accurate observations made of its characteristics in several countries. Such observations have served as a basis for establishing the parameters governing its manifestation at the surface, and the form and extent of its appearance in different conditions. This led to investigators being able to formulate both conceptual and prediction models of the subsidence process. The reliability and accuracy of prediction models have been judged on the basis of comparison with field data.

This has resulted in subsidence engineers and specialists tending to favour the more advanced forms of empirical models for prediction work. No single prediction model can give

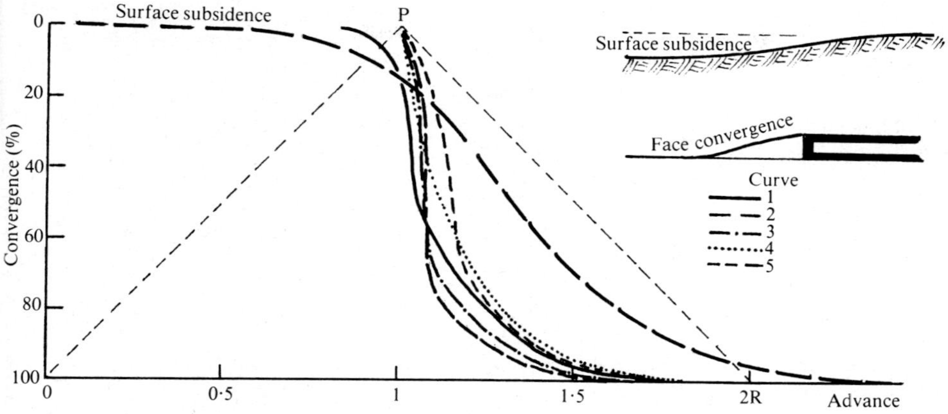

Figure 16(b) Advance of working through area of influence of surface point P, after Wardell and Webster (1957)

Data: Curve 1 according to Winstanley (1938)
Curve 2 according to Jones (1945)
Curve 3 according to Evans and Jones (1946)
Curve 4 according to Phillips and Henshaw (1942)
Curves 4 and 5 correspond to where full area of influence has been worked

the complete answer to every subsidence problem since the diverse range of factors influencing the many different mining settings still need careful engineering assessment. Prediction models of the empirical type give the most accurate and reliable assessments of anticipated subsidence. In most cases they give results possessing a high order of accuracy and sufficient to meet the engineering needs of surface structures in regions of mining subsidence particularly where longwall extraction is the method of working.

CHAPTER 3

MINING AND GROUND MOVEMENT

In order for surface subsidence to occur due to mining operations some form of underground collapse or abstraction process must precede such an event. Consequently the form of mining, degree and extent of activity coupled with geological factors play a role in causation of subsidence. Mining needs to be of a sufficient magnitude before its effects become of significance to disturbing the surface. The creation of an underground mining excavation results in a re-distribution of the stress field and this reacts with the rocks surrounding the excavation thereby inducing movements with possible collapse potential or subsequent closure that is related to time.

Stress re-distribution associated with a mine roadway

An example of the stress redistribution process and its resulting effects is demonstrated in the Memorandum on the Design of Mine Workings, NCB (1950-51) for a mine roadway which forms the basis of Figure 17. The example used is a rectangular opening in a stratified geological setting, corresponding to a coal seam development drivage in a predominantly vertical stress field. A pressure arch is formed and the beds spanning the opening tend to sag and cause bed separations; shearing forces are in strong evidence. If the excavation were left unsupported, it would possibly give rise to the strata within the pressure arch collapsing into the excavation especially where weak mudstone beds form the immediate roof. Strong roof beds, however, can give permanent natural support and in those partial extraction methods involving room and pillar working providing the mine pillars are stable, then the risk of surface subsidence becomes very small.

In the situation illustrated in Figure 17, the natural re-distribution of stress around the narrow excavation, results in generally confining the collapse process to the vicinity of the roadway providing sufficient depth of cover exists. The natural pressure arch formed allows a stress gradient to occur between the arch and the enclosed dome; if the stress gradient is sufficient to exceed the strength of the rock in its triaxial state of stress, then fracture will occur with the resulting formation of a stress relieved dome. The latter will most certainly need support to prevent subsequent collapse, at least within the life-span of the mining operations or at least the serviceable life of the roadway. Supports used in mine roadways deteriorate in the long-term (outside the life-span of the mine and may involve more than one hundred years). Consequently, in some situations, a previously limited collapse may progress to a higher level by virtue of loss of support resistance arising from such as timber rotting or steel corrosion.

Ground movement associated with a mine roadway

Figure 18 demonstrates two commonly encountered forms of ground movement associated with mine roadways, whose width is generally around 5-6m but may be as much as 10m in favourable geological conditions. There are two distinct modes of ground movement which have a significant bearing on giving rise to possible surface subsidence effects. The first involves caving of roof strata following removal of support, and the subsequent bulking of

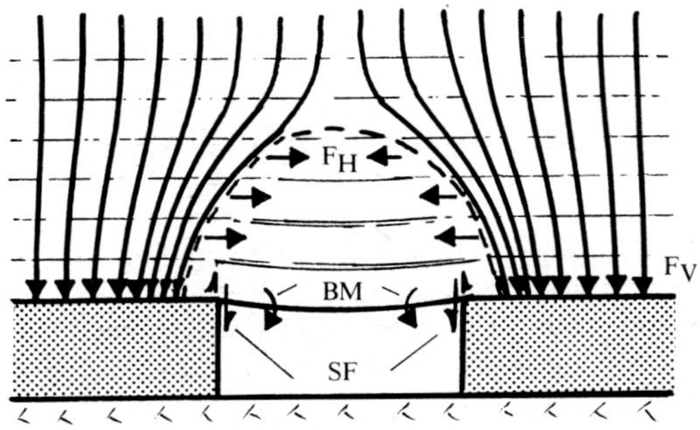

F_V = Vertical compressive forces
F_H = Horizontal compressive forces
BM = Bending moment of force
SF = Shear force

Figure 17 Distribution of forces in the vicinity of the roof of a narrow roadway according to NCB (1950-51).

fractured rock which attempts to fill or partially fill the excavated void. The second is essentially a collapse process involving unconsolidated overburden.

Figure 18(a) shows the characteristic dome shaped collapse region immediately above a mine roadway. The height of the dome is governed by three principal factors:

1. the strength of the immediate roof,
2. the bulking properties of fractured rock within the collapse dome,
3. the dimensions of the mining excavation.

Considering the strength of the roof of a mine roadway in relation to its collapse potential the most important aspect is its ability to resist fracture under the states of stress to which it is imposed; Figure 17 depicts a commonly encountered stress condition for mine roadways. The roof beds are subjected to various combinations of tensile, shear and compressive stresses. It would appear that the ability of the immediate roof to resist tensile and shear stresses is of major importance. Coupled with this, however, is the combined influence of the inherent weaknesses within the rock structure, such as slip planes, jointing, bedding planes, faulting etc, which play a significant role in triggering the collapse process. The presence of strong beds within the rock mass can discourage the upward collapse of the roof beds above a mine roadway, and such an effect is in evidence in Figure 19. This shows an ancient mine roadway (150 - 200 years old) in the Nottinghamshire Coalfield where the previous supports, if any, have long gone, and a collapse dome has formed.

It is clear from Figure 19 that the overlying sandstone bed has tended to assist in limiting the upward development of the collapsed roof of the roadway, although it would appear that some general subsidence of the sandstone itself has occurred above the mining excavation.

The basic theory of bulking of rock due to its collapse resulting in fracturing and general disturbance, has been applied in mining engineering to explain why surface subsidence is less than the excavated height. Weak rocks such as mudstones readily fracture and result in an appreciable increase in volume over their original solid state condition. The concept that broken rock will eventually fill an entire void and thereby result in restoration of a state of equilibrium within that part of the rock mass is well-founded and undoubtedly applies in

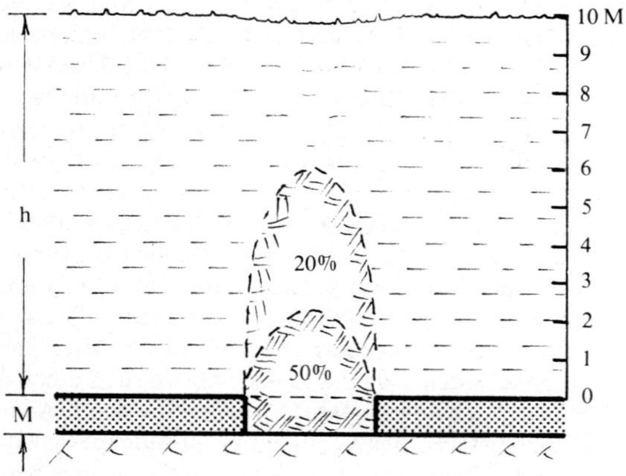

(a) Projected extent of collapse of mine roadway roof for bulking factors 20% and 50% in stratified rocks which readily cave.

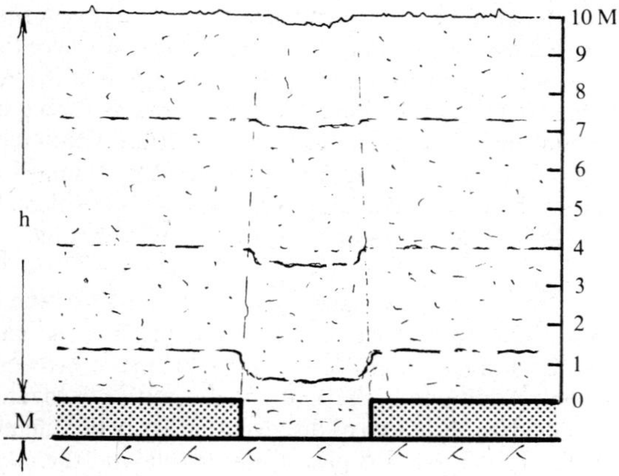

(b) Collapse characteristics of unconsolidated overburden (sands) into a mine roadway at shallow depth.

Figure 18 Ground movement characteristics above a mine roadway.

many situations. In the case of a roadway there is a greater likelihood of a well-defined collapse dome forming but as shown in Figure 19, it may not necessarily completely fill with broken rock to the extent of providing a supporting medium itself. Undoubtedly, the local geology plays a role in this respect. Sandstones are less prone to bulking in shallow mining situations than say mudstones. Referring to Figure 18(a), two levels of bulking of the collapsed roof rocks are considered, these being 50% and 20% respectively. The 50% bulking factor assumed that the height of caving will extend to twice the roadway height (from the roof level), and this appears to correspond closely with the actual situation shown in Figure 19. From the authors' own observations in many different mining conditions, the 50% bulking factor appears to be the most commonly encountered. Also shown in Figure 18(a) is the 20% bulking factor which assumes a caving height of five fold the roadway height. This would appear to be an extreme condition and one which would be rarely encountered in a normal mine roadway situation; rocks of low bulking factor tend to be strong. Consequently this would discourage upward collapse of the dome by virtue of attempting to span the opening thereby giving a natural supporting action. High collapse domes do occur, however, in particular situations but there needs to be a mechanism for removal of the previously collapsed material in order for more to subsequently form and the dome increase in height. Domes of increased height do not appear to significantly increase their width in relation to the original roadway excavation size. This aspect is discussed in greater detail in a subsequent chapter. The occurrence of subsidence at the surface above such a collapse dome in a mine roadway depends upon the depth below surface. Clearly if the depth is so shallow as to permit the dome to intercept the surface then a potential subsidence hazard exists. In deeper situations, however, where the dome may not quite reach the surface before it has choked by virtue of bulking, then no subsidence of the surface will occur if the remaining thickness of cover is adequate or it is strong enough to allow stable spanning of the disturbed ground. In some cases a small amount of subsidence may occur due to general sagging of the overlying beds.

Figure 18(b) shows the ground movement characteristics above a shallow mine roadway overlain by unconsolidated overburden, particularly sands, and where the roof is allowed to collapse. The ground movement process involves shearing above the roadway sides and a relatively confined column of material slumping downwards as shown in Figure 18(b). The effects of this form of movement can cause surface subsidence within ten fold of the roof of the roadway. But such effects generally appear to be no more than a surface depression of limited dimensions, unless there is some underground mechanism which allows removal of the caved material and thereby creates scope for the inflow into the mine.

The dimensions of the mining excavation govern the extent of the collapse dome. The width of the excavation needs to be sufficient to allow the induced stresses caused by re-distribution to exceed the natural strength of the immediate roof beds otherwise no collapse process would be initiated. Additionally the width of the mining excavation has a bearing on the potential height of the dome, although as discussed earlier in this chapter, the strength and stratified character of the roof beds also play a major role and the combined influence of these two factors usually occurs. The case of a mine roadway of relatively limited width, by comparison to a longwall extraction, permits a well-defined collapse dome to develop in time providing the rock stresses and geological characteristics are conducive to its formation. Wide excavations may not allow a well-defined collapse dome to develop owing to insufficient shearing movement at the potential dome boundary particularly at positions remote from the excavation. The extraction height has a direct influence on the potential for a collapse dome to form in the first place; it is this factor which mainly allows the scope for shearing movement to develop and produce a clearly defined collapse feature. The extraction dimensions are principal factors in giving rise to the potential for roof collapse and the possibility of effecting the surface by way of subsidence in the case of mine roadways. Other

Figure 19 Formation of collapse dome above ancient abandoned mine roadway.

factors, however, such as mining depth, rock strengths and geology may play more dominant roles in particular situations.

Ground movement associated with wide extraction

The ground movement directions associated with a wide extraction such as a longwall face have been described by Grond (1951, 1957) who summarised the ideas of German and Dutch investigators up to and around the time of his publications. Figure 20 illustrates the main points discussed by Grond concerning projected ground movements for both the transverse and longitudinal directions of advance of a longwall. Figure 20(a) is clearly based on Dutch concepts of ground movement, with general movements attracted directly towards the mined-out area with a certain amount of upward movement indicated beneath the mining horizon. Figure 20(b) is based on earlier German concepts of ground movement interacting with a pressure arch, or dome of relaxed ground above the extracted region. The idea is conveyed here that a break-line projected ahead of the face line turns sharply towards the face after having intercepted the pressure arch. Grond suggests a complex pattern of movement around the longwall face region coupled with an upward tendency for the floor beds. This ground movement pattern immediately above and ahead of the working face of the longwall has not been confirmed by later investigations, see for example Kratzsch (1983) and Whittaker, Reddish and Fitzpatrick (1985). The pattern of movement immediately below the longwall extraction as described by Grond would seem to be quite logical, although the magnitude of such movements appear to be minor in relation to those above the extraction horizon.

The ground movement pattern associated with a longwall extraction based on recent investigations is shown in Figure 21. A principal feature is the line of shearing which hades over the extracted region, and this shear line extends approximately linearly as the thickness of extraction increases, see Whittaker, Reddish and Fitzpatrick (1985). There is no indication of this shear line attempting to deviate towards or over the longwall face as suggested in the concepts discussed by Grond. Indeed, Figure 21 indicates a tendency for ground movements to approximately rotate around the coalface in the case of the advancing face line and this is accompanied by successive occurrence of shear lines which play a role in the mode of ground movement behind the face. A certain amount of differential movement between sheared blocks occurs together with the flexing of the roof beds. The transverse section of the longwall, Figure 21(a), shows the establishment of symmetrical shear lines hading over the extraction with a rotational tendency for ground movement to occur from the ribsides; these shear lines are permanent features and can remain open over appreciable distances unless, of course, the solid ribsides were removed. In the case of the shear lines shown in Figure 21(b), these are connected with the process of advancing the longwall and tend to close up again at some distance behind the face-line; this appears to take effect at about 1/4 to 1/3 (depth) from the face.

The ground movement process is linked with the re-distribution of strata pressure around the longwall. Figure 22 shows the pressure arch condition as related to sub-critical, $w/h < 1{\cdot}2 - 1{\cdot}4$ and super-critical, $w/h > 1{\cdot}2 - 1{\cdot}4$ extraction dimensions in relation to depth. Figure 22(a) indicates that natural arching across the extraction is taking place thus preventing full return to cover load pressure within the goaf area and such a process discourages full development of surface subsidence. Conversely, Figure 22(b) shows the pressure arch to intercept the surface and this allows full return to cover load pressure over part of the extracted area and this is connected with development of maximum surface subsidence.

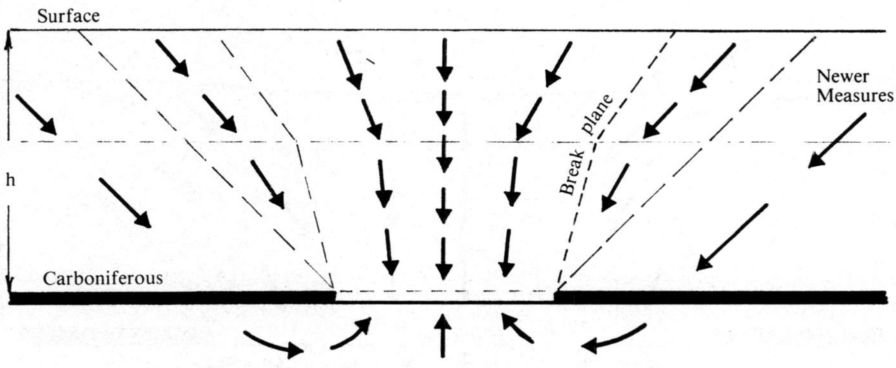

(a) Ground movement directions across longwall extraction.

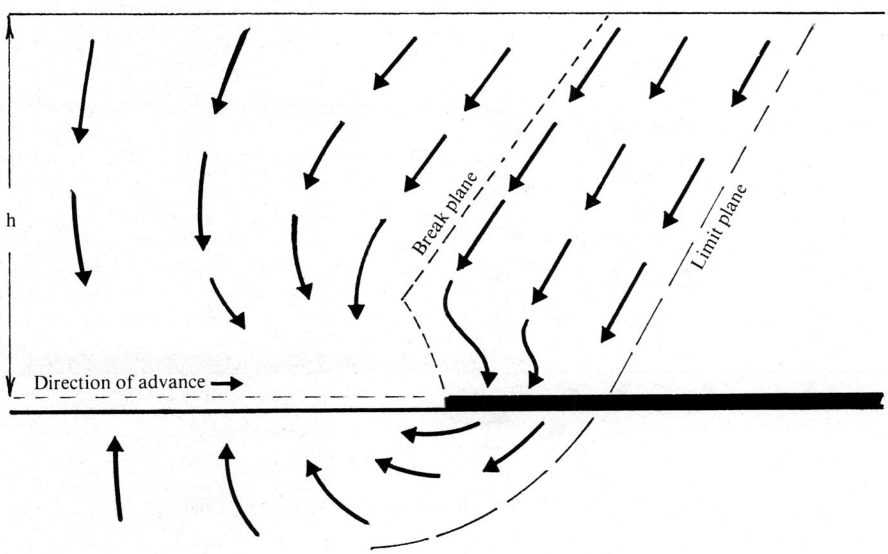

(b) Ground movement directions associated with advancing longwall face.

Figure 20 Ground movements around a longwall extraction based on Dutch and German theories according to Grond (1951, 1957).

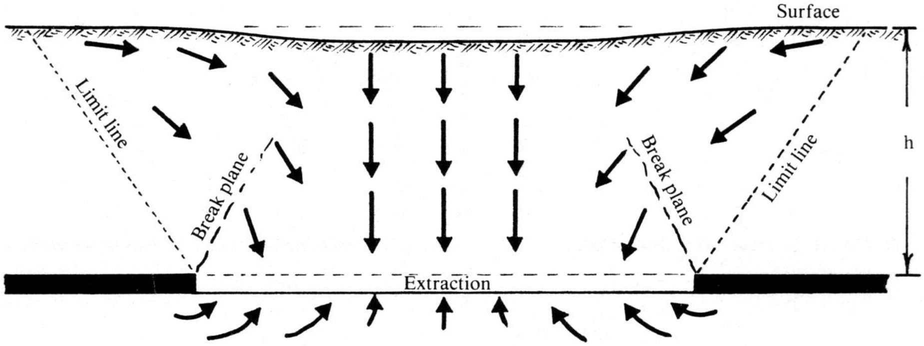

(a) Ground movement directions across longwall extraction

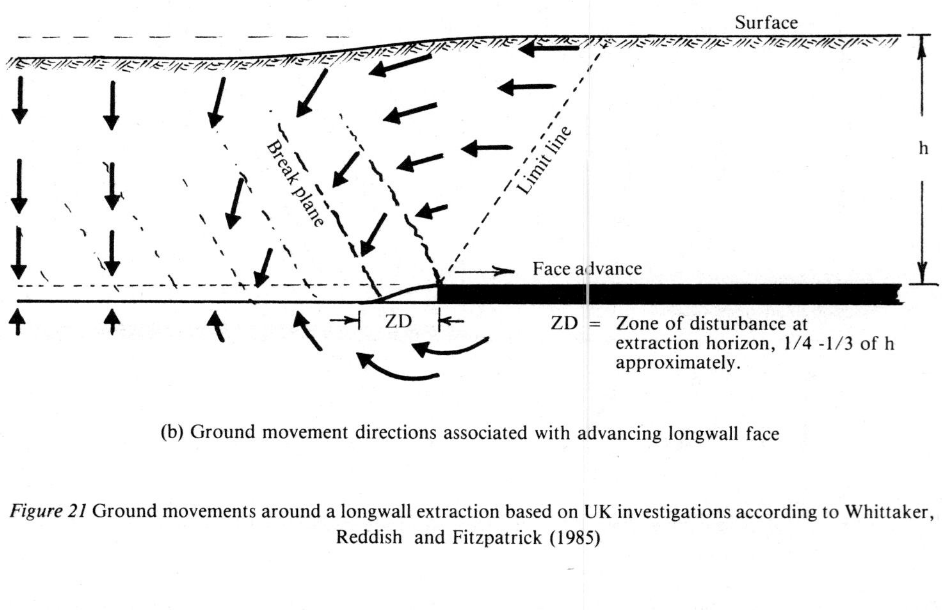

(b) Ground movement directions associated with advancing longwall face

Figure 21 Ground movements around a longwall extraction based on UK investigations according to Whittaker, Reddish and Fitzpatrick (1985)

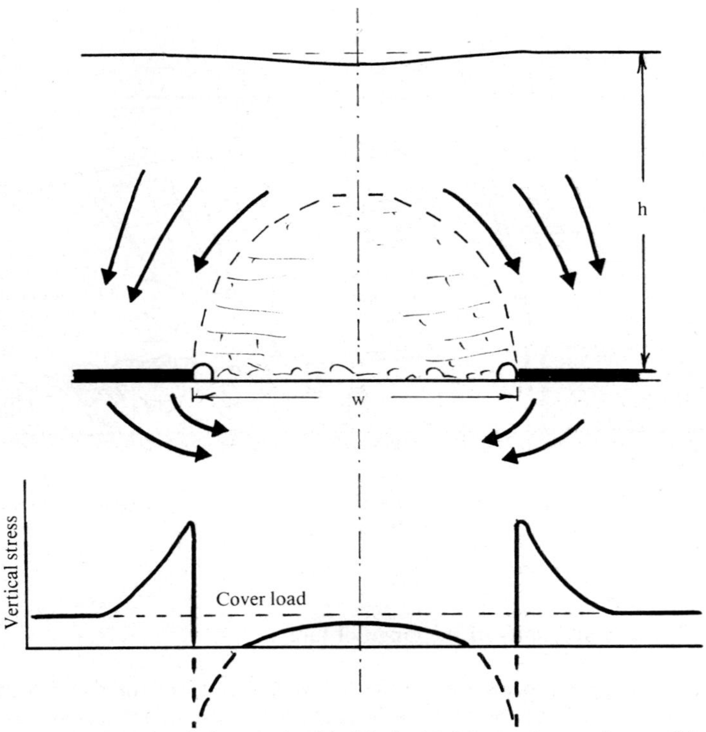

(a) Pressure arch in relation to sub-critical (w/h < 1·4) longwall extraction condition

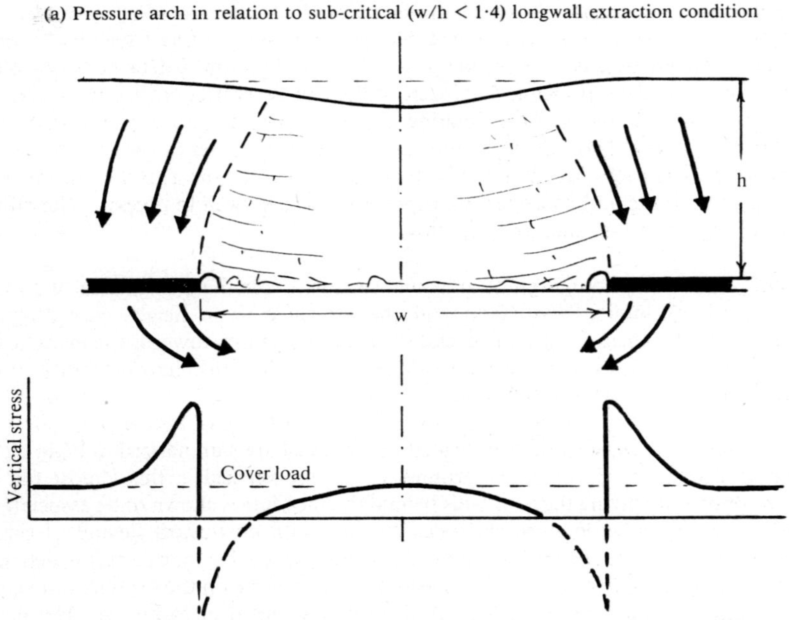

(b) Pressure arch in relation to super-critical (w/h > 1·4) longwall extraction condition

Figure 22 Illustrating arching of strata pressure across a longwall extraction in relation to subsidence according to Whittaker and Pye (1977)

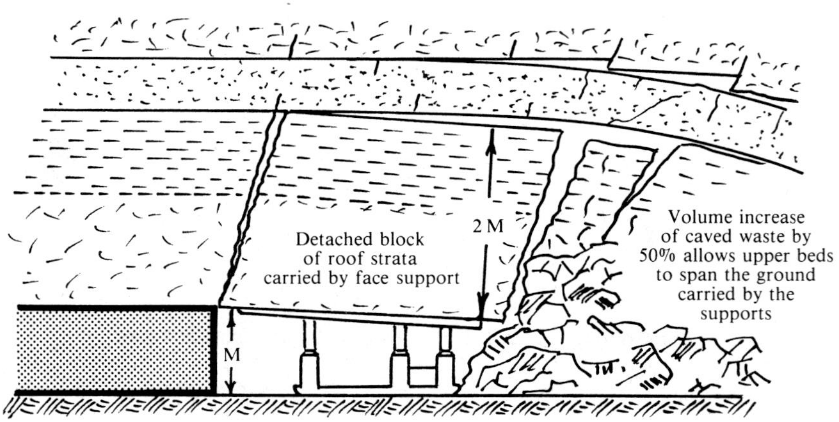

Figure 23 Ground control on longwall face and generation of support loading by detached block of roof strata, from Whittaker (1974)

Ground movement in proximity of the longwall face

The natural arching process shields the longwall face from the full effects of strata pressure due to the cover load. Support of the roof on a longwall face is mainly based on the principle illustrated in Figure 23 which shows a detached roof block being carried by the support system. The caved waste is assumed to generate self-supporting resistance by virtue of its bulking characteristics. A bulking factor of 50% would allow a detached roof block to form of height 2M. This principle has been utilised in several countries with major coal mining industries, as a means of estimating the minimum resistance of the face support system, see Whittaker (1974). The variations in geological character of the roof strata of coal seams, however, has necessitated ample factors of safety to be incorporated in the detailed design of such supports and consequently support resistances well in excess of the minimum indicated by Figure 23 are commonly employed.

Figure 24 shows a photograph of an actual section across the central area of a longwall face; the depth of mining was 600m and the extracted seam height was 0·94m. The photograph indicates the existence of several shear planes hading towards the extracted area. The ground ahead of the longwall face is relatively undisturbed especially the strata above the seam and beyond 1·5 - 2M into the solid.

The ground movement features indicated in Figure 24 are summarised in Figure 25. The main line of fracture is shown to be around 26° to the vertical in this almost flat seam. Ground movement involving the roof beds behind the face-line is shown to be associated with differential slipping between sheared blocks coupled with downward flexing of the beds. Immediately behind the face support system, the process of caving takes effect which in weak strata may simply involve the collapse of sheared blocks resting on the support or in stronger strata the falling of fractured roof beds previously suspended in cantilever. The detached rocks tend to act as a cushion above the caving-effected zone. The extent of the roof fractures depends upon the extraction height and the nature of the immediate roof beds. This latter aspect is discussed in a later chapter.

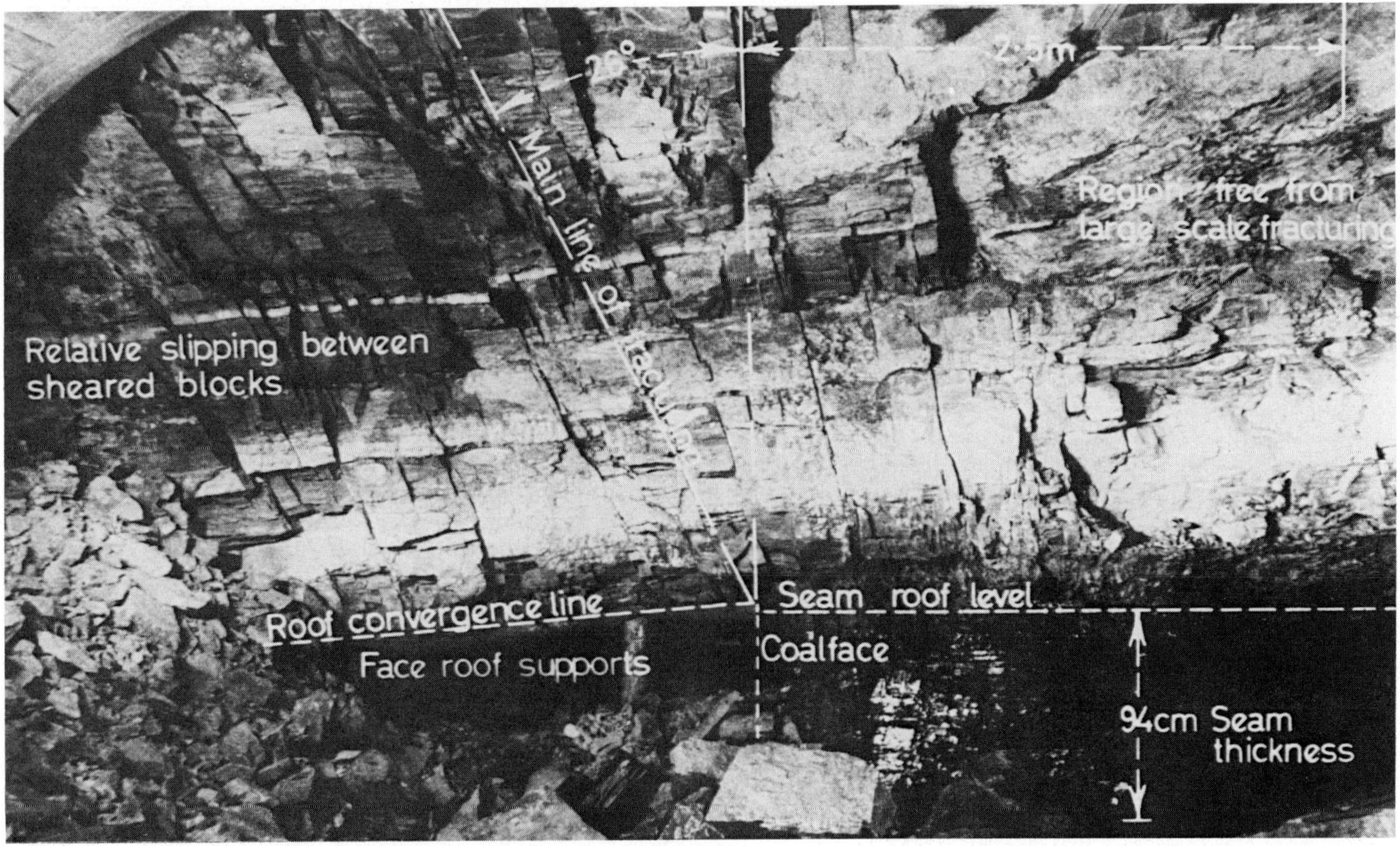

Figure 24 Roof fractures associated with longwall face

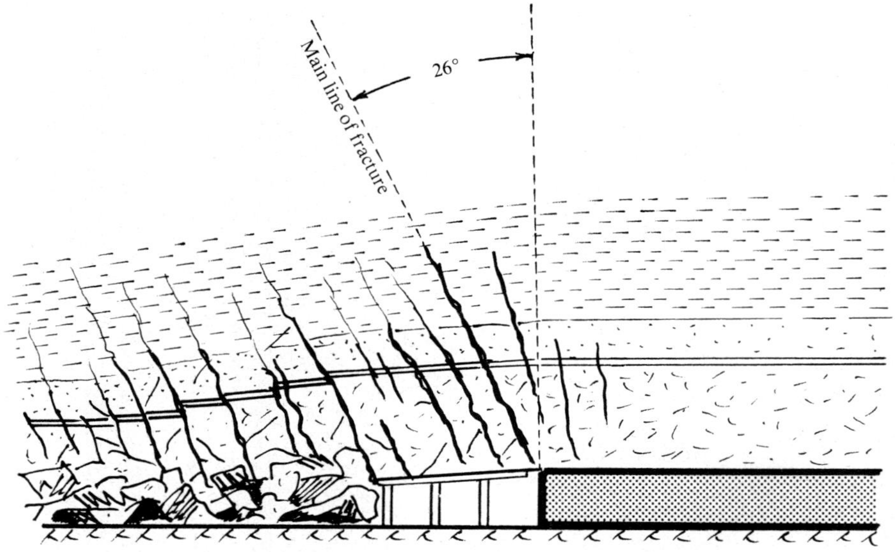

Figure 25 Illustrating principal mode of roof subsidence immediately behind a longwall face

Notes 1. Main fracture lines are based on actual observations and particularly Figure 24
2. The roof strata were Coal Measures rocks
3. Goaf control is assumed to be full caving of the waste area

Influence of type of mining operation on the mode of resulting surface subsidence.

Underground mining operations are adopted to suit the character of the mineral deposit being exploited and the geological and geotechnical conditions which prevail at the time and place of mining. The value, and distribution of values, of the mineral(s) within the deposit; the strength properties of the mineral, its host rock if applicable, and the surrounding rocks; the existence of special mining hazards; the location of the mineral deposit; the geometry of the deposit, dip and its depth below surface, are all major factors influencing the choice of mining method.

The incorporation of support within mined-out areas and particularly as a permanent feature within a working stope forms a principal component of some mining methods. This is especially so where the size and geometry of the ore body and the strength of its side-walls necessitate such support for safe practice and adequacy of ground control for effective mining.

Discussion is focussed here on particular mining operations which generally result in some form of surface subsidence. The influence of the mining method on the general character of any ensuing subsidence at the surface is briefly examined and its relevance to control is also considered. The following examples are not intended to indicate the wide spectrum of mining methods, but rather serve to bring to attention the potential problems of surface subsidence associated with some of the more common methods.

Solution mining. This method has been used extensively in the extraction of salt and Figure 26(a) illustrates the basic principle and its commonly associated collapse feature. The solution cavities are generally unknown regarding their extent and size. When such cavities

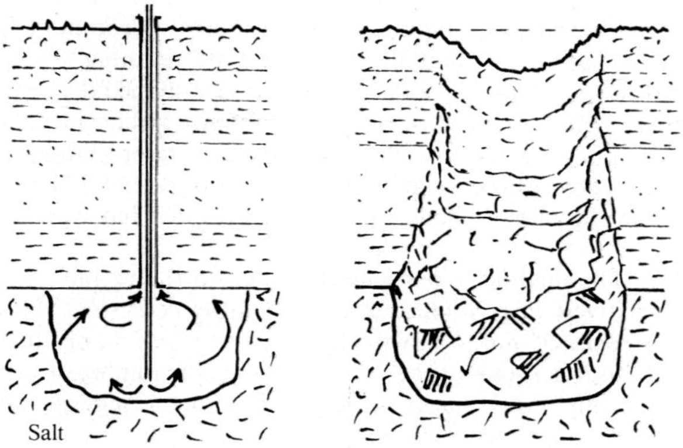

(a) Subsidence arising from solution mining

Notes
1. Size and extent of solution cavities usually unknown
2. Potential for later subsidence of surface depends on strength properties and thickness of overlying beds of rock
3. If an established subterranean network of ground-water flow exists at the pre-mined horizon, i.e. in the salt bed, then further subsidence can be anticipated in the form of a deepening depression or as a sink-hole

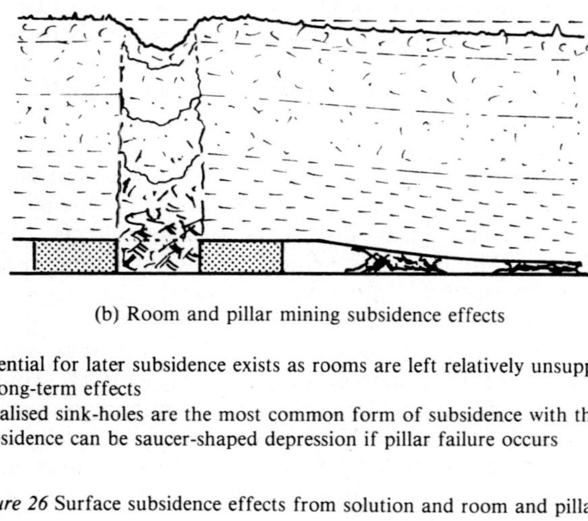

(b) Room and pillar mining subsidence effects

Notes
1. Potential for later subsidence exists as rooms are left relatively unsupported in respect of long-term effects
2. Localised sink-holes are the most common form of subsidence with this type of mining
3. Subsidence can be saucer-shaped depression if pillar failure occurs

Figure 26 Surface subsidence effects from solution and room and pillar mining

are abandoned, their unknown dimensions and extent can present a significant problem. If water is allowed to flow through inter-connecting cavities in a salt bed, then further removal of salt in solution becomes inevitable. The form of subsidence collapse is generally of the sink-hole type with a collapse chimney migrating towards the surface. Subsidence of the surface can be either as a saucer-shaped or conical depression, or the sudden appearance of a cavity as can occur with collapse of surface capping beds spanning a migrating collapse chimney.

Room and pillar mining. This is probably the oldest form of mining method and many old mine workings exist world-wide. Figure 26(b) indicates two common forms of surface subsidence which arise, the first being the sink-hole type resulting from collapsed mine junctions whilst the second is that of a widely spread saucer-shaped depression as can occur following pillar failure. In both situations, surface subsidence has occurred many years after abandonment of the mining operations. Where there is an appreciable risk of sink-holes appearing at the surface above old room and pillar operations, then the risk may continue for many years especially in the case of shallow coal mine workings having a weak overburden.

Longwall mining. The method is particularly suited to the extraction of seam type of mineral deposits of relatively large lateral extent and fairly consistent thickness. It can be employed to extract deposits of almost any thickness by a succession of lifts. It is not limited by depth of mining. Indeed it is one of the safest methods for mining at great depths where special rock stress problems pose a hazard to safe operation. Longwall mining has special advantages in respect of the following:

1. efficient extraction of mineral deposits,
2. economically attractive in relation to other methods,
3. offers high productive capacity in lending itself to concentration of mining operations, and
4. permits controlled subsidence which can be accurately pre-assessed in terms of magnitude, effects and duration.

Figure 27(a) demonstrates how limiting the width of extraction and using an intervening rib pillar, the surface subsidence can be reduced and if appropriately designed the respective subsidence profiles from each extraction can interact to produce a smoother subsidence profile. The example given is the most commonly applied method for reducing mining subsidence effects at the surface in the UK.

Where mining from two or more seams in the same vertical succession needs consideration to be given to surface subsidence, the advantageous use of rib pillars can considerably assist in this respect. Figure 27(b) shows such an example. The staggering of rib pillars and extractions within the same vertical succession permits excellent scope for interaction of surface subsidence profiles to produce surface ground strains of lesser magnitude due to a smoother resulting final profile. It should be noted, however, that the upper seams should have been mined firstly before mining the lower seams otherwise live interaction can effect the stability of current workings at higher levels. Multi-seam situations call for careful planning and phasing of layouts in order to minimise subsidence and strata pressure interaction effects between workings. Lower seams, especially where they offer a greater economical return, are sometimes mined before upper workable seams. This is entirely feasible but as stated above a carefully phased plan of extraction is required if the upper seams are to continue remaining workable.

Top-slicing. The successive removal of descending slices and the support of the ground overlying the current operation, whilst mining is completed, is applied to mineral deposits of various types and geometries. Figure 28(a) shows diagrammatically the operating principle of top slicing. As the current mining operation is in progress the ground above is supported and thereafter allowed to cave onto the top of the next slice. In many metal mining operations, the wooden supports are not recovered following withdrawal of support within the current slice and consequently allows a mat of timbers to build-up at the interface between the caved material and the next slice thereby effecting a measure of roof control and particularly assists in preventing large scale inflows of waste rock. Where top-slicing is applied to thick coal deposits, the caving process takes place immediately behind the working face, the main

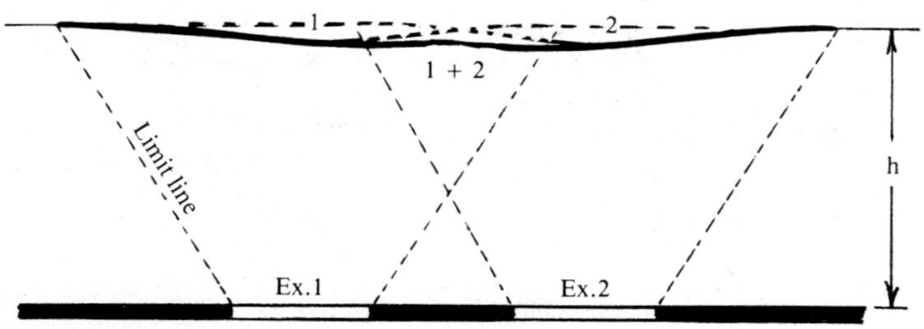

(a) Interaction of surface subsidence profiles from extractions in the same seam

Notes
1. Subsidence of surface is gradual and spread over an area exceeding that mined out
2. Surface subsidence is predictable with considerable accuracy in most mining situations
3. The duration of subsidence is predictable and is relatively short-lived by comparison to other mining methods

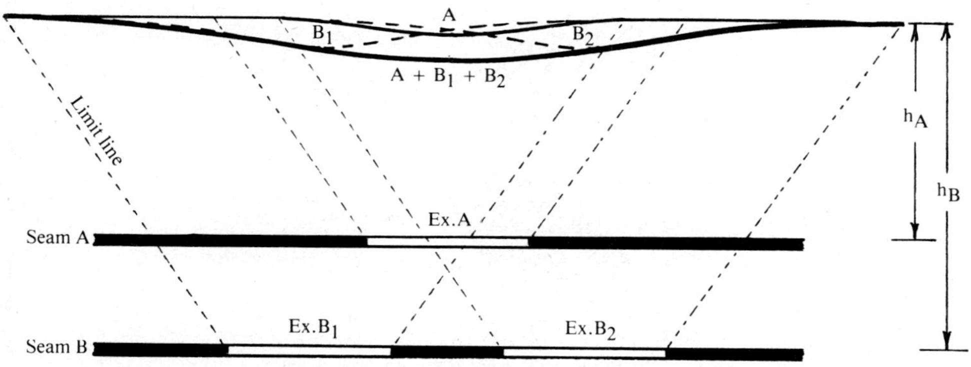

(b) Interaction of surface subsidence profiles from extractions in different seams

Notes
1. Staggering pillars under extractions can produce a smoothing effect to the resultant subsidence profile
2. The example presupposes Extraction A to have finished before Extractions B_1 and B_2 commence

Figure 27 Surface subsidence arising from longwall mining involving interaction of profiles

supports advancing forward and in many situations a layer of steel wire meshing is left at the interface between the caved rock and the top of the next slice. Several slices can be taken, and the build-up of wire meshing at the base of the caved zone controls the formation of cavities within the waste area in addition to controlling dilution of the mineral product. Top-slicing allows efficient removal of mineral deposits and is applied where the side walls are weak and present caving problems which need control. The form of mining subsidence accompanying top-slicing is the gradual formation of a surface depression with appreciable tension cracks owing to the localised concentration of strains. In some parts of the world, the subsidence depression may fill up with water, creating lakes of significant size and consequently presents the additional hazard of the risk of such water gaining access to the mine; some mines pump the water from large surface depressions overlying current operations, whilst most others rely on the overburden retaining its impermeable character.

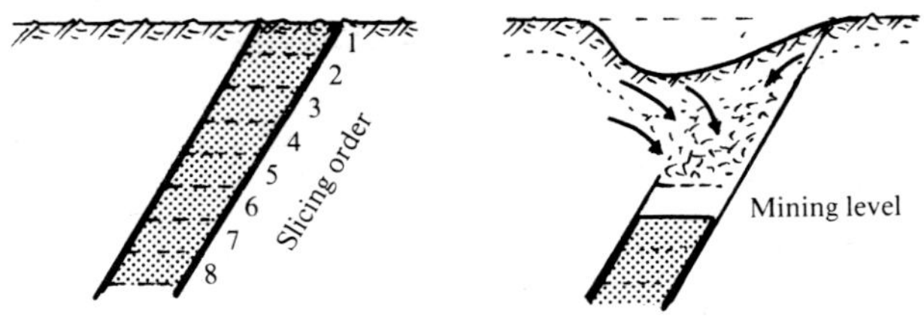

(a) Surface subsidence from top-slicing

Notes 1. Subsidence tends to be gradual with a surface depression developing as mining proceeds deeper
2. Surface extent of subsidence depression depends on nature of surrounding rocks and overlying superficials

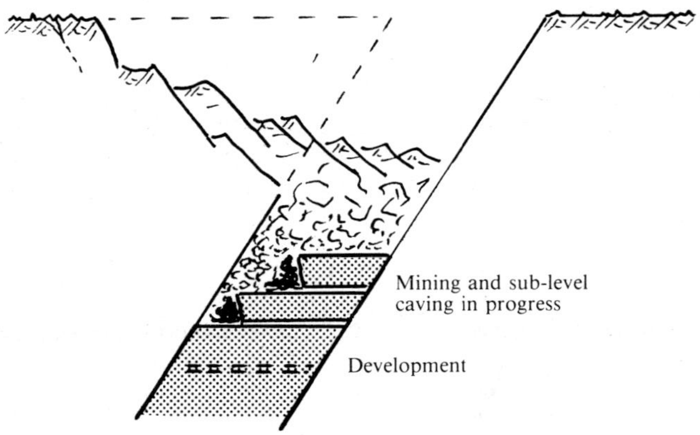

(b) Surface subsidence from sub-level caving

Notes 1. Main form of subsidence is slumping of hanging-wall into mined out area; foot-wall failures are comparatively less common but can occur in particular geological conditions
2. Progressive failure of hanging-wall tends to decrease as mining progresses deeper

Figure 28 Surface subsidence effects from top-slicing and sub-level caving methods of mining

Sub-level caving. This is a fairly popular choice of mining method where the geological conditions favour its adoption. Figure 28(b) illustrates the method of mining in a classical situation involving a strong ore body which is dipping steeply. Blasting of sub-levels is common practice. Ore loading triggers further caving and subsidence in a progressive manner. The character of surface subsidence displayed in Figure 28(b) corresponds to the strong ore situation with comparably strong side-walls; the hanging wall mainly subsides. Sub-level caving is also a successful method for mining thick underground deposits of lignite in Yugoslavia; there is often no need for blasting of the coal forming the sub-level although steel wire mesh is frequently used to control the caved mass and avoid waste rock flowing into the mined product. Surface subsidence over sub-level caving operations as observed by the authors in Yugoslavia produce large depressions with major fissuring and surface stepping.

Block-caving. The controlled caving of a block of weakened ore requires its break-down to sizes which permit its discharge through cones at the base of the block and through finger raises into underlying transport drifts. The ore needs to possess adequate natural weakness as

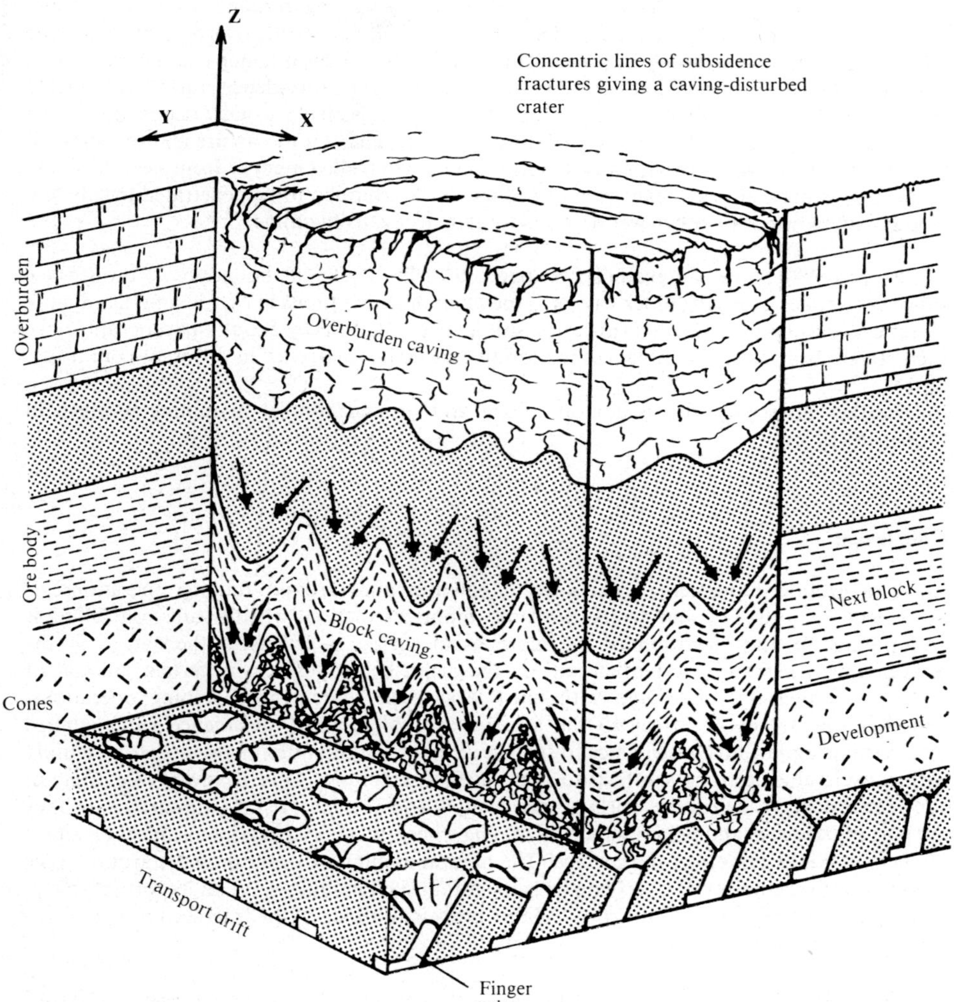

Figure 29 Effect of block caving on surface subsidence development

Notes
1. The block mainly transmits its effects to the surface vertically giving rise to a caving-disturbed crater with concentric lines of surface fractures
2. The effects of subsidence can increase laterally if rocks allow an appreciable draw to develop

to allow progressive caving without the need to resort to excessive blasting. There is the need to exercise close control over ore withdrawal otherwise ore dilution with waste rock will occur; ore withdrawal needs to be uniform from all its draw points. Subsidence of the block of ore is illustrated in Figure 29 with the movement of ore tending to funnel towards the draw-points. Controlled withdrawal of ore from the block will permit the overburden to subside with a measure of uniformity and result in a surface subsidence crater characterised by a tendency to form concentric lines of subsidence fractures. The subsidence crater has its major effects immediately over the mined-out block although subsidence usually occurs outside of this area and is subjected to an angle of draw which depends on the nature and thickness of the overburden. Should ore withdrawal occur in an uncontrolled manner from just one draw-point, then a draw ellipsoid can form and allow the overburden to be penetrated locally and possibly allow water or other wet deposits to gain access to the mine.

Stout (1980) shows an example, on page 39 of his book, of the nature of the surface subsidence crater overlying block caving in metal mining operations in North America. Figure 29 reflects closely the surface features of subsidence indicated by Stout's photograph since such craters bear close resemblances irrespective of their occurrence in different countries.

Further interesting discussion on caving and ground control in mining is given in Morrison (1976).

Conclusion

The ground movement process accompanying underground mining operations, is dependent upon the size and geometry of the mining extraction, whether any void filling is incorporated in the operation or the more usual caving is practised, the nature of the surrounding rocks and overburden and on the depth below surface. A certain amount of surface subsidence is associated with most large scale underground mining operations, although the degree of effect to the surface can range from subsidence being almost indiscernible and of no consequence to surface structures to that of a highly localised feature with considerable destructive effects to some surface structures. Mine roadways are rarely of subsidence concern unless the workings are shallow and present a risk to the surface, or they are part of the room and pillar method of mining where under certain circumstances collapses of such workings can effect the surface. The natural arching of strata pressure around mine roadways and the ability of overlying beds to sustain such arching plays an important role in limiting upward migration of the roof collapse process. Bulking of the fractured roof beds is a further important factor.

Wide extractions create the opportunity for the caving and subsidence process to attain a state of surface stability in a relatively short time span after mining has ceased and the longwall method of mining features prominently in this respect. Longwall extractions lend themselves to controlled surface subsidence with respect to both time and position.

Other mining methods give rise to particular forms of surface subsidence. A clear understanding of the mechanisms of ground movement generating such subsidence greatly assists in planning mining operations to achieve greater control over effects to the surface and avoiding the creation of potential paths for water inflow into underground mining operations.

CHAPTER 4

METHODS OF PREDICTING MINING SUBSIDENCE

The basic aim of all prediction methods is to produce an acceptably accurate assessment of mining subsidence and in some cases its associated effects on surface structures and ground conditions. Most of the studies on developing such prediction methods have been based on coal mining situations. Of particular importance here has been the fact that seams of uniform thickness and relatively large horizontal and regular extent of extraction have provided the basis on which prediction methods could be satisfactorily developed.

Efforts have been made in many coal mining countries to develop mining subsidence prediction techniques, and generally the starting point has been to ensure that the pre-calculation method follows closely the phenomenon as observed in practice. There are many publications dealing with predicting mining subsidence, some covering theoretical concepts whilst others resort to applying observed surface behaviour in the form of empirical prediction models. This chapter reviews the more important developments in methods of predicting mining subsidence and discusses the advantages and limitations of various methods in relation to particular mining conditions which prevail in some countries.

Basic characteristics of trough subsidence

The methods of subsidence prediction are virtually all directed towards considering the creation of a surface depression above an extraction of wide extent in relation to its extracted thickness. It is assumed that the extraction has progressed sufficient to allow development of maximum subsidence so that the problem can be considered in two-dimensions. Additionally only one extraction is generally examined in the development of a basic model, so that solid ribsides exist with no previous mining effects.

There are three fundamental conditions which influence the character of mining subsidence trough development at the surface:

1. **Subcritical extraction.** The width/depth (w/h) ratio of the extraction is insufficient to allow development of the full potential subsidence owing to a degree of natural arching of the strata across the extraction. It is a situation which commonly exists in deep mining layouts. For U.K. coal mining conditions this corresponds to

$$w/h < 1{\cdot}4$$

2. **Critical extraction.** The width/depth (w/h) ratio is just sufficient to allow development of the full potential subsidence. For U.K. coal mining conditions this corresponds to

$$w/h = 1{\cdot}4$$

3. **Supercritical extraction.** The width/depth (w/h) ratio allows development of the full potential subsidence. The extraction width is too great in relation to the depth to allow natural arching across the ribs on each side of the extraction. For U.K. coal mining

conditions this corresponds to

$$w/h > 1{\cdot}4$$

Figure 30 demonstrates the subcritical, critical and supercritical extraction conditions for mining subsidence in flat seam conditions. The subsidence profile (vertical component of displacement) is confined between the limit lines shown as angle of draw, with subsidence occurring outside of the plan-area of extraction.

Subsidence of the surface is accompanied by horizontal displacements which act towards the area of maximum subsidence. The magnitude of the horizontal displacements are an approximate function of the gradient of the subsidence profile.

In the regular geometrical extraction configuration shown in Figure 30, the maximum subsidence would occur over the centre of the mined-out area section, and the horizontal displacements would be zero at that point. A state of symmetry exists as displayed by Figure 30. The horizontal displacements result in differential movements which cause states of strain to occur along parts of the subsidence profile, these strain states being in both compression and tension. The subsidence profile itself can introduce significant elements of slope and curvature which can have a major influence on surface structures under certain conditions.

A fundamental assumption of all prediction methods is that the surface subsidence cannot exceed the extracted seam height. Field observations indicate that a maximum value of surface subsidence appears to be 0·9M, where the extraction conditions permit its development. There are isolated situations, however, where subsidence has exceeded the extracted seam height especially where large quantities of water have flowed out from undermined strata and in multi-seam workings particularly where old extraction areas have been undermined and given rise to reactivation of subsidence. The latter can be particularly significant where longwall operations take place under old abandoned room and pillar workings which exist at shallow depths.

The magnitude of the full potential subsidence is a linear function of the extracted seam thickness. A single extraction of a given height is assumed to produce the same subsidence as two extractions of only half of the formers' extraction thickness; this is providing the extraction width and depth below surface are virtually the same as can occur with top slicing variations. The following general expression (2) is commonly employed in subsidence prediction assessments.

$$S_{max} = a\,M \qquad \ldots (2)$$

where, S_{max} = maximum value of subsidence
M = extracted seam height
a = subsidence factor

The subsidence factor (a) is dimensionless and is dependent upon the nature of the overburden and type of goaf treatment such as total caving or some form of stowing. The value of the subsidence factor has been established by observations in different conditions and in several countries. In practice, the value of the subsidence factor (2) corresponds to 0·9 based on observations in France, Germany and the UK whilst stowing results in the factor being generally around 0·4 to 0·5 (pneumatic) and even as low as 0·12 (hydraulic sand stowing), the latter being observed in Poland, according to Bräuner (1973a). Table 5 gives a range of values for the subsidence factor as observed in different countries.

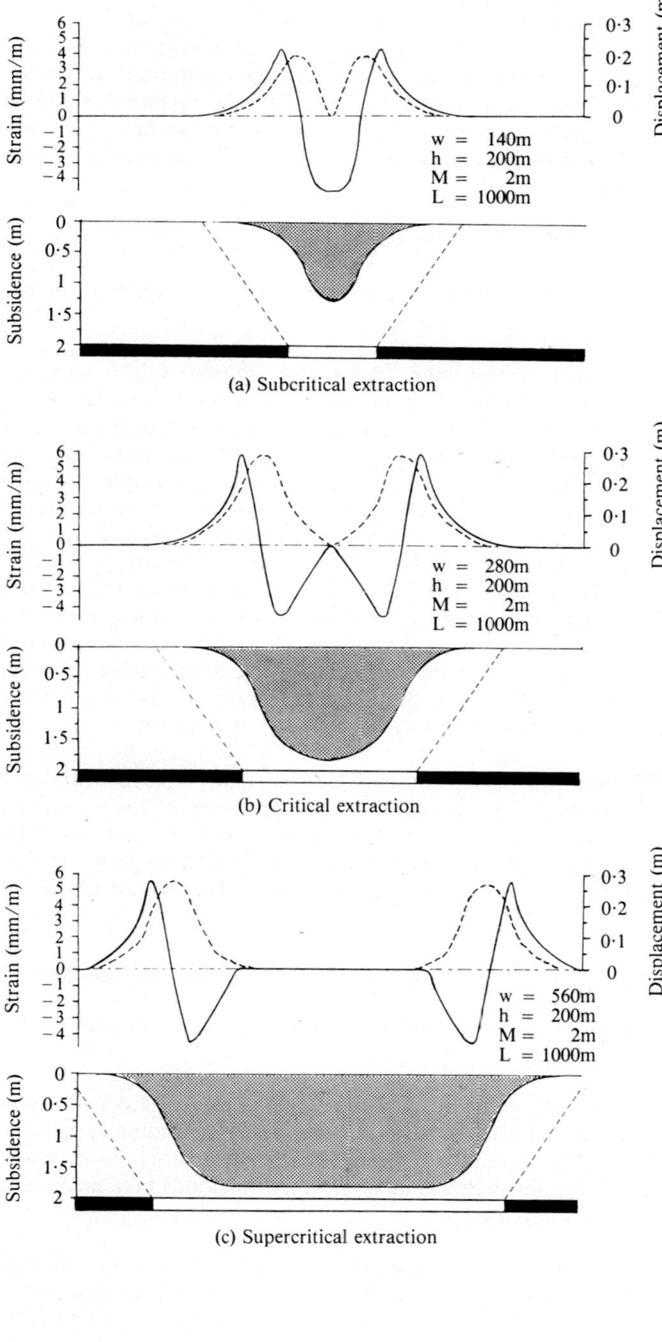

Figure 30 Comparison of subcritical, critical and supercritical mining subsidence situations

Bräuner (1973b) draws attention to the basic assumptions used in subsidence predictions as being mainly based on the principle of superposition of surface displacements. Those subsidence profiles which exhibit overlapping are assumed to produce a resultant which is the simple addition of the vertical displacements. A further important aspect is that of the principle of equivalence of extracted areas with respect to the surface and assumes that where an extraction has straight lines projected from its edges to that surface point, then extractions at different depths but extending across the limiting lines will produce the same displacements on that surface point.

Principal methods of predicting mining subsidence

The various principal methods of predicting mining subsidence can be grouped as follows:

1. **Empirically derived relationships.** Within this grouping falls a number of formulae which have been established from observed surface subsidence behaviour and consequently they generally apply specifically to the ground and mining conditions which prevail where the observations were made. Some formulae have been derived for a specific mining locality and are known to give a fairly accurate assessment of mining subsidence. In other cases subsidence observations have been collated for major mining areas or even countries, and the data rationalised for general prediction purposes. Such subsidence prediction models are factual and provide a basis for comparison with values predicted by other methods. Empirically derived relationships rely on sufficiently large numbers of observations and case studies in order to establish levels of accuracy for prediction of anticipated mining subsidence. The UK NCB empirical model (SEH) has an accuracy of generally better than 10% of the seam extraction for prediction of the amount of resulting subsidence. Such methods can be used to distinguish between the effects of different types of surface geology and overburden.

2. **Profile functions.** This group contains methods that predict profiles in set directions across an excavation. They generally use the geometry of the excavation along with equations or tables of data to predict a longitudinal or transverse profile for subsidence or strains. This approach has been used very extensively throughout the world because it is relatively easy to use and apply to new situations. The method can be subdivided into two types:

   (a) The use of equations to define the profiles in terms of mining geometry.

   (b) The use of tables of data and nomograms to define the profiles in terms of the mining geometry.

   The constants for the equations and the nomograms are all derived empirically from local observations. The methods vary in complexity depending upon how many of the more significant factors are taken into account. Simple methods may only consider depth to be significant, whereas more advanced methods take many if not all of the significant factors into account.

3. **Influence functions.** This group contains methods which calculate subsidence at surface points. It is based on the theory that there is an area of influence around a point due to extraction of a small element within the sphere of influence of that surface point. A typical influence function would look at annular zones around a point and use equations to relate the percentage extraction in each area to the subsidence of that point. The area of influence is generally assumed to be circular, although hexagonal

shapes have been employed occasionally in order to simplify subsequent calculations. The functions commonly used can be subdivided as follows:

(a) Empirically derived functions which use many arbitrary constants to produce the desired value and form of subsidence.

(b) Functions relating to the rock properties and mechanics of the problem rather than to arbitrary constants.

The second of these groupings is the more popular but is less flexible when carrying out calibration work. A major advantage of influence functions is their ability to predict subsidence movements at any point above extractions of any practical shape. The major drawback of influence functions is that they are considerably more difficult to apply than profile functions and much more difficult to check and calibrate.

Table 5

Observed values for the subsidence factor (a) according to Bräuner (1973b)

Coalfield and form of goaf treatment	Subsidence factor (a)
British coalfields (1)	
Solid stowing	0·45
Caving or strip-packing	0·90
Ruhr coalfield, Germany (2)	
Pneumatic stowing	0·45
Other solid stowing	0·50
Caving	0·90
North and Pas de Calais coalfield, France (3)	
Hydraulic stowing	0·25 — 0·35
Pneumatic stowing	0·45 — 0·55
Caving	0·85 — 0·90
Upper Silesia, Poland (4)	
Hydraulic stowing	0·12
Caving	0·70
USSR (5)	
Caving, Donets and Kizelov coalfields	0·60
Caving, Kuznetsk and Karaganda coalfields	0·70
Caving, Chelybinsk coalfield	0·90
USA, Pennsylvania (6)	
Caving	0·50 — 0·60

Data sources:

1. NCB (1966)
2. Fläschenträger (1957), Schulte (1957)
3. Grard (1969)
4. Knothe (1953)
5. GIMS (1958)
6. Maize, Thomas and Greenwald (1941)

4. **Analytical models.** This group contains those models which are based on treating subsidence as a problem which can employ the laws of elasticity, plasticity and visco-elasticity. The elastic treatment of ground movement has attracted considerable attention in the past and in a number of cases a marked degree of success has been achieved. Several mathematical solutions have been advanced in the form of sets of equations for determining subsidence; such equations generally represent the geometry of the problem in a simple two-dimensional form and require knowledge of the elastic behaviour of the rock mass. The mass behavioural properties of the overburden are frequently determined by back-analysis using known in-put data for the other subsidence and mining extraction parameters. Finite element methods are probably the most advanced form of mathematical modelling within this group but care needs to be exercised in the choice of rock mass behavioural properties since a purely isotropic elastic solution of resulting ground movements tends to be unrealistic as the mathematically derived profiles exhibit a much flatter form of surface subsidence profile. Simulation of rock failure and resulting ground movement using non-linear analysis provides a more realistic basis for mathematical prediction of surface subsidence. This method of modelling mining subsidence offers the advantages of predicting anticipated ground movements at intermediate horizons; it also allows scope for incorporating geological variations particularly those of lithology in the model.

5. **Physical models.** The use of small scale models to represent different mining situations for the purpose of observing the resulting subsidence behaviour has been employed with various degrees of success in many countries. Gravity loaded models have tended to prove more acceptable than other types although surcharge loading and centrifugal loading of models have been applied with some measure of success. A wide range of model materials has been used but particular success has been achieved using sand, gelatine and sand/plaster in weakly bonded states. A striking advantage of physical models is the fact that the actual mechanisms of deformation and failure can be observed, thus allowing scope for selection of the more important aspects of material behaviour. However in some situations it is claimed that theoretical models involve less time in performing investigations. Physical models allow some features to be readily observed which theoretical models find very difficult to represent, and that is the progressive caving of the roof behind a longwall extraction and the subsequent propagation of cracks and caving effects into the roof beds and the transition from caving to subsidence. Physical models can be applied to complex geological and mining situations. They play an important role in the prediction of caving and subsidence and are particularly advantageous to investigating new situations. They have proved useful in supplementing the other prediction methods referred to earlier.

Empirically derived relationships

The best known of empirically derived relationships is the NCB SEH method which utilises rationalised subsidence data established by observations in various UK coalfields. It is based essentially on observations in fairly flat coal seams. The generalised relationship between the principal variables is given by equation (3).

$$\frac{S}{M} = f\left(\frac{w}{h}\right) \qquad \ldots (3)$$

The ground strain values produced by the creation of the subsidence trough have been investigated and the empirical relationship given by equation (4) is used to determine the maximum values of compressive and tensile strains

$$E = f\left(\frac{S}{h}\right) \qquad \ldots (4)$$

Additionally, the maximum slope of the subsidence trough can be determined from equation (5)

$$G = f\left(\frac{S}{h}\right) \qquad \ldots (5)$$

where E = maximum value of ground strain
S = maximum subsidence
h = depth below surface
w = width of extraction
M = thickness of extraction
G = maximum slope (tilt)

Subsidence data have been established which have allowed constants to be derived for the purpose of determining the respective constants for various ratios of w/h so that maximum values of subsidence, strain and slope can be readily calculated. The NCB SEH method is discussed separately in a later chapter.

Empirical formulae have been used in the USSR for calculating anticipated subsidence over longwall mining extractions and this aspect is discussed by Akimov (1958) who reports that equation (6) has been established from observations in Russian coalfields. Equation (6) allows the maximum subsidence (S) to be determined from the extraction height (M) and knowledge of the extraction width (w) and the critical width (w_c).

$$S = \frac{2\,M\,w}{(w + w_c)} \qquad \ldots (6)$$

Bräuner (1973 b) has reported that the angle of draw for most coalfields in the USSR is around 30° for level seams and this gives a relationship between the critical width and depth of $w_c = 1{\cdot}15$ h. Equation (6) is a relatively simple relationship and was probably formulated to give an approximate value of the likely maximum subsidence. This expression gives a consistently higher value of subsidence than the NCB SEH empirical model for UK coalfields. It appears that equation (6) applies particularly to those geological conditions where the overburden and especially the surface rocks exhibit a strong measure of pliability. Such simple relationships provide general guidance; more comprehensive relationships and tables of data are required for building an overall picture of surface subsidence above longwall mining extractions. The latter is most certainly the case where advanced forms of computerised treatment of mining subsidence prediction are involved.

Profile functions

The profile function S = f (x) approach to predicting mining subsidence has probably been the most widely applied of all methods in view of its representation being conveniently expressed in mathematical terms or data-base form. Consequently reproduction of the subsidence profile is relatively straightforward. Nearly all profile functions have an empirical basis whether they use graphs or functions to define the profile. Those methods which use equations to define the profile simplify the problem but generally at the cost of accuracy and flexibility when compared to a graphical or data-base approach.

Of the several publications on subsidence prediction, Bräuner (1973 b) and Kratzsch (1983) perhaps give the most detailed reviews and discussions of the state-of-the-art, and the following section is based mainly on the former's paper. The basic subsidence profile function is shown in Figure 31 with the transition point corresponding to the position where S/2

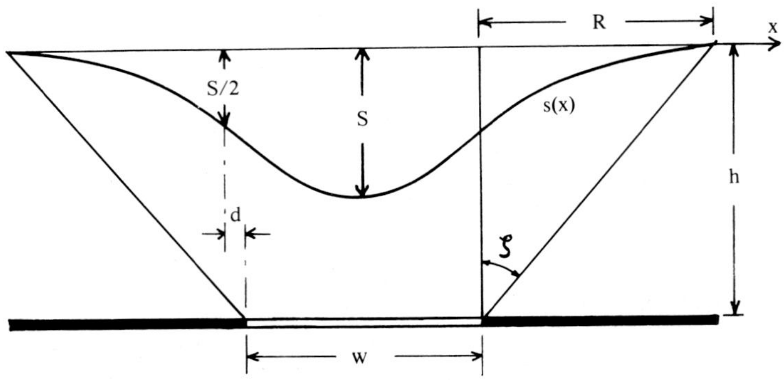

Figure 31 Basic subsidence profile function s(x)

occurs. The function can be symmetrical or asymmetrical about the transition point. In actuality observations generally indicate a non-symmetrical condition. The position of the transition point relative to the ribside needs to be established by observation, and its magnitude (d) depends upon the depth below surface. Also shown in Figure 31 is the critical radius of extraction (R), so that the value of subsidence (s) for a critical extraction at a point along the profile can be expressed as given by equation (7).

$$s = f\ (S, x, R) \qquad \ldots (7)$$

Should the extraction be subcritical, then an additional parameter, namely w/h, would need inclusion as a function.

A profile function simply defines the shape of the curve, and it requires a maximum value of subsidence and a number of constants to calculate the subsidence profile. The method is generally applied by adjusting the constants in the equation until a satisfactory shape is obtained which provides an acceptable match to the observed data. Consequently, using this calibrated equation it should then be possible to predict subsidence curves for that particular locality. Profile functions are advantageous from two standpoints:

1. Since the subsidence curve is defined by an equation, it can quickly and easily be calculated by a computer program.

2. When very little measured data are available, the method can still be applied using limited data to calculate constants.

The method has the following disadvantages:

1. The profile function may not match exactly the shape of that given by measured data and consequently introduce an error. Such an error may not be significant in respect of subsidence, but it may gain significance when tilt or curvature are calculated from the subsidence profile.

2. The shape of profiles may vary depending upon various external parameters not taken into account by the function.

The calculation of strain, or horizontal displacements using profile functions requires the use of relationships between strain and curvature. For a given location it is possible to relate strain to curvature using simple relationships. Curvature can be calculated from the profile function equation for subsidence, this being given by the second derivative of the subsidence equation. If the relationship between strain and curvature is then applied to the second derivative equation, a relationship for strain is derived. However, it will be shown later that the relationship between strain and curvature is specific to particular locations and even then the validity of the relationship comes under question when high orders of accuracy are required.

It is probably more accurate and valid to calculate the curvature for a situation and then relate the curvature directly to the surface damage, than to complicate the situation by attempting to calculate strain and then use the existing damage criteria based upon strain.

The following expressions contain a selection of some of the more commonly used profile functions. They readily fall into two groups, the first includes functions which use the transition point as the origin, whilst the second group uses the face centre as the origin.

Functions based on transition point origin

1. **Hyperbolic function:** King and Whetton (1957)

$$s = (S/2)\{1 - \tanh(\frac{2x}{R})\} \qquad \ldots (8)$$

2. **Exponential function:** Martos (1958)

$$s = S\exp\{-0{\cdot}5\left[\frac{x+R}{R}\right]^2\} \qquad \ldots (9)$$

3. **Donets trigonometrical function:** GIMS (1958)

$$s = (S/2)\{1 - \frac{x}{R} - \frac{1}{\pi}\sin(\frac{\pi x}{R})\} \qquad \ldots (10)$$

4. **Trigonometrical function:** Hoffman (1964)

$$s = S\sin^2\{\frac{\pi}{4}(\frac{x}{R} - 1)\} \qquad \ldots (11)$$

Figure 32 shows a comparison of equations (8) to (11) together with the UK SEH anticipated curve.

Equation (8) has been demonstrated by King and Whetton (1957) to provide a fairly close agreement with surface subsidence observations obtained in UK coalfields. It gives reasonable agreement with actual observations especially over the rib-side. The exponential function expressed as equation (9) appears to give an increasing degree of error over the rib-side when compared to the other functions and actual data from UK observations. Equations (10) and (11) give similar results to the hyperbolic profile function of equation (8). All the curves plotted from the profile function equations tend to give a more gentle gradient than that indicated by the SEH method.

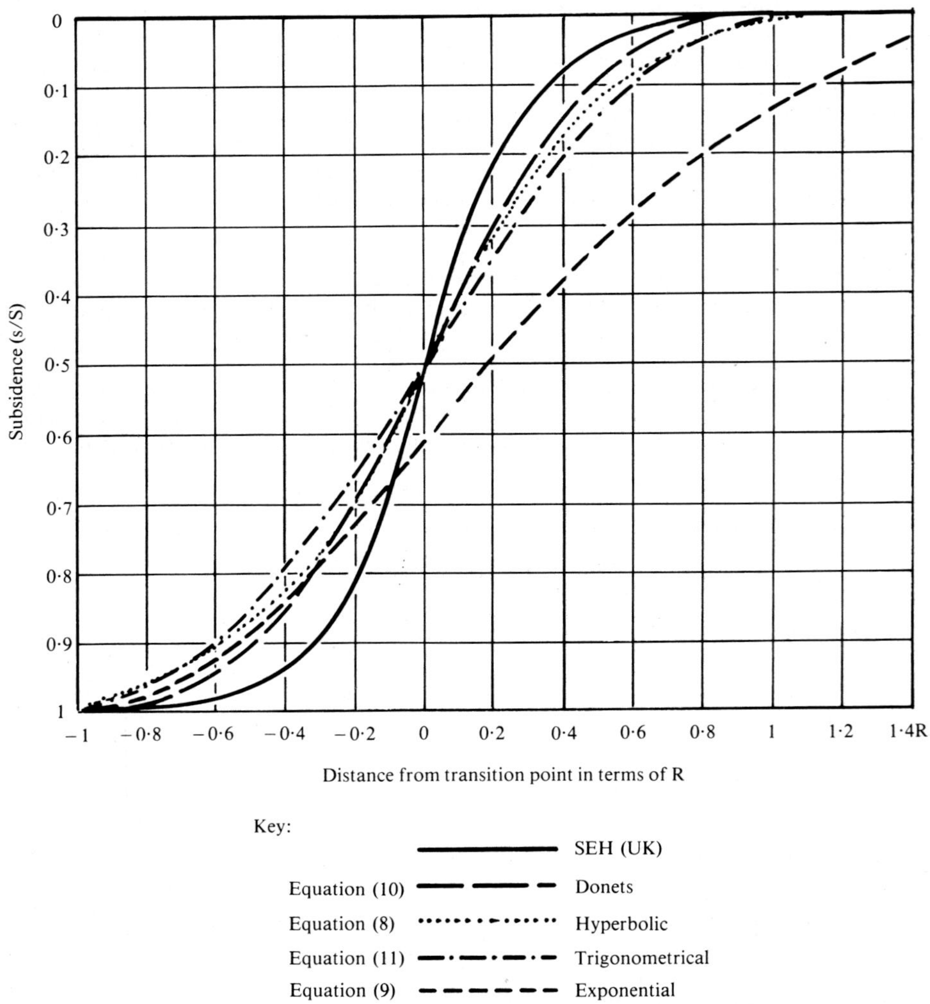

Figure 32 Profile function characteristics based on transition point origin

Notes:

1. Data used for SEH calculation: w = 140m, h = 100m, limit angle = 35°, transition point position = 0·8R from panel centre.

2. All calculations based on a critical extraction condition.

Functions based on face centre origin

1. **Polish profile function:** Kowalczyk (1966)

$$s = S\,e^{-\bar{n}x^2} \qquad \ldots (12)$$

where $\bar{n} = \dfrac{S}{R^2 c}$

c = average roof bed lowering
R = radius of critical area of extraction

2. **Hungarian profile function:** Martos (1958)

$$s = S\,e^{-\frac{x^2}{2d^2}} \qquad \ldots (13)$$

where d = distance of transition point from panel centre

3. **Niederhofer's method:** Niederhofer (1962)

$$s = S\left\{1 - \left(\frac{x}{p}\right)^2\right\}^2 \qquad \ldots (14)$$

where p = half width of subsidence profile, ie $w/2 + R$

4. **Indian profile function:** Kumar, Saxena and Singh (1983)

For subcritical extractions:

$$s = S\,e^{-\frac{nx^2}{p^2 - x^2}} \qquad \ldots (15a)$$

For critical widths:

$$s = S\,e^{-\frac{nx^4}{p^4 - x^4}} \qquad \ldots (15b)$$

where n = constant (1 to 5 generally)

x = distance from extraction centre

R = radius of critical area of extraction

p = half width of subsidence profile ie $w/2 + R$

Equation (12) was established from surface subsidence observations and appears to be mainly applicable to relatively flat stratified measures of the Upper Silesian coalfield.

Equation (13) has been derived from observations in Hungarian coalfields; the predicted subsidence profile this equation produces tends to be much flatter and wider than would be expected for UK coalfields.

Niederhofer's equation can be applied in a complex form to inclined seams; it has a degree of flexibility in its range of applications although it tends to differ markedly with the SEH curve around the ribside region.

Equations (15a) and (15b) present the Indian profile function of Kumar, Saxena and Singh. This equation tends to give curves of similar shape to the SEH method but gives a

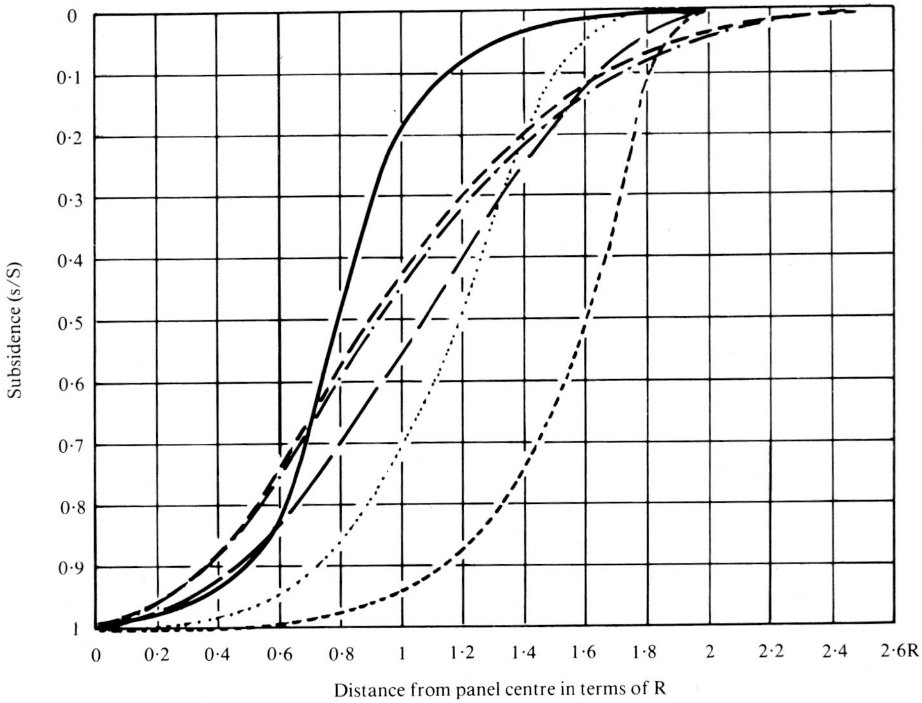

Key:

	———	SEH (UK)
Equation (12)	— — —	Polish
Equation (13)	—·—·—	Hungarian
Equation (14)	—— ——	Niederhofer
Equation (15b)	········	Indian (n = 5)
Equation (15b)	- - - - -	Indian (n = 1)

Figure 33 Profile function characteristics based on face centre origin

Notes:

1. SEH profile for w = 140m, h = 100m, transition point = 0·8R from panel centre.
2. In equation (12), used S = 0·84M (from SEH) and c = M
3. In equation (13), used d = 0·8R
4. In equation (14), p = w/2 + R, ie 2R
5. In equation (15b) p = 2R and n = 1 and 5

broader shaped trough over the extraction. There is a marked rate in reduction of subsidence as the extremities of the subsidence curve are approached. It would appear that equation (15b) is probably mainly suited to shallow mine workings in view of the steep sides it predicts for the subsidence trough.

Figure 33 shows a comparison of these profile functions in relation to the SEH prediction for UK conditions.

There are many more functions that can be applied to subsidence predictions but the above listed equations provide a representative selection of the various types available for such work. Many of the other functions either require several constants (derived empirically) or more complex calculations. If fairly simple functions achieve a satisfactory level of accuracy in subsidence prediction work, then there seems little point in making the problem more complex than necessary by using more complicated functions.

Of the profile functions discussed here it would appear that the Indian, hyperbolic and exponential equations deserve special consideration.

The profile function method of subsidence prediction lends itself to adaptation to particular coalfield conditions and gives fairly accurate results in a number of situations. However, the principal limitation lies in the method's fairly restricted application to mainly longwall extractions of consistent width.

Influence functions

The basic principle of the influence function method is the application of the law of superposition to determine the overall influence of an extraction area treated in infinitesimal parts.

Figure 34 illustrates the concept of superposition of subsidence troughs produced by extraction of infinitesimal parts of a flat seam of uniform thickness; summation of the individual elementary troughs gives the resulting subsidence of point P for the entire extraction. Each extraction element of area dA can be considered as making a contribution to the overall subsidence of the surface as expressed by equation (16).

$$k_Z = f(r) \qquad \ldots (16)$$

where k_Z = magnitude of influence of dA on P
r = horizontal distance between position of P and the infinitesimal element dA

Equation (16) is the standard form of an influence function as related to subsidence prediction. The product of dA and the influence function k_Z gives the contribution which that element makes to subsidence of a surface point. This principle is illustrated in Figure 35(a).

The position of the surface point P is chosen as the origin of r (radial variable in polar co-ordinates) as the influence function k_Z addresses the effect on P. Consequently the element directly under P contributes the most influence, so mathematically the function $k_Z(r)$ is at a maximum when $r = 0$. With reference to Figure 35(b) it is seen that when P is positioned in the centre of a critical width of extraction, then P becomes influenced by all the elements within the range, governed by the function and this results in full subsidence of P. With reference to Figure 35(a) the full subsidence (S) is given by the area under the curve $k_Z(r)$. In

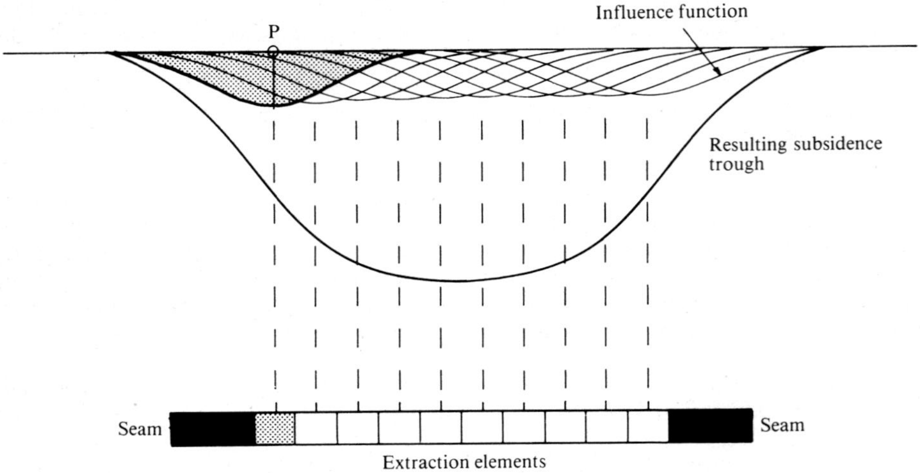

Figure 34 Illustrating superposition of infinitesimal influences to produce resulting subsidence trough, based on Bräuner (1973b)

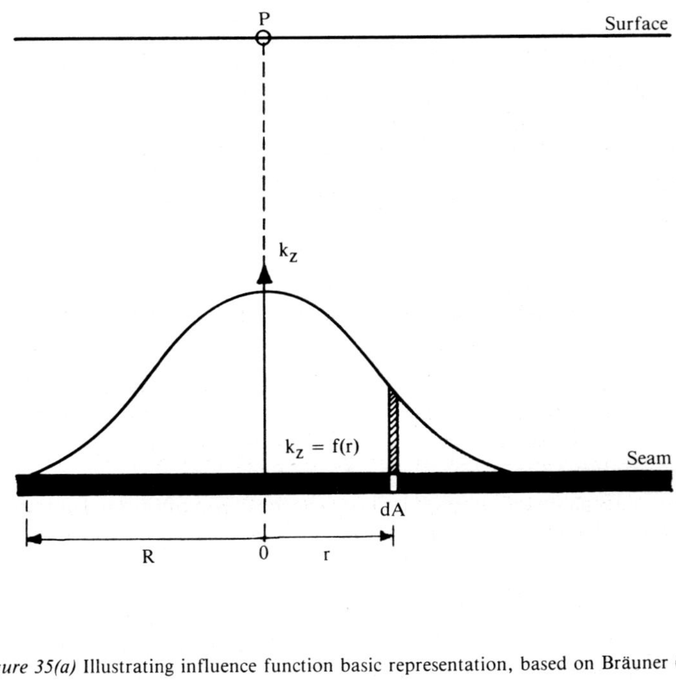

Figure 35(a) Illustrating influence function basic representation, based on Bräuner (1973b)

the situation where P is located directly over the extraction edge, then it experiences only half of the possible influence and consequently results in half of the full subsidence, see Figure 35(b). The point of inflexion coincides with the position of S/2. It is shown later that field observations indicate that the position of S/2 does not always coincide with the edge of the extraction and that a correction factor needs to be applied. However, in accordance with the principle of superposition, the extraction of an area A will result in a surface point subsiding by a magnitude s as expressed by equation (17).

$$s = \iint_A k_z(r)\, dA \qquad \text{. . . (17)}$$

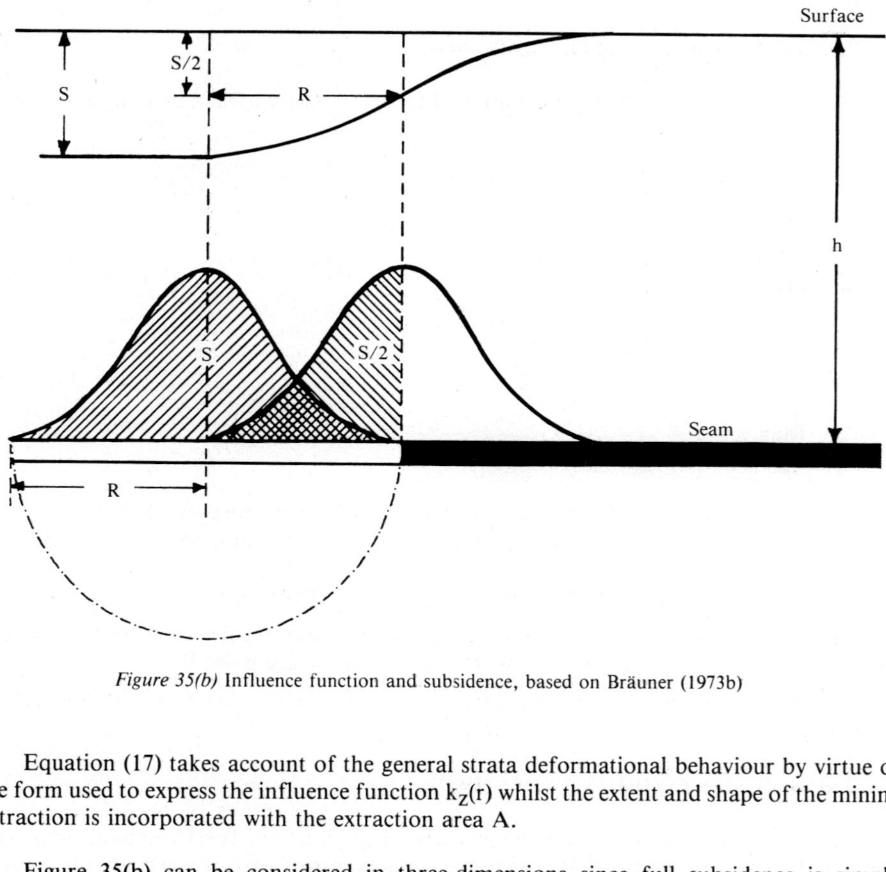

Figure 35(b) Influence function and subsidence, based on Bräuner (1973b)

Equation (17) takes account of the general strata deformational behaviour by virtue of the form used to express the influence function $k_z(r)$ whilst the extent and shape of the mining extraction is incorporated with the extraction area A.

Figure 35(b) can be considered in three-dimensions since full subsidence is simply expressed by the volume determined by rotating the curve $k_z(r)$ around the z axis. For the condition of the radius of the solid of revolution corresponding to the critical radius R, then the integration limit is that of r = R where k_z becomes zero; should k_z tend to approach zero asymptotically, then the integration limit becomes infinite. Full subsidence is expressed by one of two equations (18a) and (18b) under these conditions.

$$S = 2\pi \int_o^R r\, k_z(r)\, dr \qquad \text{. . . (18a)}$$

$$S = 2\pi \int_o^\infty r\, k_z(r)\, dr \qquad \text{. . . (18b)}$$

Influence functions can be determined from observed values of subsidence (s) resulting from an area of extraction (A). Using field data, equation (17) can be solved in terms of k_Z usually by numerical methods. Theoretical studies have produced a wide range of analytical functions, but the most commonly adopted influence function is of exponential form and is given as equation (19), this being advanced following work by Polish investigators, Knothe (1957) and Litwiniszyn (1957, 1964).

$$k_Z = \frac{S}{R^2} e^{-\pi(r/R)^2} \qquad \ldots (19)$$

Selected influence functions

1. **Keinhorst's method.** Keinhorst (1928, 1934)

Figure 36 shows the basic representation of the terminology used in deriving Keinhorst's influence function.

$$k_Z = \frac{2}{3\pi} \frac{\tan^2\beta}{\tan^2\gamma} \frac{S}{R^2} \qquad \ldots (20)$$

for $O \leqq r \leqq \dfrac{R\tan\gamma}{\tan\beta}$

$$k_Z = \frac{\tan^2\beta}{3\pi(\tan^2\beta - \tan^2\gamma)} \frac{S}{R^2} \qquad \ldots (21)$$

for $\dfrac{R\tan\gamma}{\tan\beta} \leqq r \leqq R$

where γ = angle of influence of outer zone (measured to horizontal)
β = angle of break of inner zone (measured to horizontal)
$R = h \cot\gamma$
h = depth of extraction in flat seam with level surface

The critical area of extraction is subdivided with an inner circular zone resulting in $(\frac{2}{3})S$ and an outer zone accounting for $(\frac{1}{3})S$ with respect to a surface point P.

Bräuner (1973b) has suggested employing values of $\gamma = 55°$ and $\beta = 70°$ in equations (20) and (21) to provide a working solution.

2. **Bals' method.** Bals (1931/32), Zenc (1969)

Bals' method is based upon Newton's law of gravitation. The hypothesis employed considers that the surface point P is attracted by the mass of part of the goaf with a force which is universally proportional to the square of the distance. Bals' directed his work towards finding ratios of attraction from different zones. Equations (22) to (25) show the development of the well known Bals' equation using the terminology mainly based on Zenc (1969) and Grond (1953). Figure 37 illustrates the basic concept and terminology used by Zenc (1969). Bals neglected the constants since he was only seeking the ratios of influence of different zones so that equation (25) could be used to define the influence function. The function k_Z in relation to the surface point P overlying a goaf area can be expressed as follows:

$$k_Z = \frac{\text{const.}}{R^2 + h^2} d\alpha$$

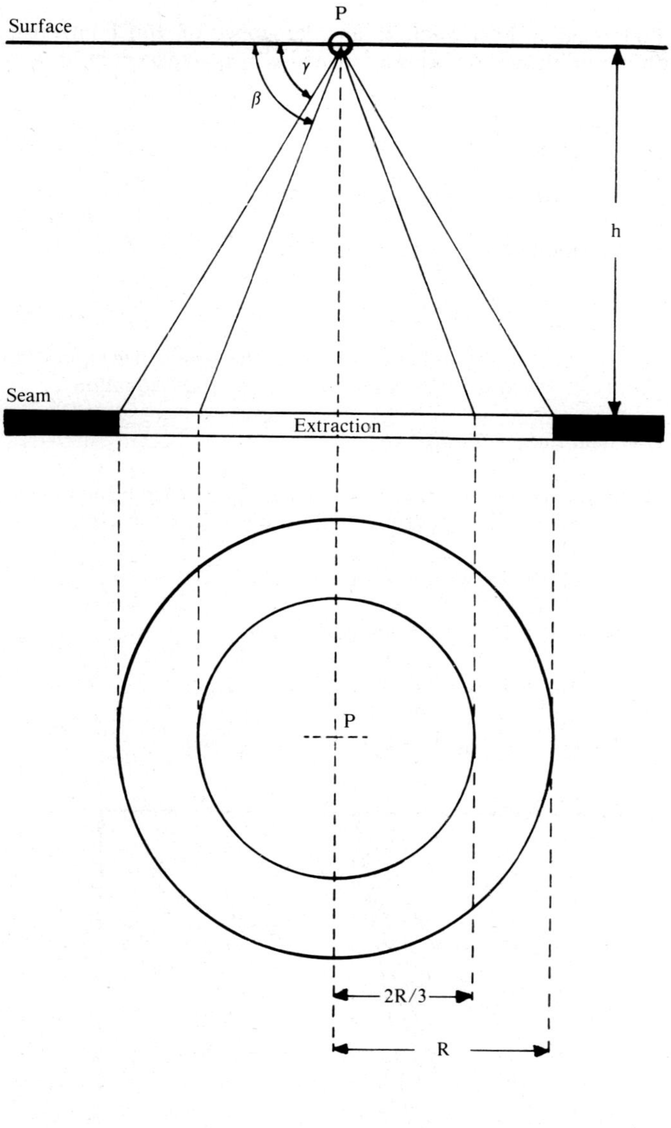

Figure 36 Illustrating basic representation of Keinhorst's influence function, based on Bräuner (1973b)

$$k_z = \frac{\text{const.}}{h^2} \cos^2 \alpha \, d\alpha$$

Equation (22) requires integration if annular zones of equal influence are to be determined, the limits of integration being $\alpha = 0$ and $\alpha = \alpha_m$ (angle of influence measured to the vertical).

$$K_z = \frac{\text{const.}}{h^2} \int_0^{\alpha_m} \cos^2 \alpha \, d\alpha \qquad \ldots (23)$$

where K_z represents the total area of the function.

Integration of equation (23) gives equation (24).

$$K_z = \frac{\text{const.}}{h^2} \; 1/4 \; (\sin 2\,\alpha_m + 2\,\alpha_m) \qquad \ldots (24)$$

Since the coefficient const./h^2 will be constant for flat seams then expression (24) can be written as equation (25), which is the well-known form of Bals' equation

$$K_z = 1/4 \; (\sin 2\,\alpha_m + 2\,\alpha_m) \qquad \ldots (25)$$

Bals divided the full area of influence into annular zones, each being of equal influence. Five zones are considered and the relevant equations are as given in (26).

$$1/5 \, [\, 1/4 \, (\sin 2\,\alpha_m + 2\,\alpha_m) \,] = 1/4 \, (\sin 2\,\alpha_1 + 2\,\alpha_1)$$

$$2/5 \, [\, 1/4 \, (\sin 2\,\alpha_m + 2\,\alpha_m) \,] = 1/4 \, (\sin 2\,\alpha_2 + 2\,\alpha_2)$$

$$3/5 \, [\, 1/4 \, (\sin 2\,\alpha_m + 2\,\alpha_m) \,] = 1/4 \, (\sin 2\,\alpha_3 + 2\,\alpha_3)$$

$$4/5 \, [\, 1/4 \, (\sin 2\,\alpha_m + 2\,\alpha_m) \,] = 1/4 \, (\sin 2\,\alpha_4 + 2\,\alpha_4) \qquad \ldots (26)$$

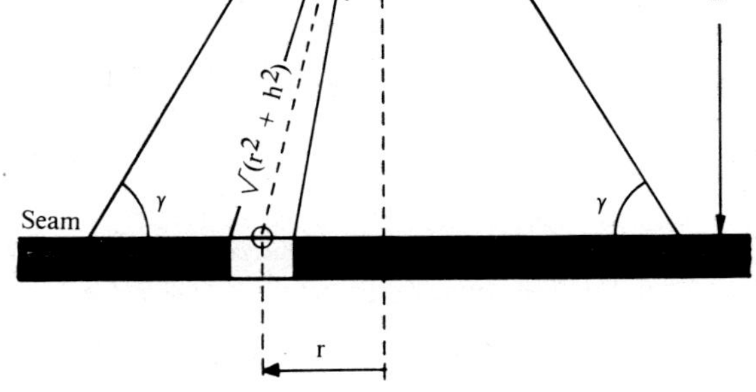

Figure 37 Basic representation of Bals' influence function, based on Zenc (1969)

Equations (26) are solved iteratively by substituting values into them until a satisfactory accuracy is obtained for α_1, α_2, α_3 and α_4. The radii of the annular zones are calculated using equation (27).

$$r_{1,..4} = h \tan \alpha_{1,..4} \qquad \ldots (27)$$

Consequently, the annular zones can then be drawn from the initial knowledge of α_m being the angle of influence (measured to the vertical), so that a graphical integration grid can be produced simply by insertion of suitably spaced radii to give annular sub-divisions of equal influence.

Figure 38 illustrates the graphical integration grid for the Bals' influence function method of determination of subsidence of a surface point P.

3. **Knothe's method.** Knothe (1953, 1957), Zenc (1969).

Knothe's work is based on surface subsidence observations in the Upper Silesian Coalfield, from which he formulated the influence function given by equations (28a) and (28b) in cartesian and polar co-ordinate forms respectively.

$$k_Z\ (x,\ y) = \frac{S}{R^2} \exp\left[-\frac{\pi}{R^2}\ (x^2\ +\ y^2) \right] \qquad \ldots (28a)$$

$$k_Z\ (r) = \frac{S}{R^2} \exp\left[-\frac{\pi r^2}{R^2} \right] \qquad \ldots (28b)$$

where, x, y are plane rectangular co-ordinates

r is polar co-ordinate, $r^2 = x^2 + y^2$

R is radius of influence

Integration of equation (28a) within the limits of A gives the subsidence of the surface point P under consideration.

$$s = \iint_A k_Z\ (x,\ y)\ dA$$

$$s = \frac{S}{R^2} \iint_A \exp\left[-\frac{\pi}{R^2}\ (x^2 + y^2) \right] dA \qquad \ldots (29)$$

Equation (29) is difficult to handle in its present form but if the underground extraction is considered sufficiently extensive in one direction for the ends to be regarded of no influence, equation (30) can be derived.

$$s(a,h) = \frac{S}{R} \int_a^{\infty} \exp\left(-\frac{\pi x^2}{R^2} \right) dx \qquad \ldots (30)$$

where, a is the distance from the excavation edge (positive outwards).

Equation (30) requires numerical integration. The polar form lends itself to simpler analysis where annular zones require to be determined.

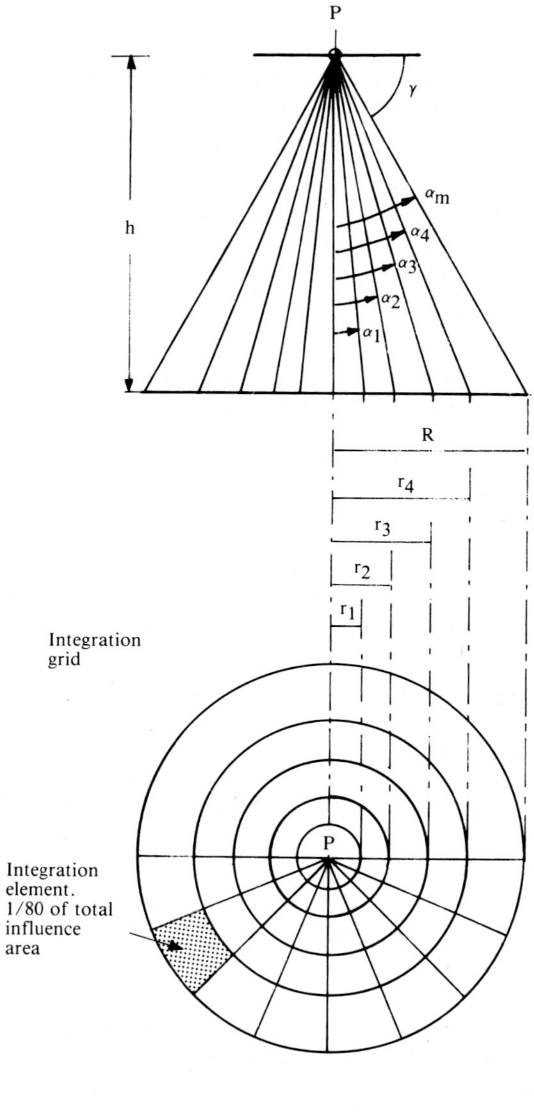

Figure 38 Illustrating Bals' influence zone subsidence calculation method, based on Zenc (1969)

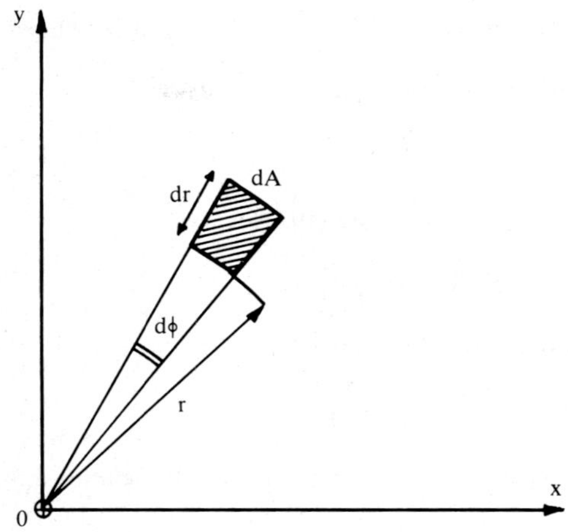

Figure 39 Polar co-ordinate representation, based on Zenc (1969)

Considering the polar form of integration, reference is made to Figure 39. This shows the basic representation of dA in polar form and can be expressed as equation (31).

$$dA = r\,d\phi\,dr \qquad \ldots (31)$$

Equation (29) can be now written in the form given by equation (32) and using $r^2 = x^2 + y^2$.

$$s = \frac{S}{R^2} \iint_A \exp\left(-\frac{\pi r^2}{R^2}\right) r\,d\phi\,dr \qquad \ldots (32)$$

Extraction of an annular area of radii r_1 and r_2 where $r_1 > r_2$ will produce subsidence (s) as given by equation (33).

$$s = \frac{S}{R^2} \int_o^{2\pi} d\phi \int_{r_2}^{r_1} \exp\left(-\frac{\pi r^2}{R^2}\right) r\,dr \qquad \ldots (33)$$

The integrated form of equation (33) is as follows, namely equation (34).

$$s = S\left[\exp\left(-\frac{\pi r_2^2}{R^2}\right) - \exp\left(-\frac{\pi r_1^2}{R^2}\right)\right] \qquad \ldots (34)$$

The latter expression has a feature that requires special consideration. This function requires areas outside the normal radius of influence to be taken into account. In order to ensure an accuracy of ± 1% it is necessary to employ a radius of influence (R) 1·21 times larger than normal to allow satisfactory compensation. Also, owing to the highly curved nature of the function it would appear appropriate to use at least 10 annular zones.

4. **Other particular influence functions**

The following list contains some of the more commonly applied influence functions.

1. **Beyer's method.** Beyer (1945) and Niemczyk (1949)

$$k_z = \frac{3S}{\pi R^2}\left[1 - \left(\frac{r}{R}\right)^2\right]^2 \qquad \ldots (35)$$

2. **Sann's method.** Sann (1949, Bräuner (1973b)

$$k_z = \frac{2\,S}{\pi\sqrt{\pi}\,R}\ \frac{1}{r}\ \exp\left[-4\left(\frac{r}{R}\right)^2\right] \qquad \ldots (36)$$

3. **Litwiniszyn's method.** Litwiniszyn (1957)

$$k_z = \frac{n\,S}{R^2}\ \exp\left\{-\,n\,\pi\left(\frac{r}{R}\right)^2\right\} \qquad \ldots (37)$$

where n = an independent parameter and in this relationship its value seems to be usually 1.

Equation (37) was firstly established from field and experimental observations in addition to deduction from probability considerations. Litwiniszyn demonstrated that the theory of stochastic rock movement could be used to verify equation (37).

Ehrhardt and Sauer (1961) have investigated equation (37) and confirmed its validity in respect of West German coalfields.

A similar approach has been adopted by Kochmanski (1959) who suggested a slightly modified version of equation (37).

Practical application of influence functions

Influence functions form the basis of a graphical calculation method for determining subsidence of a surface point overlying level seam workings. An integration grid is normally used, which involves selection of an appropriate influence function.

Figure 40(a) shows a selection of influence functions drawn as circular and annular zones of equal influence (20%) for all except that of Keinhorst's. All the functions are drawn as circles to a maximum diameter of 2R with the exception of Knothe's influence function which exceeds 2R.

There are similarities between the methods of Erhardt and Sauer, Knothe and Beyer in respect of giving anticipated maximum subsidence and the form of subsidence profile. Bals' method gives subsidence predictions which are not distinctly different from these three methods, although the Bals' profile tends to be more uniform. Sann's method predicts a subsidence profile having a deeper central area than the other methods and consequently in partial extraction layouts can produce a higher value of subsidence. Keinhorst's influence function gives an over simplified representation of the subsidence profile. The other five methods are in wide general use in several countries. Zenc (1969) has demonstrated very favourable comparisons between Bals' and Knothe's methods.

Figure 40(b) gives an example of applying the influence function integration and to a mining subsidence problem, and Figure 40(c) compares results given by different methods. The comparisons are consistent with the comments made above.

Figure 40(a) Comparison of influence function zone areas for subsidence calculation by graphical methods of integration

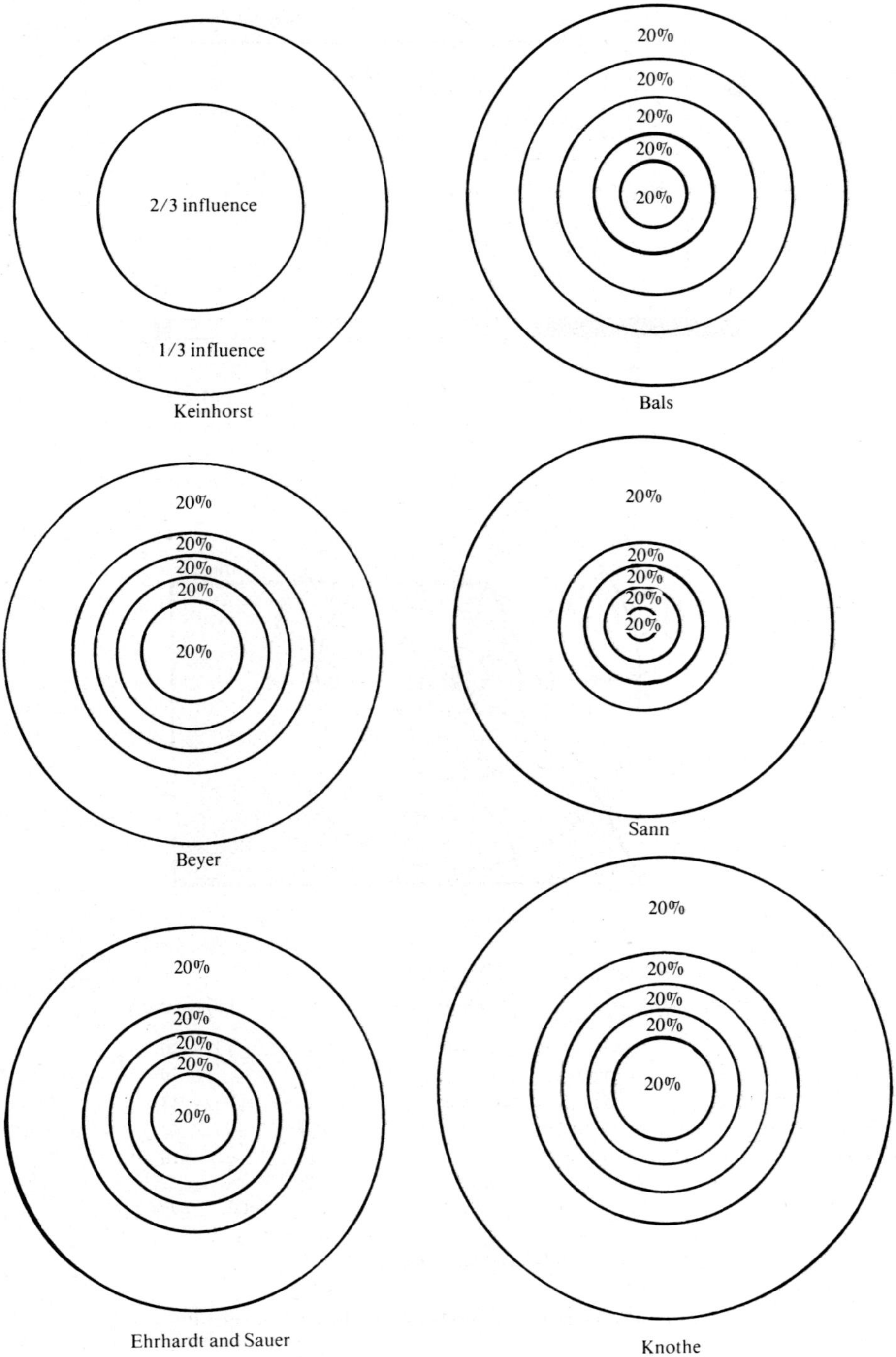

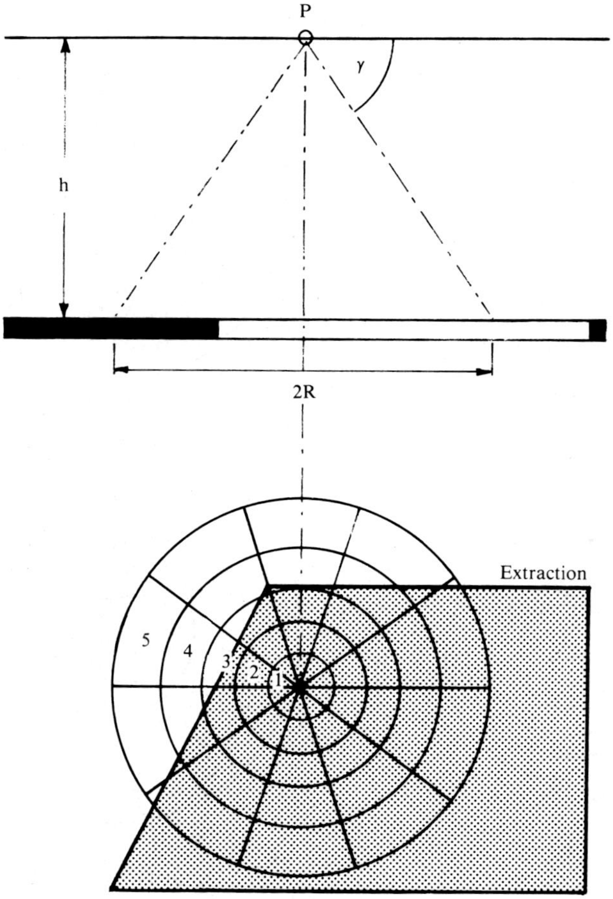

Example:

Annular zone	Number of influence elements	Influence (%)
1	10	20
2	10	20
3	9	18
4	6	12
5	5	10
	Total	80%

Subsidence at point P = 0·8 S

Figure 40(b) Example of calculation of subsidence of point P

Marr (1975) discussed the application of the zone area system to prediction of mining subsidence in UK coalfields. He drew attention to how the effect of coal pillars can be adequately taken into account, and concluded there was scope for increased use of such prediction methods in the UK in view of the reasonable standard of accuracy achieved and high degree of reliability which such methods provide.

Comparison of profile functions and influence functions

The following discussion makes a comparison between profile and influence functions. The points listed below refer specifically to profile functions:

1. Field observations can be easily employed to derive profile functions.

2. Profile functions can be readily applied to a wide range of subsidence problems, including subcritical situations.

3. Profile functions are easier to calibrate to field data and are generally more accurate in subsidence predictions.

4. The range of subsidence problems to which profile functions can be applied is mainly limited to relatively straightforward two-dimensional mining extraction situations involving essentially rectangular geometrical configurations.

5. Profile functions can be adapted to take into account the effects of small pillars in modifying the surface subsidence profile.

The following comments refer specifically to influence functions:

1. Shape of extraction is not a limiting factor in respect of influence functions, since they can be applied to any shape of extraction in one or more seams.

2. The method is more mathematically rigorous than that involved with profile functions, and consequently requires more detailed assessment when performing subsidence predictions.

3. Influence functions are based on the principle of superposition and consequently assume complete validity of applying the principle to any mining situation.

4. Since influence functions depend mainly upon the radius of influence (R), they can be readily adapted to take different geological conditions into account.

5. Accuracy in subsidence prediction with influence functions tends to be reduced in preference to increased flexibility of application particularly regarding extensive extractions of irregular configuration.

6. Prediction of displacement with influence functions tends to be more complicated than with profile functions; for increased accuracy of prediction of subsidence and displacements, corrections require to be made to the basic method in the form of edge zones of incomplete closure around extraction areas.

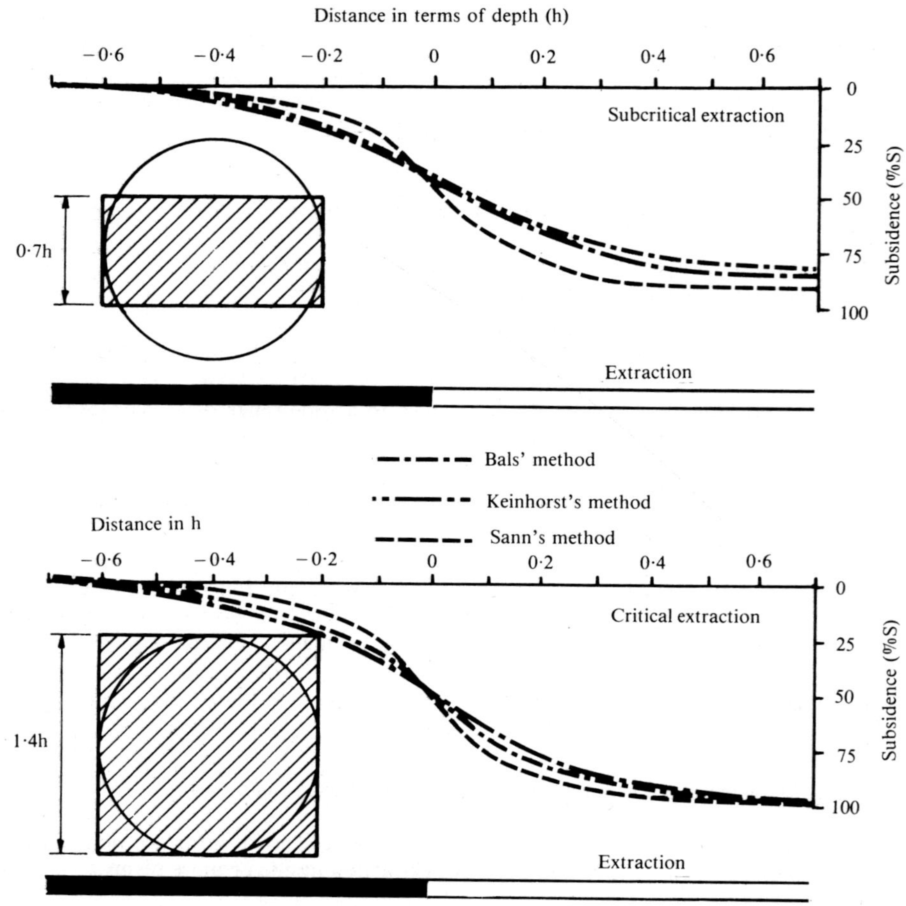

Figure 40(c) Comparison of subsidence results calculated by different influence function methods for critical and subcritical situations

Analytical models

Specific reference is made to stochastic, elastic, visco-elastic and plastic treatments of ground behaviour prediction.

Stochastic medium. Litwiniszyn (1957) examined the notion of a stochastic medium in an attempt to explain ground displacement behaviour associated with mining subsidence. This form of mathematical model differs from those based on the mechanics of continuous media, owing to its non-exclusion of discontinuous transformations of the medium. The stochastic medium consists of many elements possessing multiple degrees of freedom, thus making it impossible to solve by methods of classical mechanics, although solution of the problem can be achieved by other methods. The geometrical structure of the problem governs the laws controlling the set of elements in the stochastic medium. The structural properties of the problem are incorporated in the form expressed by coefficients of the differential equation for the medium. Litwiniszyn points out that in reality the natural ground only approximately resembles the stochastic model, but it does bear some resemblance to models based on the mechanics of continuous media. Further work was performed at the Polish Academy of Science in Krakow by Litwiniszyn (1964, 1972) using experimental models consisting of dry sand, and this gave confirmation of the stochastic medium in relation to describing the behaviour of ground movement due to under-mining.

Berry (1964a) has discussed the stochastic medium in relation to ground movement and suggests that for the theory to have practical use three conditions should be satisfied.

1. the theoretical medium properties must be similar to the natural ground,
2. the behaviour under analysis must be of practical relevance, and
3. the results given by the theory must provide a means of determining expected effects.

Berry concluded that the stochastic medium theory was rather limited especially in relating dry sand behaviour to the large regions of ground movement associated with mining operations.

Elastic analysis of rock movements. The initial concept of elastic treatment of the ground movement process in seeking an explanation of subsidence due to mining probably arose from research studies at the University of Leeds, King and Whetton (1957). They demonstrated that a physical model subjected to elastic behaviour gave a fairly accurate insight into ground deformation above a thin horizontal excavation. Hackett (1958-59) used the concept of representing a longwall extraction as a crack in an elastic medium, and this probably initiated a programme of theoretical studies of ground movement due to mining subsidence by Berry at the Department of Mining Engineering, University of Nottingham. Salamon (1963-64) also performed extensive theoretical work into the extraction of tabular deposits.

Sonntag (1958, 1960) also considered theoretical treatment of the ground movement process, but abandoned the more usual isotropic assumption in his analyses of uniformly stratified ground. Sonntag assumed homogeneity of the ground under analysis as a principal factor in providing results for practical purposes.

Berry (1963) discussed the application of the equations of classical elasticity to the ground movement phenomenon, and reported that the treatment for isotropic ground, gave results significantly at variance with field measurements of subsidence in British coalfields. This prompted work to be directed towards the development of a transversely isotropic treatment of the problem. Berry recognised that the rock disturbance process due to mining such as caving and that of geological discontinuities would influence the validity of any form of elastic treatment of the ground mass, but he assumed that such effects would be small in relation to the whole region under the influence of mining. In the earlier work, Berry ignored the effects of time and assumed that the stress states determined were those reached after completion of the ground movement process.

Two-dimensional treatment of isotropic ground. An early assumption made by Berry (1963) was that the ground could be considered isotropic and homogeneous. He distinguished three post-excavation states:

1. *Non-closure* meaning no part of the roof and floor is in contact.

2. *Partial closure* meaning that some part of the roof and floor is in contact.

3. *Complete closure* meaning that every part of the roof and floor is in contact.

Berry suggests that the third state is physically impossible but still makes a contribution to consideration of the problem. These three states lead to different solutions of the problem. The complete closure solution has the property that the surface displacements are independent of the elastic constants (under isotropic conditions). Berry demonstrated that the isotropic theory produced surface subsidence results that were too shallow and projected too far laterally in relation to observed subsidence data; in actuality, surface subsidence troughs were deeper and much more confined than could be explained by the isotropic elastic treatment of ground movements.

Figure 41 illustrates the isotropic case shortcoming in comparison to UK mining subsidence field data. King (1964) has pointed out that at extremely low values of width/depth ratios incomplete closure would be experienced owing to the tendency for free span to occur and it is questionable that the graph would begin at the origin shown in Figure 41.

Transversely isotropic ground. The recognition of the limitations of the isotropic theory led Berry and Sales (1961) to consider extending the elastic treatment to transversely isotropic ground. Five constants govern the behaviour of transversely isotropic material. These constants, c_{11}, c_{12}, c_{13}, c_{33}, c_{44}, appear in the stress-strain relationships given by equation (38), Berry (1977):

$$\begin{aligned}
\sigma_x &= c_{11}\varepsilon_x + c_{12}\varepsilon_y + c_{13}\varepsilon_z,\\
\sigma_y &= c_{12}\varepsilon_x + c_{11}\varepsilon_y + c_{13}\varepsilon_z,\\
\sigma_z &= c_{13}\varepsilon_x + c_{13}\varepsilon_y + c_{33}\varepsilon_z,\\
\tau_{yz} &= 2c_{44}\gamma_{yz},\\
\tau_{xz} &= 2c_{44}\gamma_{xz},\\
\tau_{xy} &= (c_{11} - c_{12})\gamma_{xy}.
\end{aligned} \qquad \ldots (38)$$

where σ_x, σ_y and σ_z are the direct stresses in directions x, y, and z.
ε_x, ε_y and ε_z are the strains in directions x, y, and z.
τ and γ represent shear stress and strain respectively in relation to the indicated directions.

The solution of the transversely isotropic problem involves using two parameters α_1 and α_2 which are roots of equation (39).

$$c_{11}c_{44}\alpha^4 + \{c_{13}(2c_{44} + c_{13}) - c_{11}c_{33}\}\alpha^2 + c_{33}c_{44} = 0 \qquad \ldots (39)$$

Berry suggests that it is sometimes more convenient to use the parameters k_1 and k_2 which are expressed in (40).

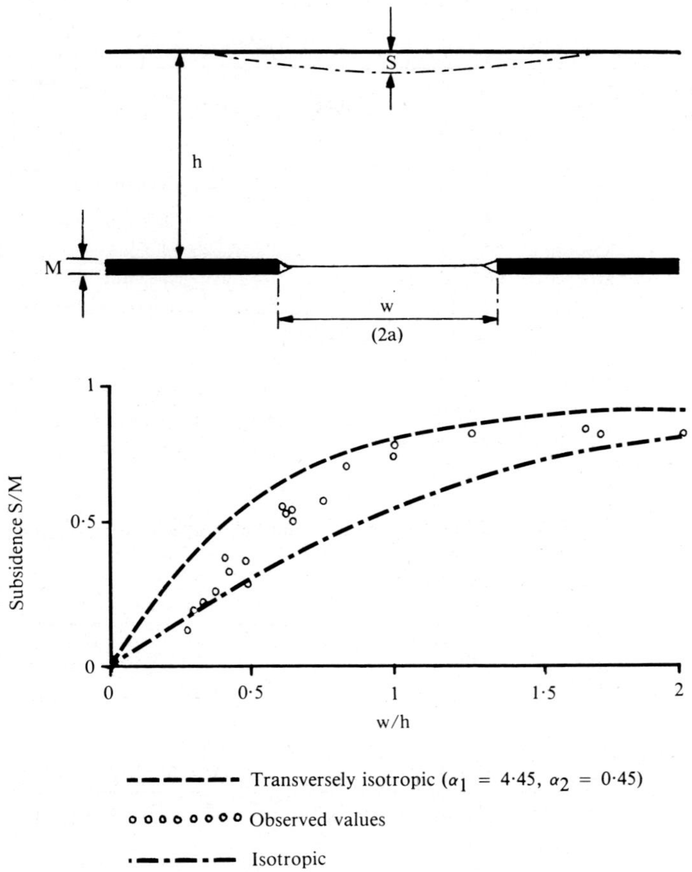

Figure 41 Comparison between theoretically derived and observed subsidence values, after Berry (1977)

$$k_1 = \alpha_1\alpha_2 = (1-\nu_1{}^2)^{\frac{1}{2}}(E_1/E_2-\nu_2{}^2)^{-\frac{1}{2}}$$

$$k_2 = (\alpha_1{}^2+\alpha_2{}^2)/2 = \{E_1/2[G-\nu_2(1+\nu_1)]\}(E_1/E_2-\nu_2{}^2)^{-1} \quad \ldots (40)$$

where E_1 and E_2 are Young's moduli (tensional) in the horizontal and vertical directions.
ν_1 and ν_2 are Poissons' ratio (tensional strain), corresponding to $-\varepsilon_x/\varepsilon_y$ and $-\varepsilon_z/\varepsilon_y$ respectively.
G is shear modulus in the vertical plane.

With the extraction considered to have advanced beyond the zone of influence, then a state of plane strain can be assumed to exist. For a flat seam of uniform depth h and extraction width 2a*, and extraction thickness t*, then for complete closure, the vertical displacement $v_o(x)$ and horizontal strain $\varepsilon_o(x)$ are given by equations (41) and (42) respectively, for real values of α_1 and α_2 (Berry 1963, 1977).

*The symbols used by Berry have been retained in these equations: t corresponds to M for extraction thickness whilst 2a corresponds to w.

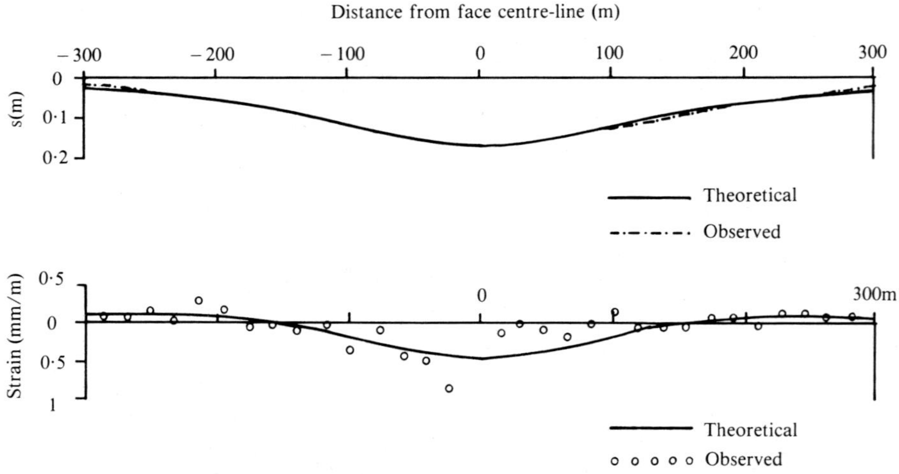

Figure 42 Comparison between theoretical and observed subsidence and strain values according to Berry (1977)

$$v_0(x) = -\frac{t}{\pi(\alpha_1 - \alpha_2)}\left\{\alpha_1 \tan^{-1}\frac{2ah_1}{x^2 - a^2 + h_1^2} - \alpha_2 \tan^{-1}\frac{2ah_2}{x^2 - a^2 + h_2^2}\right\} \quad \ldots (41)$$

$$\varepsilon_0(x) = \frac{2a\alpha_1\alpha_2 t}{\pi(\alpha_1 - \alpha_2)}\left\{\frac{x^2 - a^2 - h_1^2}{(x^2 - a^2 - h_1^2)^2 + 4h_1^2x^2} - \frac{x^2 - a^2 - h_2^2}{(x^2 - a^2 - h_2^2)^2 + 4h_2^2x^2}\right\} \quad \ldots (42)$$

where $h_i = h/\alpha_i$ $(i = 1, 2)$.

The mass behavioural properties of the ground are influenced by α_1 and α_2, or by k_1 and k_2. Although, equations (41) and (42) relate to complete closure, Berry comments that nevertheless closure is never complete and this factor does not affect the shape of the subsidence and strain profiles significantly when the depth (h) is large although their magnitudes will differ. In considering the three-dimensional case Berry and Sales (1962) found a similar dependence on the quantities α_1 and α_2. Berry (1963) has used equations (41) and (42) to find a reasonable fit with field data from a site in the Nottinghamshire Coalfield; he used data from the Clumber line and obtained a good fit with values of $\alpha_1 = 4{\cdot}45$, $\alpha_2 = 0{\cdot}45$ (or $k_1 = 2$ and $k_2 = 10$). The comparison is illustrated in Figure 42.

The Clumber comparison used by Berry has appeared in several of his papers. Little attempt appears to have been made in making more comprehensive comparisons between field data and theoretical predictions as most of the papers have been concerned with discussing the theoretical aspects of the work. Salamon (1964) expressed reservations on just using a pair of constants (k_1 and k_2) to describe quantitatively the subsidence phenomenon in British coalfields but Berry (1964b) produced additional supporting data to confirm the validity of his original conclusion.

Salamon (1963/64, 1974, 1977) has put forward ideas on elastic analysis of displacements and stresses resulting from mining of tabular deposits. He discusses homogeneous, isotropic and transversely isotropic models and makes similar observations to Berry. Salamon (1963) describes the face element principle approach to elastic treatment of ground movement and gives examples of its solution by numerical integration. Two other theoretical models are discussed by Salamon (1963/64) and are referred to as the *frictionless laminated model* and the *multi-membrane model.* Both these latter models were aimed at incorporating a 'stratified' characteristic within their treatment of the problem. Reservations have been expressed regarding the degree of simplification used in arriving at the basic equations used in these two models, Berry (1977).

Concluding remarks on elastic analyses. The general lack of homogeneity of the natural ground overlying mining extractions is a limiting factor to the application of elastic theory to the prediction of ground movements caused by mining. Even if the elastic properties of individual layers are accurately known, there are still problems in matching of stress and displacement components across interfaces (Berry 1977). The finite element method offers special advantages in this respect since the elastic constants can be selected in respect of particular elements so that a layered model can be readily assigned appropriate elastic constants.

Visco-elastic treatment of ground movement. The incorporation of time into the theoretical treatment of ground movement has been discussed by Marshall and Berry (1966) and Astin (1968). Such work demonstrated the feasibility of applying visco-elastic analyses to ground movement but Berry (1977) concluded that it gave no certain indication that it was appropriate especially in view of the general lack of rheological data available on Coal Measures rocks.

Plastic treatment of ground movement. Consideration of ground movement obeying the laws of plasticity has been undertaken by Cherny (1966) and he derived a hyperbolic tangent function to describe the resulting subsidence. The general application of plastic models has not attracted any significant attention. It is probable that this type of model has limited application to shallow depth situations where the overburden exhibits plastic flow characteristics as can occur with some stiff clays, silty deposits and unconsolidated deposits, possessing a fairly high water content. Plastic models are rarely used probably because of the difficulties in establishing reliable mass behavioural properties of the overburden. There has not been sufficient evidence published to indicate that models based on the laws of plasticity are adequately valid or produce results comparable with other methods of subsidence prediction.

Physical models and prediction of subsidence

Physical models have played an important role in the development of knowledge pertaining to understanding ground behaviour in relation to mining subsidence. Litwiniszyn (1957) and Knothe (1957) have demonstrated the importance of using physical models to confirm or justify theoretical concepts in establishing quantified treatment of mining subsidence problems in Poland. King and Whetton (1957) reported the findings of mining subsidence investigations using physical models constructed of solidified gelatine solution and all the experiments were conducted within the elastic limit of the medium employed; these authors ably demonstrated the mechanics of mining subsidence from which important fundamental reasoning and theoretical developments emerged.

Sand/plaster physical models are the most commonly used since other types do not readily provide the advantages given by such modelling materials. In particular, sand/plaster types easily allow the principal variations in geology to be built into the model. Additionally, the propagation and growth of cracks caused by caving and subsidence can be observed and studied in relation to other factors. Surface subsidence can be studied as a separate feature, although it is usual to take account of sub-surface behaviour with such models. Multi-seam workings and pillar configurations can be incorporated into the model. Consequently, sand/plaster models increase the scope of the subsidence investigation and allow many other important factors to be observed simultaneously. This technique has been used in several countries, notably the USSR, West Germany, USA, India, Australia and the UK. It forms a separate chapter later in this book.

Conclusions

The principal conclusions drawn are listed as follows:

1. Prediction of mining subsidence in European coalfields is based on experience covering many years of extensive field and laboratory measurements in addition to detailed theoretical studies. Empirically based prediction methods continue to find the greatest favour in view of the sheer volume of reliable subsidence data available and which have been incorporated in such prediction methods.

2. Profile functions serve an important role and can greatly assist in producing rapid assessments of mining subsidence in those situations which employ rectangular extractions in relatively simple layouts. Influence functions increase the scope of the profile function method especially in allowing complex mining layouts to be assessed in relation to anticipated surface subsidence. The use of zone area methods with graphical integration is expected to continue to be a popular choice for subsidence prediction especially in view of its convenience and relative speed of calculation attributes.

3. Theoretical developments have tended to be used to explain ground behaviour; predictions of mining subsidence by such methods have needed to be confirmed with empirical observations to judge their validity and adequacy.

4. Finite element models and physical models have special advantages in investigating particular problems. These methods allow an insight to be made into anticipated ground behaviour between the surface and the mining horizon.

CHAPTER 5

OBSERVED BEHAVIOUR OF MINING SUBSIDENCE

Knowledge of the character of mining subsidence in relation to underground exploitation of minerals has been most reliably established from detailed surveying observations at the surface overlying and adjacent to such mining activities. From such observations patterns of ground movement behaviour have been identified. This has led to the emergence of reliable methods of subsidence prediction.

The purpose of this chapter is to present a number of observed cases of mining subsidence, many of which have been incorporated in the UK SEH empirical model. These examples of case studies of mining subsidence serve to demonstrate behavioural trends. They also effectively exhibit the influence of the main controlling parameters, particularly depth and extraction dimensions of mining.

Beevers and Wardell (1954-55) in commenting upon particular differences between European experiences regarding subsidence, and especially in contrasting Dutch and German with British mining subsidence knowledge as existed during the early 1950's made the following points:

1. There was wide general agreement on the Continent pertaining to the maximum possible subsidence in relation to the form of goaf treatment such as packing and stowing, and recognition of the area of influence.

2. There was wide agreement on the uniformity of the angle of draw over wide extractions; there were larger magnitudes of draw reported than by comparison to UK observations.

3. The significance of conclusions pertaining to the nature, extent and disposition of horizontal ground movements were widely accepted.

4. It was recognised that horizontal ground movements played a major role in relation to damage to surface structures.

Beevers and Wardell (1954-55) suggested that the degree of accuracy adopted in levelling when taking subsidence observations would make a significant difference to the observed angle of draw. They put forward a comparison of observed subsidence over a ribside by ordinary and precise levelling methods, and their findings are given here in Figure 43. As can be seen, their results indicate that by ordinary levelling methods the angle of draw was observed to be 18·5° but by precise levelling was 35° for this longwall mining situation. These results imply that between 18·5° and 35° the effects of mining on the surface are almost indiscernible for this case history.

Shadbolt (1986) who was involved extensively with subsidence observations over a period of thirty years since the 1950's has given the following explanations for increased accuracy in respect of mining subsidence:

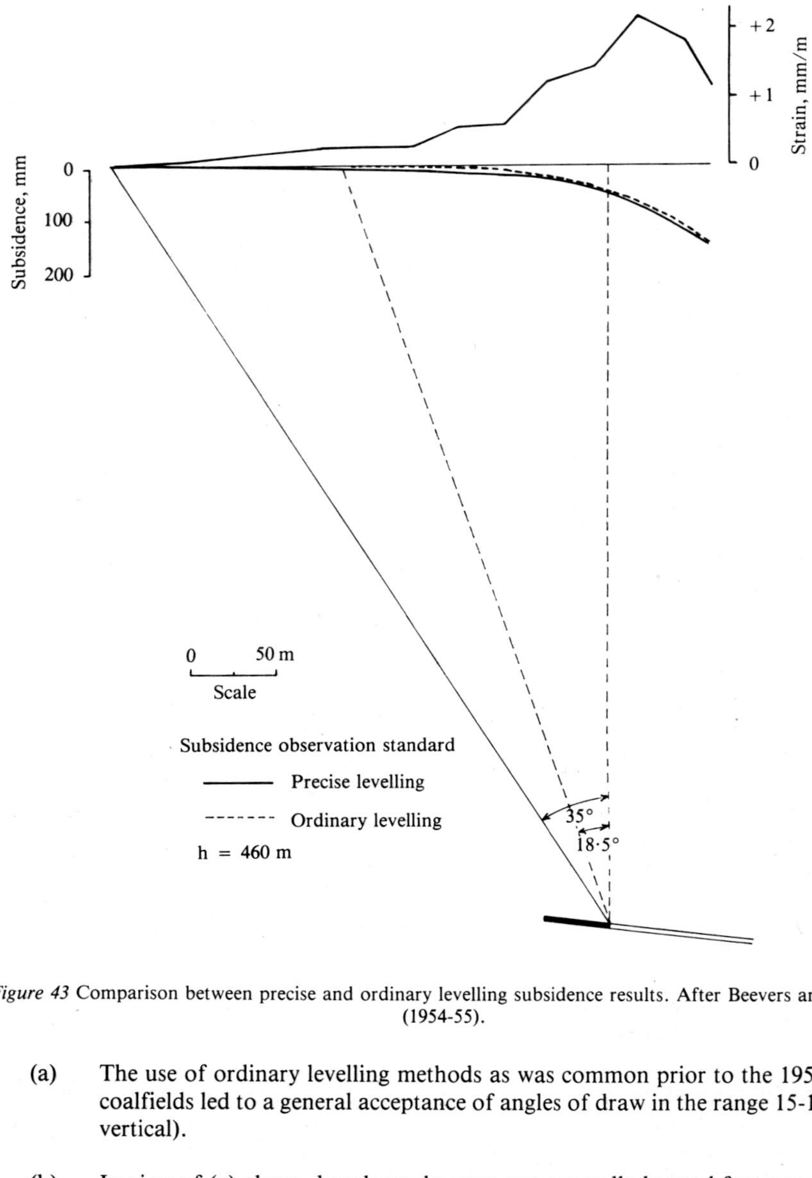

Figure 43 Comparison between precise and ordinary levelling subsidence results. After Beevers and Wardell (1954-55).

(a) The use of ordinary levelling methods as was common prior to the 1950's in UK coalfields led to a general acceptance of angles of draw in the range 15-18° (to the vertical).

(b) In view of (a) above, bench marks were not generally located far enough outside the area of influence; unstable bench marks in such situations would naturally lead to indicating a reduced value for maximum subsidence. Consequently the question of the influence of increased accuracy of levelling methods was compounded by the bench marks being possibly located in areas within the influence of mining.

(c) The general introduction of precise levelling methods in the 1950's resulted in a three-fold increase in accuracy; the experience gained in UK Midland coalfields indicated that ordinary levelling methods could achieve an accuracy of ± 1·5 mm whereas the introduction of precise levelling at that time achieved accuracies of ± 0·5 mm.

(d) The extracted seam height influenced the angle of draw by virtue of its effects being more easily measured at the surface.

Yorkshire Coalfield mining subsidence observations

Beevers and Wardell (1954-55) have reported the following general comments regarding subsidence observations in the Yorkshire Coalfield, UK.

1. The maximum subsidence from longwall extractions employing strip packing was observed to be 70-85% (of M). The degree of goaf packing influenced the maximum subsidence as did the depth of mining where strip packing was employed.

2. Angles of draw were found to be remarkably consistent over the whole coalfield, with the range being 32-38° even where the overburden contained Permian limestones and Triassic sandstones.

3. Subsidence development was in relation to the proportion of the area of influence affected, and maximum subsidence only occurred if the full area of influence was extracted.

4. The symmetry of a transverse subsidence profile can be distorted by previous adjacent workings, and this applies also to the displacement and strain curves.

5. The maximum displacement and change from compressive to tensile ground strain occur at the position of maximum gradient on the subsidence curve. The subsidence observations reported by Beevers and Wardell tend to show the maximum tension occurring above the ribside; see Figures 44, 45 and 46 which show the observation lines in relation to the extraction together with results for the transverse and longitudinal (advancing) observed profiles.

6. The greatest effect from the travelling strains occurs when the extraction width approaches the diameter of the influence area, and consequently is more prone to occur at relatively shallow depths.

7. The travelling strains displacement and subsidence profiles exhibit a time-dependency effect as illustrated in Figure 47(b) which shows results from the observation line 1-27 of Figure 47(a). There was a tendency for slight continued subsidence after the face stopped in this case study which resulted in a small shifting of the peak tensile strain displacement and transition point positions. The subsidence limit point was not influenced by this time dependency phenomenon, and the effects of time were the least at the outer limits of subsidence.

8. Partial extraction layouts involving narrow longwalls separated by intervening wide rib pillars can significantly limit surface movements due to undermining. The case study illustrated in Figures 48 and 49 represents a partial extraction scheme where panel widths of 72 m and rib pillar widths of 96 m gave a 40% extraction. The depth was 690 m and extracted seam height was 1·7 m whilst the

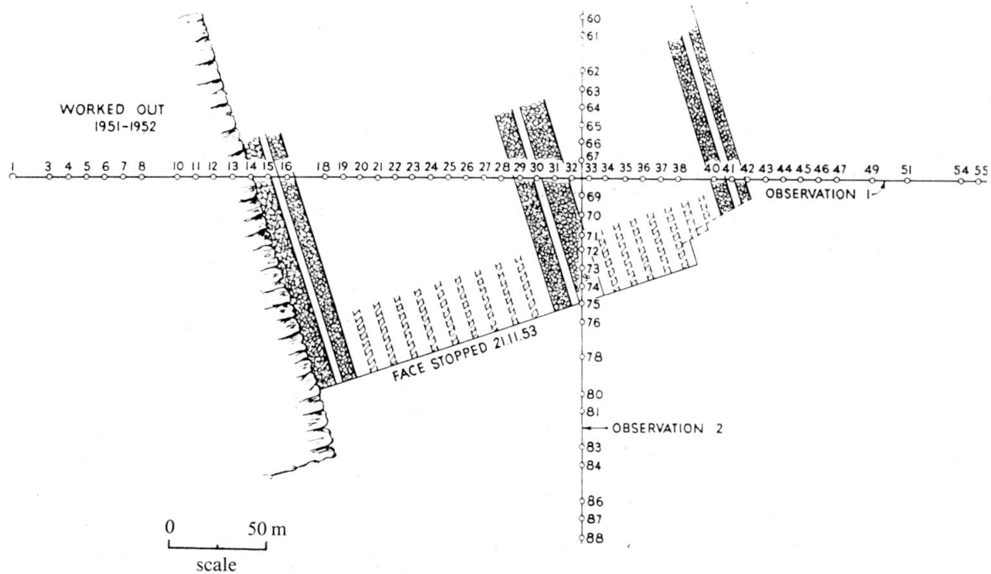

Figure 44 Location of observation lines referred to in Figures 45 and 46. After Beevers and Wardell (1954-55).

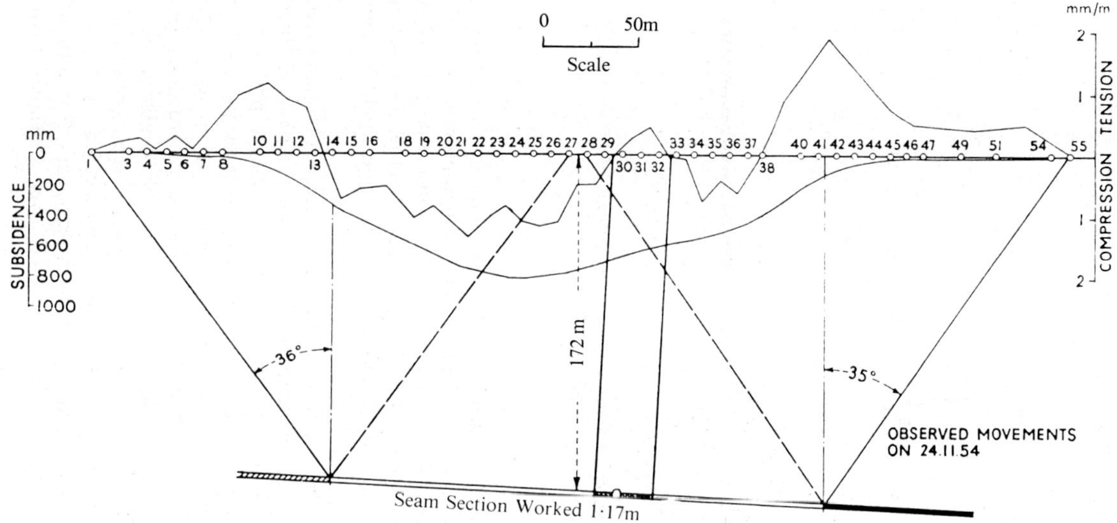

Figure 45 Transverse subsidence and strain profiles for observation line 1-55. After Beevers and Wardell (1954-55).

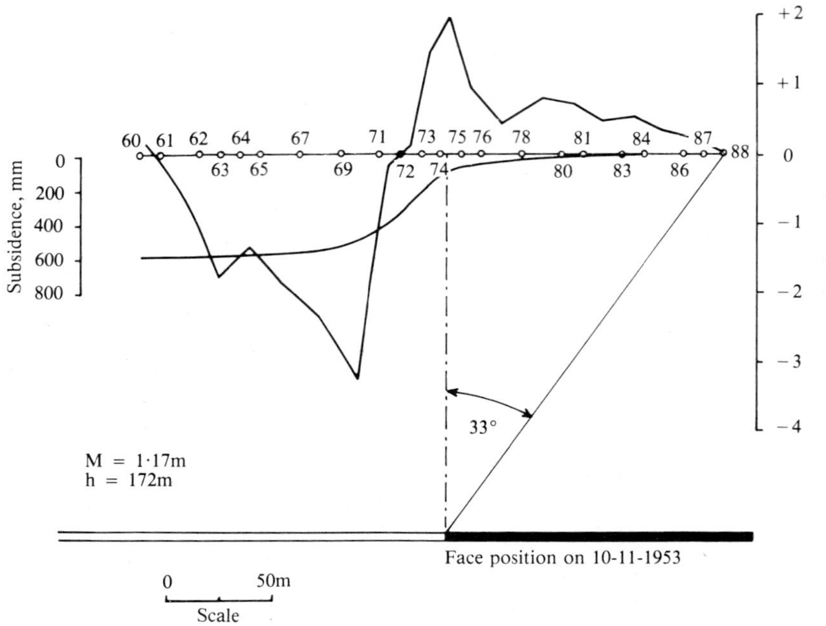

Figure 46 Subsidence development curve showing travelling strain profile in relation to Figure 44. After Beevers and Wardell (1954-55).

maximum resulting subsidence was around 100 mm. This case study clearly demonstrated that pressure arching across the longwalls was taking place and thus resulting in reducing the maximum subsidence, whilst interaction of the individual subsidence profiles produced a smoothing out effect of the resultant profile so that the effects of surface subsidence in this situation would be minimal.

9. The main cause of damage at the surface is the effect of ground strain, tension or compression, which accompanies subsidence; this is highlighted by the fact that the greatest damage does not necessarily occur at the position of maximum subsidence, but rather where the greatest curvature is encountered. The degree of damage sustained by a surface structure is not necessarily related directly to the magnitude of strain but more on its shape, size and construction together with the orientation relative to the underground workings. Noticeable damage (cracking of internal plaster, doors and windows sticking) was observed when the tensile strain exceeded 0·4 mm/m and the compressive strain exceeded 0·8 mm/m. Damage for compressive strain is less frequent than with tensional effects although it is often more severe. However, buildings located over tensional fissures can be appreciably damaged.

Webster (1951-52) drew attention to the differential variations being of major importance rather than just the question of subsidence. He demonstrated the importance of interaction of surface strain profiles in producing a resultant of greatly reduced significance in respect of causing surface damage.

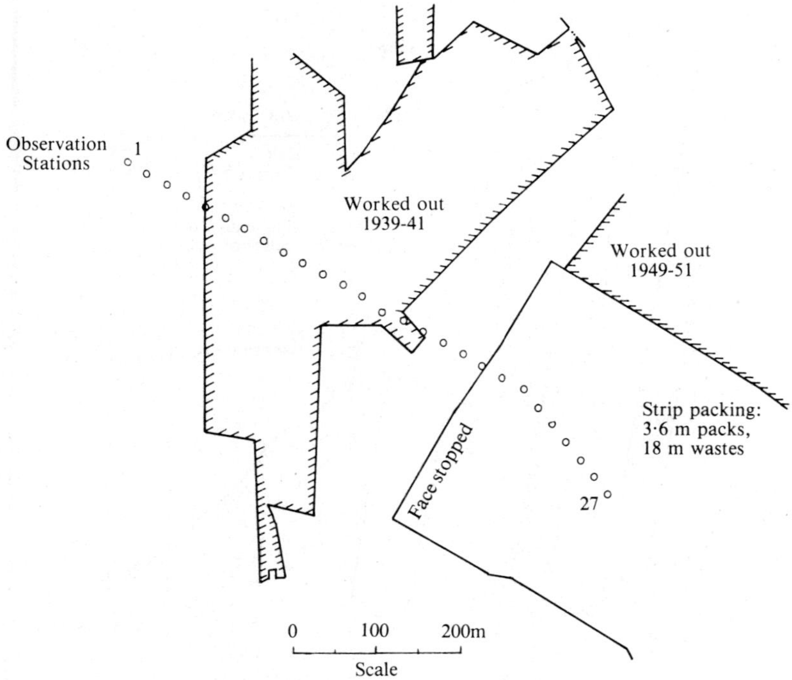

Figure 47(a) Observation line and workings in relation to subsidence results presented in Figure 47(b). After Beevers and Wardell (1954-55).

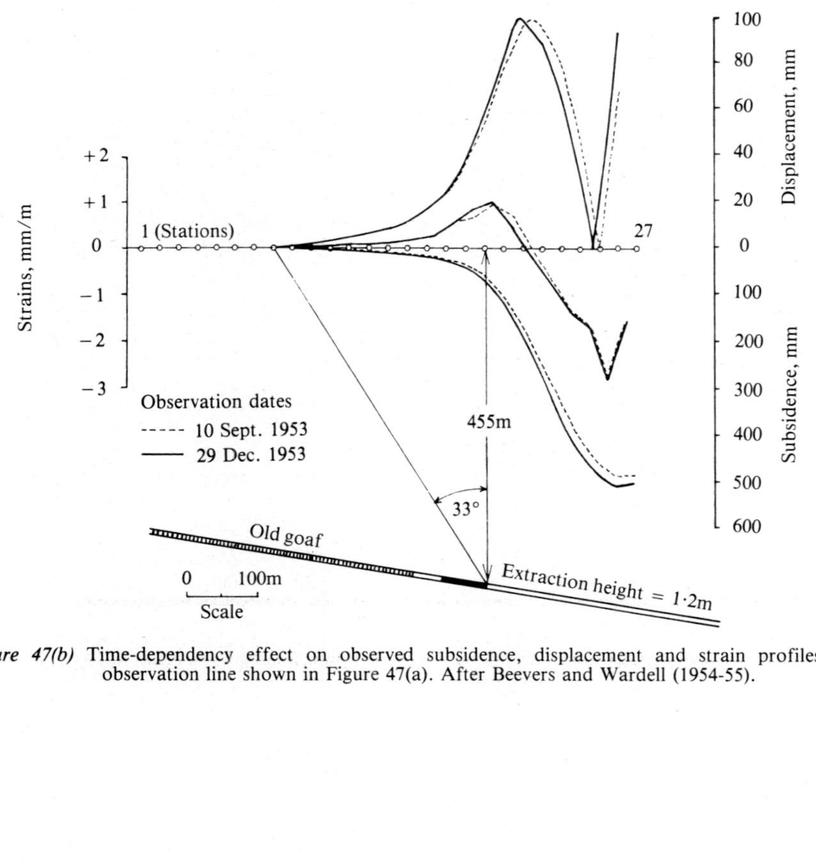

Figure 47(b) Time-dependency effect on observed subsidence, displacement and strain profiles on the observation line shown in Figure 47(a). After Beevers and Wardell (1954-55).

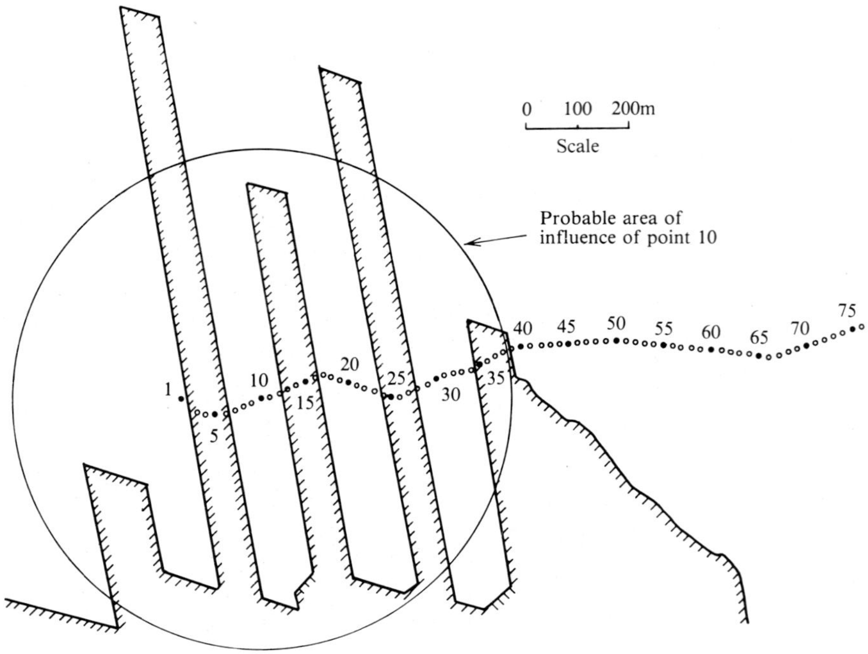

Figure 48 Observation line over partial extraction workings. After Beevers and Wardell (1954-55).

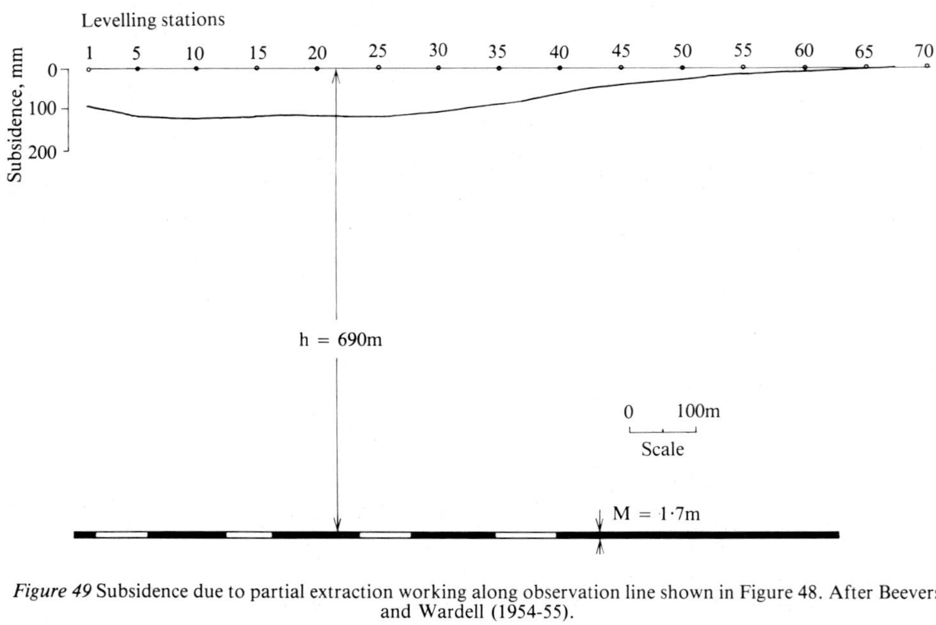

Figure 49 Subsidence due to partial extraction working along observation line shown in Figure 48. After Beevers and Wardell (1954-55).

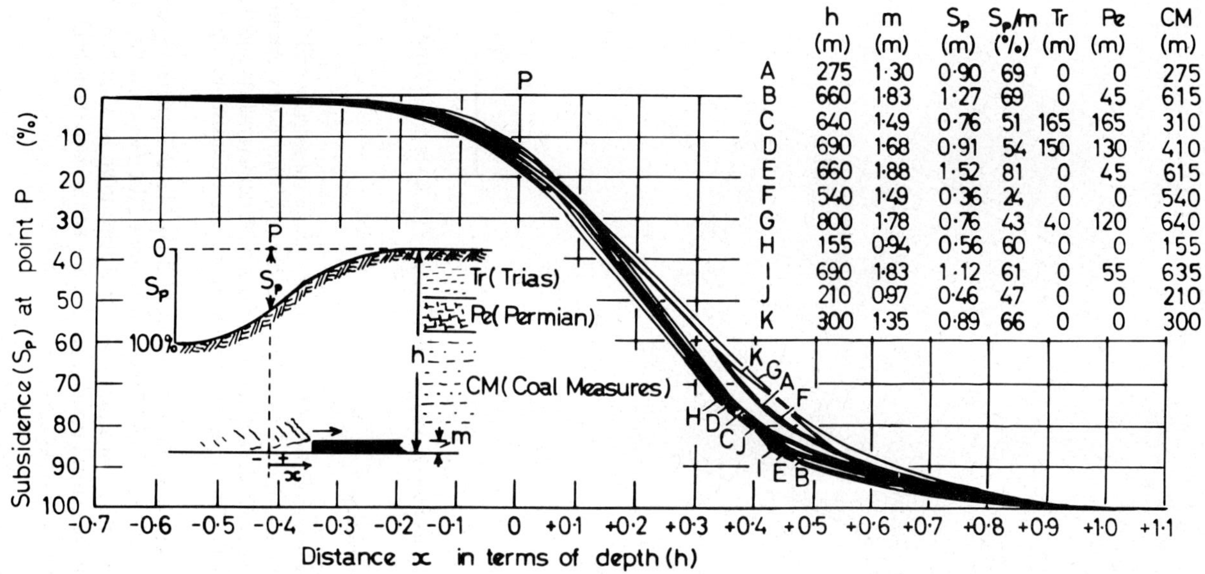

	h (m)	m (m)	S_p (m)	S_p/m (%)	Tr (m)	Pe (m)	CM (m)
A	275	1·30	0·90	69	0	0	275
B	660	1·83	1·27	69	0	45	615
C	640	1·49	0·76	51	165	165	310
D	690	1·68	0·91	54	150	130	410
E	660	1·88	1·52	81	0	45	615
F	540	1·49	0·36	24	0	0	540
G	800	1·78	0·76	43	40	120	640
H	155	0·94	0·56	60	0	0	155
I	690	1·83	1·12	61	0	55	635
J	210	0·97	0·46	47	0	0	210
K	300	1·35	0·89	66	0	0	300

Figure 50 Development of subsidence in relation to face position (revision of Wardell's original results, Wardell (1953-54) and Forrester and Whittaker (1976).

Figure 51 Location of subsidence line shown in Figure 52. After Orchard (1957b).

Development of mining subsidence

Wardell (1953-54) investigated the development of mining subsidence and his results are shown as Figure 50. Of special importance is the consistency of the results. Although a wide range of depth has been used, the development curves are particularly close and demonstrate the same characteristic shape. The subsidence limit points show consistency also.

A further important feature exhibited by Figure 50 is the fact that different combinations of Coal Measures, Permian limestone and Triassic sandstone exist within the data used. These results suggest that within the combinations of overburden shown by Figure 50, there is no significant departure from the general trend regarding any of the rock types being present on the development of mining subsidence.

Effect of gate-side packing on surface subsidence

A common practice in previous longwall mining layouts, particularly during the 1950's and earlier was the incorporation of access roadways within the goaf area behind the longwalls. Present-day practice is generally to construct the gate roadways at the rib-sides,

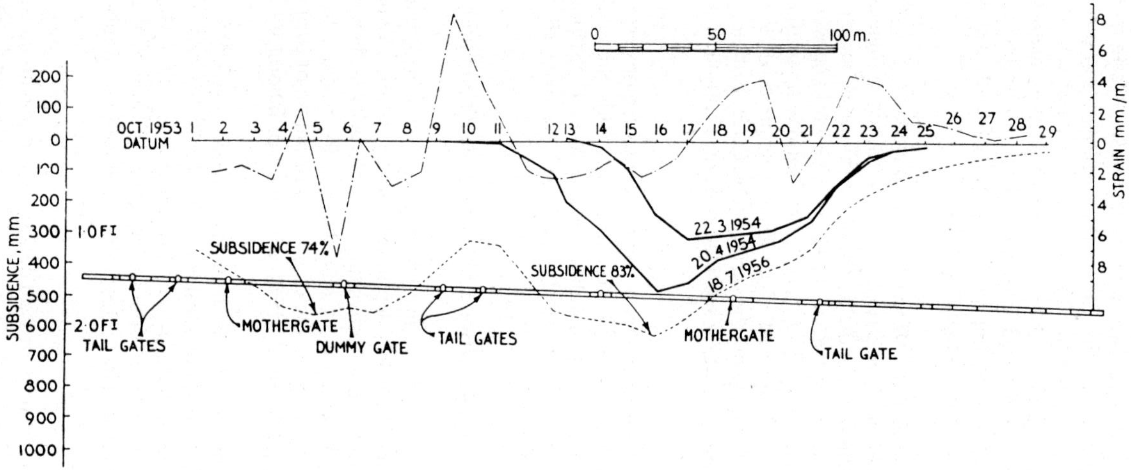

Figure 52 Showing hump effect in subsidence and strain profiles influenced by gate-side packs.

although with some forms of advance layout roadways may be located within a goaf region and be subjected to the full effects of subsidence.

Whether the gate roadways are located at the rib-sides or towards the centre of the face, they require a measure of protection against the effects of caving and subsidence behind an advancing longwall face. Such protection is usually in the form of gate-side packs. The construction of the packs was predominantly by hand using waste rock although pneumatic stowing was widely employed in the 1950's. More recent gate-side packing construction methods have extended beyond pneumatic placement to that of scrapers, ram-packers, pump packing, hydraulic methods and combinations of two or more methods. The basic aim has been to improve efficiency and speed in gate-side pack construction. With some methods, however, it appears that seeking improved strength of pack construction has been the principal objective so as to promote gate stability.

Orchard (1957b) examined the effect of gate-side packing of roadways serving longwall advancing faces and a case study described by him is shown in Figures 51 and 52. This case study is taken from the Durham Coalfield. The depth (h) was 60 m, extracted seam height (M) = 0·76 m and the panel width (w) was about 122 m. The workings were caved apart from the gate-side pack constructions shown in Figure 51. The subsidence profile clearly shows a hump where gate-side packing was employed. The tail gates demonstrate that supporting action of the packs is taking place and reducing the surface subsidence. It should be noted however that this packing effect on reducing surface subsidence locally is probably also associated with the shallow depth of only 60 m.

The case study shown earlier in Figure 44 and 45 also demonstrate the hump effect in the surface subsidence profile due to the construction of gate-side packs adjacent to a centre gate located in the goaf area behind a longwall advancing face. The depth below surface in this case study was 171 m, which is still classified as being relatively shallow in relation to the development of overburden pressure.

Observations on the relationship between subsidence, depth and horizontal strain

Orchard (1957b) has argued that in view of curvature of the ground surface being more pronounced over workings of shallow depth, and that curvature is an indication of the degree of horizontal displacement, then horizontal strain intensity will be related to both depth below surface and the magnitude of subsidence. Horizontal strain is proportional to subsidence and inversely proportional to depth. Confirmation of such relationships had been already indicated by King and Smith (1954) as a result of laboratory investigations.

Orchard (1957b) examined 40 cases from seven different coalfields in order to investigate the relationship between maximum subsidence (S), depth (h) and maximum tensile strain (+E); the basic data used by Orchard are presented in Table 6 whilst Figure 53 shows a graphical representation of the relationship between +E and S/h. It was pointed out by Orchard that the w/h ratio would influence the relationship, and this is discussed further in Chapter 6.

Orchard extended his analysis of field data to that of maximum compressive strain (−E) in order to investigate its relationship with the S/h ratio. He again emphasised that the w/h ratio would play a role in influencing the nature of the relationship. Table 7 presents Orchard's data for the 16 cases he used where w is less that R, whilst Figure 54 shows −E plotted against S/h as giving a reasonably linear relationship for the cases considered.

The results presented in Figures 53 and 54 were further examined and rationalised by

No.	Coalfield	Depth h (m)	Width/ Depth w/h	Max. Subsidence S (m)	Max. Sub. Depth S/h	Max. Tensile Strain (+E) Fracture Peak	Max. Tensile Strain (+E) Fracture Equivalent Normal	Max. Tensile Strain (+E) Normal
1	North Staffordshire	823	0·18	0·68	0·00083			0·00072
2	,,	541	0·375	0·38	0·0007			0·00056
3	,,	506	0·60	0·53	0·00115	0·00128	0·00083	
4	,,	549	0·42	0·61	0·00111			0·00111
5	,,	579	0·26	0·30	0·0005			0·00033
6	,,	616	0·32	0·76	0·00122			0·00083
7	,,	409		0·56	0·0014			0·001
8	,,	375	0·729	0·75	0·002	0·00344	0·00167	
9	,,	396	0·62	1·04	0·0026			0·00194
10	,,	335	0·755	0·99	0·00295			0·0025
11	,,	249	0·776	0·91	0·00367	0·00444 0·00156	(Dipside) mean 0·003 (Riseside)	
12	,,	320	0·755	1·68	0·00524			0·00367
13	,,	107	1·540	1·07	0·010			0·00683
14	Durham	127	0·568	0·32	0·0025			0·0014
15	,,	127	1·17	0·60	0·0047			0·0026
16	,,	46	1·97	0·18	0·004			0·002
17	,,	267	1·14	0·65	0·0024			0·0012
18	,,	64	2·26	0·61	0·0095			0·005
19	,,	107	1·5	0·59	0·0055			0·0025
20	Yorkshire	112	0·33	0·09	0·0008			0·0005
21	,,	77	1·4	0·62	0·008			Goafside 0·005 Ribside 0·006
22	,,	23	4·3	0·60	0·026			0·018
23	,,	101	1·27	0·55	0·0054			Goafside 0·001 Ribside 0·0036
24	,,	41	2·44	0·63	0·0153		0·0058	Goafside 0·0058 Ribside 0·0092
25	,,	114	0·8	0·81	0·007	0·013		
26	,,	101	1·0	0·64	0·0063			Goafside 0·0031 Ribside 0·0042
27	,,	177	0·57	0·42	0·0024			0·00169
28	,,	207	1·3	0·52	0·0025			0·001
29	,,	454		0·52	0·00114			0·00095
30	,,	107	0·83	0·53	0·005			Goafside 0·0009 Ribside 0·000379
31	,,	171	1·5	0·80	0·0047			Goafside 0·00185 Ribside 0·0019
32	Warwickshire	591	0·47	1·55	0·0026			0·002
33	,,	396	0·68	0·76	0·0028			0·0026
34	,,	463	0·53	0·87	0·0012			0·0008
35	Nottingham-shire	349	0·4	0·23	0·00066			0·000555
36	,,	213	1·5	0·58	0·0027			0·001
37	Lower Rhine (Left Bank)	175	1·9	0·87	0·005			0·0035
38	,, ,,	170	0·53	0·46	0·0027			0·00173
39	,, ,,	200	0·90	0·37	0·00186			0·00133
40	Czechoslovakia	65	0·86	1·6	0·0245			0·0126

Table 6 Subsidence data relating to maximum tensile strain (+E) After Orchard (1957b)

Case	Depth h (m)	Maximum Subsidence S (m)	S/h	Width/ Depth w/h	Maximum Contraction (−E)	
					Peak	Equivalent Normal when fractured
1	249	0·91	0·00367	0·776	0·00778	0·00625
2	541	0·38	0·0007	0·375	0·00111	0·00111
3	549	0·26	0·00049	0·330	0·001	0·001
4	412	1·17	0·00285	0·624	0·004	0·004
5	396	1·04	0·0026	0·620	0·00693	0·00417
6	375	0·75	0·002	0·729	0·0033	0·0033
7	354	0·40	0·00114	0·56	0·0015	0·0015
8	381	0·91	0·0024	0·52	0·0033	0·0033
9	683	0·44	0·00064	0·22	0·0009	0·0009
10	546	0·34	0·00063	0·37	0·0014	0·0014
11	579	0·30	0·0005	0·26	0·00067	0·00067
12	570	0·21	0·00036	0·23	0·0007	0·0007
13	793	0·31	0·0004	0·31	0·0005	0·0005
14	506	0·58	0·00115	0·60	0·0014	0·0014
15	488	0·46	0·00094	0·60	0·00111	0·00111
16	409	0·56	0·0014		0·00167	0·00167

Table 7 Subsidence data relating to maximum compressive strain (−E). After Orchard (1957b)

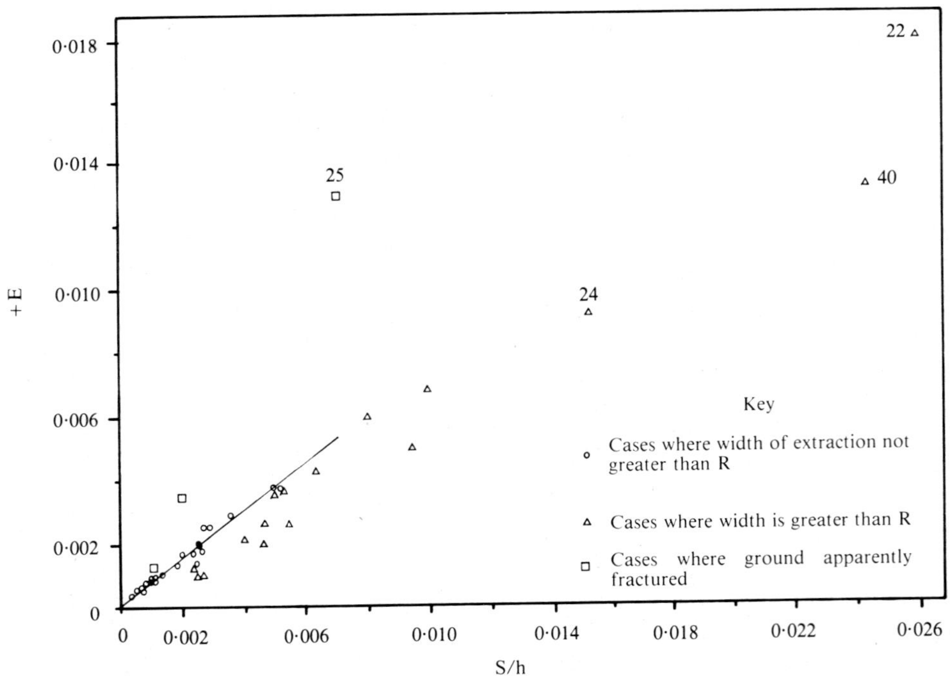

Figure 53 Relationship between +E and S/h. After Orchard (1957b).

Orchard in order to arrive at a more precise understanding of the role played by the w/h ratio; attention was also focussed on deriving a similar relationship for deducing the maximum tilt due to mining subsidence. These results were then embodied in the SEH empirical model for determining subsidence strains and tilt.

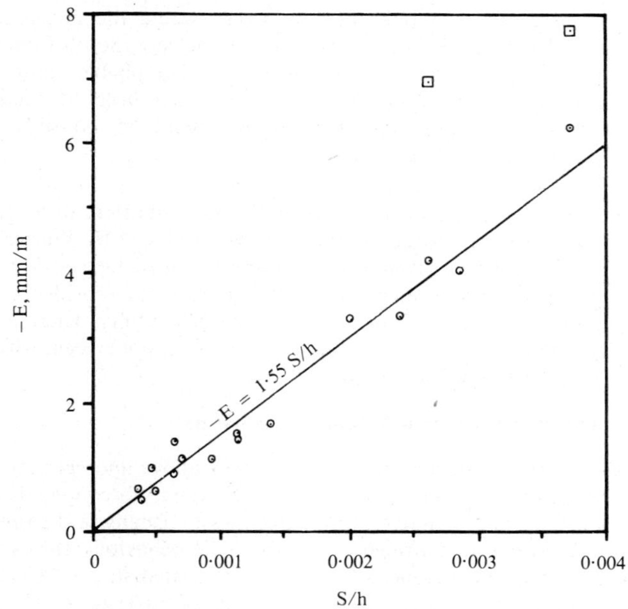

Figure 54 Relationship between – E and S/h. After Orchard (1957b).

Depth of working, h (m)	Thickness of non-Carboniferous (m)	Thickness of Carboniferous (m)	Angle of draw (from the vertical) (deg.)	Dip (deg.)
598	340	258	45	7
572	344	228	45	8
564	343	221	50	14
419	99	320	35	8
370	236	134	45	12
297	232	65	45	10
251	195	56	45	14
224	93	131	35	6
74	18	56	35	7
43	12	31	35	17
38	15	23	30	17
30	16	14	35	17

Table 8 Subsidence data for South Limburg showing influence of depth on angle of draw. After Drent (1957).

Subsidence observations in the South Limburg Coalfield

Drent (1957) has drawn attention to the non-Carboniferous overburden associated with the South Limburg Coalfield, The Netherlands, having a significant effect on increasing the area of influence of mining subsidence.

Table 8 shows subsidence data published by Drent where the angle of draw was predominantly 35-40° but attained 50° (to the vertical) in one case. South Limburg is noted for its fairly large areas of influence of mining subsidence. The pliable nature of the near surface rocks coupled with frequent occurrence of saturated beds of sand and other unconsolidated materials have been considered to be mainly responsible for such a phenomenon in this particular European coalfield.

Drent (1957) examined seven case histories from South Limburg in order to investigate their comparison with the subsidence development curve produced by Wardell. The basic comparison of the subsidence data is given in Table 9 where four surface positions of point P are considered. There is general agreement on the development of subsidence between the South Limburg results and the average curve for the Wardell results. There appears to be more variation with those results from South Limburg and this is not evident with those from the Yorkshire Coalfield, see for example Figure 50.

Subsidence observations in the Austrian Tertiary Coal Basins

Perz (1957) has reported subsidence observations taken above underground workings in the Austrian Tertiary coal basins which have been predominantly brown coals occurring in seams of 1 to 20 m thickness and with overburden thicknesses usually in the range of 100 to 300 m. The nature of the overburden influences its subsidence behaviour; the overburden has a pliable character with frequent presence of weak and saturated strata. Perz has reported angles of draw of 41-45° (to vertical); he also reports the occurrence of surface fractures located at around 18·5° (to vertical). In one case having a flat surface and 90 m of overburden, a subsidence factor of 85% was recorded and this is about the average value encountered in these Austrian coal basins. Perz reports that subsidence is also immediate and almost fully developed after extraction of the critical area.

Surface effects over the edges of workings

Where one side of a longwall panel has been previously worked, then additional subsidence can be expected on that side by comparison to over a solid ribside. Wardell and Webster (1957) have considered this aspect and Figures 55 and 56 reproduce their results to demonstrate the observed additional subsidence over an adjacent goaf of a previous extraction. An important feature is the reduced ground strain tension and the tendency for the peak tensile strain to be displaced further from the current extraction and indeed appreciably over the adjacent old goaf area.

Effects of previous workings

Wardell and Webster (1957) examined the effect of a large underlying coal pillar in relation to the character of surface subsidence from a current extraction; Figure 57 shows their results which Wardell and Webster suggest should be compared to Figure 56 in view of the similarity of depth and dimensions of extraction. The underlying pillar appears to have resulted in an intensification of the ground movement effects especially in relation to increasing the magnitude of the tensile and compressive strains.

Narrow rib pillars are occasionally left for underground convenience of working close to

South Limburg Case No.	Thickness of non-Carboniferous overburden (m)	Thickness of Carboniferous (m)	Depth (m)	Distance of P relative to face-line −0·5R	0	+0·5R	+R
1	340	258	598	3	33	82	96
2	99	320	419	5	18	77	96
3	93	131	224	1	15	74	93
4	15	73	88	1	4	50	92
5	18	56	74	1	6	46	89
6	20	31	51	1	4	31	88
7	15	23	38	3	15	62	92
Average				2·1	13·6	60·3	92·3
Wardell Curve, UK		Average curve taken for comparison of results		1·5	15	70	95

Table 9 Subsidence data for South Limburg showing development of subsidence compared with Wardell's results. Data derived from Drent (1957) and Wardell (1953-54).

Notes

1 The distance of the observed point P at the surface when at −0·5R signifies that P has not yet been undermined.

2 R corresponds to the radius of influence.

3 The results from the Wardell curve are for subsidence observations in the Yorkshire Coalfield, the detailed values being given in Figure 50.

a previously extracted panel. Wardell and Webster (1957) showed quite effectively how such a narrow rib pillar (coal rib) flanked by gate-side packs and two roadways can influence the surface subsidence profile particularly in giving rise to a hump effect and high localised ground strains. However, due account should be taken of the relatively shallow depth in this case study since at greater depths such a phenomenon would be diminished; the effects would probably be undetectable in the subsidence profiles of many deep workings. Similar observations have been reported by Aynsley and Hewitt (1961) who considered subsidence behaviour over shallow workings and the effects of pneumatic stowing.

Subsidence observations in the Nord-Pas-de-Calais Coal Basin, France

Dejean and Martin (1973) have discussed subsidence aspects of the Nord-Pas-de-Calais Coal Basin where for horizontal seams the angle of draw is of the order of 35°. They report that the maximum subsidence (S) can be expressed as given by equation (43)

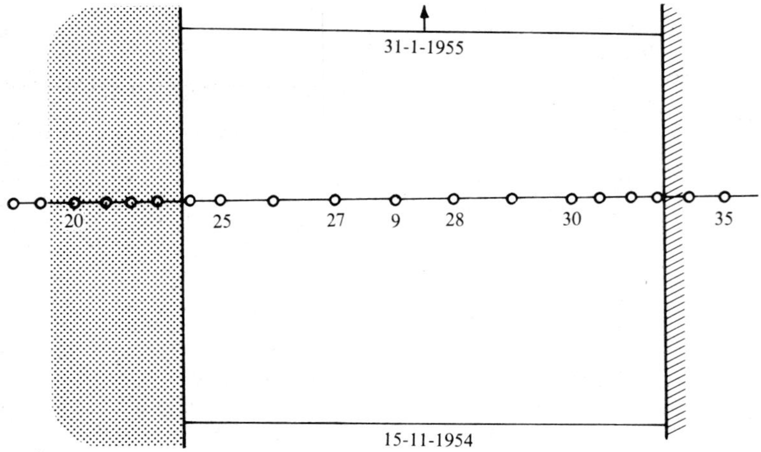

mining data:

$h = 41$m ; Strip packing

$w = 100$m ; Coal Measures (45% Sandstone)

$M = 0{\cdot}93$m ; No other seams worked

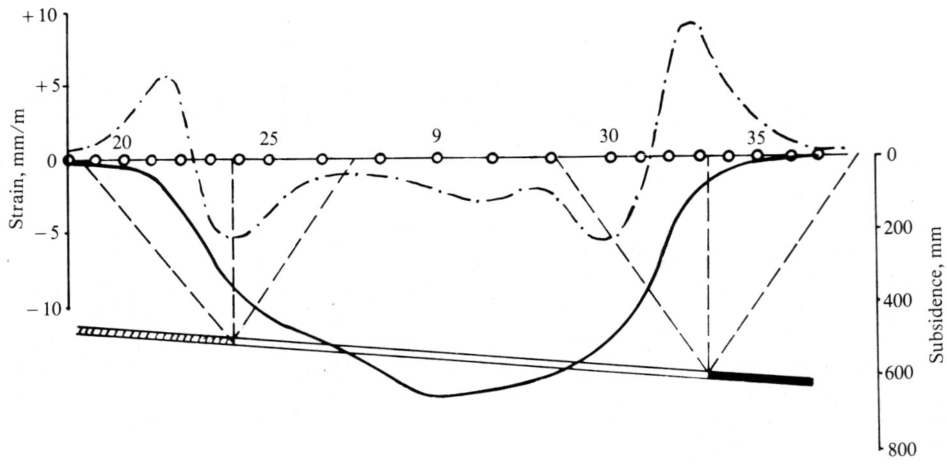

Figure 55 Surface effects over edges of workings. After Wardell and Webster (1957).

$$S = D\,q\,M \qquad \text{. . . (43)}$$

where D = a depth factor

q = a stowing factor

M = extracted seam height.

Values quoted by Dejean and Martin for the stowing factor are:

Caving	0·85 to 0·9
Pneumatic stowing	0·45 to 0·55
Hydraulic stowing	0·20 to 0·35

The depth factor is reported to increase with depth from about 0·7 at a depth of 50 m whilst reaching its maximum of 1·0 at about 500 m.

Dejean and Martin reported that from their researches the shape of the subsidence curve is largely retained at different horizons between the surface and the working horizon, although more concentrated nearer the seam and with the exception of the last few metres near to the roof of the workings which are disturbed by localised caving effects. They also concluded that the maximum subsidence varies with depth.

Arcamone *et al* (1986) have drawn attention to the importance of taking due account of the nature of the overburden especially in new mining developments since they report that this factor has a controlling influence during the early stages of subsidence.

Mining subsidence observations in South African Coalfields

Wagner and Schümann (1985) have presented a comprehensive review of mining subsidence observations and interpretations above longwall coal mine workings in South African coalfields, and conclude that predictable patterns of surface movement are possible in view of the consistency of the results obtained so far.

A recent paper by Van der Merwe (1986a and b) has reported studies of the effects of longwall mining at Sigma Colliery, South Africa, where two coal seams have been worked in close proximity. The depth of the floor of these two seams is 115m and 127m respectively, whilst each seam is about 3m thick; a combination of siltstone, mudstone and sandstone separate the two seams giving an intervening section of about 9m thickness. Longwall widths of 200m are generally employed. Extraction of the Sigma Double Seam produces a maximum surface subsidence of some 2·6m (for super-critical extraction dimensions) whereas a comparable South African single seam extraction would give about 1·5m and the UK SEH prediction 2·7m maximum subsidence. The corresponding maximum tilt is 55mm/m, 35mm/m and 58mm/m; the maximum compressive strain 20mm/m, 10mm/m and 11mm/m; the maximum tensile strain 15mm/m, 10mm/m and 14mm/m; and the angle of draw 65°, 0 to −30° and 64° for these three respective conditions. It is concluded that the working of the second seam at Sigma Colliery results in a subsidence pattern with strong similarities to the SEH prediction, owing mainly to the first seam working weakening the overlying rocks.

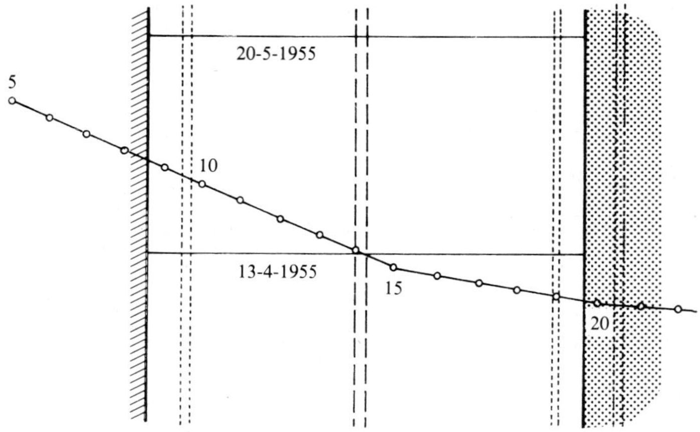

mining data:

h = 101m	;	Mechanical stowing
w = 101m	;	Coal Measures strata
M = 1·04m	;	5 seams worked below

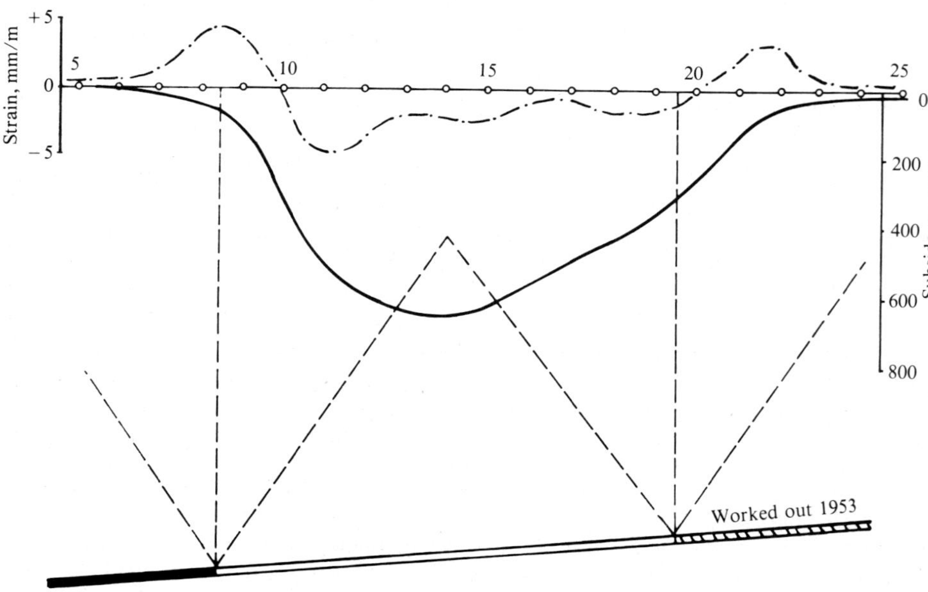

Figure 56 Surface effect over edges of workings. After Wardell and Webster (1957).

mining data:
h = 112m ; Strip packing
w = 91m ; Coal Measures strata
M = 1·04m ; 5 seams worked below

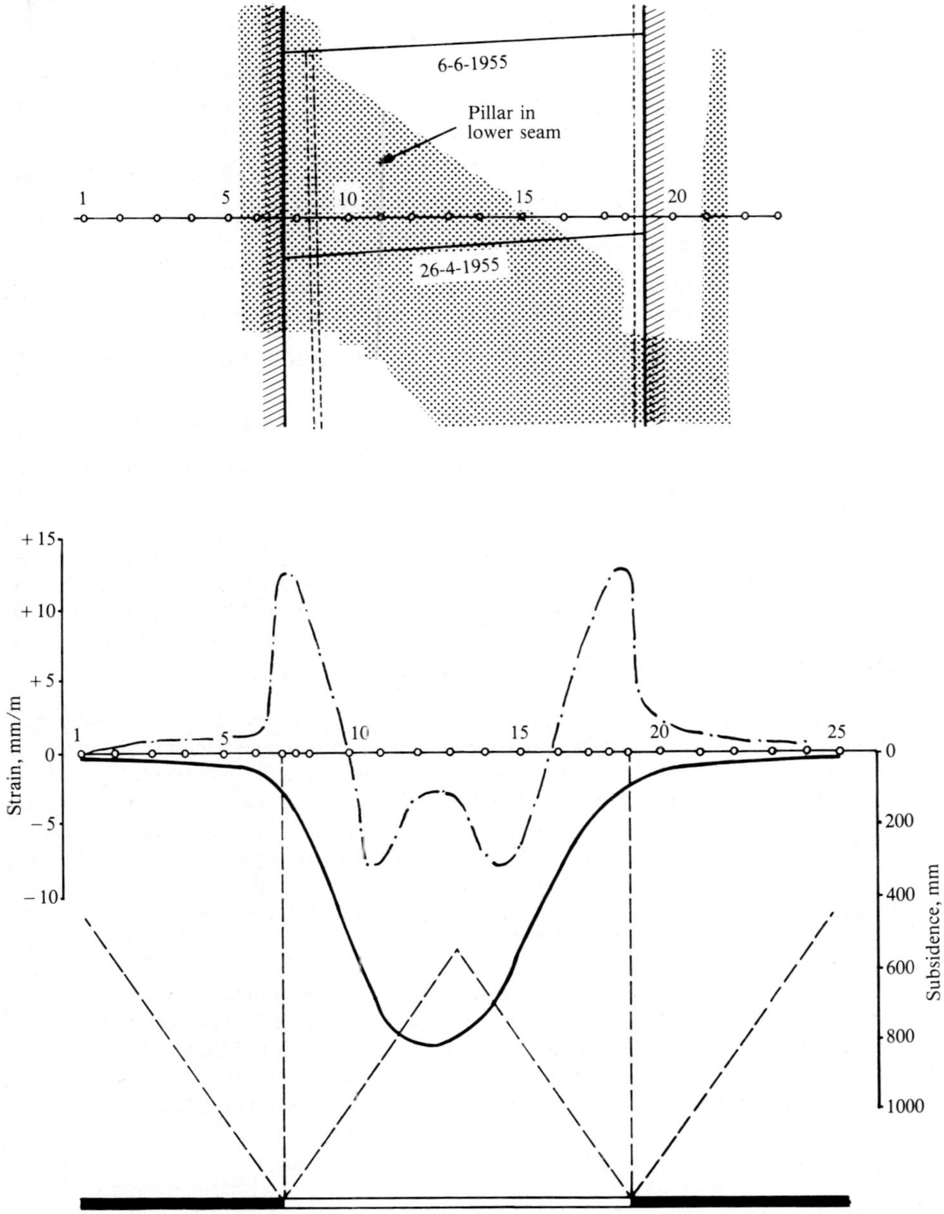

Figure 57 Illustrating effect of previous working on subsidence from current working. After Wardell and Webster (1957).

Schümann (1983, 1984) has discussed mining subsidence investigations at South African collieries where the depth range has been 100 to 200 m with typically 200 m wide longwalls separated by rib pillars of 20 to 30 m width. The extraction mining height has been generally 2·5 to 3 m. Extensive longwall mining has taken place under competent dolerite sills of 30 to 70 m thickness and this has had a significant influence on the caving behaviour of the overburden, Galvin (1982). The following comments, however, are based on Schümann's investigations where massive dolerite sills were not present in the overburden:

1. The transverse and longitudinal subsidence profiles indicate an angle of draw of about 11° (to vertical), as no significant movement is shown to occur outside of 0·2 h from the ribside.
2. Only a small amount of subsidence occurs over the solid ribsides, with not more than 0·04 S/M occurring over the edge of the extraction.
3. The maximum subsidence is less than 0·5 S/M.
4. The transition point for both longitudinal and transverse subsidence profiles occurred at about 0·3 to 0·4 h from the edge of the extraction.
5. Some 80% (or more) of the subsidence had occurred at a distance of 0·5 h from the edge of the extraction.
6. Mining subsidence had almost ceased at a distance of 1·0 h from the face line.
7. Around 90 to 95% of subsidence had occurred after about three months since the time of undermining.

The 1986 Symposium on the effect of underground mining on the surface held at Sandton, South Africa received a number of contributions on recent observations on subsidence particularly above coal mining operations SANGORM (1986). Developments in monitoring, computation and data analysis of surface subsidence were dealt with comprehensively by Schümann (1986).

A number of case studies of surface subsidence over collapsed bord and pillar workings have been described by MacCourt, Madden and Schümann (1986), and their results show that above 75% extraction, in excess of 60% (S/M) was being observed in some cases. The surface effects of shallow undermining on the East Rand has been discussed by Cameron-Clarke (1986) who draws attention to the observed depths of potholes (4 - 10m) to be surprising with a reported mining depth of around 50 - 60m. He attributes the increased height of underground collapses (beyond 10 times the mining thickness of about 3m) to be possibly associated with a low bulking factor for the Karoo Shales, perhaps increased thickness of mining, or the effect of water ingress into the old workings.

Further interesting papers were included on subsidence prediction by Oravecz (1986), and Gürtunca and Schümann (1986) whilst Van der Merwe (1986 a and b) presented important papers on panel pillar aspects in relation to subsidence and the effect of longwall operation at 50m below surface on the resulting subsidence which was found to be 0·49 (S/M) for a super-critical extraction indicating that complete consolidation was not taking place.

Mining subsidence behaviour of thick lignite deposits in Yugoslavia

The Velenje Mine in Yugoslavia is well-known regarding the progressive extraction of a lignite deposit of up to 120 m thickness and located 250 to 450 m below the surface. Underground mining is by a combination of slicing and sub-level caving thus creating subsidence of the surface. Typically a longwall is driven taking a 2·5 m extraction height, whilst a vertical thickness of 7·5 m of lignite is left above the face and this is caved behind the current longwall thus effecting a 10 m extraction. The overburden consists mainly of clays, silts, sands, marls and gravels and has a high natural water content of 20 to 40% generally and exhibits weak strength properties, 1·5 to 4 MPa (UCS) whilst the lignite has an UCS* of 10 -20 MPa. The deposit is of Pliocene age and the maximum surface subsidence has reached up to 50 m depth, Sovinc *et al* (1982).

The underground operations at Velenje Mine have resulted in a large trough-shaped subsidence depression and as observed by the present authors, it was allowed to fill with water thus forming a significant lake of around 400 - 500 m dimensions and around 40 - 50 m depth. The angle of draw appears to be in the range 33 - 42° (to vertical). The surface subsidence depression exhibits significant shear steps and can be 1 - 2 m vertically, rather than a continuous smooth change in gradient as is commonly observed in many European longwall operations. The weak overburden allows the beds to deform and subside uniformly although with some localised steps. The formation of sink-holes is prevented during the mining operation by the extensive use of steel meshing; this prevents inflow of caved overburden into the sublevel caving (of lignite) operations taking place behind the longwalls. Although subsidence and caving are effectively controlled, the present authors observed the release of significant quantities of gas from the strata and this bubbled through the lake at the position where large fissuring occurred within the tensile strain region of the subsidence depression.

Surface subsidence associated with Japanese coal mines

Coal mining operations in Japan are virtually entirely by the longwall method. Hiramatsu *et al* (1979) have reported general characteristics of mining subsidence in the northern district of Kyushu. Table 10 presents subsidence observations published by these authors for the Kyushu district and summarises geological and working conditions and subsidence data for 16 coal mines. Figure 58 illustrates the terminology. Additionally, Figure 59 shows a relationship between the subsidence (S/M) plotted against plan width of the longwall (l/h). Figure 59 shows the dependence of subsidence development in relation to the longwall plan width/depth below surface (l/h) ratio. The general relationships indicated are fairly consistent with subsidence observations in European coal mining countries. Hiramatsu *et al* summarise their subsidence data in relation to goaf treatment and the main findings are given in Table 11; the anticipated effects of the form of goaf treatment are similar to European observations.

Surface subsidence observations in Indian coalfields

Subsidence observations have been reported above underground mining operations in Indian coalfields by Kumar and Singh (1967) and Kumar *et al* (1973) and several field investigations have been subsequently performed, Singh (1983). The general subsidence behaviour as observed at Indian coal mines indicates the following by comparison to European experiences:

1. smaller magnitudes of maximum subsidence,

2. areas of substantially less influence due to subsidence.

---

*UCS — Unconfined (or Uniaxial) Compressive Strength

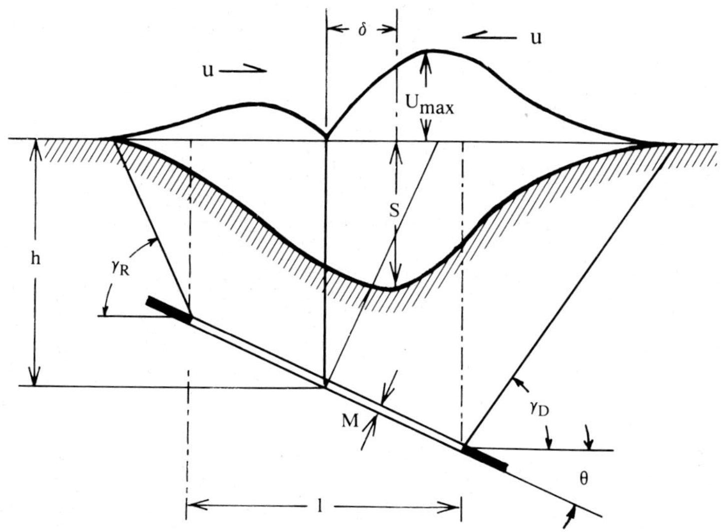

Figure 58 Illustrating symbols used to describe subsidence data presented in Table 10. After Hiramatsu *et al* (1979).

Table 10 Summary of the measurements of surface subsidence at Japanese Coal Mines. After Hiramatsu *et al* (1979)

Coal mines (pits)	Geological conditions				Working conditions				Subsidence data				
	θ (deg)	h (m)	d (m)	presence of faults	M (m)	∫ (m)	treatment of goaves	presence of old workings	γ_R (deg)	γ_S (deg)	γ_D (deg)	S (m)	δ (m)
Kaho (1)	4	71	4	no	1·30	258	50% packing	no		44		0·70	
Yoshinoura	4	95	<1	no	0·63	290	50% packing	no		45		0·38	
Hino	3	255	<1	no	1·20	192	caving	no		60		0·73	
Kogayama	5	392	7	yes	1·70	570	caving	no		65		1·52	
Hojo (1)	3	485	100	no	1·55	424	caving	yes		60		0·78	
Sagara	12	180	<1	no	1·80	60	caving	yes	45	43	40	0·52	20
Tokiwa	12	730	15	no	2·25	530	caving	no	65			0·90	97
Hojo (II)	14	160	<1	no	(a)2·10 (c)1·30	260	caving	yes	65	60	57	2·30	27
Yoshikuma	14	615	7	yes	(a)0·92 (b)1·80	240	caving	no	70		60	0·58	77
Shinnyu	15	660	15	no	(a)1·25 (c)1·70	590	caving	no	65	60	55	1·42	90
Futase (I)	15	135	10	no	2·10	105	caving	no	65		55	0·93	30
Futase (II)	16	350	10	no	2·15	135	caving	no	65			0·72	72
Kaho (II)	19	80	4	no	1·65	80	100% packing	yes	50	45	40	0·32	18
Lizuka	20	225	10	no	(a)1·30 (c)2·00	175	caving	no	62		55	1·78	50
Futase (III)	20	190	10	no	1·80	160	caving	no	63		54	0·91	58
Futase (IV)	23	210	10	no	1·70	110	caving	no	63		55	0·56	75

θ: Dip angle of coal seams, h: mean depth of workings, d: thickness of Quaternary strata, M: thickness of extraction, ∫: width of workings measured horizontally, S: maximum subsidence, γ_R, γ_S and γ_D: limit angles of subsidence in the directions of rise, strike and dip respectively, δ: deflection of the position of the maximum subsidence, that is the horizontal distance from the centre of mined area to the point of the maximum subsidence, (a), (b) and (c): upper, middle and lower seams respectively.

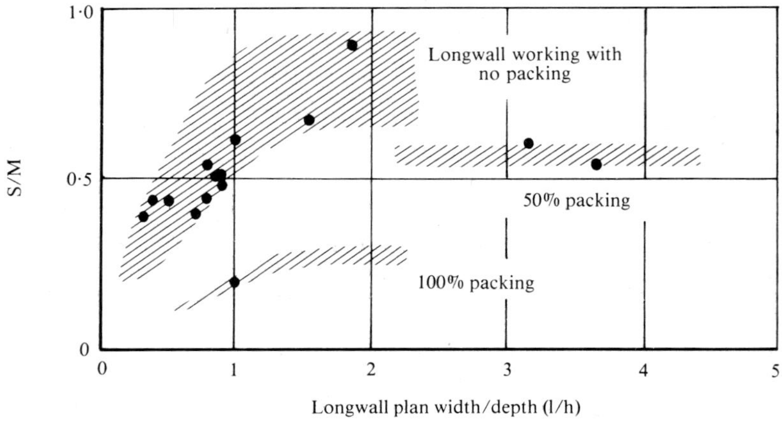

Figure 59 Relationship between maximum subsidence (S/M) and plan width/depth (l/h) ratio. After Hiramatsu *et al* (1979).

Form of goaf treatment	Maximum subsidence (S/M)
Total caving	0·7 - 0·9
50% partial packing (stowing)	0·55
100% packing (stowing)	0·3

Table 11 Influence of form of goaf treatment on the development of subsidence for longwall workings of supercritical dimensions at Japanese coal mines, from data given by Hiramatsu *et al* (1979).

Kumar *et al* (1973) refer to subsidence observations in the Jharia, Raniganj, Karanpura and Korba coalfields where results have been obtained at mines in the depth range of 68 to 165 m. These authors express the view that the strong nature of the overburden and particularly the frequent occurrence of large proportions of sandstones (43 to 88%) within it contribute significantly to the relatively small amounts of maximum subsidence and its rather limited lateral extent at the surface. The present authors having observed subsidence in India in relation to the type of geology encountered in the overburden agree with this viewpoint of Kumar and his colleagues.

The subsidence results of Kumar *et al* were obtained at eleven mines and they report that no subsidence was observed in two cases where room and pillar working was practised. Where longwall mining with hydraulic sand stowing was practised the maximum subsidence was 2 to 5·6% (S/M); longwall working with caving gave maximum observed subsidence values of 3·2 to 107% (S/M) with other individual values of 7%, 32% and 61% being quoted. The 107%

appears to be due to unusual conditions as the authors refer to a false roofing effect and is probably a non-representative value.

The travelling angle of draw was found to be 4 to 21° in representative cases for the strike side, whilst for the static situation the angle was 13 to 28° again on the strike side all being relative to the vertical. Although subsidence over the edge of the extraction was observed to be less than 50% (S/M), the subsidence troughs were still strongly indicative of an elastic beam behaviour as these authors went on to discuss.

There is a tradition to use stowing of the goaf areas in some mining situations, especially in very thick seams where mining takes place in several lifts, in order to promote greater safety against the possible effects of heavy caving of the strong sandstone overburden. The application of robust and heavy duty hydraulic roof supports in Indian longwall workings has in a number of cases eliminated the need for stowing where the immediate concern was that of controlling movement and caving of the roof strata. However, the likelihood of surface subsidence will undoubtedly increase where stowing operations are discarded in such circumstances.

Observed behaviour of mining subsidence in Australia

Kapp (1973a, 1973b) has reviewed aspects of subsidence behaviour in Australian mines, and particularly those of coal workings, and reported that subsidence movements can be predicted. He comments that methods developed outside of Australia will give reliable guidance to subsidence movements in the Southern Coalfield. Kapp (1973b) refers to subsidence observations above longwall extractions at Kemira Colliery, New South Wales, where extracted seam heights of 1·65 to 1·9 m and face widths of 160 to 189 m have been taken in a depth range of 146 to 292 m; the overburden consisted predominantly of interbedded sandstones with some shales. Rib pillars were used between some of the workings and their widths were in the range of 12 to 45 m. The following general comments are made by Kapp on these subsidence observations and their interpretations:

1. The maximum observed subsidence was less than that predicted by the UK SEH method; the Kemira results varied from 60 to 80% of SEH. This is most probably due to the high proportion of sandstone within the overburden at Kemira.

2. The barrier (rib) pillars influenced the maximum subsidence over each extracted panel. It would appear that the barrier pillars created a greater spanning effect of the overburden than would be expected by the SEH prediction model and this is indicative of the influence of high proportions of sandstone being present in the overburden.

3. The topography influences the nature of the resulting lateral displacements and strain profile.

4. Considering a levelling accuracy of 6 mm the limit angle of subsidence at the surface was indicated to be in the range of 9 to 35°.

Considering Kapp's results for observed slope and strain, there appears to be a reasonable degree of consistency with SEH predictions. It appears that rib pillars play a greater role in controlling surface subsidence at Kemira than would be predicted by SEH, and this phenomenon indicates that substantially less flexing of the roof beds is taking place on each side of the pillar.

Kapp (1977) has investigated subsidence behaviour in the Northern Coalfield, New South Wales, where he reported upon the effects of shortwall mining at Lambton and Burwood collieries around the Newcastle area. The mining depth range was 102 to 127 m, and extraction widths were in the range 49 to 71 m giving a w/h ratio range of 0·43 to 0·66; the extracted seam height range was 2·04 to 2·4 m. The maximum observed subsidence (S) ranged from 52 to 168 mm giving a range of S/M of 2·2 to 8·0%. The overburden was mainly composed of sandstone beds. Kapp reports that in such residential areas where the w/h ratio is around 0·6, the support left between successive panels should be equivalent to the combined effect of two rows of pillars 11 by 13 m dimensions (assuming a pillar height of 2 to 2·4 m). It was concluded from this work that this form of mining layout (panel and pillar system) reduced subsidence and associated effects in the residential areas around Newcastle and that the geological conditions of the overburden allowed fairly high recovery of coal consistent with low orders of resulting subsidence.

Kapp (1982 and 1985) has strengthened the conclusions he has drawn in earlier investigations by further work on subsidence behaviour especially in relation to the significant role played by strong and massive sandstones which are frequently encountered in the overburden above coal mines in Australia.

Frankham and Mould (1980) and Holla (1985a, 1985b and 1986) have also reported observations of coal mining subsidence in New South Wales. An important feature of these observations has been that full subsidence (S/M) develops at a width/depth (w/h) ratio of around 1·4 which is consistent with the UK SEH observed relationship. However, whilst the SEH curve levels off at around 0·9 (S/M), these authors report that for New South Wales coalfields the corresponding value is about 0·6 (S/M) with data spread between 0·5 and 0·7 (S/M). At a w/h ratio of 0·5, the Northern Coalfield subsidence observations indicate <0·05 (S/M) whereas the Southern Coalfield observations indicate around 0·3 (S/M). Conversely at a w/h ratio of 1·0 the subsidence observations indicate similar values of around 0·5 (S/M). This would suggest that the different overburden conditions are probably playing a role at w/h ratios of 0·5 and less.

Holla (1986) has made the following specific recommendations following his analysis of subsidence data from the Newcastle Coalfield:

1. The maximum value of subsidence is of the order 0·45 - 0·6 (S/M).

2. The maximum value of tensile strain (+E) is 0·4 (S/h).

3. The maximum value of compressive strain (−E) is 0·6 (S/h).

4. The maximum value of tilt (G) is 1·8 (S/h).

5. The angle of draw varies widely in the Coalfield and in some cases corresponded to zero. The 35° limit angle is regarded as very conservative for determining pillar sizes for subsidence protection. Where possible, the angle of draw should be established from local observations, but in their absence it appears that 26·5° is more realistic than 35° in respect of this Coalfield.

Mining subsidence observations in US coalfields

Dunrud (1976), Dunrud and Osterwald (1978a, 1978b) have reported on observations of coal mining subsidence in Utah, Colorado and Wyoming. Attention is drawn to depressions, cracks, pits and troughs overlying underground workings affecting the use of the surface for development. Reference is made to fires starting by spontaneous combustion in those abandoned coal mines where air can gain access to seams prone to such a natural hazard. This can induce further surface subsidence in time.

Gentry and Abel (1978a and 1978b) have described mining subsidence characteristics at York Canyon Mine, Raton, New Mexico where interest was focussed on the response of the overburden to undermining by a totally caved longwall panel. The extraction thickness was 3 m, the depth below surface ranged from 73 to 134 m whilst the face width was 107 m and it travelled 762 m. The overburden consisted mainly of beds of strong sandstone, siltstone and shale; the overburden comprised of about 30% sandstone. These authors reported the following general findings from the subsidence investigation:

1. The topography appreciably influenced both the magnitude and distribution of the resulting surface movements caused by the longwall mining operation.

2. Topographic lows resulted in a decrease in the angle of draw; the observed average angle of draw was 15·5° (to vertical).

3. The onset of surface subsidence occurred when the longwall was around −0·2 h from P and essentially complete at +0·84 h and observed to be complete at around +1·0 h beyond P.

4. Operation of the longwall resulted in almost instantaneous surface movement; conversely appreciable surface subsidence ceased virtually on cessation of longwall mining operations.

5. The predicted subsidence values for the ridge tops and topographic lows were 15 -21% and 55% greater respectively than the measured values.

6. The maximum subsidence observed over this single longwall was 2·16 m (or 0·7 S/M).

These authors extended their observations and findings to the establishment of a model aimed at general prediction of subsidence in York Canyon Mine conditions, Abel and Gentry (1978). Their preliminary subsidence model exhibits similarities in shape to that predicted by the UK SEH model, although with differences in detail.

Wade and Conroy (1980) have presented the results of a rock mechanics study of a longwall panel of Old Ben Mine No 24 located at Benton, Illinois, where the face width was 140 m and 529 m long, whilst the extracted seam height was 2·1 m. The depth below surface was 190 m. The maximum observed surface subsidence over Panel 1 was 1·34 m which corresponded to 0·63 (S/M). The angle of draw observed on three sides of the longwall gave values of 14°, 30° and 23° respectively.

Subsidence observations at Old Ben Mine No 24 in Illinois have also been discussed by Hood *et al* (1983) who report angles of draw as high as 43° for transverse profiles and 17·5° in the longitudinal direction. These authors report that the measured values of subsidence at Old Ben Mine No 24 compare favourably with the UK SEH predicted values for the mining dimensions in this particular situation. The results presented by the authors show that the UK SEH method predicted the order of maximum subsidence fairly accurately for panel nos 1 and 2 and gave a fairly good prediction for the trough shapes. However, the observed subsidence showed a steeper gradient over the rib edges than that predicted by SEH, and these authors report that the peak strains were observed to be around four fold higher than predicted. Khair *et al* (1987) have reported similar subsidence characteristics for transverse profiles observed in the Northern Appalachian region in West Virginia, and concluded that the caving behaviour of the overburden is a major factor to subsidence development in relatively shallow (less than 100 m) depth conditions as exist in this mountainous region above the Redstone Coalbed.

A number of investigators have directed their attention towards evaluating and revising subsidence prediction methods for US coalfields and in particular the work of Adamek and Jeran (1981) and O'Rourke and Turner (1981) deserve special mention. It was recognised, however, that more field observations were required in order to establish an adequate basis from which prediction models with increased scope could emerge. Karmis *et al* (1981) examined the potential of the zone area method for subsidence prediction in the Appalachian Coalfield and presented observed data on surface behaviour in relation to subsidence. Their observations above longwall workings in the Appalachian Coalfield are summarised as follows:

(a) The observed angle of draw ranged from 12 to 34°.

(b) Where w/h exceeded 1·2 the angle of draw tended to maintain a consistent value of around 28°.

(c) The maximum subsidence appears to be attained when w/h reaches around 1·2.

(d) The point of inflexion occurs at about 0·2 h inside the ribside (similar situations in the UK would correspond to 0·14 h according to SEH).

(e) With regard to maximum subsidence (S) the upper limit and mean value for the Appalachian Coalfield are indicated to be 0·6 to 0·65 (S/M) and 0·45 to 0·5 (S/M) respectively.

Karmis and co-workers (1983, 1984, 1987) have discussed further aspects of observed subsidence behaviour due to longwall mining in the US Eastern coalfields. Earlier views expressed by Karmis *et al* (1981) have been re-confirmed in his later works. He draws attention to higher values of horizontal strain being indicated in relation to curvature for Appalachian coalfield subsidence observations when compared to UK SEH (1975) predictions. The results of Peng and Geng (1982), Geng and Peng (1983), Gentry *et al* (1981) and Bauer and Hunt (1981) are used by Karmis to highlight this feature.

Subsidence observations in relation to a longwall panel in the Pittsburgh Coalbed (seam) operating at a depth below surface range of 230 - 305 m in Western Pennsylvania, have been discussed by Moebs and Barton (1985) and Trevits and King (1986) and report two particular categories of subsidence limit angle. They show an angle of 15° (to vertical) as the limit of major surface deformation, and an angle of 25° (to vertical) as the limit of detectable surface deformation.

Subsidence observations in Czechoslovakia

Zilavy (1968) has discussed the general subsidence characteristics of the surface overlying longwall operations in the Ostrava-Karvina Coalfield of Czechoslovakia. He describes a number of case histories where the angle of draw ranges generally between 25° and 39° (to vertical) but most values appear to be around 25-30°. The maximum subsidence is reported to be 0·9 (S/M); the subsidence profiles presented are similar to those observed in Western European coalfields.

Conclusion

Subsidence of the surface above longwall coal mining operations has been observed in many countries and observed data are regarded as the most reliable means for establishing a basis from which to predict associated movements. Observations in different countries indicate that the same principal parameters control the resulting subsidence, although dissimilar geological settings produce detailed differences in the magnitude of subsidence and extent of area of influence. Observed subsidence data in the UK are sufficiently consistent as to provide the foundation for reliable and accurate empirical prediction methods.

CHAPTER 6

SUBSIDENCE PREDICTION BY THE UK EMPIRICAL MODEL

The UK SEH (Subsidence Engineers' Handbook) empirical model is based on observations at around 200 sites in several coalfields. The subsidence data were mainly collated during the 1950-65 period and were rationalised by National Coal Board specialist staff and subsequently published, NCB (1965, 1975). These two publications have been employed extensively throughout UK coalfields and have served as excellent means of assessing the anticipated response of the surface to underground coal mining operations. Additionally, these publications have been employed widely in other countries and are frequently used as a standard basis on which to compare local subsidence observations.

The SEH empirical model consists essentially of a series of tables of data, charts and graphs from which subsidence data can be derived and which relate to specific mining situations. The general geological conditions are not considered as an individual parameter, since most of the observations relate to overburden comprised of either entirely Carboniferous Coal Measures, or in combination with younger rocks, namely Permian limestones or Triassic sandstones. A detailed discussion on the influence of such rocks being present in the overburden is given in a later chapter which deals specifically with geological factors.

Scope of the SEH empirical model

The subsidence data refer to longwall mining operations of fairly consistent extraction dimensions. It is generally assumed that the width (w) of longwall extraction is fairly constant throughout the area of influence, and generally the same applies to extraction height (M). Although the subsidence charts and nomograms are directed at caving operations, other forms of goaf treatment are taken into account by simply applying a coefficient to the maximum subsidence (S).

The approach to subsidence prediction employed by the SEH method is to produce assessments of the maximum subsidence and associated maximum values of strain, tilt and displacement. Having established the maximum values this then allows longitudinal (advancing or dynamic) and transverse profiles of the subsidence, strain, tilt and displacement to be drawn using data from specially prepared charts. Transverse and longitudinal profiles of subsidence and associated parameters meet many general situations, and these profiles can be combined by appropriate interpolation procedures to allow a plan of subsidence contours to be presented in relation to the longwall workings.

Irregular (in plan) longwall extractions are not considered by the SEH prediction charts, nor are different thicknesses of extraction within the same longwall panel. However, the SEH approach can be adapted to cater for such situations, as is discussed in the next chapter.

Subsidence prediction terminology

The terminology presented in Table 12 is based on that given by SEH since it has been largely adopted internationally. One major change has been introduced by the symbol representing

Table 12 Subsidence prediction terminology based on SEH (1975)

Symbol	Term	Explanatory Comments
γ	Angle of draw	Also referred to as limit angle. The angle subtended between the horizontal and a line joining the extraction edge with the limit of subsidence.
ζ	Angle of draw	Also referred to as limit angle. Refers to the angle subtended between the vertical and line joining the extraction edge and limit of subsidence.
R	Critical radius	$R = h \tan\zeta$. Radius of critical area.
ϱ	Curvature	Radius of curvature of ground induced by subsidence.
h	Depth	Depth below surface of workings.
w	Panel width	Refers also to face length being the width of extraction of the longwall.
M	Extraction height	Thickness of extraction taken on longwall.
p	Pillar width	Refers to the shortest dimension of a rectangular pillar; width of a rib pillar.
L	Panel length	Panel dimension along line of advance.
d	Distance	Any horizontal distance.
a	Subsidence factor	Ratio of complete subsidence to extracted seam height, S/M.
s	Subsidence	Vertical movement anywhere within a subsidence trough.
S	Maximum subsidence	Maximum subsidence occurring within a subsidence trough.
e	Strain	Change in length relative to an original dimension; can be shortening (−) or lengthening (+).
E	Maximum strain	Maximum strain encountered within the subsidence trough.
g	Slope	Slope of any part of a subsidence trough.
G	Maximum slope	The maximum slope encountered within a subsidence trough.

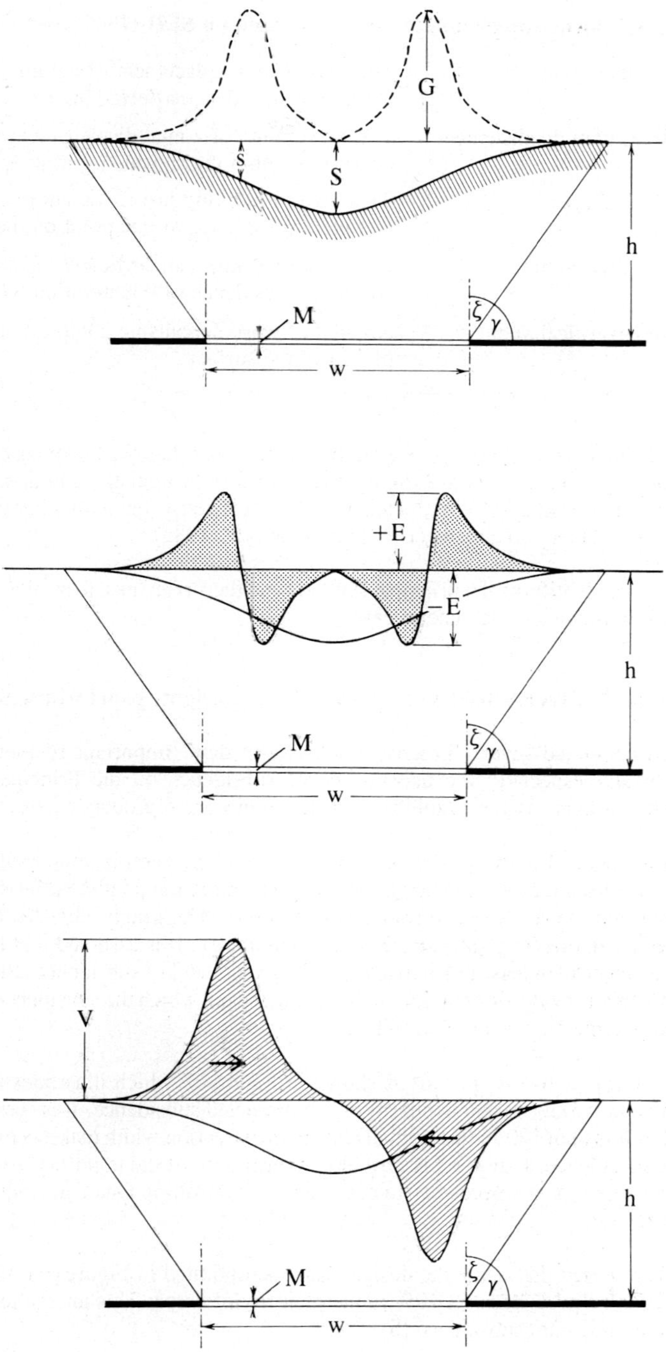

Figure 60 Illustrating the principal symbols in subsidence prediction terminology, SEH (1975)

Table 12 Subsidence prediction terminology based on SEH (1975) (continued)

v	Displacement	The horizontal displacement of a point occurring within the subsidence affected area.
V	Maximum displacement	The maximum horizontal displacement of a point occurring within the subsidence affected area.
	Critical area	That area of working just sufficient to cause complete subsidence of one point on the surface.
	Sub-critical area	An area of working not sufficient to cause complete subsidence of one point on the surface.
	Super-critical area	An area of working causing complete subsidence of part of the surface.

extracted seam height (M); the upper case is adopted here rather than the lower case as the latter has in the authors' opinion led to confusion in publications and calculations in distinguishing it from the metre symbol. Using M for extracted seam height is consistent with other publications, see for example Whittaker and Breeds (1977) and Kratzsch (1983).

In order to assist identifying the principal symbols and their relevance to mining subsidence, Figure 60 has been drawn for the general case.

Basic relationships between subsidence, extracted seam height, panel width and depth

SEH (1965) contained several graphs of subsidence data. Important relationships were presented graphically especially the dependence of subsidence on the principal extraction dimensions. Some of these earlier graphical representations are reproduced here.

Maximum subsidence. Figure 61 shows the relationship between maximum subsidence (S) expressed in terms of seam extraction height (M) and the extraction width/depth (w/h) ratio for the solid stowing and caving (or strip packing) conditions. The graph indicates that the w/h ratio can be used as an effective subsidence control parameter. It is assumed that the longwall has advanced a distance of at least 1·4 h to allow full development of subsidence. Figure 61 was generally used to obtain a value for maximum subsidence, from which the subsidence profile and other associated parameters could be calculated.

A further development of Figure 61 is shown in Figure 62 which illustrates the effect of increasing extraction width (w) and depth (h) on the predicted subsidence. It is seen in the case where the depth is less than 100 m that a small change in extraction width makes a marked effect on the resulting subsidence. Figure 62 is further demonstration of the important role played by depth of working although it is necessary to consider this controlling factor in conjunction with the width of extraction.

SEH (1965) reported that using the design basis incorporated in Figures 61 and 62 would allow subsidence to be predicted to ±10% in the majority of cases. This undoubtedly referred specifically to the maximum subsidence (S).

Limited advance. It was recognised by SEH (1965) that in those situations where a limited amount of face advance occurred namely less than 1·4 h, the maximum subsidence predicted by Figures 61 and 62 would need reducing in relation to the limited advance. Figure 63 shows a

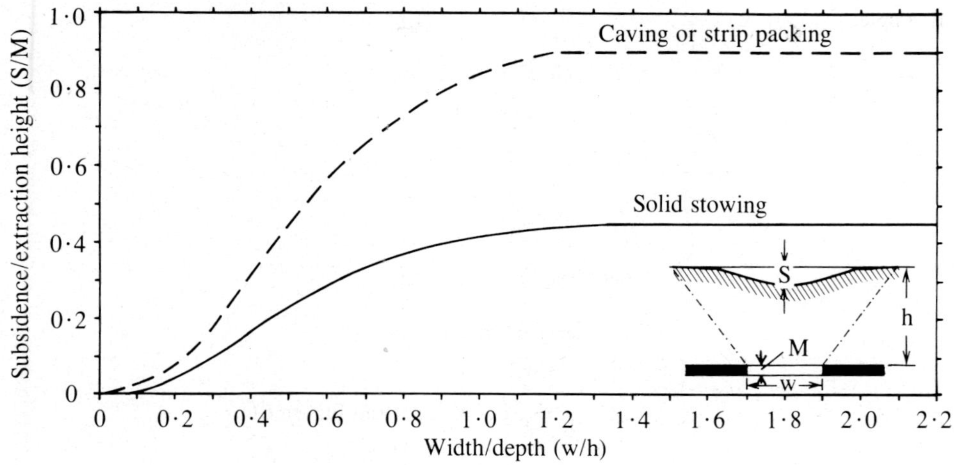

Figure 61 Relationship between S/M and w/h, after SEH (1965)

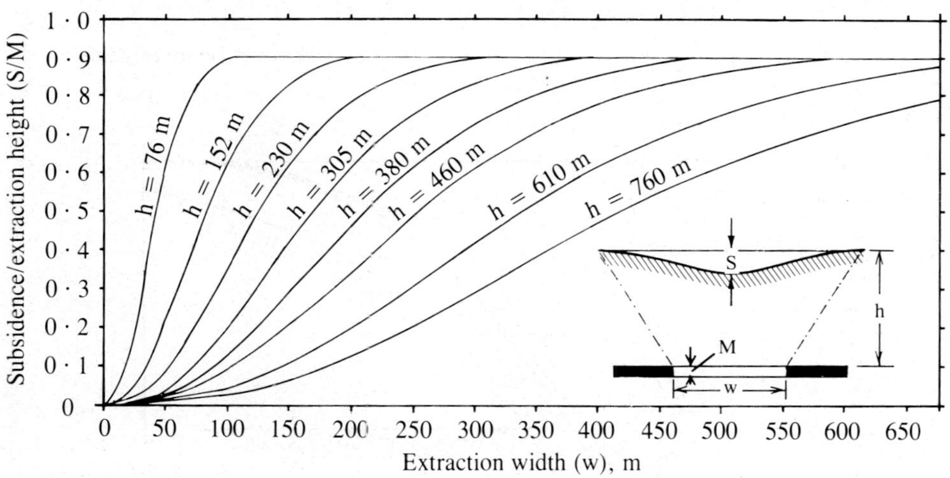

Figure 62 Illustrating the effect of increasing depth and extraction width on the resulting subsidence, after SEH (1965)

correction graph which was produced for such situations and the following example demonstrates its application.

Example:

Given parameters:

$M = 1{\cdot}5$ m

$h = 300$ m

$w = 200$ m

$L = 150$ m

Goaf treatment = caving

Derived parameters:

$w/h = 0{\cdot}67$

$S/M = 0{\cdot}64$ (from Figure 61)

$S = 0{\cdot}64\,(1{\cdot}5)$ m

$= 0{\cdot}96$ m for $L \geqslant 1{\cdot}4\,h$

$L/h = 0{\cdot}5$

$s/S = 0{\cdot}5$ (from Figure 63)

Partial subsidence (s) = $0{\cdot}5\,(0{\cdot}96)$ m

$= 0{\cdot}48$ m (occurring over centre of goaf)

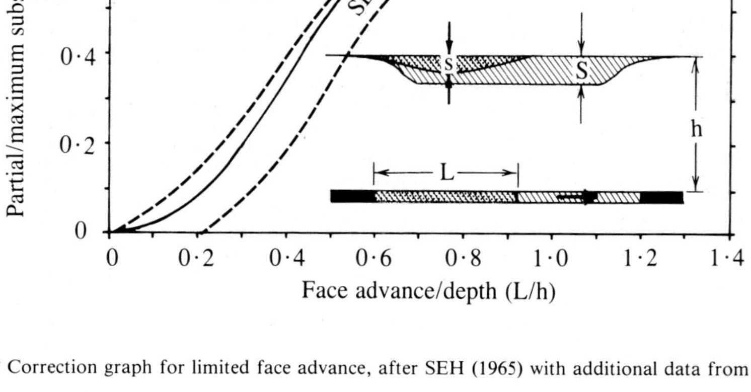

Figure 63 Correction graph for limited face advance, after SEH (1965) with additional data from SEH (1975)

Included in Figure 63 are revised graphical relationships presented in SEH (1975); depths of 900 m and 100 m are considered. These results indicate depth to be a factor in determining the response of the overburden to the subsidence phenomenon. It would appear that the SEH (1965) curve gives a fairly representative average result for the 200−900 m depth range; the 100 m depth curve suggests that the lower overburden pressure is allowing the roof beds to hang behind the extraction for greater distances before subsiding.

The first edition of SEH (1965) drew attention to the importance of the subsidence development curve in relation to face position and this is shown in Figure 64. The relationship shown in Figure 64 corresponds to the vertical movement of the surface point P as it is progressively affected by subsidence from onset to cessation of movement.

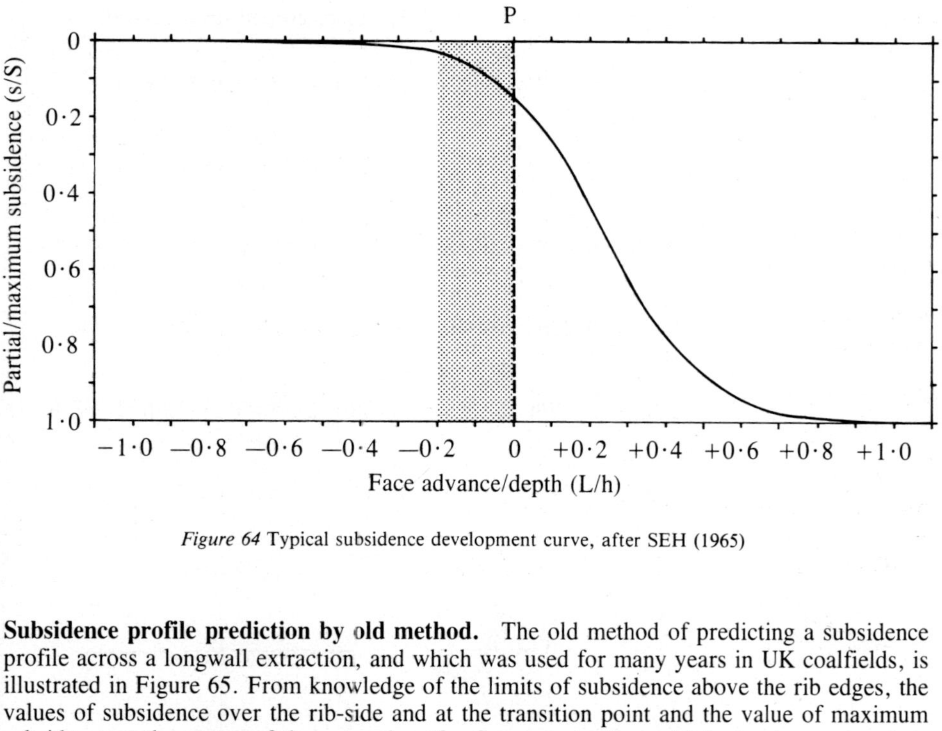

Figure 64 Typical subsidence development curve, after SEH (1965)

Subsidence profile prediction by old method. The old method of predicting a subsidence profile across a longwall extraction, and which was used for many years in UK coalfields, is illustrated in Figure 65. From knowledge of the limits of subsidence above the rib edges, the values of subsidence over the rib-side and at the transition point and the value of maximum subsidence at the centre of the extraction (for flat seams), then sufficient points were given through which the subsidence profile could be drawn. A general knowledge of profile shape was expected in producing complete subsidence profiles by this method. It was convenient in providing a fairly quick and reasonably accurate assessment of the transverse subsidence curve, although it did depend to some extent on the skill and knowledge of the user.

Subsidence and strain profile prediction by SEH (1965). The approach adopted earlier to subsidence and strain profile prediction in SEH (1965) was to use dimensionless charts from which subsidence and strain data were simply read off. The w/h ratio, and the values of S, +E and −E were required as a pre-requisite to the data preparation operation (see Table 13). This was followed by reading off the distances, in terms of depth and relative to the transition point, of particular pre-determined values of subsidence and strain. It was of course necessary to have pre-knowledge of the position of the transition point in order to fix the positioning of the subsidence and strain profiles relative to the longwall extraction.

Table 13 Values of maximum subsidence, surface ground strains and tilt due to longwall mining. Compiled from data given in SEH (1965, 1975) and modified after Whittaker and Breeds (1977)

Longwall width/depth (w/h)	0·2	0·25	0·33	0·5	0·75	1·0	1·4
Max. subsidence/extracted seam height S/M; caved wastes	8%	12%	22%	45%	70%	84%	90% (max.)
Coefficients for deducing magnitude and position of maximum ground strains and tilt							
Max. strain due to compression (−E)	2·2S/h	2·15S/h	1·9S/h	1·35S/h	0·75S/h	0·55S/h	0·5S/h
Position of −E from centre line (−Ex)	0	0	0	0·02h	0·10h	0·20h	0·39h
Max. strain due to extension (+E)	0·5S/h	0·65S/h	0·75S/h	0·8S/h	0·65S/h	0·65S/h	0·65S/h
Position of +E from centre line (+Ex)	0·49h	0·42h	0·34h	0·32h	0·40h	0·51h	0·70h
Max. ground tilt (G) (at transition pt.)	2·2S/h	2·6S/h	3·15S/h	3·35S/h	2·85S/h	2·75S/h	2·75S/h
Position of G from centre line (Gx)	0·32h	0·27h	0·22h	0·21h	0·26h	0·37h	0·56h

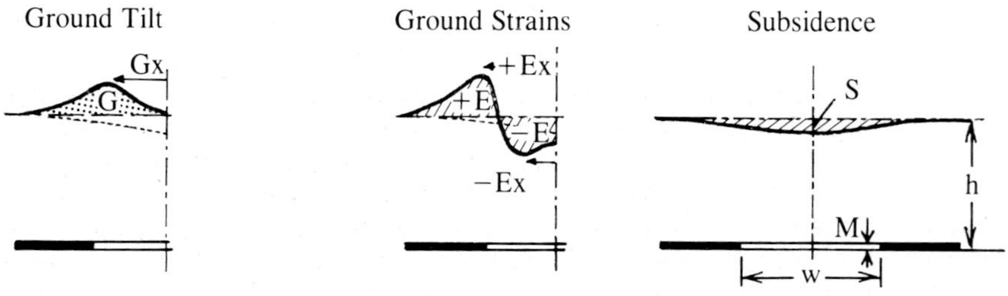

Example: Longwall width = 200 m, depth = 400 m, extracted seam height = 2 m

Since w/h = 0·5, max. subsidence (S) = 45% (2 m) = 0·90 m at the centre line

Max. strain due to compression (−E) $= 1{\cdot}35 \left(\frac{0{\cdot}9}{400}\right) =$ 0·0030 (or 3·0 mm/m) at 8 m from centre line.

Max. strain due to extension (+E) $= 0.8 \left(\frac{0{\cdot}9}{400}\right) =$ 0·0018 (or 1·8 mm/m) at 128 m from centre.

Max. ground tilt (G) $= 3.35 \left(\frac{0{\cdot}9}{400}\right) =$ 0·0075 (or 7·5 mm/m) at 84 m from centre line.

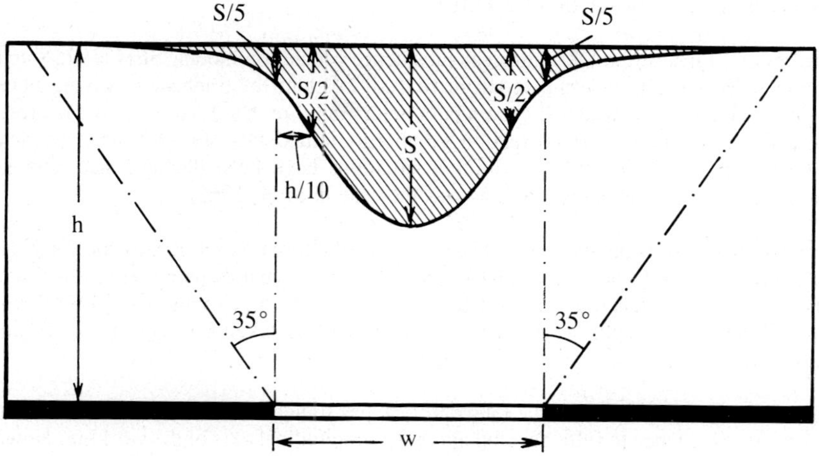

Figure 65 Old method of predicting a subsidence profile, after SEH (1965)

An example of the SEH method of determining the maximum values of subsidence, strain and tilt is included in Table 13 which shows a condensed compilation of data for a range of width/depth ratios. The maximum values of strain and tilt are referred to the centre line of the panel, however, which is consistent with SEH (1975).

The multiplication constant values, which are shown in Table 13 as K(S/h), have been taken from Figure 66. This illustrates its variations with the width/depth ratio.

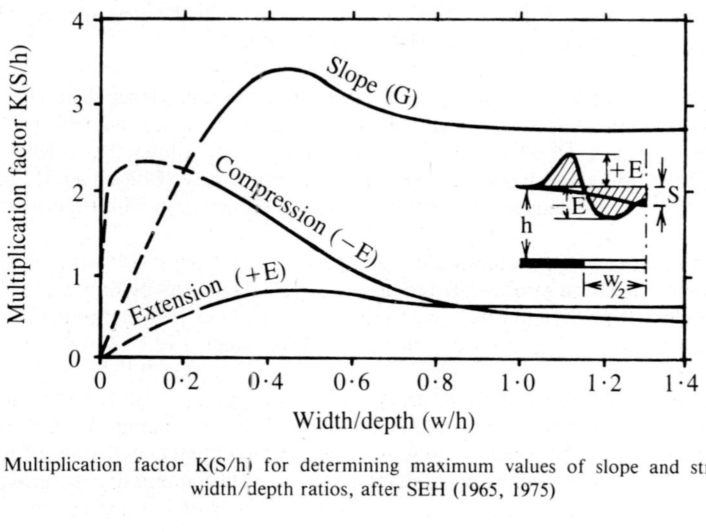

Figure 66 Multiplication factor K(S/h) for determining maximum values of slope and strain at various width/depth ratios, after SEH (1965, 1975)

Subsidence and strain prediction by SEH (1975)

The revised form of the UK empirical subsidence prediction model, SEH (1975) included a number of different features compared to the first edition, with the emphasis on giving increased refinement and scope to consider the main parameters in more detail. An example is presented in Figure 63 which shows the correction graph for limited face advance whereby shallow (h = 100 m) and deep (h = 900 m) mining situations have been distinguished between as opposed to just using an average curve as was the case in SEH (1965).

Prediction of maximum subsidence. The revised method for predicting maximum subsidence has been to consider longwall face width (w) and depth (h) as separate parameters and these have been used to plot subsidence in terms of S/M values as shown in Figure 67. This allows any particular influence of depth on resulting subsidence to be taken into account in respect of extraction width.

Figure 67 again assumes that the longwall face has travelled a distance of at least 1·4 h in order to allow subsidence to fully develop along the longitudinal axis of the working. However, where the longwall does not achieve 1·4 h of advance, a correction needs to be applied as discussed earlier in this chapter, when referring to Figure 63.

Consequently to summarise, Figure 67 is used to determine the maximum subsidence S; should the face advance be less than 1·4 h then the derived value for S will need correction using Figure 63.

The maximum subsidence possible is shown by Figure 67 to be 0·9 M which corresponds to the caved condition. The reduced magnitude of subsidence achieved by different forms of stowing of the goaf area has been discussed earlier; the maximum value of subsidence given by Figure 67, and possibly also after correction for limited advance, would need further adjustment to take any special goaf treatment into account if applicable.

Prediction of subsidence profile. After having deduced the maximum subsidence (S) as explained in the preceding section, the subsidence profile is drawn from data obtained from Figure 68. This presents a design graph of w/h against d/h which allows the position to be determined at which specified magnitudes of subsidence occur. Basically, a horizontal line is drawn corresponding to the w/h ratio and the interception points with the lines of equal subsidence are read off to give the distance from the centre of the panel.

Prediction of strain profile. The first step in determining the subsidence data with which to plot the ground strain profile, is to evaluate the maximum strain values, namely +E and −E. This is done using Figure 66 which for the particular w/h ratio allows the constant K to be established as a multiplication factor for +E and −E respectively. Consequently, numerical values for +E and −E are deduced which can be used in conjunction with Figure 69.

Referring to Figure 69, this shows a design graph of w/h against d/h with lines of equal horizontal strains. The strain profile data are obtained from Figure 69 by drawing a horizontal line corresponding to the w/h ratio value, and the interception points with the strain lines are used to read off the appropriate distances from the centre of the panel.

The strain profiles given by Figure 69 are of three basic types, as is shown in Figure 70. At small values of the w/h ratio eg w/h = 0·42, only one peak value of compressive strain occurs and this is in the centre of the panel. With w/h ratios of around unity, two areas of maximum compressive strain are evident, and as the w/h ratio increases the compressive strain at the panel centre progressively decreases until achieving zero at w/h ratios of 1·4 and greater.

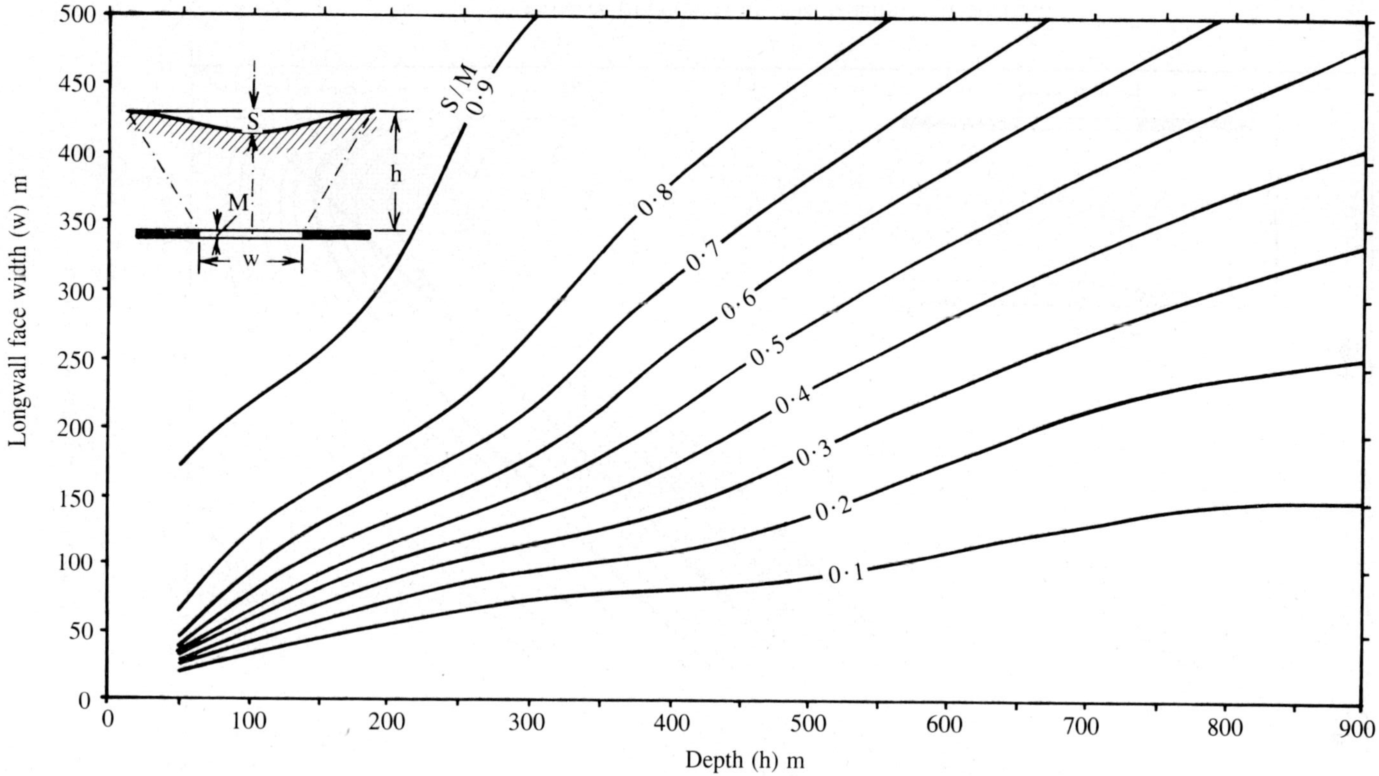

Figure 67 Relationship of subsidence to extraction width and depth, after SEH (1965).

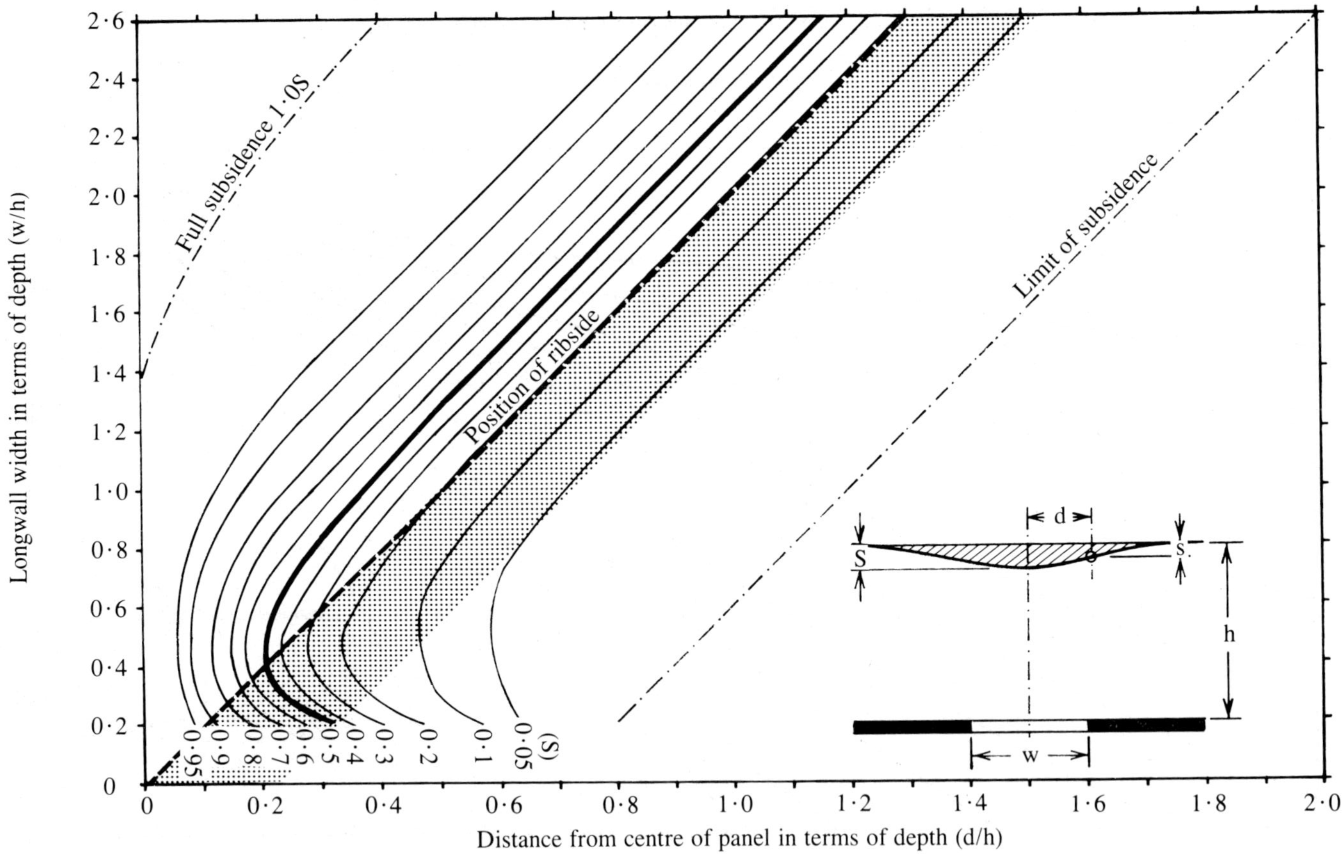

Figure 68 Design graph for prediction of subsidence profiles, after SEH (1975).

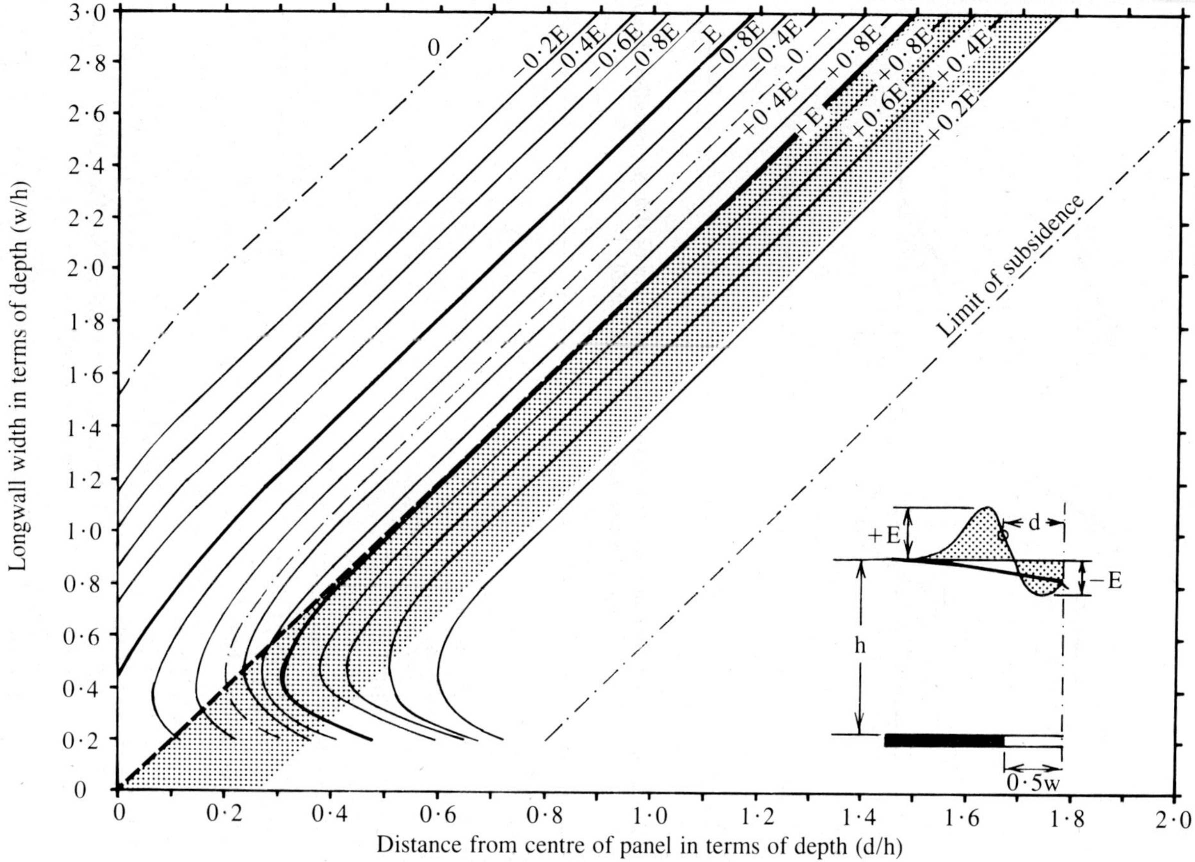

Figure 69 Design graph for prediction of strain profiles, after SEH (1965).

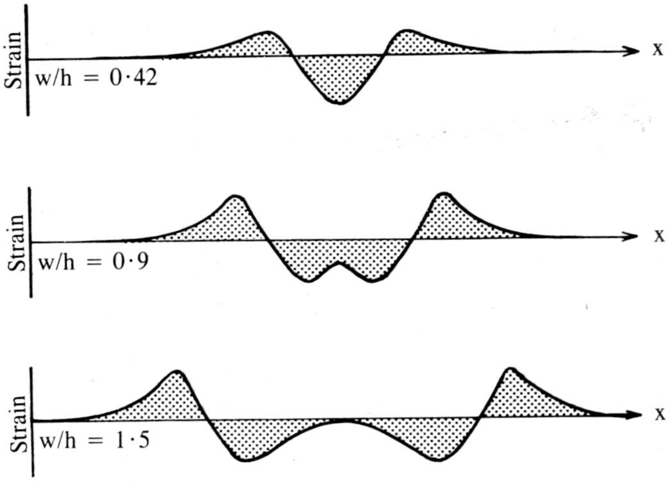

Figure 70 Illustrating the three principal types of strain profile, after SEH (1975)

Table 14 Example of data preparation for subsidence and strain profile presentation due to longwall extraction using SEH (1975)

Basic mining data:

M	=	1·5 m
h	=	400 m
w	=	200 m
L	=	1000 m
Overburden	:	Coal Measures
Goaf treatment	:	Caving

Derived subsidence data:

w/h	=	0·5	
S/M	=	0·47	From Figure 67
S	=	0·71 m	
L/h	=	2·5	No correction required, see Figure 63
+E	=	0·8 S/h	From Figure 66
	=	1·42 mm/m	
−E	=	1·4 S/h	From Figure 66
	=	2·49 mm/m	
G	=	3·3 S/h	From Figure 66
	=	5·86 mm/m	

Table 14 Example of data preparation for subsidence and strain profile presentation due to longwall extraction using SEH (1975). (continued)

Subsidence profile data: From Figure 68

Subsidence		Distance from panel centre	
In terms of (S)	(m)	(d/h)	(m)
0	0	0·95	380
0·05	0·04	0·59	236
0·1	0·07	0·47	188
0·2	0·14	0·34	136
0·3	0·21	0·28	112
0·4	0·28	0·24	96
0·5	0·36	0·21	84
0·6	0·43	0·17	68
0·7	0·50	0·15	60
0·8	0·57	0·12	48
0·9	0·64	0·08	32
0·95	0·68	0·06	24
1·0	0·71	0	0

Strain profile data: From Figure 69

Strain		Distance from panel centre	
In terms of (E)	(mm/m)	(d/h)	(m)
0	0	0·95	380
0·2 (+E)	+0·28	0·60	240
0·4 (+E)	+0·57	0·51	204
0·6 (+E)	+0·85	0·43	172
0·8 (+E)	+1·14	0·38	152
+E	+1·42	0·32	128
0·8 (+E)	+1·14	0·27	108
0·4 (+E)	+0·57	0·24	96
0	0	0·21	84
0·4 (−E)	−1·00	0·16	64
0·8 (−E)	−2·00	0·08	32
−E	−2·49	0·02	8
0·8 (−E)	−2·00	0	0
0·6 (−E)	−1·49	0	0
0·4 (−E)	−1·00	0	0
0·2 (−E)	−0·50	0	0

Example of determining subsidence and strain profiles

An example of subsidence and strain profile determination by SEH (1975) is given in Table 14. The first step is to derive the subsidence data from the basic mining data. Then the subsidence profile data are prepared using Figure 68 whilst the strain profile data are obtained from Figure 69. In the latter situation the corresponding values of +E and −E need to be deduced from Figure 66.

The profile data from Table 14 are shown in graphical form in Figure 71.

This method of producing subsidence and strain profiles is the most widely used world-wide, most probably because it is based on observed subsidence phenomena. The results plotted in Figure 71 refer to a transverse profile although longitudinal profiles can be produced by considering the advancing face line as a half profile. Due account should also be taken of Figure 64 which shows the typical subsidence development curve as indicating that 15·5% (S) occurs over the face line but the design graph (Figure 68) shows that slightly more subsidence is likely at this position. The reason for this is that the development curve is likely to experience slightly more subsidence over the rib edge due to residual subsidence effects.

Extraction of rib pillar between adjacent longwall extractions

The presence of a rib pillar between adjacent extractions produces a significant influence on the resulting surface subsidence. Where the rib pillar is 1·4 h width or more then there will be no interaction between the respective subsidence profiles providing the angle of draw does not exceed 35°. Rib pillar widths of progressively decreasing magnitude below 1·4 h result in increasing interaction of profiles with correspondingly greater subsidence over the pillar area. Profile interaction can be considered as a separate entity using standard subsidence computational procedures.

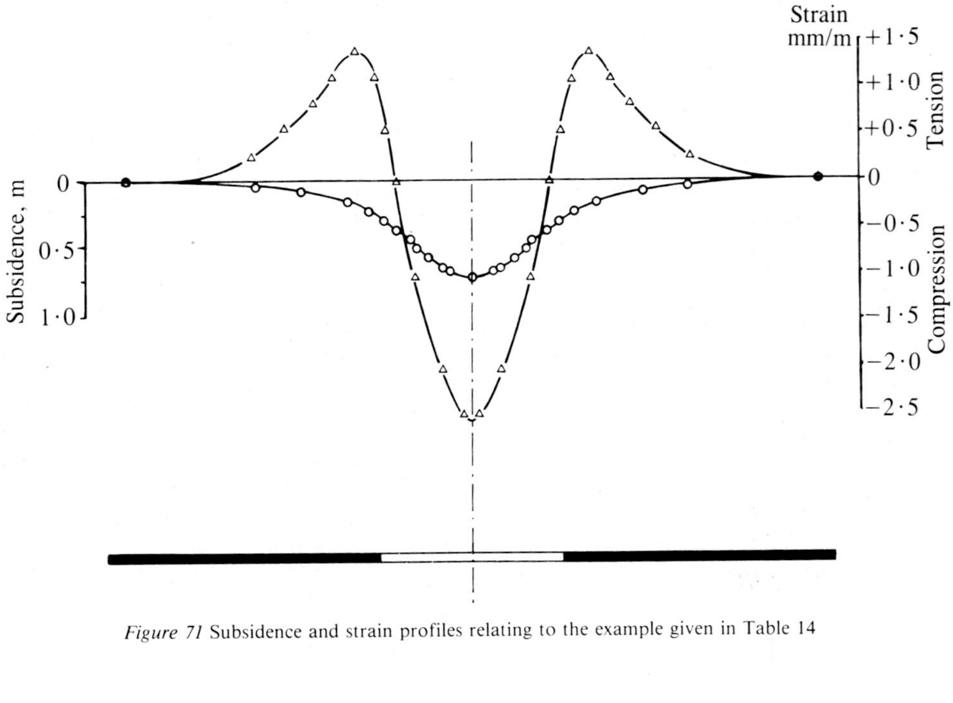

Figure 71 Subsidence and strain profiles relating to the example given in Table 14

Figure 72 shows a graphical relationship for the subsidence over the solid coal rib pillar. The rib pillar width is considered in terms of the depth, whilst the resulting subsidence over the centre of the pillar is based on that which would occur if the pillar were not present.

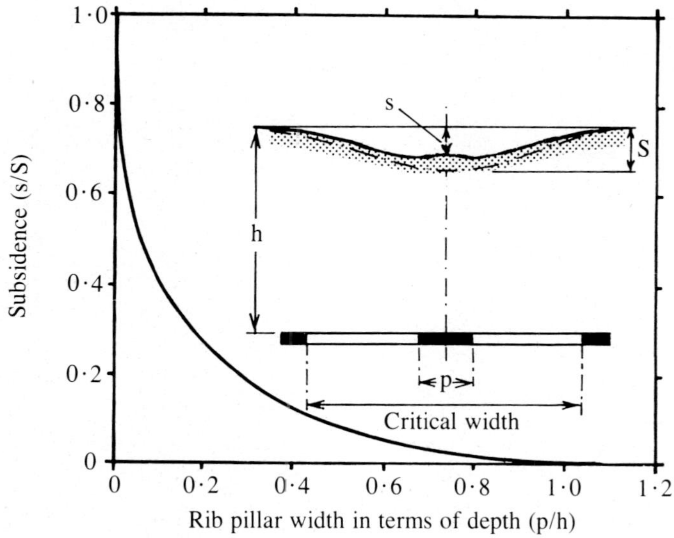

Figure 72 Subsidence over solid coal rib pillars, after SEH (1975)

Conclusion

The SEH empirical model has proved to be a reliable method of prediction owing to its approach being based on observed subsidence phenomena. The method has been thoroughly tested against many different longwall mining situations and is itself used as a means of testing the validity and degree of accuracy offered by other prediction methods and associated theories.

A striking advantage of the SEH empirical model is its convenience and adaptability for wide ranging applications in subsidence engineering. Some of these applications are referred to in subsequent chapters of this book, and a notable example is the ability of the SEH data to be applied in computerised form to more complex situations.

CHAPTER 7

COMPUTER BASED PREDICTION OF MINING SUBSIDENCE EMPLOYING EMPIRICAL DATA

The previous chapter discussed the UK SEH (Subsidence Engineers' Handbook) method of subsidence prediction. It demonstrated how subsidence and ground strain profiles can be produced for longwall mining extractions. The method essentially involves direct reading from graphs and design charts of previously established empirical data.

In many circumstances, however, the determination of just two subsidence profiles, namely in the transverse and longitudinal directions of longwall mining, is insufficient for complex extraction layouts particularly those involving several faces in one or more seams. Such profiles possess appreciable accuracy but difficulties can arise when extrapolating them to contoured plans of subsidence and strains. This is important in respect of the random occurrence of certain features such as pipelines crossing the path of a longwall working.

This chapter examines the application of a computer based prediction method which employs UK SEH empirical data for calculation of surface subsidence and strains due to longwall mining. It readily allows a more comprehensive assessment to be made of the subsidence and strains over the whole surface area of influence. Attention is focussed here on methods of employing SEH data to producing more accurate presentation of subsidence displacement and ground strain data by computerised methods.

Empirical computer model

The computer program developed at the Department of Mining Engineering, University of Nottingham for the calculation of subsidence displacements and surface ground strains arising from longwall mining extraction layouts has been described by Whittaker and Reddish (1984). The program essentially uses a series of graphs and tables of data from SEH (1975) to form the basis of the calculations for obtaining subsidence and horizontal strain profiles for the transverse and longitudinal directions of the longwall working. The input parameters of panel width (w), depth (h), distance of travel (L) and thickness of extraction (M) are interactively checked by the program to see if they are within the range of the program. These basic parameters are used to calculate the maximum subsidence and strains, from which the subsidence and strain profiles are subsequently determined. At this stage the profiles can be partially calibrated to local conditions by applying multipliers to the subsidence and strain values.

An important feature of the calculation sequence is the treatment of the longitudinal profiles. Calculation of the longitudinal profiles relates independently to each end of the longwall panel and requires further routines to produce two-ended profiles. In some cases, all that is required is to simply reverse the data and place it at the other end of the profile, thus leaving a gap in the centre. When the gap length is small, overlap needs to be eliminated from the profile data. This feature is illustrated in Figure 73 (a) and (b).

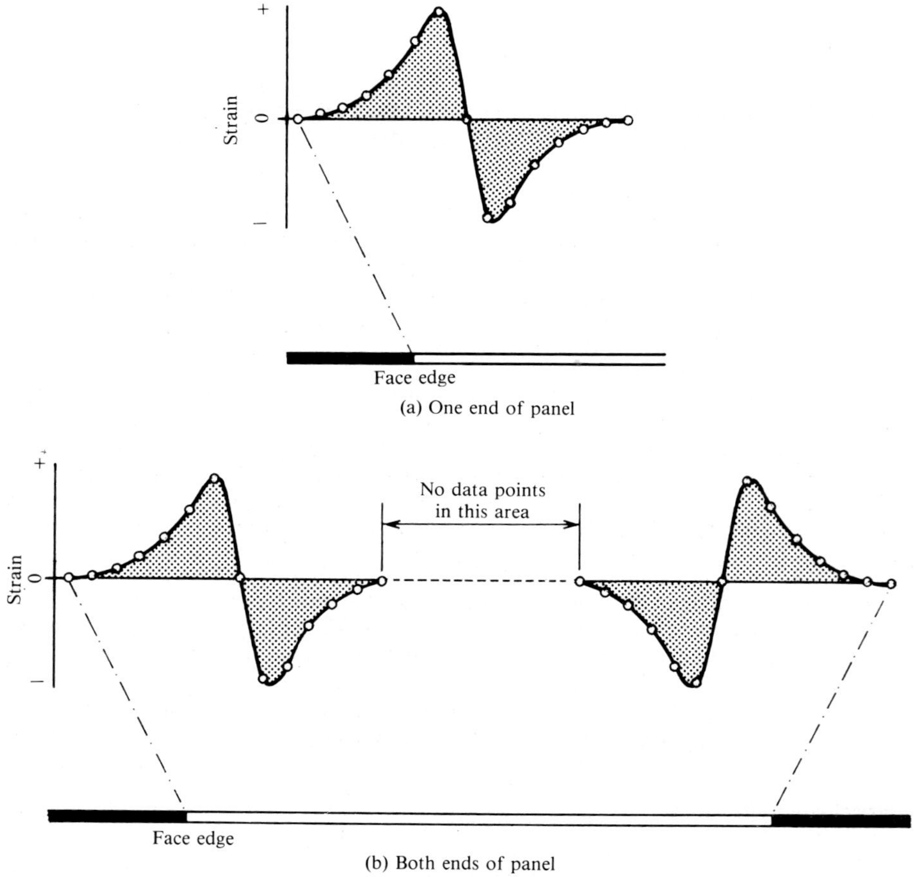

Figure 73 Longitudinal strain profile data distribution

Calculation of horizontal displacement profiles from strain

Up to this stage the program has calculated the usual profiles for the transverse and longitudinal directions as would be obtained by standard manual operation of the SEH method. However, the horizontal displacement profiles need to be calculated for further progress towards detailed treatment of surface ground strain patterns. In determining the ground strain pattern correctly, it is firstly necessary to convert the strain data into displacement data, since vector addition of strains is difficult, with a minimum of three strains being required. Conversely vector addition of displacement offers convenience and ease of performance, Whittaker, Reddish and Fitzpatrick (1985).

The displacement profiles are established by straightforward calculation on the strain profile data. The average strain in each bay is calculated firstly and then converted into the change in length for that bay by multiplication of the bay length. The displacement relative to one end is then simply equal to the sum of the displacements between that point and the end.

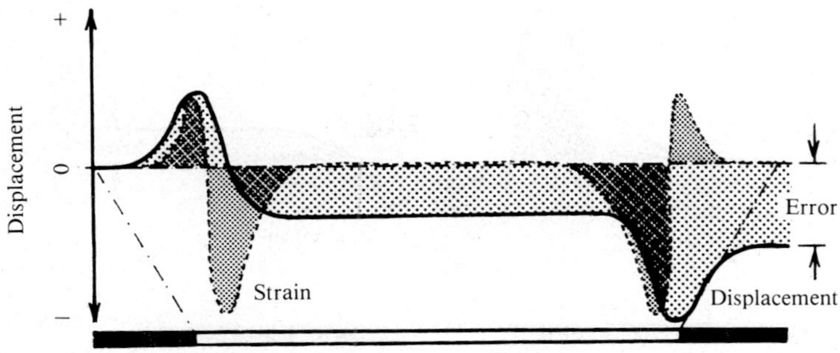

(a) Displacement derived from strain

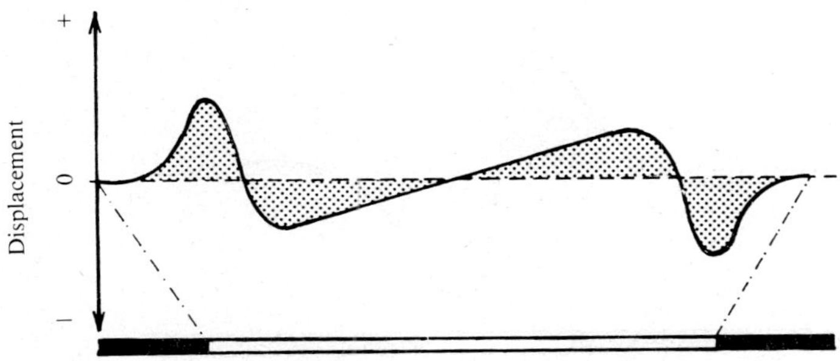

(b) Displacement after linear error distribution

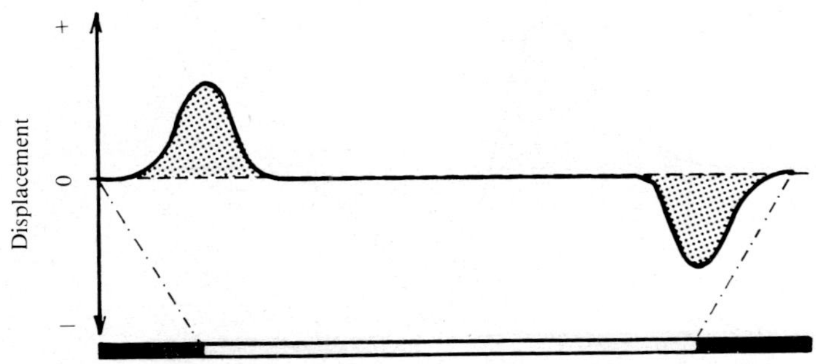

(c) Displacement after error distributed in proportion to strain

Figure 74 Longitudinal displacement profile correction

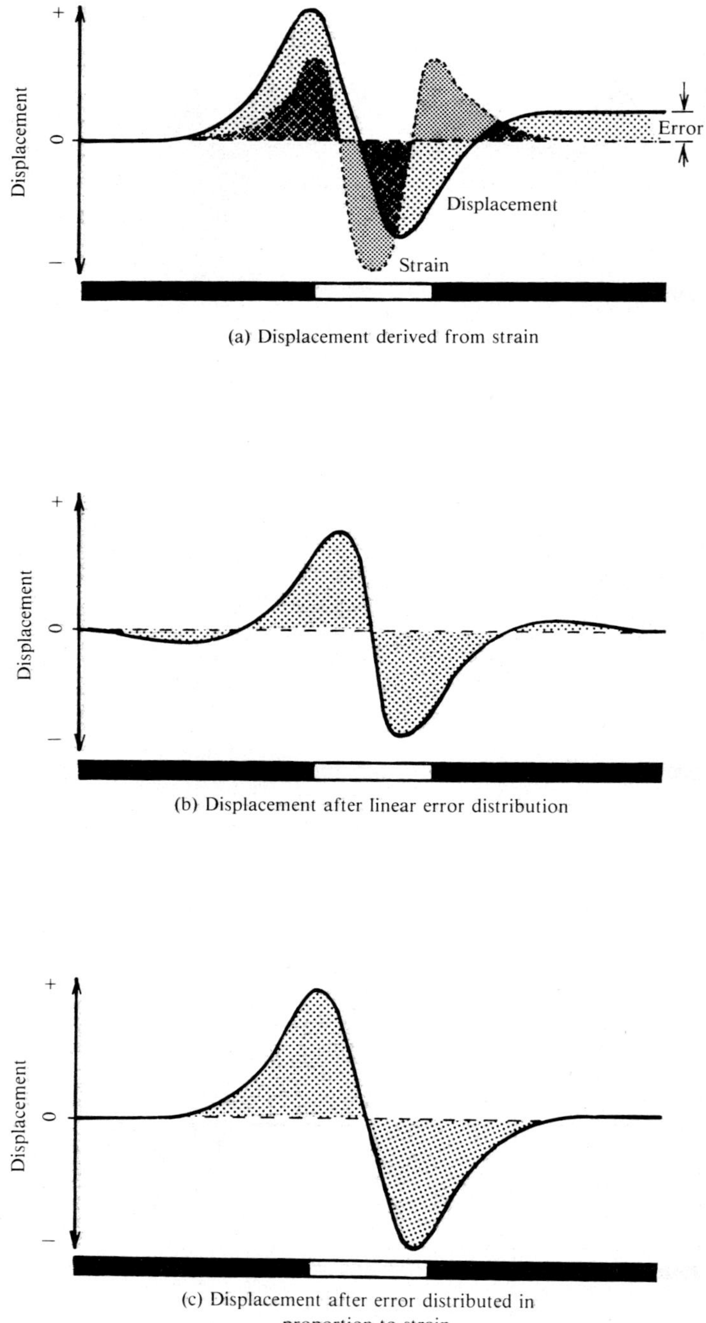

(a) Displacement derived from strain

(b) Displacement after linear error distribution

(c) Displacement after error distributed in proportion to strain

Figure 75 Transverse displacement profile correction

However, the displacement profile produced does not necessarily start and finish at zero displacement which is a common feature of empirically derived displacement curves. This profile needs to close and this requires the application of an appropriate correction. The method of correction adopted here is to distribute the displacement closure error in proportion to strain, because where the strain is high then measurements are most likely to be in error. This procedure produces the correct profile shape but has a small error in symmetry, which can be eliminated by averaging values on either side of centre.

The displacement profile correction methods are illustrated in Figures 74 and 75.

The form of error discussed above essentially arises from the way in which the strain values presented in the SEH were obtained by idealising measured curves. If the displacement curves had been actually measured (in every case), they would undoubtedly have been balanced. The strain curves, however, being difficult to check, were not balanced before incorporation in the SEH.

The subsidence, strain and displacement values are all determined in the same positions using a specially adapted curve fitting routine. The special routine deals with large data gaps as can occur in longitudinal profiles. A curve fit would normally extrapolate into data gaps and possibly cause large errors. In this case, however, data gaps are bridged linearly, as illustrated in Figure 76. The profiles are then produced as output either graphically or in tables.

An example of output for a single panel that can be requested at this stage in the program, is illustrated in Figure 77.

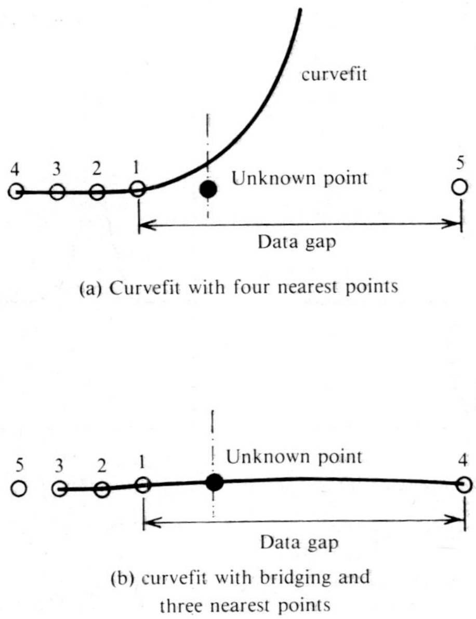

Figure 76 Curve fitting routines

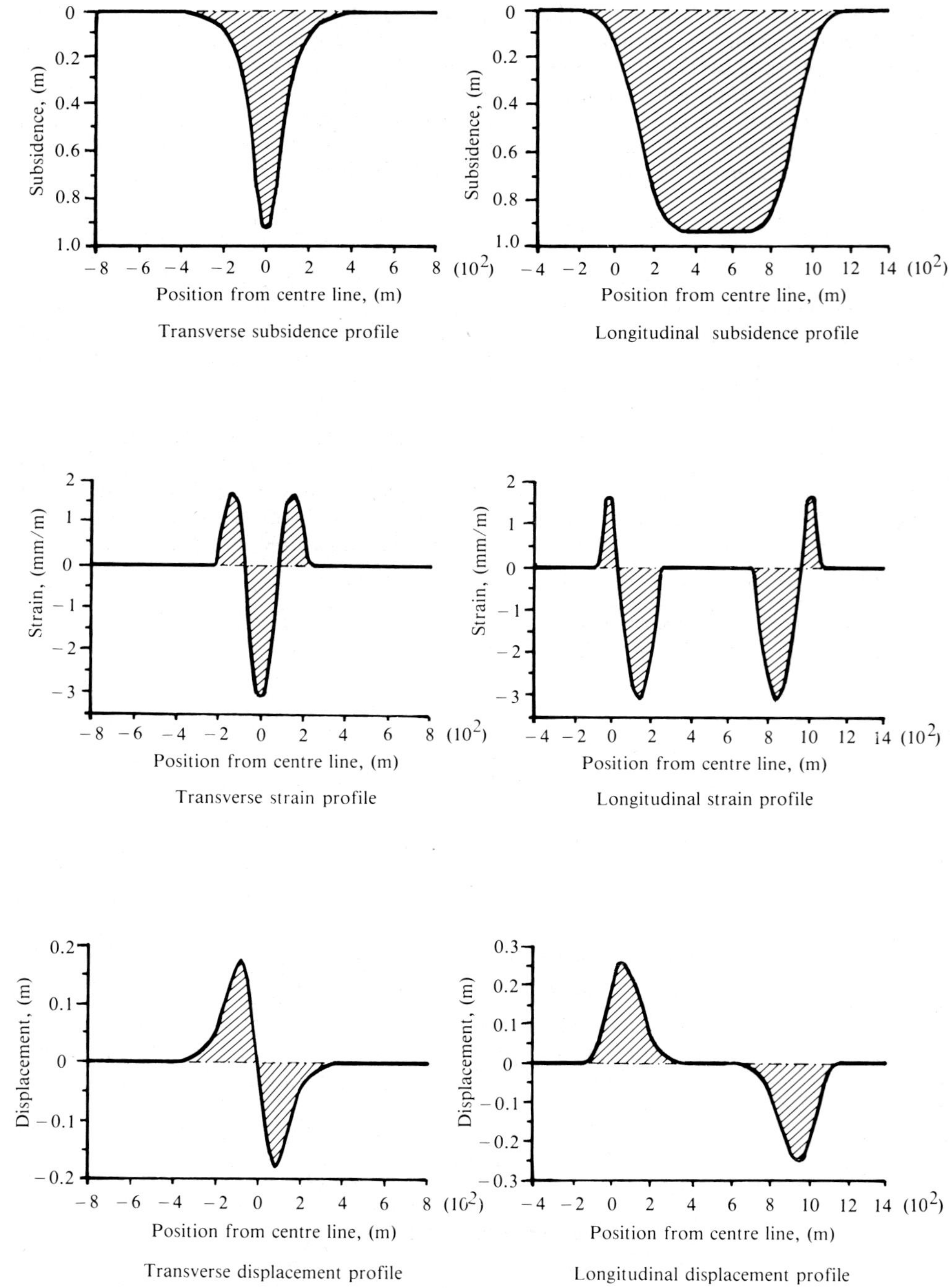

Longwall extraction details: w = 200m ; M = 2m
h = 400m ; L = 1000m

Figure 77 Single panel profiles

Generation of grids of data over whole surface area of influence

Up to this stage in the computer model, only profiles along two lines have been generated. The next step involves employing these two lines of data to generate a grid of data over the whole area of influence. In respect of subsidence, the grid is generated in the knowledge that both profiles will reduce in proportion to each other. The central value of the transverse profile reduces with the longitudinal profile so giving a grid of values with maximum subsidence in the centre and zero subsidence at the extremities.

Figure 78 illustrates the principle of reducing subsidence profiles over a longwall panel in relation to transverse profiles.

Longitudinal displacement reduces with transverse subsidence, and similarly transverse displacement with longitudinal subsidence. This allows a grid of values to be established of displacements in two directions which cover the plan surface area influenced by subsidence. The spatial movements of individual surface points are correspondingly deduced.

It follows that from knowledge of the spatial displacements of surface reference points, it is possible to determine changes in surface length between various points and consequently specify strain values. A striking feature of this method of treating the subsidence data by firstly calculating spatial displacements is that it readily allows surface ground strains to be calculated along any line across the longwall extraction.

Having established the three displacements for any point above a panel, it is relatively straightforward to calculate horizontal displacements and total displacements if they are required.

The grids of data can be used for a number of purposes, the most common of which is for the drawing of contour diagrams of the various movements. Figure 79 presents examples of contour diagrams of subsidence, transverse displacement and longitudinal displacement.

An alternative form of surface displacement representation can be as output as a vector diagram as illustrated in Figure 80. Conversely spatial displacements can be presented as shown in Figure 81.

Where the strains are required along a specific line across the longwall panel extraction, then routines can be called which refer to the grid, and thereby calculate relative displacements along the line. Special curve fits to the grid are required for the purpose of selecting out the curved nature of the grid of data.

Calculation of principal strains

Knowledge of the strains in three directions is a pre-requisite for calculation of the principal strains. The displacement data grid can be directions at any point within the grid. The relationships presented in equations 44 and 45 can be used to calculate the principal strain magnitudes and their corresponding directions from three known strains.

The main parameters referred to in equations 44, 45 and 46 are shown graphically in Figure 82. The principal strains over the whole of a longwall panel extraction can be calculated using these equations and the knowledge that the largest known strain must be within 45° of the largest principal strain. The principal strains are subsequently plotted as demonstrated in Figure 80(b).

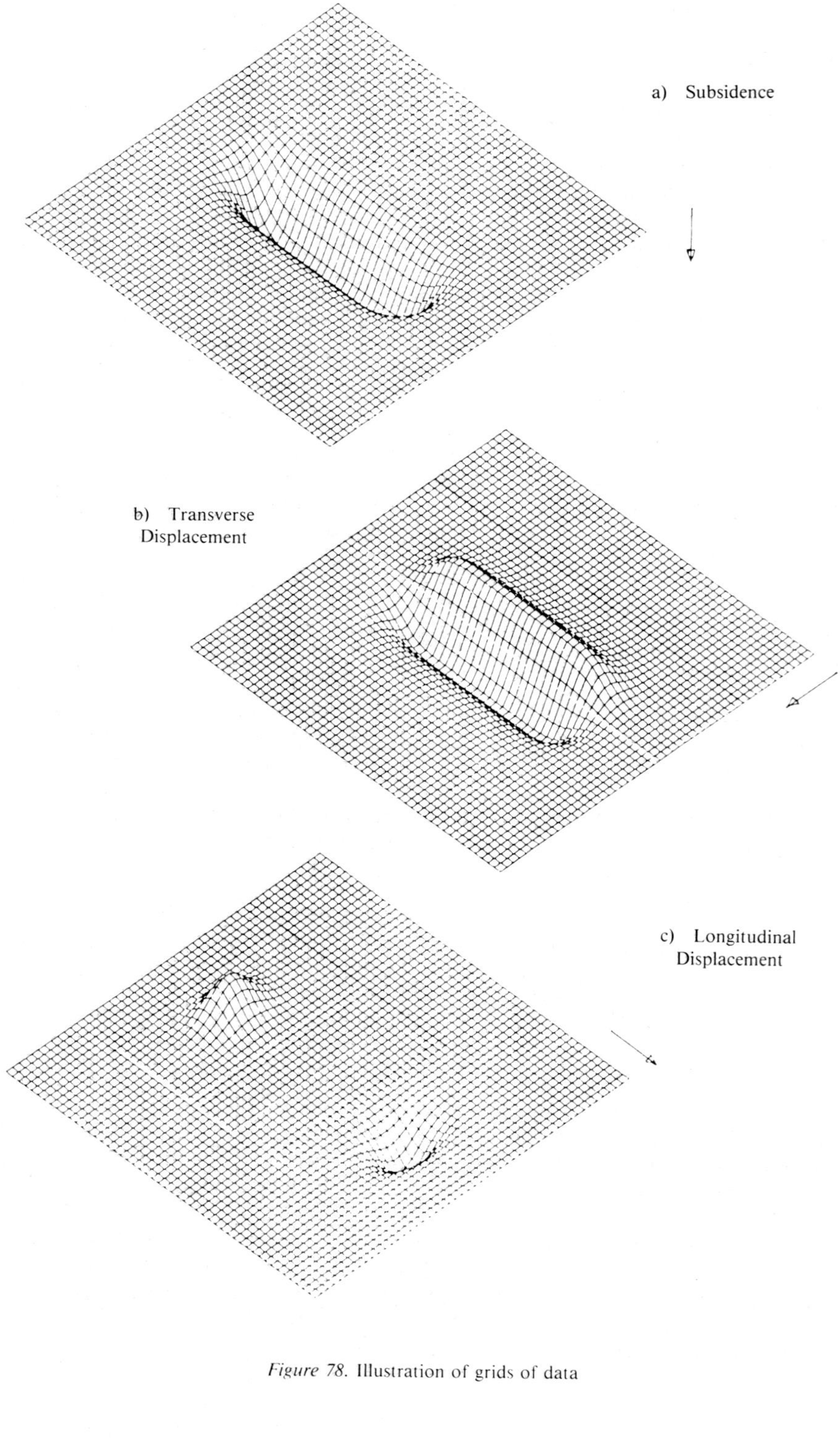

Figure 78. Illustration of grids of data

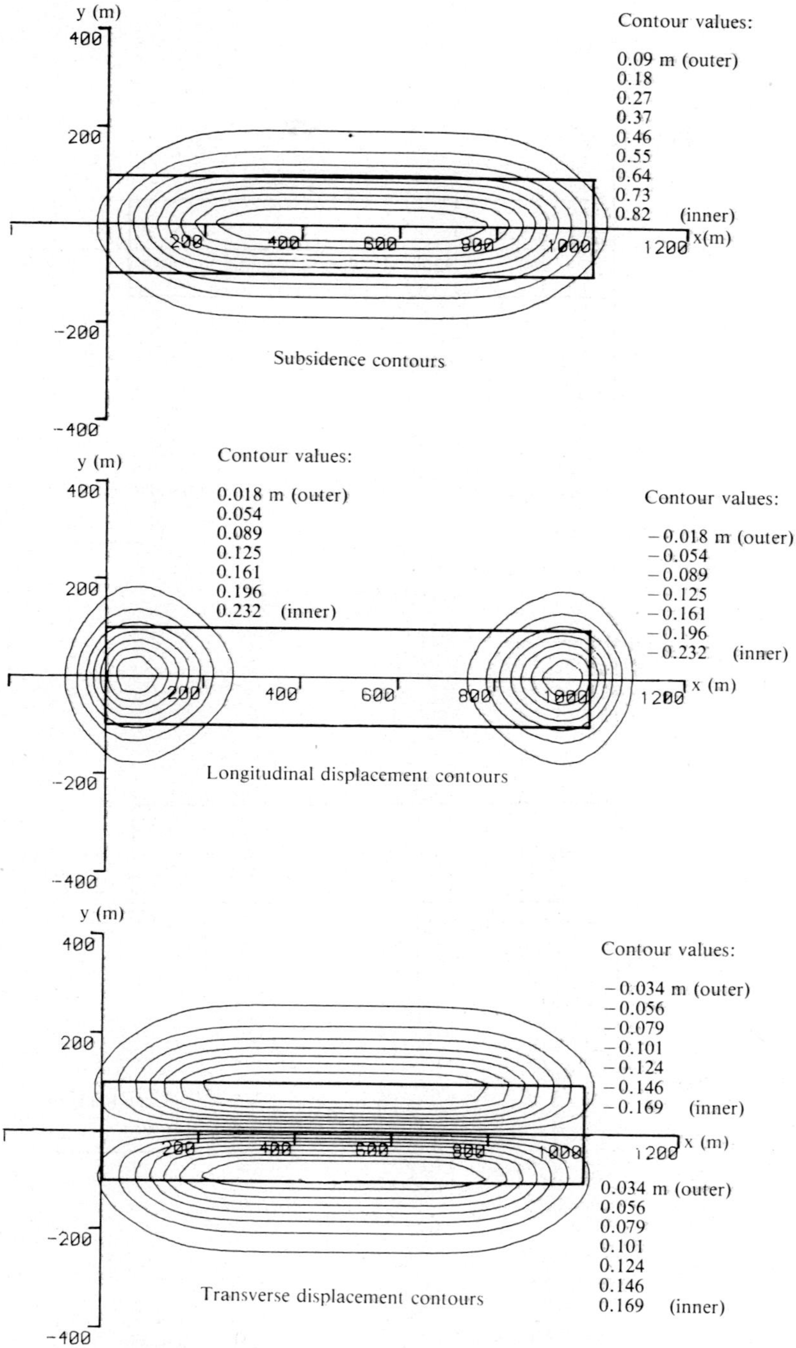

Longwall extraction data: w = 200m ; M = 2m
h = 400m ; L = 1000m

Figure 79 Contour diagrams of ground movements.

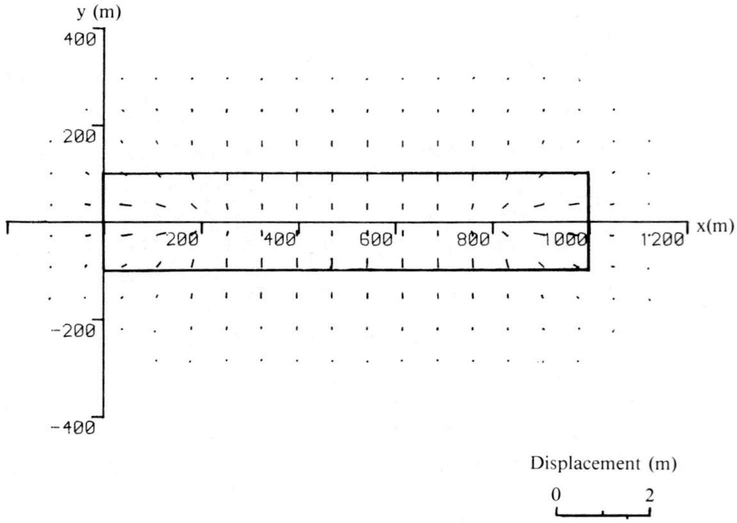

(a) Displacement vectors

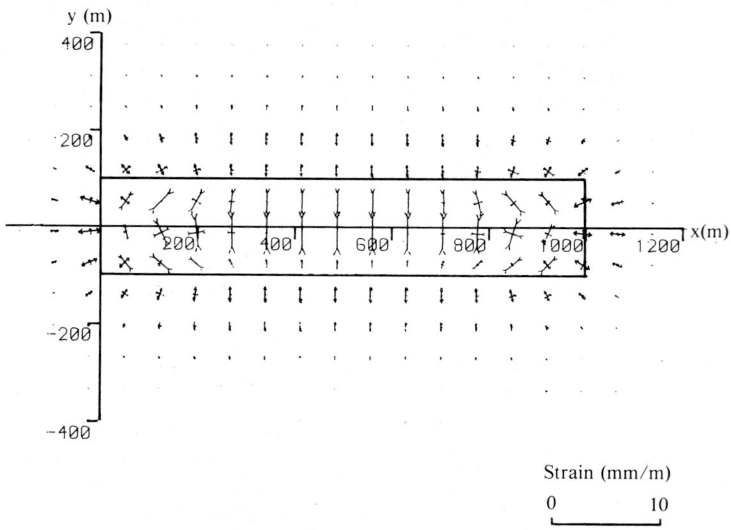

(b) Principal strains

Longwall extraction data: w = 200m ; M = 2m
h = 400m ; L = 1000m

Figure 80 Displacement vector and strain graphical representations.

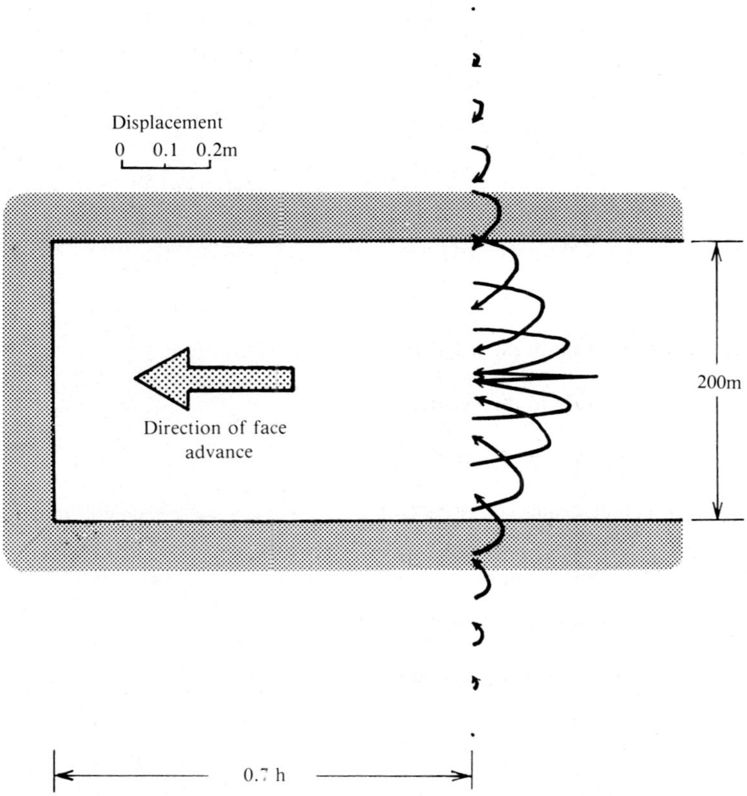

Longwall extraction data: w = 200m ; M = 2m
h = 400m ; L = 1000m

Figure 81 Spatial movement of points on a transverse line

Treatment of multi-panel and multi-seam situations

Up to this stage discussion of computerised treatment of subsidence, displacements and strains has been concerned with single longwall panel extraction situations. A further program has been developed which takes the grids of data from the single panel calculations and relates the coordinates and subsidence data to a global set of axes. The corrected grids of data are then regridded onto a global grid. The principle of superposition is assumed to apply to the data to produce a combined overall result. This allows multi-panel and multi-seam subsidence predictions to be carried out.

$$\tan 2A_1 = \frac{2(e_2 - e_3 \sin^2 v_2 - e_1 \cos^2 v_2)}{(e_1 - e_3) \sin 2v_2} \quad \ldots (44)$$

$$2E_1 = e_1 \left[1 + \frac{1}{\cos 2A_1}\right] + e_3 \left[1 - \frac{1}{\cos 2A_1}\right] \quad \ldots (45)$$

$$E_2 = e_1 + e_3 - E_1 \quad \ldots (46)$$

where e_1 = known strain in direction 1

e_2 = known strain in direction 2

e_3 = known strain in direction 3

E_1 = principal strain

E_2 = principal strain

A_1 = angle of principal strains to strain e_1

v_2 = angle between strain e_1 and e_2

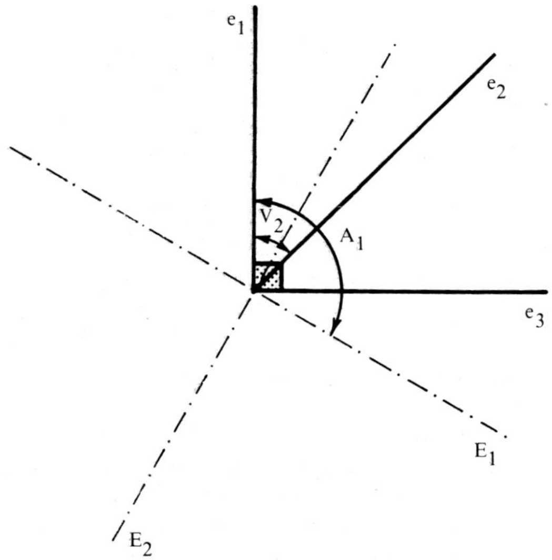

Figure 82 Illustrating determination of principal strain

The following list indicates, some of the capabilities of the program:

1. Prediction of strains along any line across one or more panels.

2. Prediction of principal strain patterns over one or more panels.

3. Prediction of the displacement of any point over one or more panels.

4. Good graphical output of the predictions.

Figures 83 and 84 illustrate subsidence contours and principal strain outputs for multi-panel and multi-seam longwall mining layouts.

Computerised predictions of mining subsidence and displacement using influence function methods

A flexible influence/zone area approach to subsidence and horizontal displacement prediction has been described by Ren, Reddish and Whittaker (1987), where particular emphasis has been placed on the applicability of the method to computerised treatment. A favourable comparison was made with SEH (1975) which considered ground movement prediction over a full panel.

Appendix 1 presents details of the findings of Ren, Reddish and Whittaker (1987) in respect of comparing the results of influence functions and empirical methods.

Irregular extraction shape. As an example of the program capability of the influence function method adopted by these authors, Figure 85 is included to illustrate application to an irregular-shaped tabular extraction. The forms of computer output shown here are subsidence contours, principal strain configuration and surface displacement vectors.

Partial extraction layout shape. A comparison was made between the influence function method computer calculations, the SEH prediction and measured values for a case study observation line adjacent to Nottingham University and discussed on pages 90-92 of SEH (1975). The results are presented in Figures 86 and 87. Predicted subsidence contours and principal strains for the three panel extractions are compared in Figures 86 (a) and (b) for the influence function method and computerised SEH empirical prediction respectively. These two sets of results compare favourably. Comparing the results presented in Figure 87(a), the SEH prediction and influence function method indicate a value of maximum subsidence of around 150-160 mm whilst the observed maximum subsidence was around 90-100 mm. It should be noted, however, that the subsidence values are particularly small (less than 10% of the extraction height) and consequently the difference between the predicted and measured subsidence values given in Figure 87(a) are not of great significance here. However, the predicted and measured transverse strain profiles shown in Figure 87(b) compare very favourably especially over the ribsides of the two outer faces.

Irregular orebody example. A further interesting example is given by Ren, Reddish and Whittaker (1987), this being an extraction of irregular plan shape and also having different thicknesses of extraction. The shape is more akin to an irregular orebody situation than a longwall extraction. Figure 88 shows the extraction example in section and plan. The influence function method referred to in Appendix 1 has been used to predict the subsidence contours and surface horizontal movements also shown in Figure 88.

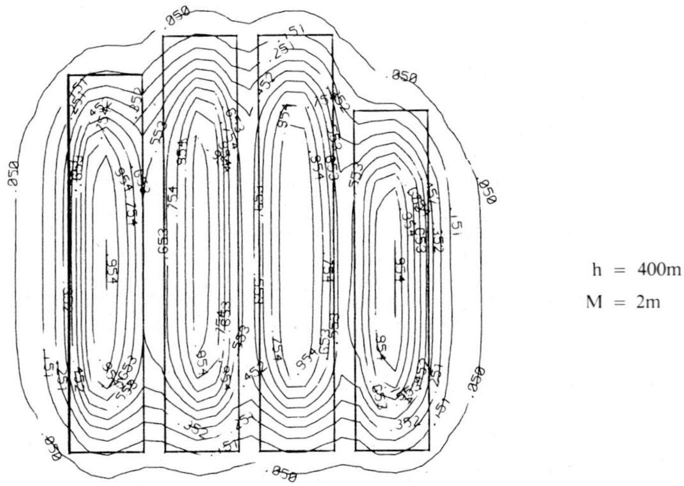

(a) Multi-panel subsidence contours

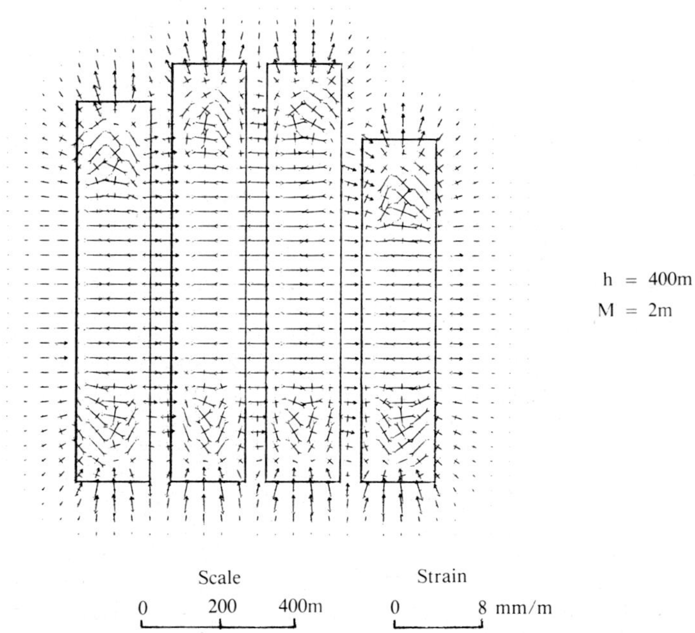

(b) Multi-panel principal strain predictions

Figure 83 Example of subsidence and strain output for partial extraction layout

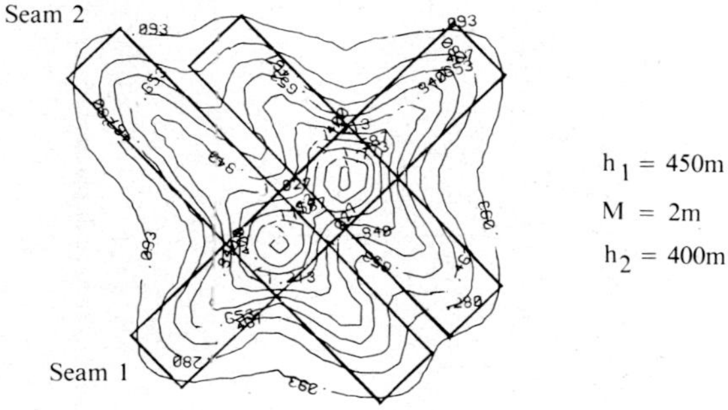

Multi-panel subsidence contours

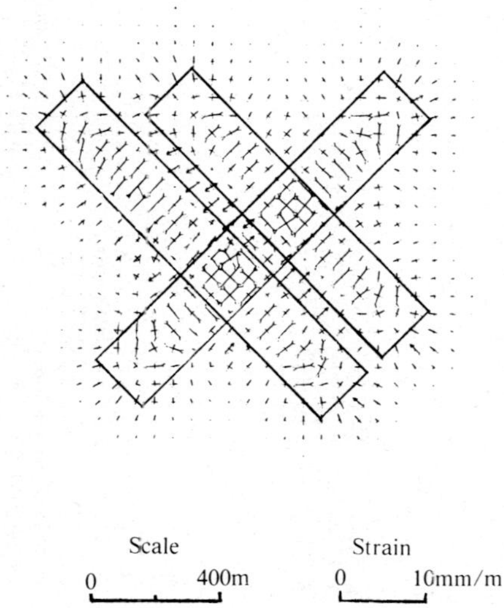

Multi-panel principal strain predictions

Figure 84 Example of subsidence and strain output for longwall extractions in different seams

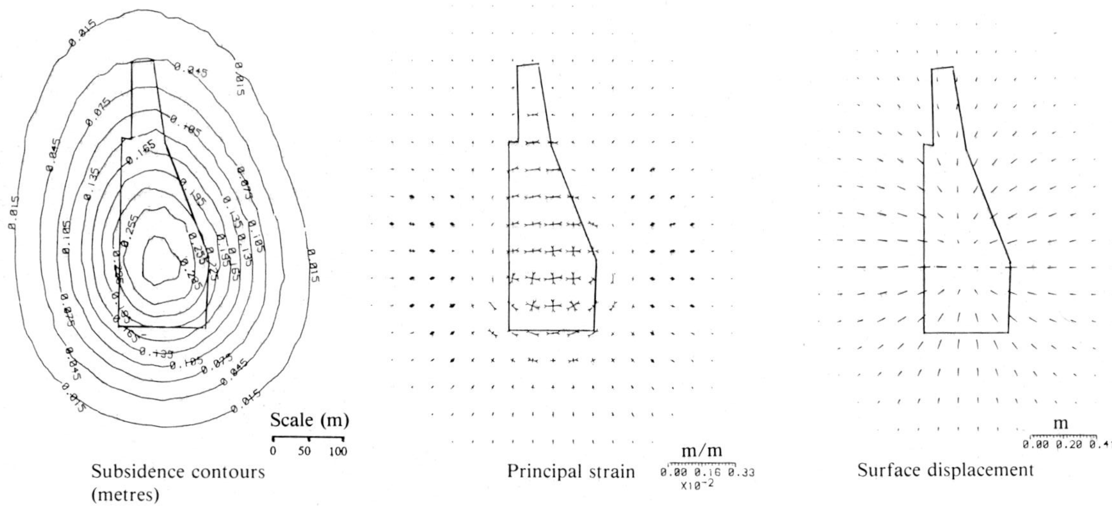

Mining data: M = 1.524m ; h = 535m

Figure 85 Subsidence, strain and displacement predictions for an irregular tabular extraction.

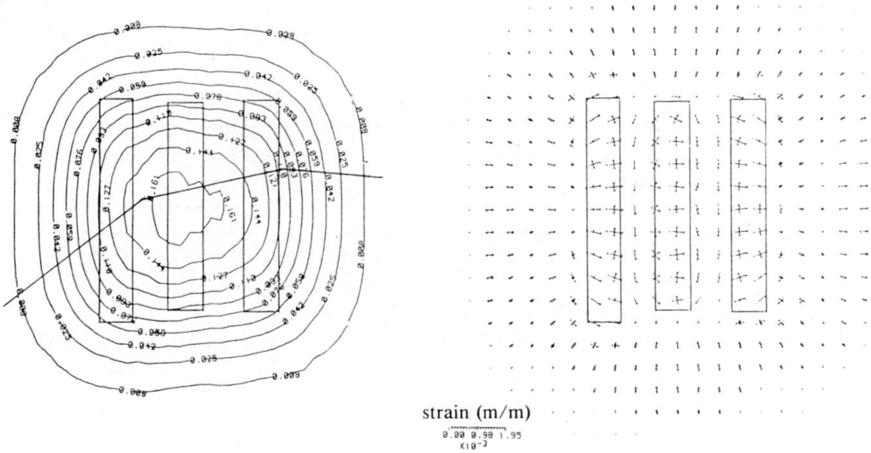

(a) Influence function method prediction.

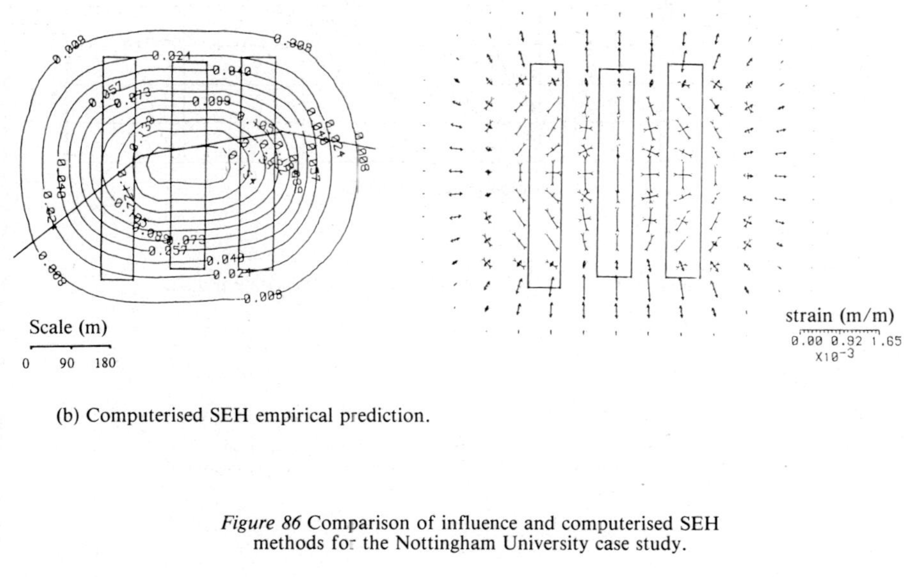

(b) Computerised SEH empirical prediction.

Figure 86 Comparison of influence and computerised SEH methods fo: the Nottingham University case study.

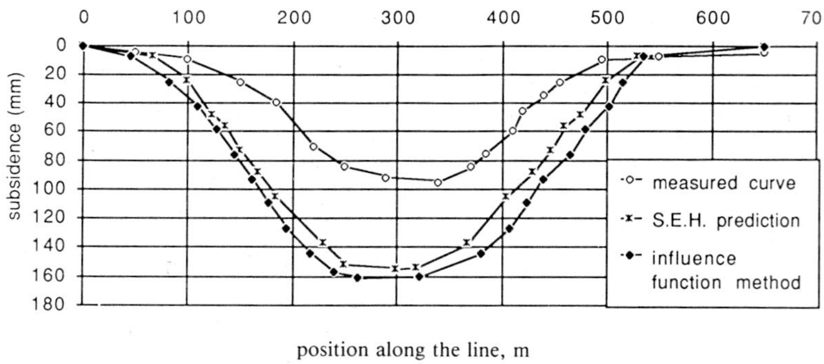

(a) Comparison between influence function method, computerised S.E.H. empirical prediction and the measured profile.

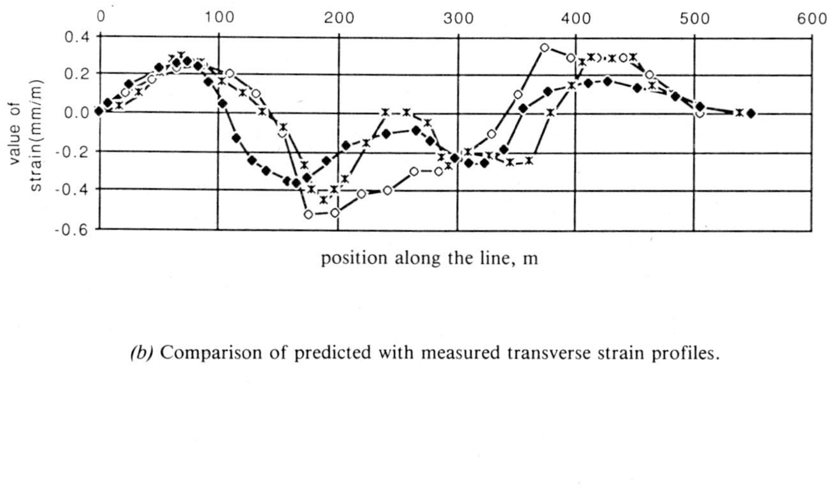

(b) Comparison of predicted with measured transverse strain profiles.

Figure 87 Comparison of predicted subsidence and strain profiles with measured values for the Nottingham University case study.

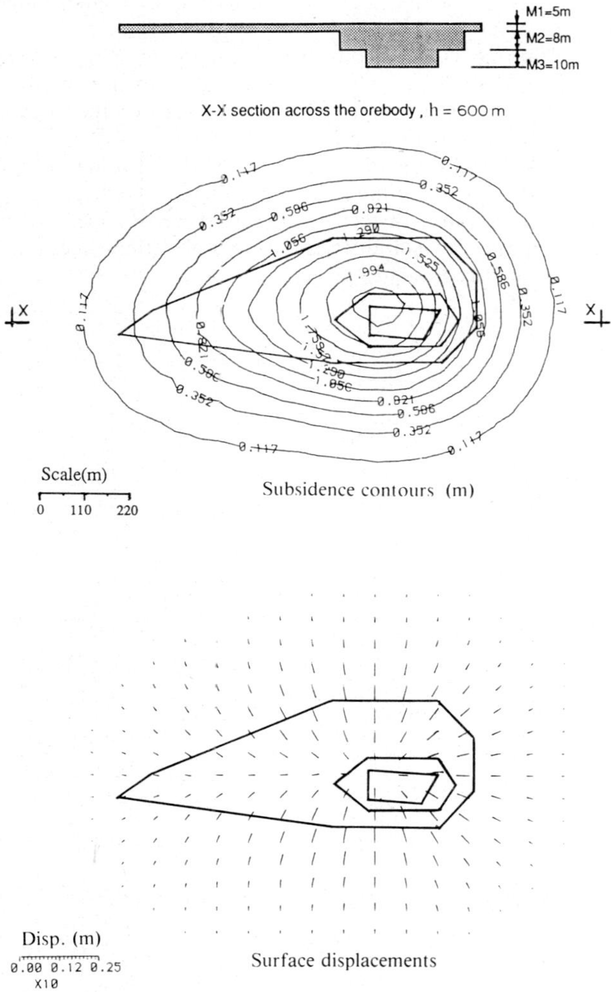

Figure 88 Predictions of subsidence contours and surface horizontal movements for an irregular orebody.

Conclusions

The basis of the SEH design method is a profile function based upon several hundred field observations covering a wide range of mining geometries. The method is well proven and is particularly aimed at predicting strains along the longitudinal and the transverse centre lines. Using a computerised step by step procedure, the SEH data can be adapted to produce ground movement predictions for first single and then multi-extraction longwall layouts.

The influence function/zone area method can be applied and adjusted to produce results comparable with the UK SEH (1975) conditions for both subsidence and displacement. This chapter has emphasised comparison with the SEH very strongly on the basis of availability of reliable results in significant quantity, but it could just have easily been calibrated to conditions in any coalfield world-wide. Even as calibrated to SEH data, the ability to alter the maximum possible subsidence (S) and the radius of influence (R) means that the geology of the site can be taken into account. In fact by altering these two factors, the program can probably be adjusted to a wide range of different geological settings since it is known that the radius of influence can vary according to geology.

CHAPTER 8

SUBSIDENCE ASSOCIATED WITH NEAR-SURFACE TUNNELS

Nature and occurrence of subsidence due to tunnelling

The occurrence of subsidence during the driving of near-surface tunnels, especially those connected with urban transport schemes, is a fairly common feature, Attewell *et al.* (1986). The construction of metro schemes in major cities world-wide has resulted in measurable subsidence, some of which has been sufficient to cause appreciable damage to surface structures. This chapter excludes subsidence above tunnels which occurs due to support deterioration as in ancient mine workings since this is discussed in a later chapter.

Influencing factors. Subsidence (also referred to as settlement in this context) above such tunnels is related to the following main factors:

1. The type of ground between the tunnel and the surface.
2. The method of tunnel drivage, together with phasing and degree of permanent support.
3. The dimensions of the tunnel.
4. The depth of the tunnel below the surface.
5. The stress conditions.

Ground conditions. Shallow tunnels are usually driven in soft ground* conditions which necessitate special considerations to be given to temporary support. The generally weak character of the ground gives rise to proneness for roof falls to occur unless a highly effective and efficient means of temporary support is employed, and even where such well-designed support schemes exist the occurrence of significant quantities of water can lead to cavities forming which could result in subsidence, O'Reilly & New (1982), Rankin (1987).

Donovan Jacobs (1975) has discussed tunnel failures particularly in respect of ground collapses which resulted in subsidence holes appearing at the surface, and draws attention to experiences with a serious collapse encountered in the construction phase of a highway tunnel in Hawaii. In this latter case it was demonstrated that tunnel excavation in soft ground demanded substantially increased attention to safety precautions as compared to tunnels of equivalent size in solid rock. The Wilson Tunnel, Hawaii was driven through a volcanic mountain ridge separating Honolulu and Kaneohe; the tunnel excavated width was 10·4 m whilst the height between the floor and crown centre of the arch-shaped profile was 7·93 m. When the tunnel was being driven through a volcanic clay material, three major underground collapses resulted in sink-holes appearing at the surface immediately above the tunnel. The third sink-hole was essentially a significant enlargement of the second since the third underground collapse occurred at the site of the second inflow. The first and third of these sink-holes measured about 22 m diameter at the surface and were 9 - 15 m deep at their maximum development. The second sink-hole was, however, much smaller being around 9 m diameter and 3·5 m deep. Water appeared to play a role in causing the tunnel roof instability

* Soft ground includes drift materials such as clay, sands and gravels which constitute engineering soils with every likelihood of being subjected to groundwater effects, in addition to weathered rocks which may possess a measure of self-supporting ability in the short-term. Soft ground implies the need for effective supporting action during both the temporary and permanent phases of the project.

to occur which led to fairly rapid appearance of these sink-holes at the surface and is a particular characteristic of collapse behaviour of unconsolidated and semi-unconsolidated clay materials.

Drivage method. The method of tunnel drivage whether hand excavated, drill and blast or machine driven usually results in some delay between excavation and support application unless a shield allows complete elimination of this excavation-support delay phase of the drivage operation. Irrespective of how well tunnel drivages are planned and carried out, there is always a strong likelihood that the operations themselves will allow the opportunity for some subsidence to occur if the other factors are favourable to its development. However, in the majority of situations the accompanying subsidence is likely to be of no significance in respect of its effect on the surface. where a major roof fall occurs in a shallow tunnel, there is a risk of collapse propagating and a conical depression or even a well-defined subsidence hole appearing at the surface. There are well-proven cavity- filling techniques which aim to prevent roof fall cavities enlarging and resulting in upward development.

Excavation size. Tunnel excavated dimensions play a major role in influencing the likelihood of subsidence developing at the surface. The greater the tunnel width, the greater the imposition placed on the immediate roof to offer self-supporting ability. Since the tunnel experiences its largest convergence rate within around one diameter of the face, the problem of convergence control becomes greater as the tunnel width increases. Effective application of temporary support measures decreases with increasing size of tunnel, and consequently calls for special care during the drivage of such tunnels in weak ground.

Tunnel depth. The depth below surface is a major influencing factor on the likelihood of subsidence occurring at the surface above a tunnel. Clearly, if the tunnel is sufficiently deep, then even the effects of major roof falls may peter out before reaching the surface. However, as is the case with mining extractions, the depth should be considered in relation to the dimensions of the tunnel.

Stress field. The nature of the *in situ* stress field can play a role in influencing the stability of near-surface tunnels and consequently their potential for the occurrence of subsidence. In most situations, however, the state of stess will be governed by the overburden loading and groundwater conditions thereby giving rise to a predominantly vertical stress field. In some circumstances high lateral stresses can exist by virtue of the geological history of the rocks through which the tunnel is driven and could give rise to stability problems in high tunnels. The conditions favouring high lateral stress fields are generally encountered in semi-competent and competent rock situations. High lateral stresses in shallow tunnels can result in premature failure of natural beams of rock spanning recently excavated tunnels; this can lead to surface subsidence occurring unless tunnel support is effective.

To summarise, therefore, of major importance is the nature of the ground through which the tunnel is driven, the depth of the tunnel below surface, and the tunnel size and mode of drivage coupled with the method and degree of support employed. This discussion relates to the possible occurrence of subsidence at the surface as a result of the tunnel drivage operations. Circumstances may arise many years later, however, where surface subsidence could occur due to the process of support deterioration taking place. Consequently, support integrity, strength and durability are important aspects in this respect.

Form of ground movement around near-surface tunnels

The general picture of ground movement around a typical near-surface tunnel is shown

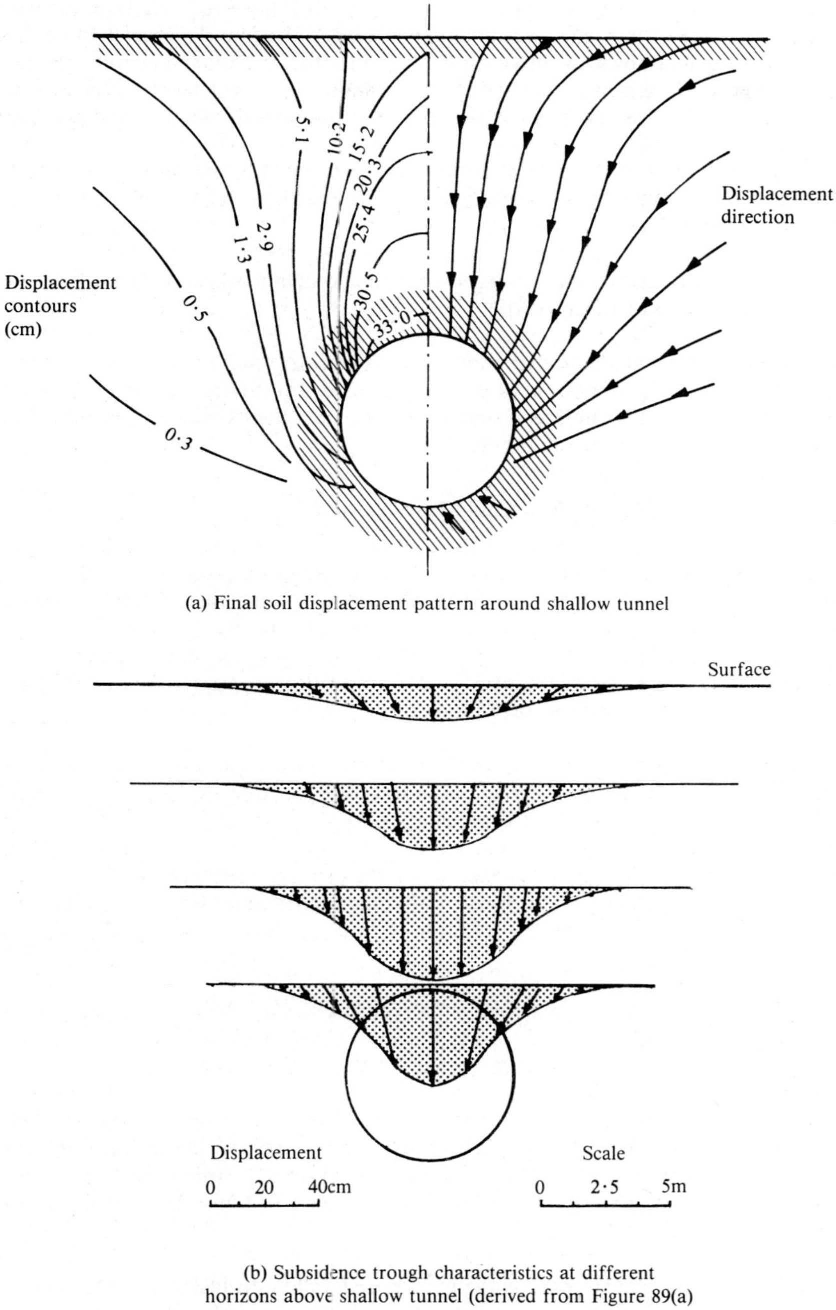

Figure 89 Ground movement behaviour around shallow tunnel, after results from Hansmire and Cording (1974)

in Figure 89(a) and is taken from Hansmire and Cording (1974) who investigated a soft ground tunnel on the Washington Metro. This shows the final observed pattern of movement. The greatest intensity of movement is in the upper third of the tunnel's excavated periphery. Displacement of ground towards the tunnel is shown to display a curved characteristic. This form of ground displacement suggests that tunnel width is an important factor as it has a major influence on the vertical height attained by the contours of equal total displacement.

The type of ground forming the overburden associated with this tunnel was Pleistocene river terrace deposits consisting of sands, gravels, silts and clays. No instability was reported in respect of water inflow.

The tunnel excavated diameter was 6·4m and the depth below surface of the tunnel centre was 15m at the instrumented site.

Figure 89(b) shows an extrapolation made by the present authors of these results in the form of a generalised representation of subsidence curves for different horizons between the tunnel and the surface. The subsidence movements become more concentrated both laterally and vertically at horizons near the tunnel.

Behaviour of a mine surface drift in Coal Measures strata

Underground mining developments are accessed by shafts or drifts, the latter of which can be in the form of simple tunnels as is often the case with shallow room and pillar workings in undulating terrain. Inclined surface drifts (also known as inclined mining tunnels) are a major means of access to underground coal mine workings; typical inclination angles for such drifts are around 10° dip. Drifts are popular where the coal seams are generally no deeper than 300m below surface, especially because of ease of drivage coupled with economy and continuity of subsequent transport operations.

Construction practice firstly involves the formation of an open-cut at the surface entry point for the drift; a monolithic concrete portal structure is then generally built to a depth of about 10 - 12m below surface. The exposed rock face is normally reinforced to promote ground arching around the intended excavation. In UK Coal Measures conditions it is usual practice for drift drivage operations to begin when the depth of cover consists of not less than 6 -8m of relatively competent strata. Most roof control difficulties tend to arise at the drift face during this critical stage of the construction operations, that is at shallow depths of cover of around less than 30m.

Drift site details. Subsidence behaviour of a typical UK Coal Measures surface drift has been reported by Whittaker and Pye (1977), which involved an arch-shaped steel girder lined tunnel dipping at 1 in 5·8 (9·8°) from an almost flat surface area. The tunnel excavated dimensions were 5·5m width at floor level and 4·0m height to the crown. At the point where the main subsidence investigation was carried out, the first 3m of cover was consolidated clay whilst the rock cover down to the crown of the drift (at 13·4m below the surface) consisted of a layered structure of siltstone, sandstone, mudstone and coal; the immediate floor and sides of the tunnel at this point were predominantly mudstones. The tunnel crown was at 13·4m and floor at 17·4m below the surface.

Subsidence measurements were made by precise levelling to instrumentation boreholes positioned above the path of the surface drift. Additionally a grid of surface pegs was used to observe subsidence for plotting transverse and longitudinal profiles. Five boreholes with strain wires and extensometer instrumentation were used to investigate sub-surface ground movement.

Development of transverse roof movements. The development of strata movement at a horizon 10m below the surface is presented in Figure 90 and this shows relative displacements in each hole plotted against tunnel advance. These results have been plotted relative to the standpipe at the mouth of the borehole. All the results indicate apparent uplift ahead of the advancing tunnel. Those plugs (Nos 2, 3 and 4) immediately over the tunnel underwent rapid subsidence until the drift was about 9 m past the observation position. It is shown by these results that the immediate roof exhibited significant subsidence at around 1 m after undermining and this phase of major movement continued until the tunnel face was some 6 m past this point. This represents the critical zone for immediate and effective roof control and virtually all occurs within 1 - 1·5 diameters of the tunnel face in this case. The shallow nature of the site and general weakness of the overburden are undoubtedly playing major roles in this particular situation. The ground is responding more in relation to load transfer by dead-weight principle rather than by natural arching as is evident in deep mining tunnels. Increased movement was observed at Station No 2 owing to localised instability of the immediate roof which became apparent during blasting. Only minor disturbances were experienced by Station Nos 1 and 5 located over the tunnel sides during the period of undermining, and these movements also ceased within 1 - 1·5 diameters behind the tunnel face.

Development of roof movement along axis of drift. Borehole No 3 results are presented in Figure 91(a) to demonstrate the similar pattern of movement experienced by the three observation plugs located at different horizons. Again relative displacements are used here to highlight the apparent uplift experienced equally by the three plugs, but after undermining as would be expected the lowest plug showed the greatest movement. The three observation horizons all tended to cease moving at the same position of tunnel advance. Inter-strata separation occurred virtually immediately after undermining the observation points.

Figure 91(b) shows the absolute vertical movements of the observation plugs of Borehole No 3, the relative displacements of which are given in Figure 91(a). The precise levelling indicated discernible uplift was occurring at 16 - 18m ahead of the tunnel face at this position, with a maximum displacement of −4mm being observed. An important feature shown by Figure 91(b) is that although the relative displacement between the plug horizons tends to cease at 8 - 9m behind the face, the surface continued to subside until the tunnel face was about 16 -18m past the observation point.

Dynamic subsidence development curve. In order to examine the dynamic subsidence development curve along the axis of the drift, five independent sets of results were obtained in the depth range of 17·4m to 20·2m (relative to the tunnel floor). The results are presented in Figure 92(a) and show fairly close consistency. A further presentation is made of these results in Figure 92(b), but subsidence is expressed in terms of its maximum observed value. Only two sets of results show discernible uplift ahead of the tunnel face. Some of these results are from ground peg observations and consequently possibly did not possess the same sensitivity as the securely grouted standpipe of Station No 3. Surface subsidence began at around 20m ahead of the tunnel face, that is a distance beyond its depth. Figure 92(b) shows that surface subsidence was complete at about a similar distance behind the tunnel face.

A comparison of the observed mean dynamic subsidence curve for these five sets of results is made with the UK SEH deep mining subsidence curve in Figure 93. Surface subsidence ahead of the drift face starts significantly earlier than would be anticipated by the SEH prediction and also it tends to exert its influence over a greater distance. This suggests a greater angle of draw is involved with the surface drift situation which probably indicates that the weak overburden and unconsolidated deposits are playing a predominant role.

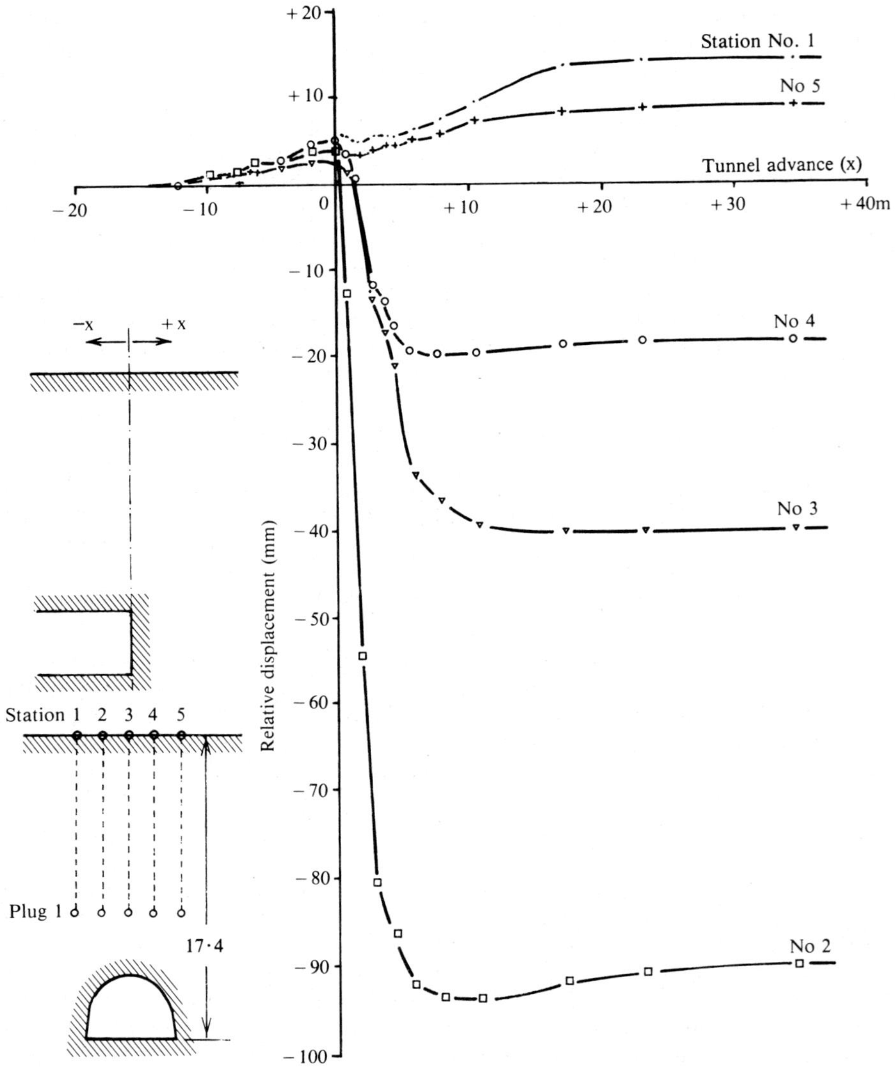

Figure 90 Illustrating development of strata movement at a horizon 10m below the surface due to tunnel excavation, according to Whittaker and Pye (1977).

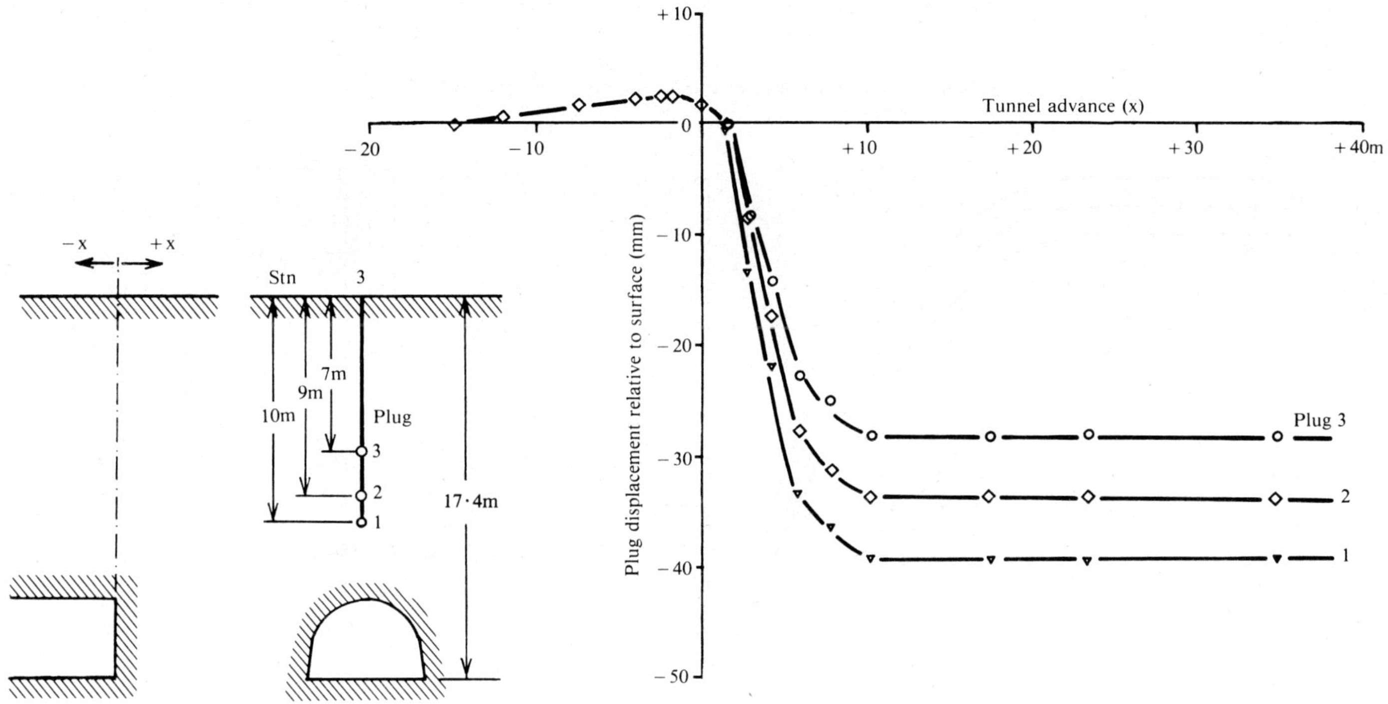

(a) Development of roof movement along axis of tunnel.

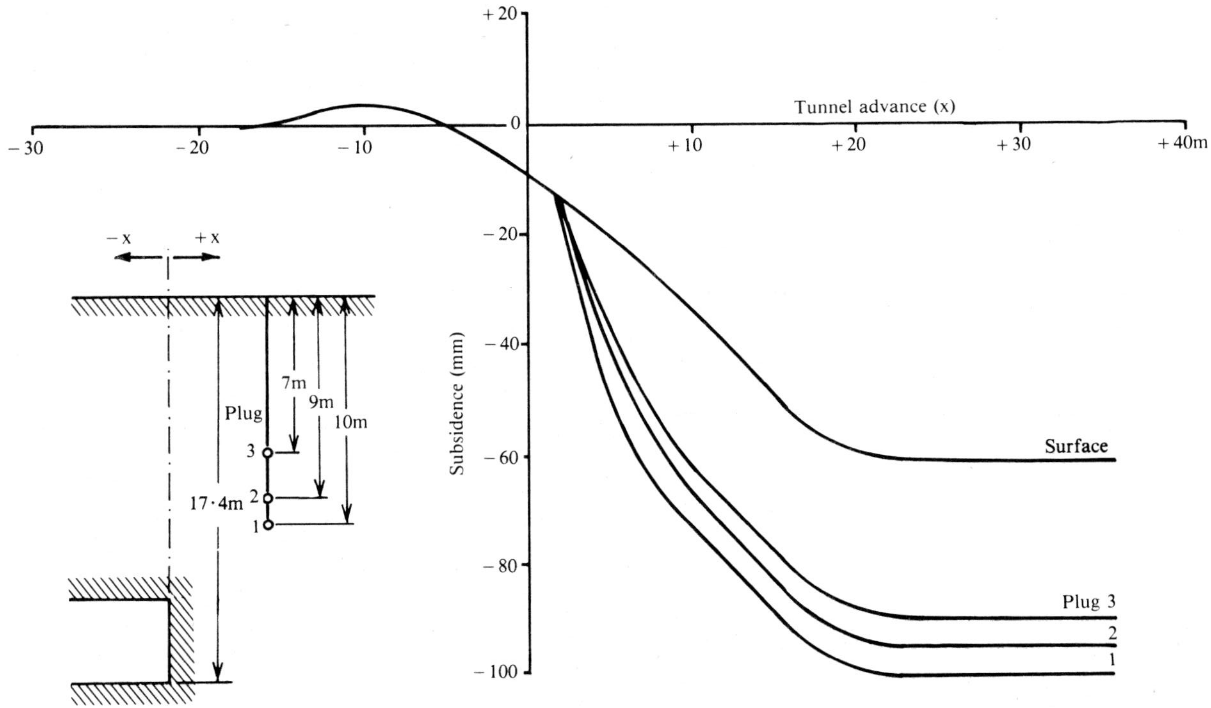

(b) Absolute vertical movements corresponding to Figure 91(a)

Figure 91 Ground movement along axis of tunnel, according to Whittaker and Pye (1977).

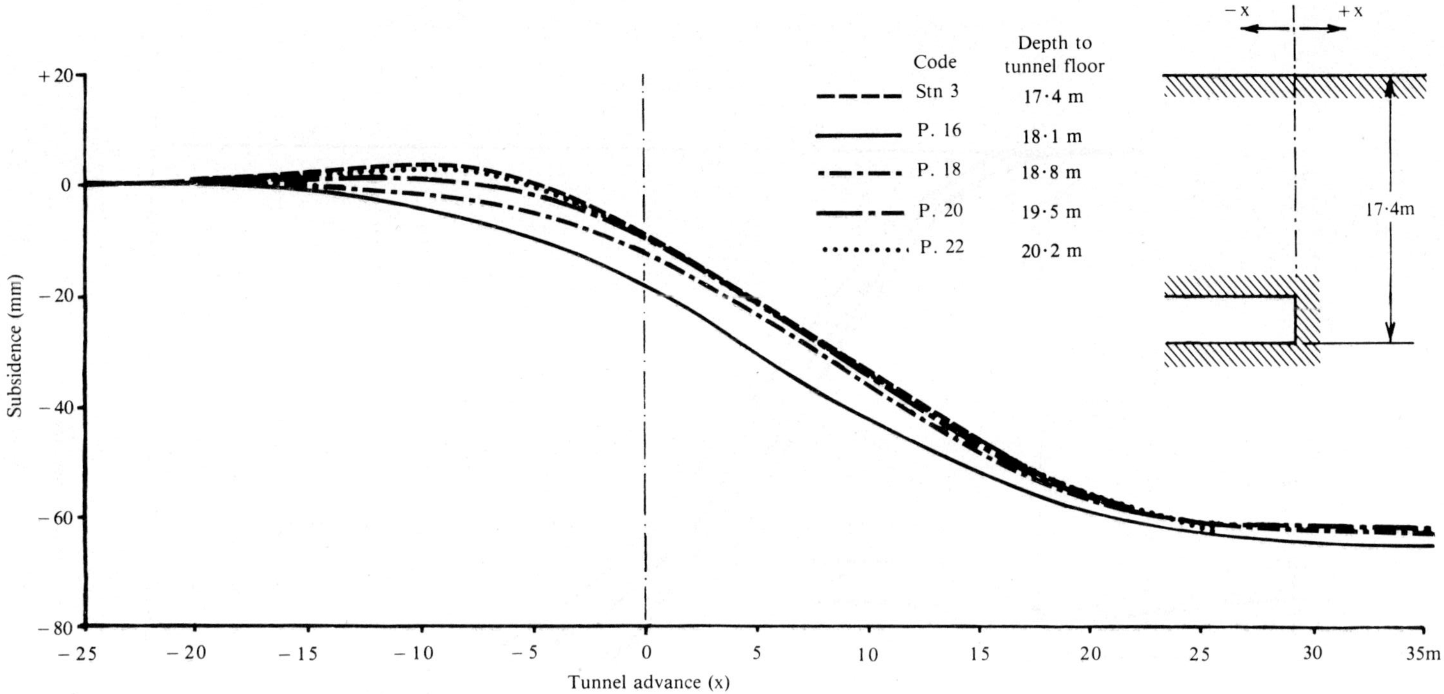

(a) Surface subsidence development curves

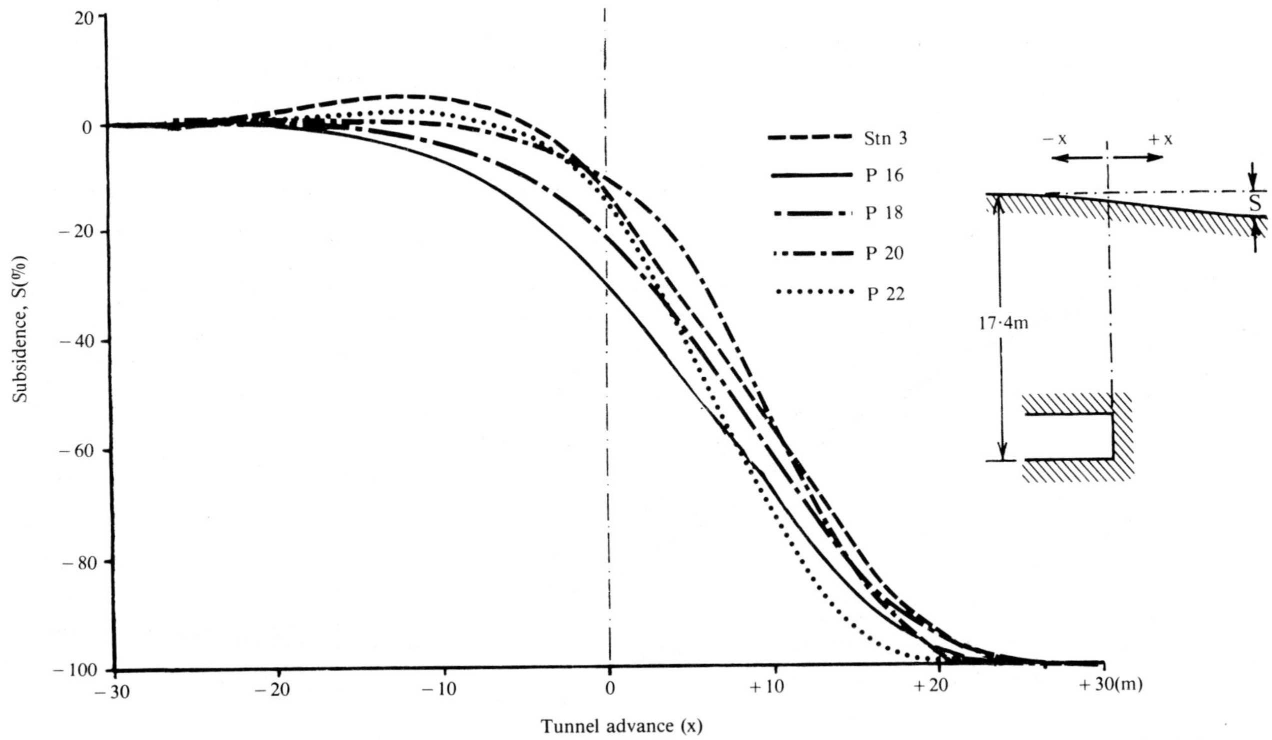

(b) Surface subsidence development curves in relation to maximum subsidence

Figure 92 Surface subsidence development curves above shallow tunnel, according to Whittaker and Pye (1977)

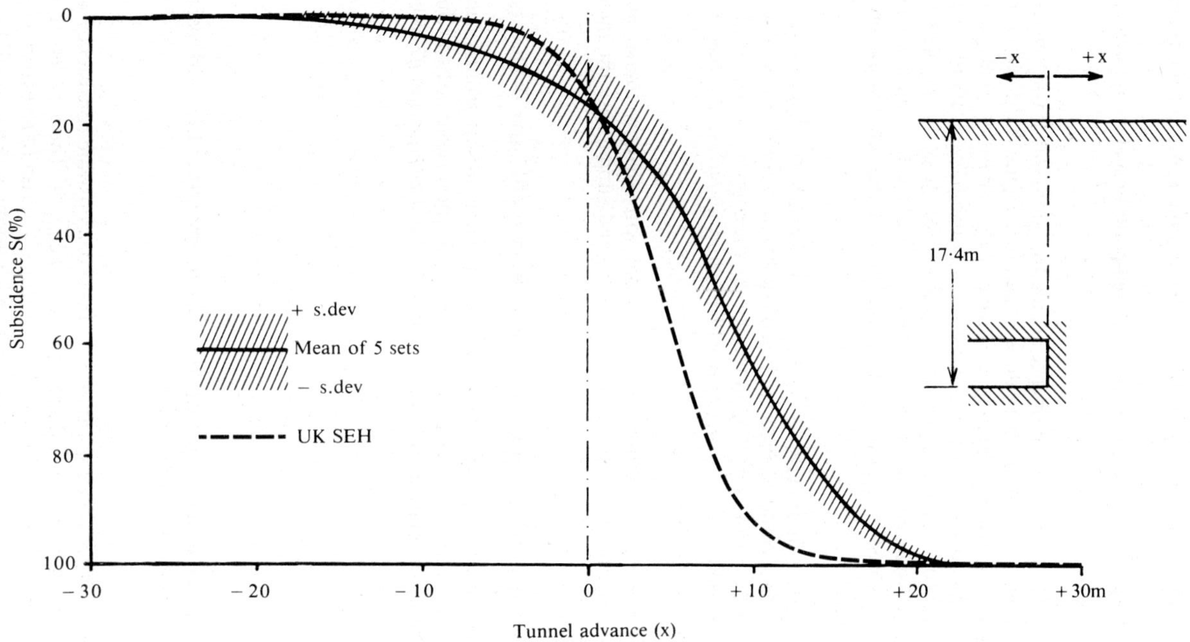

Figure 93 Comparison of subsidence result for shallow tunnel with UK SEH prediction, according to Whittaker and Pye (1977)

Transverse subsidence curve. The nature of the subsidence curve across the line of advance of the drift is demonstrated in Figure 94(a) which shows observed data, whilst Figure 94(b) shows the results expressed in terms of the maximum subsidence and is compared to the deep mining SEH predicted curve. The limit angles have been considered relative to the floor of the arched shaped excavation. This suggests a limit angle of 45° for these results. Subsidence above the drift obviously has a greater area of influence according to this comparison with the deep mining situation. The transition point is located beyond the ribside of the tunnel.

Influence of depth on the potential for subsidence to develop above shallow tunnels and drift workings

The UK SEH method of predicting subsidence arising from longwall extractions has been considered in respect of its application to shallow situations. Figure 95 shows a tentative relationship between the anticipated maximum subsidence from a narrow tabular extraction plotted against depth below surface. This extrapolated relationship considers two extraction widths, namely 5m and 10m, and uses the basic assumption that the extracted region corresponds to a situation where convergence is allowed to take place; this could be equivalent to a caved condition or even where stowing is employed. It assumes that natural arching of the strata is taking place.

The relationships given in Figure 95 indicate that the critical depth above which significant surface subsidence could be expected is around 40m in respect of tunnel widths up to 10m. This assumes that control of ground movement by the support system is maintained and that no large-scale collapses have occurred which could allow a sink-hole or conical depression to appear at the surface.

Table 15 represents observed subsidence data associated with shallow tunnel and drift workings. It would appear that the methods of tunnelling and support are the main overriding factors.

Mathematical representation of transverse surface subsidence profile over shallow tunnels

The transverse surface settlement profile for a shallow tunnel in soft ground conditions has been reported by Peck (1969) and Schmidt (1969, 1974) to conform to a shape similar to the Gaussian error function as illustrated in Figure 96. Schmidt uses equation 47 to estimate the trough width:

$$\frac{i}{a} = \left(\frac{z_o}{2a}\right)^{0\cdot 8} \qquad \ldots(47)$$

where
i is the standard deviation of the error function and corresponds to the width parameter,
z_o is the depth below surface to the tunnel centre-line,
$2a$ is the tunnel diameter.

Figure 96 includes the Kiveton Park results and these demonstrate consistency with the Gaussian error function used in the method as proposed by Schmidt. The complete transverse surface subsidence profile calculated area = $0\cdot76\text{m}^2$; using $i = 4\cdot1\text{m}$ (direct from Figure 96) gives a subsidence profile area of $0\cdot66\text{m}^2$ whilst Schmidt's formula (equation 47) gives $i = 6\cdot26\text{m}$ and a subsidence profile area of $1\cdot00\text{m}^2$. The results used by Schmidt to derive his formula had a fairly wide scatter and probably explain the difference between the actual and predicted results using Schmidt's formula for i as given here.

Table 15. Surface subsidence associated with shallow tunnel and drift workings, revised after Whittaker and Pye (1977)

Case study No.	Excavated width (W) (m)	Depth (h)* (m)	W/h	Maximum subsidence S_{max} (mm)	Excavation shape	Excavation height (m)	Type of ground	Remarks and source reference	Method of formation
1	5·5	15·4	0·36	0·64	Arch	4·0	Carboniferous Coal Measures sediments	Kiveton Park. Drift Mine, Whittaker & Pye (1977)	Drill and blast
2	5·0	8·5	0·59	16·4	Circular	5·0	Lower Chalk	T.R.R.L. Experimental tunnel at Chinnor, Oxfordshire, McCaul *et al.* (1976)	Tunnelling machine
3	4·15	29·3	0·14	6·0	Circular	4·15	London clay	London Transport Fleet Line tunnel, Attewell & Farmer (1974)	Shield tunnelling
4	5·3	11·0	0·48	21·5	Circular	5·3	Granular soil	San Francisco Mission Line BART, Peck (1969)	Digger shield; air pressure = 90 kN/m^2
5	5·2	10·5 -13·5	0·49-0·38	101·0	Circular	5·2	Granular soil	Toronto subway, Peck (1969)	Shield and hand mined
6	11·9	37·5	0·32	244·0	Arch	11·9	Hard clayey till	Large railway tunnel, GNRR, Seattle, Peck (1969)	Hand mined; small drifts with centre
7	11·0	36·9	0·30	43·0			Hard clay (clay-shale)	Garrison Dam test tunnel, Peck (1969)	Full face blasted
8	11·0	13·1	0·84	9·0			London clay	Heathrow Cargo tunnel, Peck (1969)	Shield and hand mined
9	7·3	23·5	0·31	82·0	Arch		Stiff clay	Chicago subway, Peck (1969)	Hand mined. Face benched. Air pressure = 103 kN/m^2
10	7·0	22·6	0·31	183·0	Circular	7·0	Sensitive clay	Tokyo subway, Peck (1969)	Shield with hand mining and breasted face. Air pressure = 55 kN/m^2
11	3·0	18·3	0·16	6·0	Circular	3·0	Leda clay, sensitive	Ottawa sewer, Peck (1969)	Digger shield. Air pressure = 30 kN/m^2
12	7·9	17·1-22·0	0·46-0·36	610-732			Sensitive clay	Norwegian tunnel, Peck (1969)	Shield with hand mining. Air pressure = 120-160 kN/m^2

* Depth measured to tunnel axis.

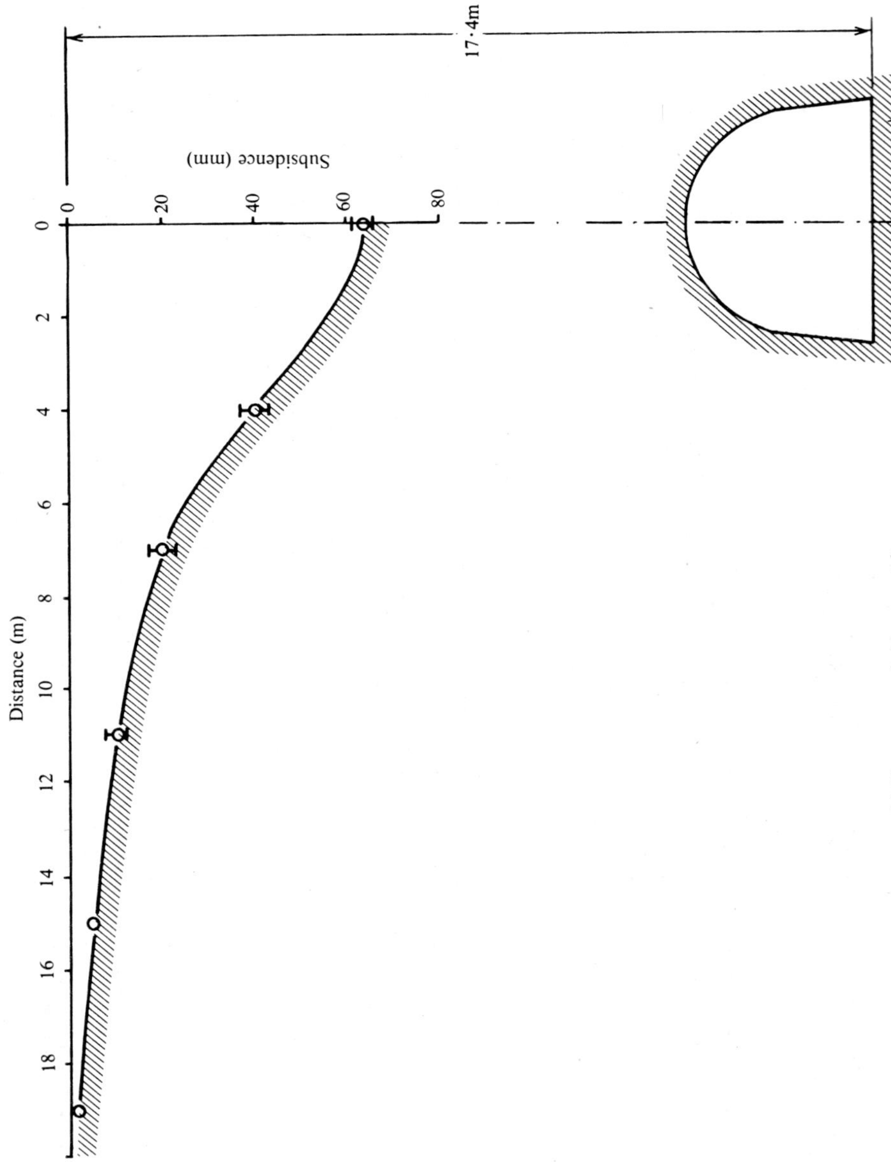

(a) Transverse subsidence results

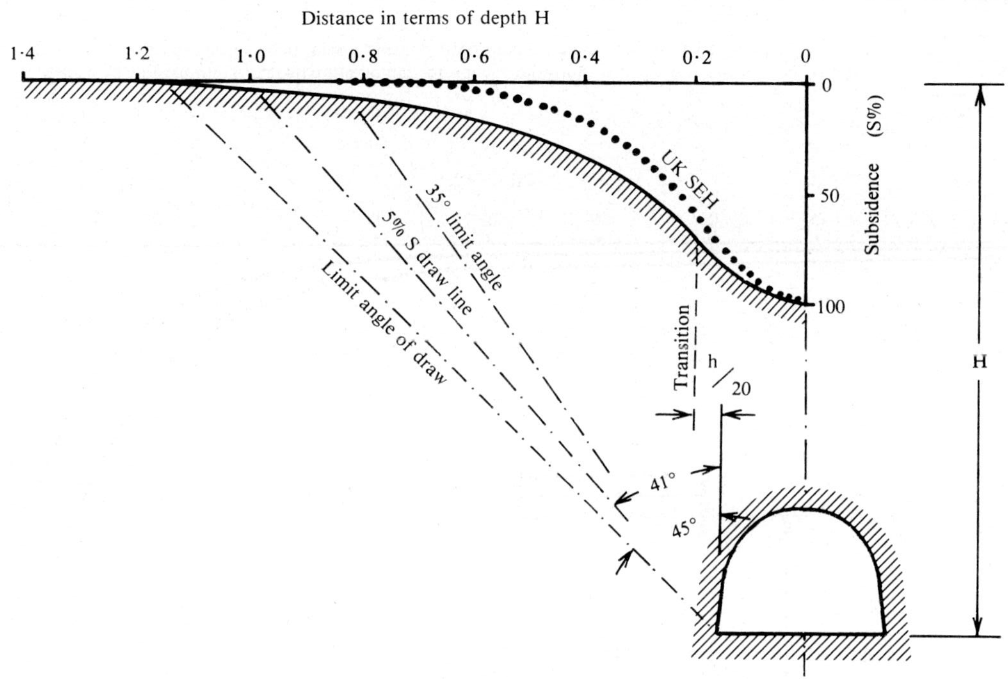

(b) Comparison of shallow tunnel transverse subsidence curve with UK SEH prediction for deep mining

Figure 94 Transverse subsidence curve, according to results from Whittaker and Pye (1977).

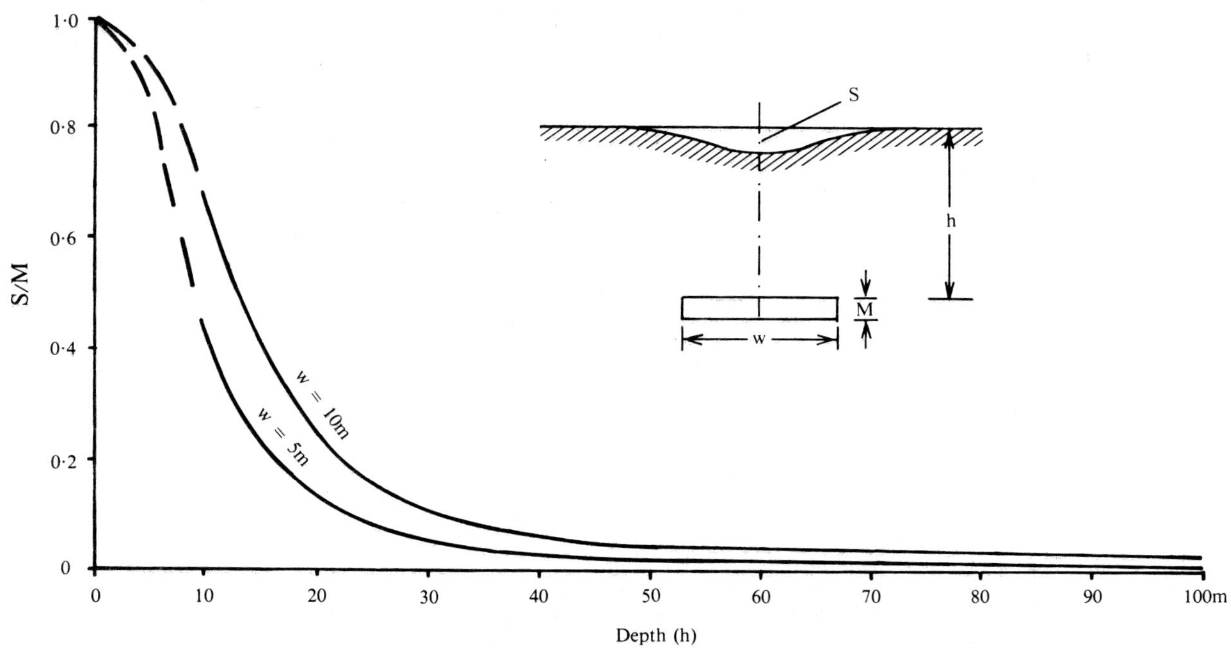

Figure 95 Illustrating anticipated maximum subsidence resulting from narrow tabular extractions in relation to depth below surface, according to Whittaker and Pye (1977)

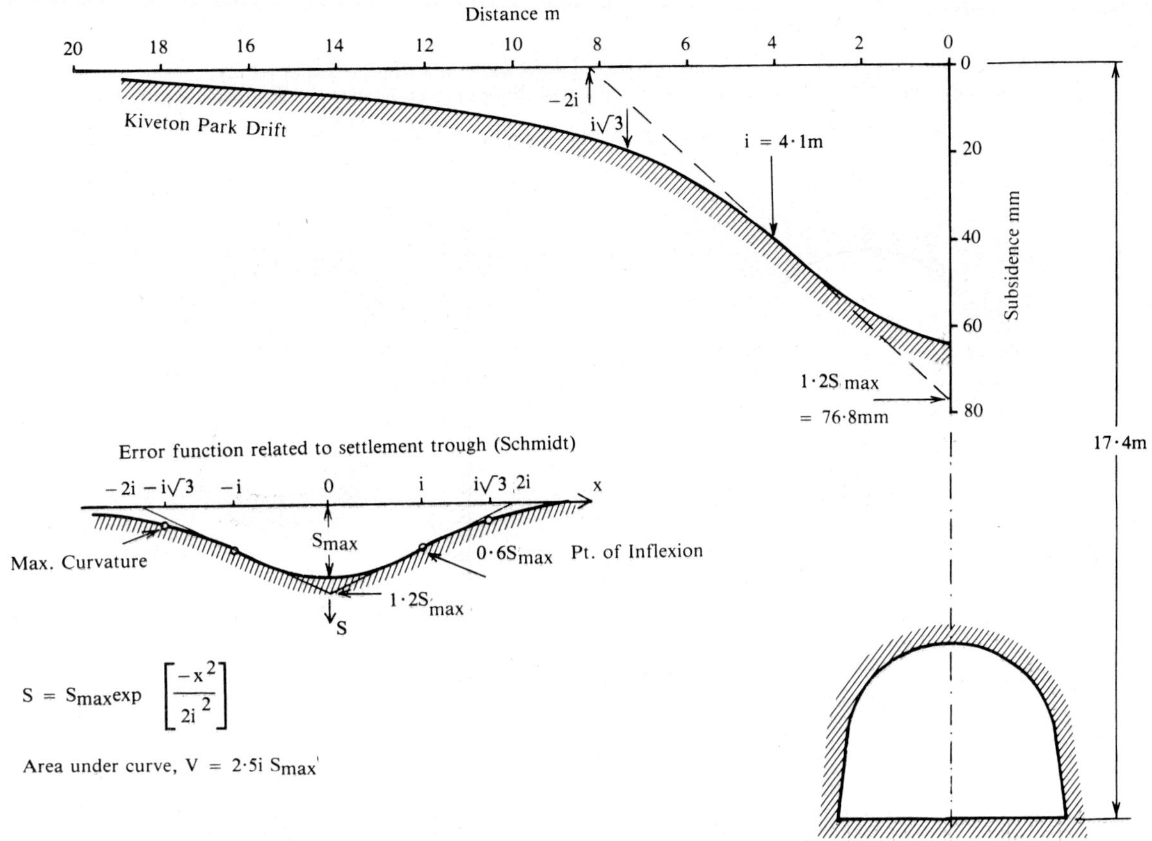

Figure 96 Error function related to Kiveton Park Drift subsidence profile, according to Whittaker and Pye (1977)

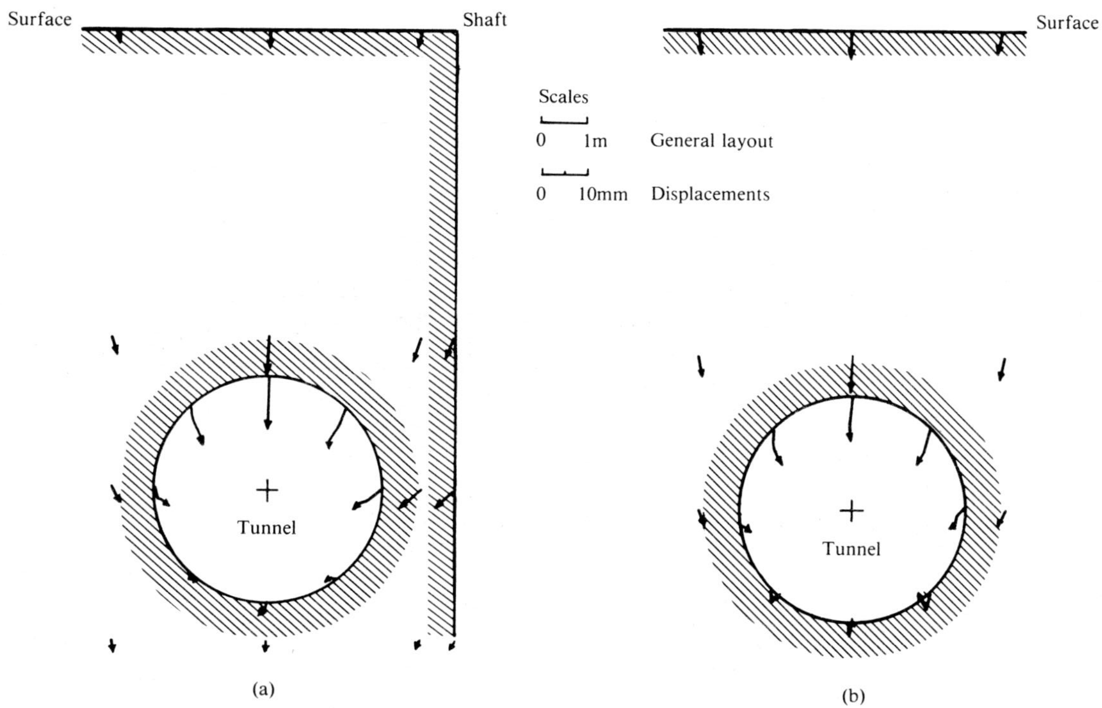

Figure 97 Comparison of displacements on vertical section across fully advanced tunnel: (a) in line with shaft, (b) 7·1 m beyond shaft centre line, after Breckels, McCaul and Whittaker (1980)

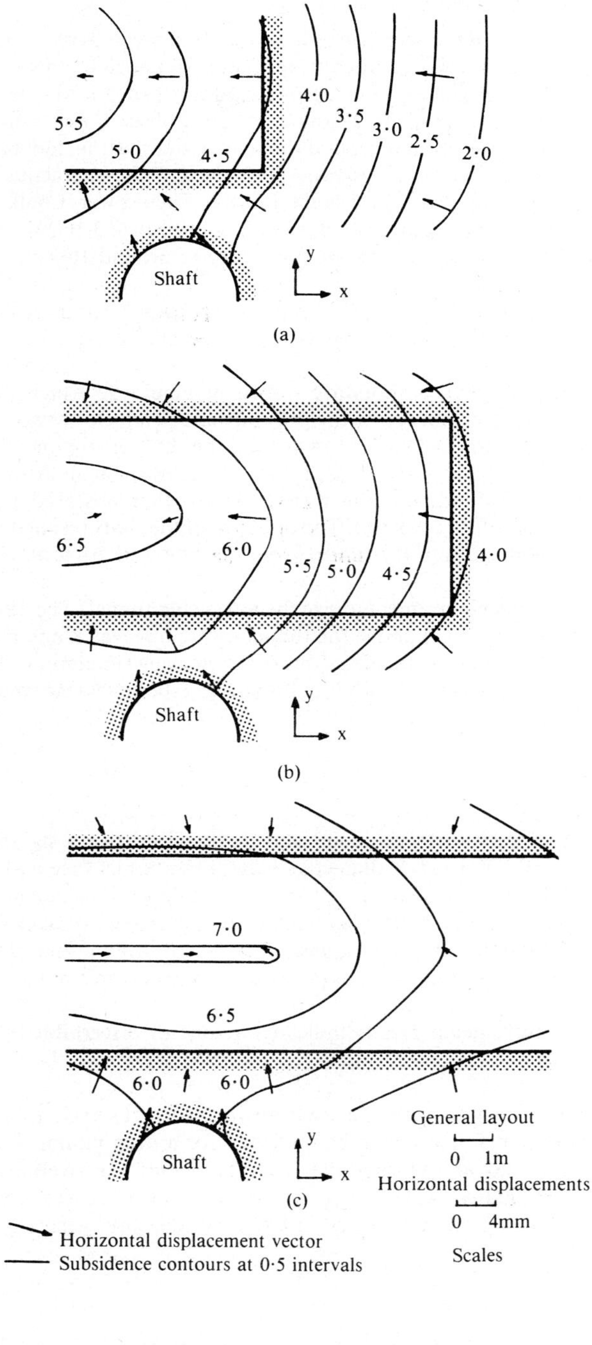

Figure 98 Surface movement above advancing tunnel in Lower Chalk after Breckels, McCaul and Whittaker (1980)

Subsidence Behaviour of a Tunnel in Lower Chalk

Breckels, McCaul and Whittaker (1980) have reported a comparison between measured and simulated displacements which resulted from shaft and tunnel interaction in Lower Chalk at Chinnor, Oxfordshire, England. An experimental tunnel of 5 m diameter was driven in close proximity to a vertical 3 m diameter shaft. Detailed observations were made of actual deformation behaviour of the ground around the tunnel and shaft including the character of surface subsidence. Finite element analysis was performed of the interactive situation for the purpose of comparing simulated with actual results. The Lower Chalk was tested and reported to have an unconfined compressive strength in the range 5-10 MPa, an average bulk density of about 2150 kg/m^3 and a saturated moisture of around 16%.

The shaft was excavated to a depth of 23·4 m and the tunnel was driven within 2 m of the shaft wall. The tunnel axis was approximately 8 m below the surface.

Results published by these authors are shown in Figure 97 which present simulated displacements on a vertical section across the fully advanced tunnel for two positions, namely (a) in line with the shaft, and (b) some 7·1 m beyond the shaft centre line. There is an initial downward movement of the invert indicated, followed by slight uplift after the face has passed. The pattern of ground movement is similar to that observed by Hansmire and Cording (1972) in the case of Figure 97(b). The presence of the shaft resulted in an appreciable increase in ground movement into the tunnel from the side-wall adjacent to the shaft.

Figure 98 shows surface movement above the advancing tunnel. The observed maximum vertical displacement (subsidence) above the tunnel centre-line was 8 mm and this compares favourably with that of 7 mm as indicated by finite element simulation. The effect of the presence of the shaft was indicated to result in broadening the subsidence movements in close proximity to the shaft.

Conclusion

The roof strata of near-surface tunnels and drifts experience a significant subsidence effect within the distance of 1 - 1·5 diameters behind the tunnel face and is regarded as a critical zone which requires immediate and effective support. The quality and timing of temporary and permanent support can play a major role in respect of achieving control over subsidence in such situations. It would appear that where drilling and blasting formation methods are used effective temporary support assumes even greater importance.

The surface drift subsidence results indicated a zone of discernible uplift immediately ahead of the drift face.

The subsidence associated with the surface drift was observed to have an area of influence greater than that which would be predicted for a deep mining situation involving wide extractions. It would appear that the weak nature of the overburden and greater significance of unconsolidated surface deposits results in an increased angle of draw, for example up to 45°, than would be expected in UK Coal Measures generally.

CHAPTER 9

SUBSIDENCE ASSOCIATED WITH ROOM AND PILLAR MINING

This chapter reviews the forms of subsidence behaviour observed with room and pillar mining of stratified mineral deposits. The factors governing surface subsidence behaviour are examined and design procedures are discussed. The longterm aspects of subsidence occurring above such mine workings are commented upon and general discussion focussed on instrumentation and remedial measures available to assist in monitoring and controlling subsidence development.

Introduction

Surface instability above previous room and pillar mining operations is a problem in many countries owing to the apparently unpredictable occurrence of subsidence holes several years after mining ceased. A later chapter deals with the problems of surface instability aspects associated with abandoned mines. Attention is devoted here to the character of subsidence associated with this form of underground mining.

Surface subsidence associated with room and pillar mining of stratified mineral deposits can present special problems. Since this method of mining is applied mainly in relatively shallow (usually less than 300m below surface) conditions, it can result in subsidence features occurring at the surface both during the life of the mine and many years after mining has ceased.

The formation of mining rooms which are designed and supported for stability during the mining operations can, however, provide significant potential for subsidence holes appearing at the surface some fifty years or so after mining ceased. There are many room and pillar mines where surface subsidence has not been a problem, and consequently the geological and mining factors are seen to play a predominant role in this respect.

This chapter reviews the forms of surface subsidence observed over room and pillar mine workings and examines the main factors. Special comments are made on mine design, geological setting and geotechnical features relating to the likelihood of subsidence occurring at the surface above such mines.

Geometrical character of room and pillar mine layouts

The basic principle of room and pillar mining is the systematic extraction of minerals from either naturally or artificially supported rooms with the transfer of cover loading to mineral pillars which are an integral feature of this system of mining. The rooms are generally only supported, if not free standing, for the duration of the mining operations. The pillars are usually designed for effective support of the overburden.

A common early form of room and pillar layout is shown in Figure 99(a), which involved the formation of independent mining districts which permitted isolation if required either during mining or on completion of the extraction operations in that part of the mine. Access to the working districts involved roadways which were required to possess an appreciable

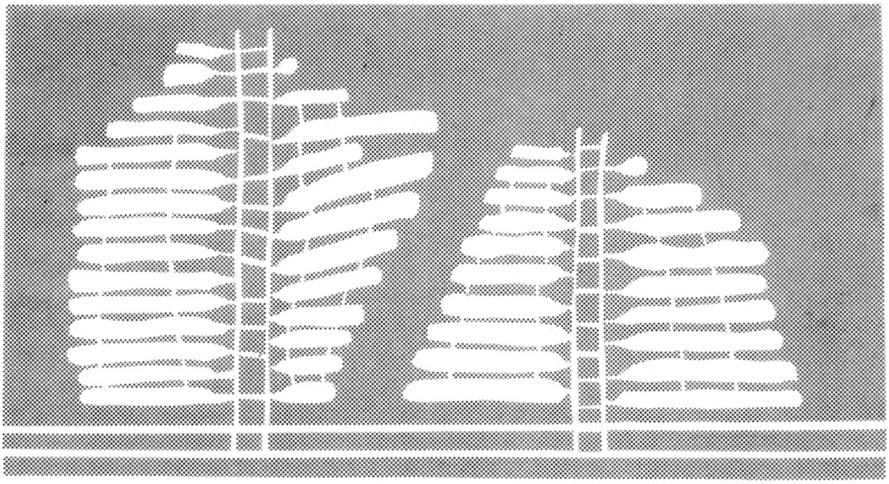

(a) Wide room form of layout.

Note: Rooms arranged so they could be isolated in due course, and each district in turn could be completely isolated within the mine.

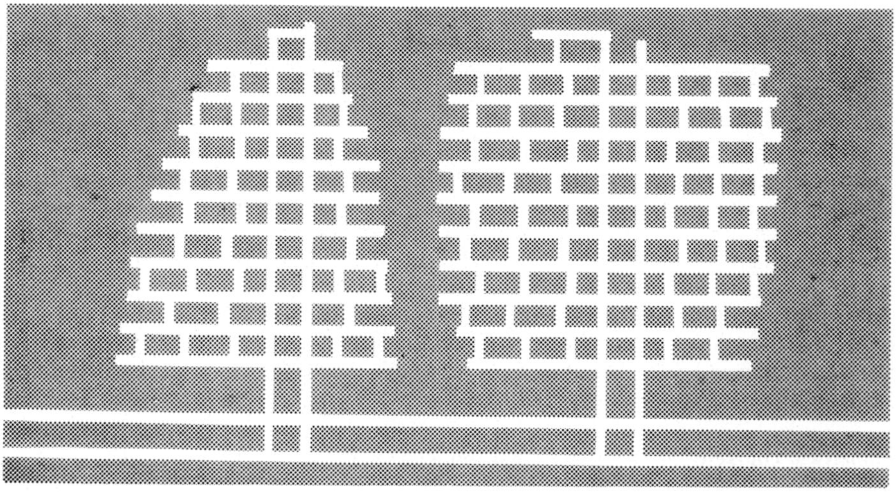

(b) Conventional room and pillar mining districts.

Note: Districts arranged to possess the ability to be completely isolated from the rest of the mine.

Figure 99 Earlier forms of conventional room and pillar layouts.

measure of stability and consequently they were usually narrower than the working rooms. The room widths were usually determined on a trial basis, and the maximum width which could be tolerated consistent with adequate roof control was used. The rib pillars separating the adjacent rooms were not generally designed taking long-term stability into account.

The long-term effect on the potential for surface subsidence above mining layouts similar to Figure 99(a) has been for subsidence features to tend to cluster over the room extraction areas. Where the rib pillars have been sufficiently thin to result in yield then subsidence troughs have frequently occurred. Subsidence holes have developed at the surface over such rooms in circumstances where the depth of cover has been shallow and the room dimensions sufficiently large.

Figure 99(a) is illustrative of a form of room and pillar layout which was very common in coal mining operations earlier than 1950, but the authors have seen it still in use in Australasia and South America during the 1980's.

Figure 99(b) illustrates a form of layout based on that given in Figure 99(a) but shows a greater regularity in width of rooms and roadways in order to cater for increased mechanisation of the mining operations. Development of the mining districts still tended to be of a fairly irregular form, with geological faulting and other disturbances and variations frequently dictating the limit of many districts. The principle of isolating mining districts however was still retained. Pillars between mining rooms tended to possess greater stability than in the form of layout shown in Figure 99(a). Subsidence effects on the surface have been mainly in the form of holes developing over the junctions.

Other common forms of room and pillar layout are shown in Figure 100(a) and (b), with rhomboidal shaped pillars in the former and square pillars in the latter. Both layouts have been designed to facilitate the application of highly mechanised mining operations. The rhomboidal shaped pillar layout assists the negotiation of corners by large capacity mineral transport vehicles en route to discharge points on conveyor networks. Figure 100(b) shows the use of square pillars but these could have been rectangular in order to reduce the number of junctions.

Subsidence holes are the main form of surface instability with such room and pillar layouts, although trough subsidence is fairly common also.

Figure 101(a) shows the staggering of pillars to produce 3-way ('T' type) junctions as opposed to 4-way types since the latter is at appreciably greater risk to potential roof instability owing to the greater span and higher concentration of rock stresses in the immediate roof. Consequently 4-way junctions represent greater surface subsidence potential than 3-way types.

Figure 101(b) shows irregular pillar configurations which can be common with certain types of minerals where there are major changes in the quality and thickness of the mineral bed being mined. Those areas with the greatest exposed roof spans, limited pillar support and having large junctions represent areas of greatest potential subsidence. Surface stability over such room and pillar layouts can take the form of holes appearing over junctions or a more uniform subsidence depression over a wide area owing to pillar failure, or failure of the immediate roof causing the pillars to punch into the overlying beds.

Small pillars in such irregular pillar configurations can yield and the strata loading thereby transfers to adjacent larger pillars possessing significant stability. The smaller width of these irregular pillars plays a major role regarding overall strength.

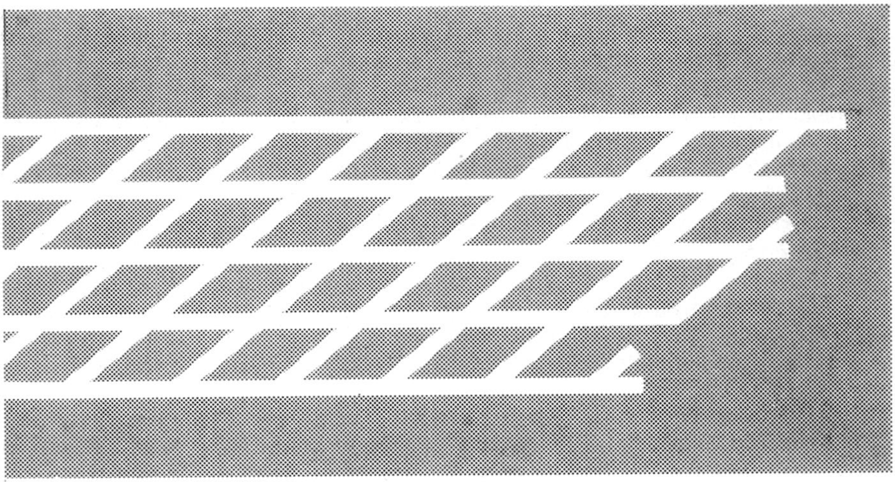

(a) Plan view of rhomboidal pillar layouts to facilitate negotiation of corners by large mining vehicles. Note: With weak minerals, the acute pillar corners would readily fail, thus increasing the area of exposed roof at junctions. This form of layout has mainly found favour in strong mineral beds with appreciable working height.

(b) Systematic development of room and pillar districts from main access roadways involving square pillars and 4-way junctions to maximise operational efficiency.

Figure 100 Common forms of room and pillar mine layout.

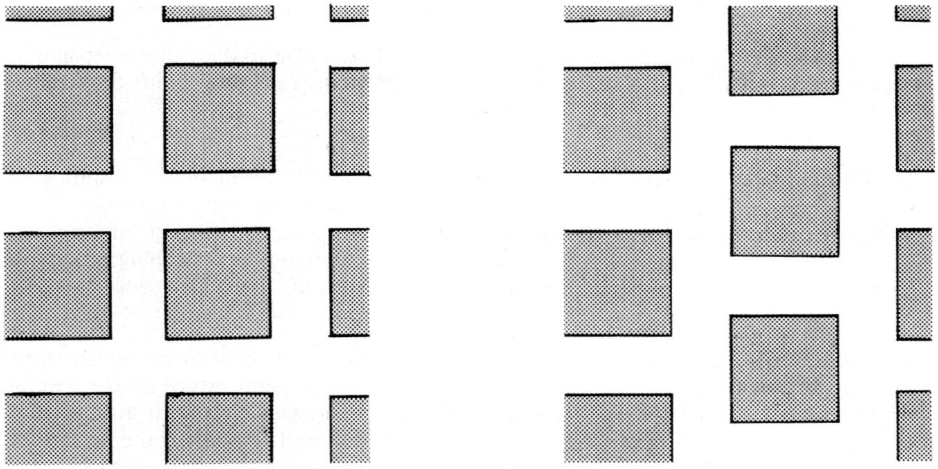

(a) Comparison of symmetrical and staggered mine pillar layouts.

Notes: 1. The 4-way junction layout pattern possesses ease of mining operation but increases the potential for subsidence above junctions.

2. Staggering of pillars to produce 3-way junctions promotes increased stability of room intersections and decreases the potential for subsidence.

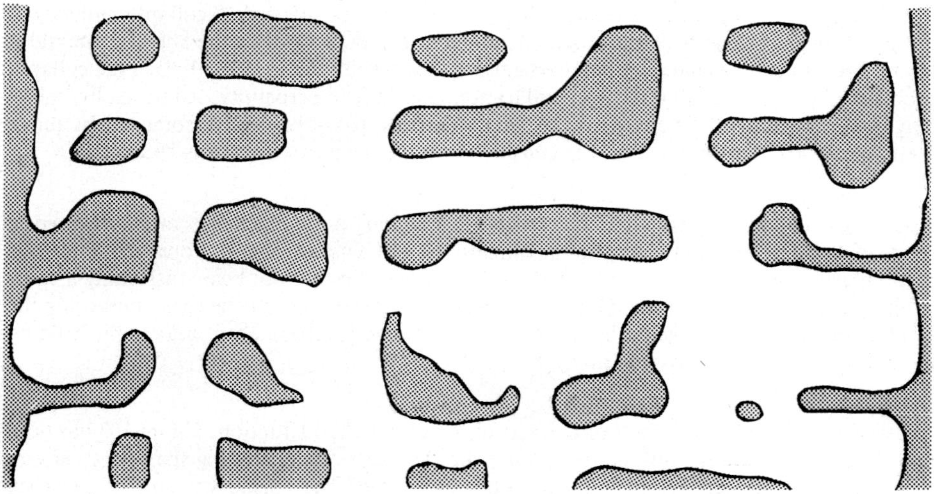

(b) Irregular pillar configurations.

Note: These are common with certain types of minerals where quality variations are significant and major variations occur in thickness of the mineral bed.

Figure 101 Staggering of junctions and irregular mine pillar configurations.

Factors influencing the stability of room and pillar mine workings

Roof collapses in room and pillar mine layouts are related either to the degree of span of a room exceeding the natural stability of the roof or to instability caused by failure of the mine pillar(s).

Roof beam failure. Figure 102 shows an example of roof beam failure resulting in subsequent collapse of the overlying rock. It is common practice to retain a thickness of the mineral bed at roof level to achieve improved support where a weak rock exists immediately over the mineral being mined. Roof failure is frequently initiated by a slabbing process in sedimentary bedded mineral formations and this can weaken the immediate roof beam.

The roof beam failure process is considered to be mainly as a result of the tensile stress magnitude generated in the roof beam by excavation of the room, exceeding the natural strength of the rock constituting the beam. Figure 102(a) shows the result of a failed roof beam causing weaker rock above the mineral bed to fail also and accompany the fallen roof beam. The weak rock overlying the mineral bed appears to fail mainly as a result of weak shear strength and low cohesion in the case of overlying clays.

Examination of several such cavities by the authors, has revealed a fairly circular collapse region with an approximately flat roof. Progressive upward movement of the collapse process is triggered by shear failure at the corners coupled with detachment of rocks within the central area of the developing hole.

The progress of upward movement of the failure process of such a collapse chimney is governed by the strength of the rocks encountered en route towards the surface. Should a strong bed of massive character be intersected as is shown in Figure 102(b), then the collapse process can be halted at that horizon and can remain there in perpetuity. Additionally, there may be sufficient depth of cover for the collapse process to choke itself at some height above the mineral bed. This latter aspect is considered in the next section of this chapter.

Roof beam stability assessment. The roof beam spanning a mining room can be assessed in terms of stability taking into account its dimensions and tensile strength properties. The basis of comparison is the calculated tensile stress induced within the roof beam, assuming it to be subjected to the laws of elasticity, and the natural tensile strength of the rock composing the beam. For a single drivage without room intersections, the roof beam maximum bending moment is as expressed by equation (48).

Using equation (51) in respect of a practical room and pillar situation, Figure 103 has been produced to illustrate the influence of roof beam thickness on the limiting span (w) to ensure that the maximum tensile stress (σ_{tens}) does not exceed the tensile strength of the rock forming the roof beam. A range of rock tensile strength values, as determined by laboratory testing, is shown for a limestone host rock for the mineral type considered here. However, the tensile strength values shown are typical values for roof types in a wide range of sedimentary mineral deposits.

Figure 103 shows that as the roof beam thickness decreases below around 1 m, then there is a considerable steepening in the rate of reduction of roof beam span to satisfy stability.

$$M_{max} = qw^2/12 \qquad ...(48)$$

$$\text{where} \quad qw = (\gamma b)\, dw \qquad ...(49)$$

$$\text{and} \quad \sigma = M\, y/I \qquad ...(50)$$

$$= \left(\frac{\gamma b\, dw}{12}\right) w \; \frac{12}{b\, d^3} \; \frac{d}{2}$$

$$\text{giving} \quad \sigma_{tens} = \frac{\gamma w^2}{2\, d} \qquad ...(51)$$

The symbols given in equations (48) to (51) are explained as follows:

M_{max} = the maximum bending moment for an elastic beam assuming the built-in condition,
q = the unit load of the roof beam,
w = the roof beam span (beam length),
qw = the total load of the roof beam
σ = the tensile stress in the beam due to bending
y = the position of the tensile stress value (σ) relative to the neutral axis of the beam,
I = the moment of inertia of the beam which is given by $bd^3/12$ for a rectangular beam about the axis considered here,
σ_{tens} = the limiting tensile stress for stability in relation to the dimensions of the roof beam,
γ = density of the roof beam material,
d, b = cross-sectional dimensions of the rectangular roof beam: d = depth of beam whilst b = breadth of beam.

Pillar strength aspects. Pillar strength is a major consideration in the design of room and pillar mine layouts in view of the need to satisfy safety against large scale failure which could arise from over-loading the pillars. The strength of coal pillars has been investigated by Salamon (1967) in South Africa, and his formula is given by equation (52).

$$\text{Pillar strength} = \frac{7180\, p^{0 \cdot 46}}{M^{0 \cdot 66}} \qquad ...(52)$$

where

pillar strength unit = kN/m^2,

p = side length of square pillar in metres,

M = room height in metres.

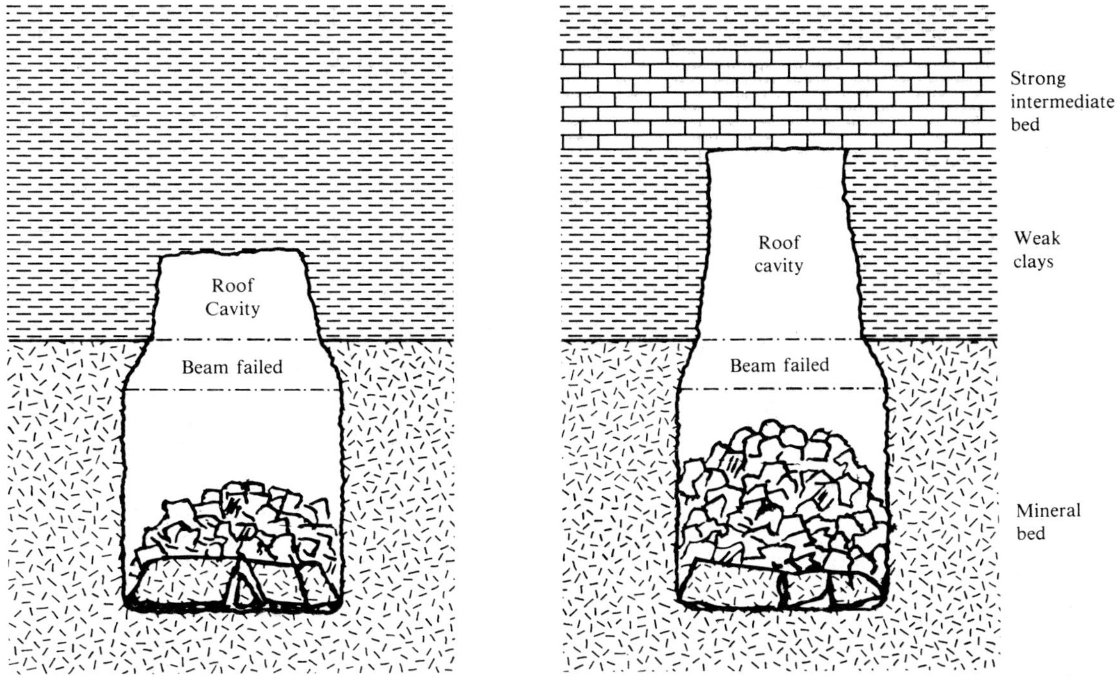

(a) Collapse of roof beam accompanied by fall of weak overlying clay.

(b) Progress of upward failure arrested by strong intermediate bed.

Figure 102 General characteristics of collapsed roof beam and subsequent cavity development.

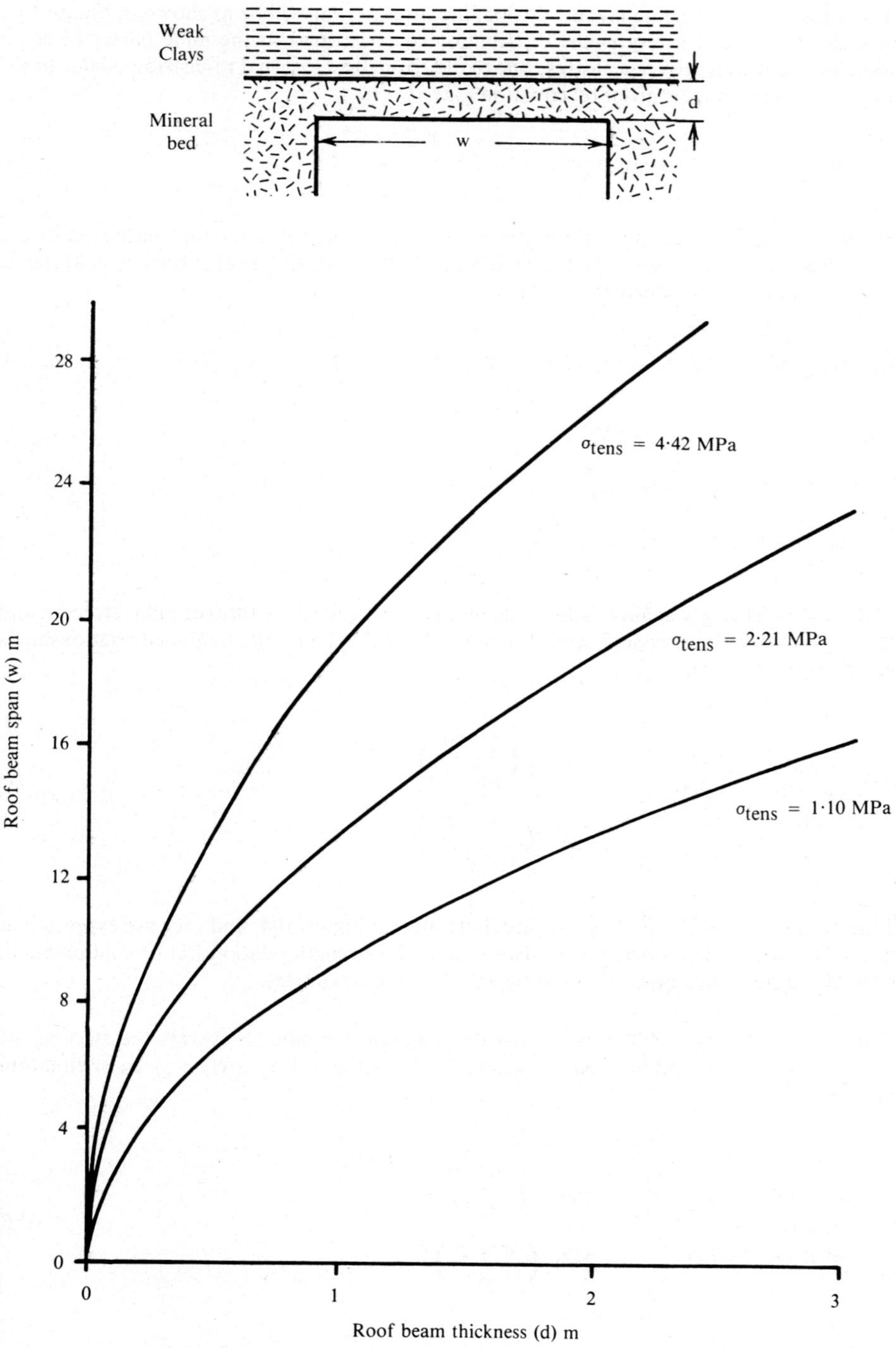

Figure 103 Self-supporting characteristics of British ironore in relation to beam span and thickness for different tensile strength values.

For a square pillar mining layout, the loading of one pillar will be as shown in Figure 104, and this assumes uniformity in the distribution of cover-load to the mine pillars of equal dimensions. The average vertical stress due to the cover-load and that re-distributed due to the presence of the rooms, is given by equation (53).

$$\text{Average vertical stress on pillar} = \gamma\, g\, h \left(\frac{w + p}{p}\right)^2 \quad \text{...(53)}$$

In case of the coal bearing sedimentary rocks an average value for the density (γ) of the cover rocks is 2350 kg/m^3. Using this latter value for density and a value for g of 9·81 m/s^2, then equation (53) can be expressed as (54).

$$\text{Average vertical stress on pillar} = 23\, h \left(\frac{w + p}{p}\right)^2 \ \text{kN/m}^2 \quad \text{...(54)}$$

where

p = pillar side length

w = width of mine room.

Pillar factor of safety against failure. Salamon has expressed the ratio of pillar strength and pillar stress to provide an assessment of the factor of safety, and this combined relationship is given by equation (55).

$$\text{Factor of safety against coal pillar failure} = \frac{7180 \left(\dfrac{p^{0{\cdot}46}}{M^{0{\cdot}66}}\right)}{23\, h \left(\dfrac{w + p}{p}\right)^2} \quad \text{...(55)}$$

All the symbols given in equation (55) are illustrated in Figure 104, and they are expressed in metres. This formula was derived for determining the strength relationship of coal pillars in South Africa but it has been found to be valid in other countries.

Whittaker and Smith (1987) have used the Salamon formula to express the strength of British ironore pillars, and their basic relationship for the factor of safety is given by equation (56).

$$\text{Factor of safety against failure of ironore pillar} = \frac{19450 \left(\dfrac{p^{0{\cdot}46}}{M^{0{\cdot}66}}\right)}{23\, h \left(\dfrac{w + p}{p}\right)^2} \quad \text{...(56)}$$

The value of the constant in the pillar strength part of the relationship has been changed on the basis of strength tests in order to allow its extension to assessing the strength of ironore pillars as opposed to the coal pillars for which it was first established.

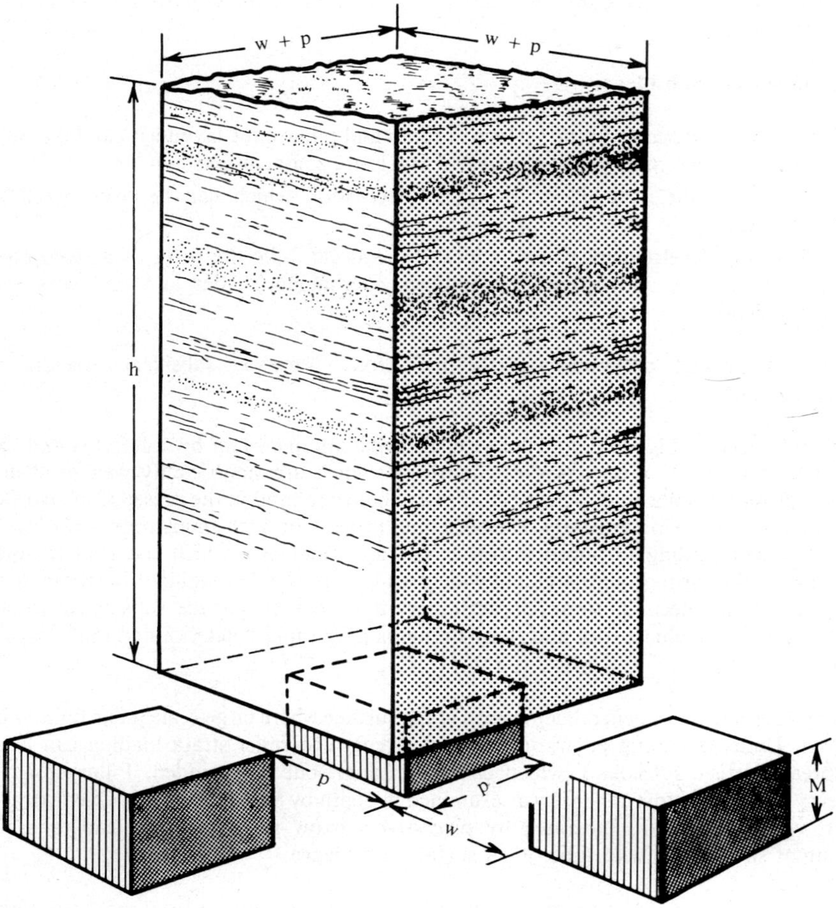

Load on pillar = $\gamma gh\ (w + p)^2$

Pillar vertical stress = $\gamma gh\ (w + p)^2\ /P^2$

Pillar strength = $7180\ p^{0 \cdot 46}/M^{0 \cdot 66}(kN/m^2)$

Figure 104 Pillar stress and strength relationships for room and square pillar workings in coal.

Figure 105 presents a graphical representation of equation (56) to indicate how the factor of safety against pillar failure is influenced by the magnitude of the square pillar side dimension for two values of depth below surface.

The factor of safety against pillar failure generally used is in the range 1·3 to 1·5 for long term stability. Lower values of factor of safety can be employed when the parameters used are accurately known and where the rib pillars (or barrier pillars) are employed to carry some of the cover load.

Forms of Surface Subsidence

Where subsidence occurs at the surface in room and pillar mine layouts it can take various forms. There are two main forms, however, which are characterised by:

(i) a general saucer-shaped or elongated depression which can be over a relatively extensive area, and

(ii) sink-hole development which is localised and can take the form of a steep conical depression or a hole with vertical (sometimes overhanging) sides and on occasions of appreciable depth.

Figure 106 presents a summary of the main trough forms of subsidence together with relevant comments.

With reference to Figure 106(a), many coal seams are underlain by seatearths and clays, and some immediate roofs consist of mudstones which are significantly weakened in the presence of water. Some under-clays can rapidly change to mud in the presence of significant quantities of water. Consequently, the softening process of such mudstones and clays can result in pillars tending to punch into the roof and floor and which can lead to surface subsidence in the form of a general and widespread depression. Flooding of coal mines after abandonment has been in some circumstances a source of surface subsidence in those conditions where the local roof and floor strengths have rapidly deteriorated in the presence of water.

Figure 106(b) shows a general depression at the surface where large-scale pillar failures have occurred. Under-designing pillars in terms of strength to resist strata loading can lead to widespread surface subsidence which can be gradual, but also sudden. Pillar design for longterm stability against collapse was examined in depth by Salamon (1967). Over extracting layouts where rooms are separated by successive narrow rib pillars can lead to gradual crushing of such pillars with subsequent surface subsidence.

A common cause of subsidence in room and pillar layouts is localised roof failure, especially over mine junctions, resulting in migration of rock over the pillars into the mined-out excavations, see Figure 106(c). Substantial widespread surface depressions can occur over the affected areas.

The most common form of surface subsidence associated with room and pillar workings is that of sink-hole development, and is illustrated in Figure 107. The mine room intersections are the locations at greatest risk to roof collapse owing to the creation of high induced stresses in the immediate roof beds and the fact that the largest expanses of exposed roof occur here also.

Consequently, it is more common to observe roof instability at 4-way junctions than 3-way (T) junctions whilst the rooms themselves do not generally give rise to collapses which can

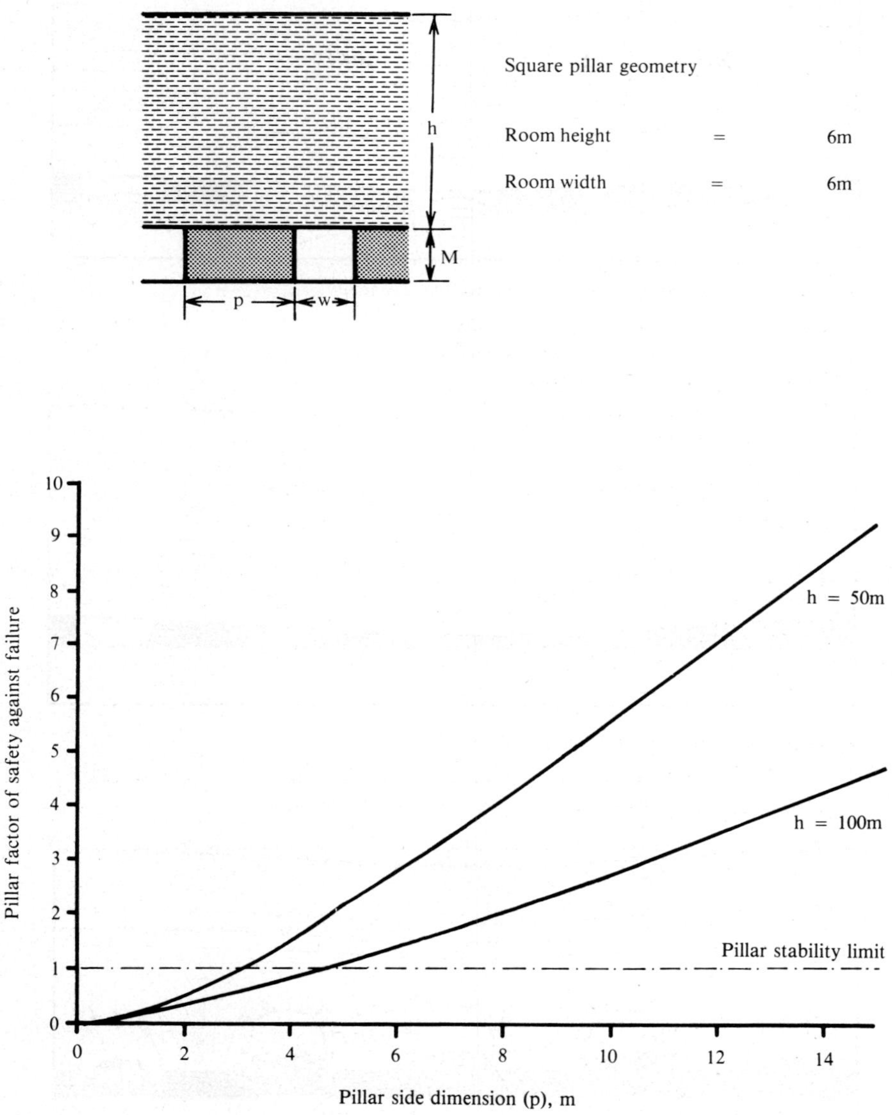

Figure 105 Illustrating influence of pillar side dimension on pillar factor of safety against failure for British ironore, after Whittaker and Smith (1987).

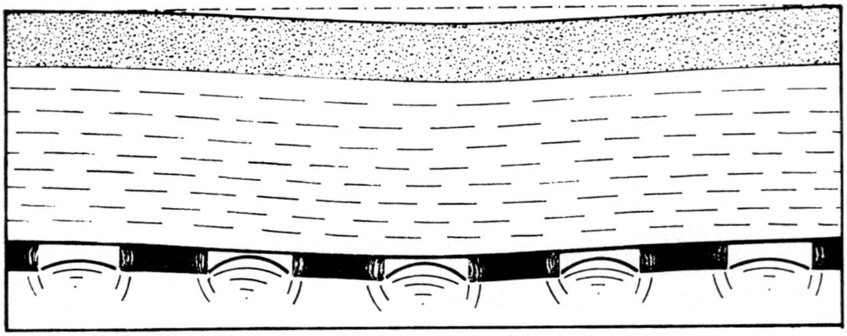

(a) Surface subsidence trough formation due to floor instability.

Notes: 1. Water affects under-clay mine floor.

2. Resulting floor softening encourages punching of pillars into floor (and occasionally the roof).

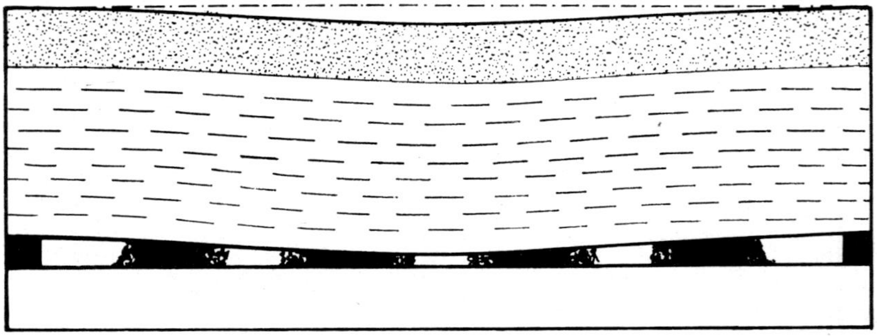

(b) Subsidence trough formed by pillar failure.

Notes: 1. Progressive spalling of pillars and gradual crushing can give risk to subsidence.

2. Such effects generally exhibit long term subsidence development.

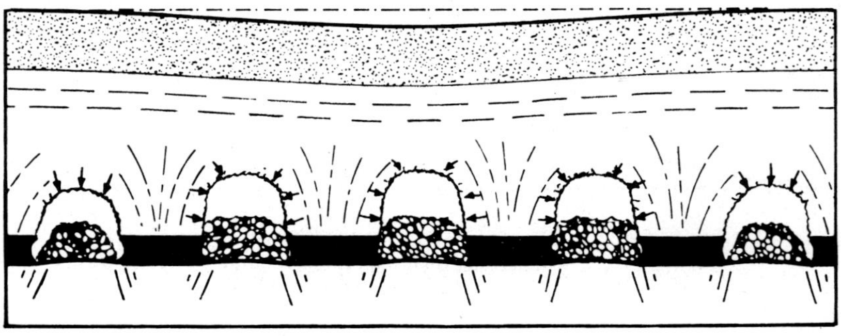

(c) Subsidence trough resulting from migration of failed roof strata.

Notes: 1. Weak rock overlying mined out area migrates into rooms.

2. Failure process is generally preceded by roof failure at room intersections resulting in loss of confinement of rock above the pillars.

Figure 106 Occurrence of trough formation over room and pillar mine workings.

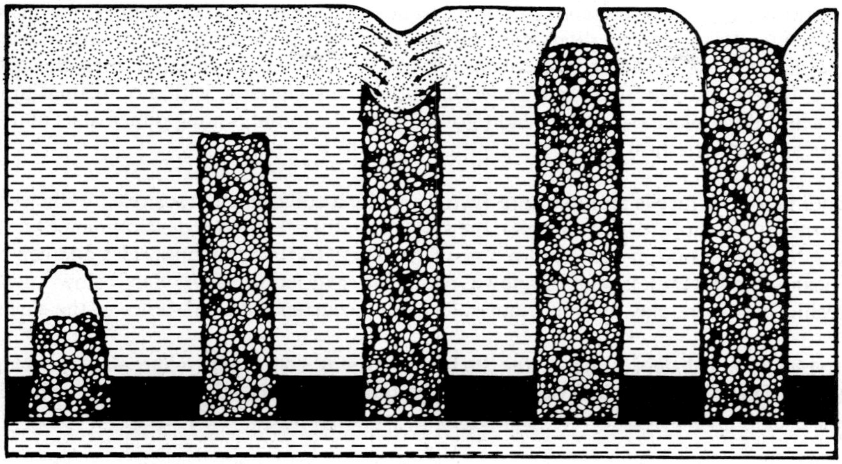

Figure 107 Sink-hole development over room and pillar mine workings

Condition	Comments
1.	Collapsed mine junction results in creation of void into which further rock can fall, providing it is sufficiently weak and stress conditions high enough to induce failure.
2.	Collapse chimney has resulted in stabilisation by self-choking; the natural bulking of the caved rock has allowed cessation of the upward collapse process within the depth of cover.
3.	The collapse chimney has reached the surface by either (a) shallow depth of cover or (b) caved material has flowed into the mine. Conical depressions of this form are very common.
4.	The collapse chimney has developed beyond that shown in condition 3, owing to shallow depth or inflow into the mine. Daylighting of the collapse chimney can result in sudden appearance of a surface cavity with overhanging sides.
5.	This can be a more mature stage of condition 4 with accompanying erosion of the sides, or it can be the result of weak surface materials slumping into the subsidence hole.

result in sink-holes appearing at the surface, except in very shallow conditions close to outcrops. Collapsed junctions can, if the overlying beds are sufficiently weak, promote an upward movement of the collapse-chimney as a draw phenomenon. Naturally, if the collapse-chimney intersects a massive sandstone horizon, then the likelihood is that upward movement will be arrested. If, however, the collapse process is allowed to continue then the natural bulking of the broken rock can result in subsequent choking with no further upward movement taking place beyond a height governed by the initial mining dimensions and local geological characteristics.

Depth below surface is an important factor. Where a collapse-chimney reaches the surface it is termed a sink-hole (or crown-hole); the surface can be affected by development of a sharp conical depression or appearance of a hole with vertical (or overhanging) sides and can be of significant depth (15m is not unknown for a 6m extraction at 70-90m below surface). Geotechnical factors play an important role in the occurrence of sink-holes at the surface.

A typical conical depression at the surface overlying room and pillar mine workings is shown in Figure 108. Conical depressions are most likely to occur where an appreciable thickness of top-soils and unconsolidated materials exist at the surface.

Conversely, Figure 109 illustrates a recently formed sink-hole with 4 - 5m of sand and top-soil overlying the bed rock. The sink-hole has sheared through to the surface with a tendency to allow a certain amount of overhang to occur.

A mature sink-hole is shown in Figure 110 where the almost circular hole enveloped trees and carried them into the mine (70m below). The surface maintained its almost vertical sides owing mainly to the very thin layer of top-soils and unconsolidated materials. The fairly competent bed rock can be clearly seen within the sides of the sink-hole. Water had played an important role in washing out caved rock within the collapse chimney and this resulted in the sudden appearance of a hole at the surface.

Sink-hole appearances commonly take the form shown in Figure 111 where shallow conditions exist and there is insufficient depth for the collapse chimney to become choked by natural bulking of caved rock. The sudden appearance, or daylighting, of the collapse chimney frequently gives rise to the stepped nature of the subsidence hole shown. The size of the hole appearing at the surface is related to the dimensions of the mining rooms.

Figure 112 also shows a sink-hole which suddenly appeared at the surface. The hole was 15m deep and about 30m across at its widest point on the surface; it was funnel shaped since the surface soils readily slipped into the sink-hole. An interesting feature is the presence of water in the sink-hole and was accounted for by the generally low-lying region where the hole appeared and the water table being very close to the surface. The mine was still operating when this sink-hole appeared at the surface, and it indicates that the collapse-chimney must have sealed itself otherwise the pond formed in the sink-hole and seen in the photograph would have flowed into the mine.

Consideration of Height of Caving of Collapse-Chimney above Room Intersection

When seeking assessment of the height to which the cavity above a collapsed junction may extend, the basic problem can be considered with reference to Figure 113 which shows the relevant stages of development of the caved roof strata. The diameter (D) of the collapse-chimney is considered as firstly $D = w$, and secondly $D = w\sqrt{2}$.

Figure 108 Conical depression at surface overlying room and pillar mine workings, Humberside, UK.

Figure 109 Recently formed sink-hole overlying room and pillar mine workings, Humberside, UK.

Figure 110 Fully developed sink-hole of approximately 20m x 20m maximum dimensions, Humberside, UK.

Figure 111 Common form of sink-hole appearance in shallow conditions, Lincolnshire, UK.

Figure 112 Sink-hole formation with immediate surface ponding.
Courtesy of British Steel Corporation.

Basic symbols relating to mine junctions

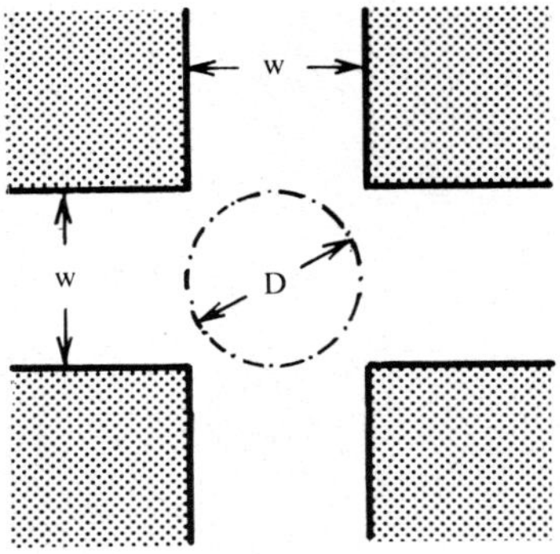

Chimney diameter assumptions

(a) $D = w$

(b) $D = w\sqrt{2}$

Development of caving above junction

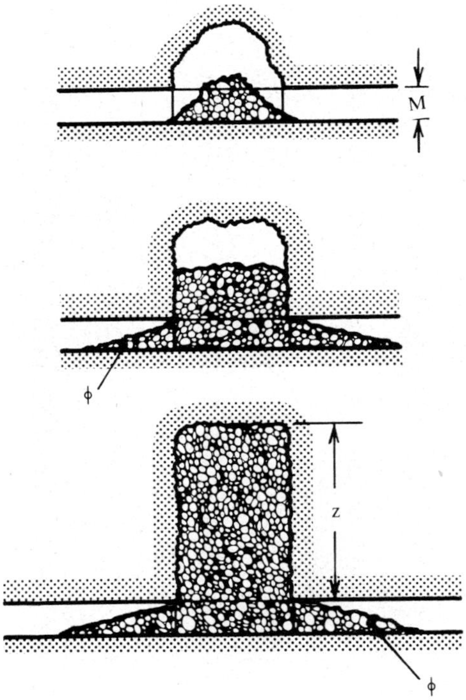

1. Initial collapse condition of mine junction.

2. Progressive development of roof collapse with caved rock spilling into adjoining rooms to roof level.

3. Caved rock fills remainder of chimney during self choking process.

Figure 113 Basic considerations in development of caving above 4-way junction

The initial collapse of the mine juction leads to caved rock spilling into the adjoining rooms forming an angle of repose (ϕ). On reaching roof level, the caved material fills the remainder of the collapse-chimney by virtue of its natural bulking characteristics.

$$V_{caved} = k z \pi D^2/4 \qquad (57)$$

$$V_{space} = 4(\tfrac{1}{2} w M^2 \cot \phi) + M w^2 + z \pi D^2 / 4 \qquad (58)$$

Equating (57) and (58)

$$z = \frac{4}{(k - 1)\pi D^2} \left\{ 2 w M^2 \cot \phi + M w^2 \right\} \qquad \ldots(59)$$

where, V_{caved} = volume of caved rock from collapse-chimney

V_{space} = volume of available space to receive caved roof rock

k = bulking factor, considered to be in the range 1·33 to 1·5

z = height of collapse-chimney

D = diameter of collapse-chimney

w = width of mine rooms

M = excavated height of mine rooms

ϕ = angle of repose of caved rock within mine rooms adjoining collapse area.

In equations (57) and (58), the rock density in its solid state is assumed to remain uniform and the bulking characteristics are also assumed to be consistent and remain unaffected by loading created during the extension in height of the collapse-chimney. It is considered that the increased loading on broken material at the base of the chimney will produce insignificant change in volume during the process of caving to the full height of the chimney.

The diameter of the chimney has been observed by the authors to be generally in the range of $D = w$ and $D = w\sqrt{2}$.

Figure 114 shows data plotted for a practical range of room width (w) using equation (59). Considering room widths to be in the range of 4 - 8m, the maximum height (z) of the collapse chimney is indicated to be in the range 3 - 9 (extraction height). The height of extraction used in this example is 3m and corresponds to a common operating height. Figure 114 suggests that narrow rooms could result in an increased height of caving, but this must be considered in relation to the fact that a narrower room is less likely to collapse. The most common room width in coal and most other stratified mineral deposit room and pillar mines is 6m, and Figure 114 indicates a caving height of 3·5 - 7M (extraction height).

Figure 115 shows graphical representations of some of the important variables influencing the height of the collapse chimney. Figure 115(a) indicates that the caving height (as a multiple of M) increases appreciably with increasing extracted mine room height. The collapse-chimney diameter (D) is seen to be a significant factor regarding the projected height (z) of the caving.

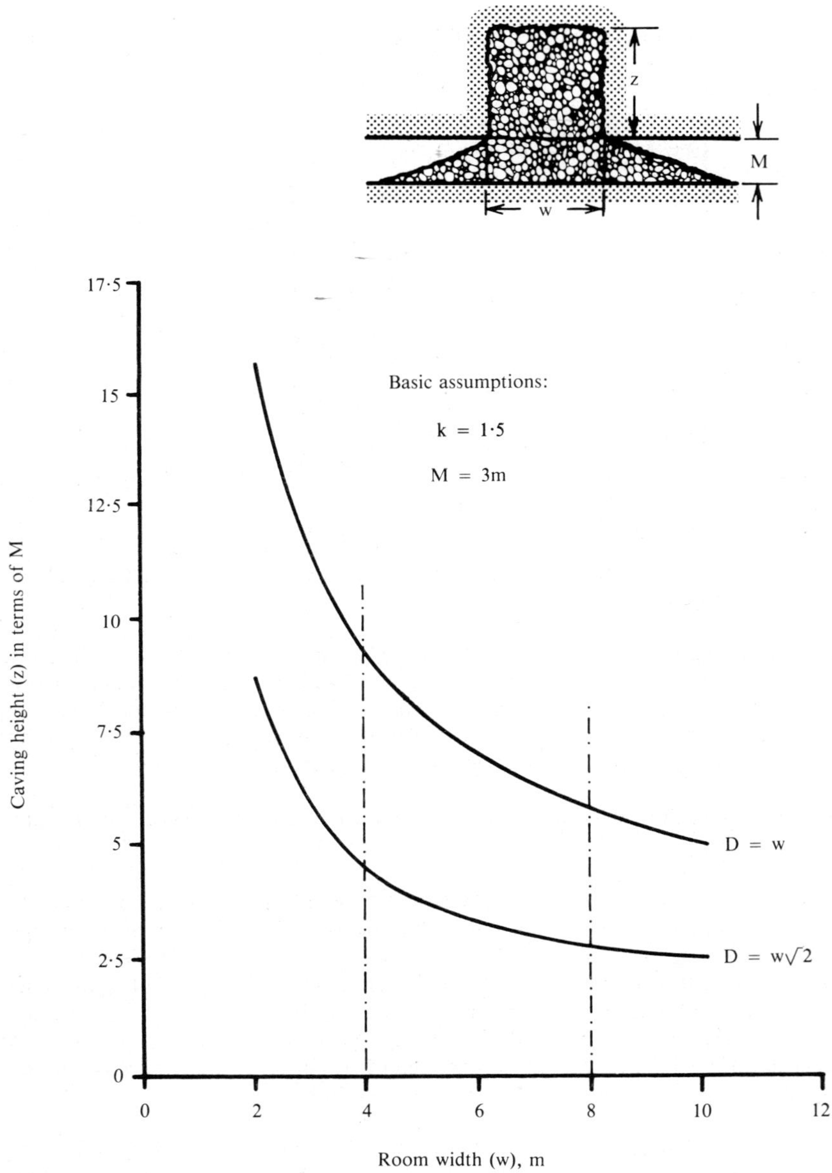

Figure 114 Relationship between caving height and room width for 4-way junction

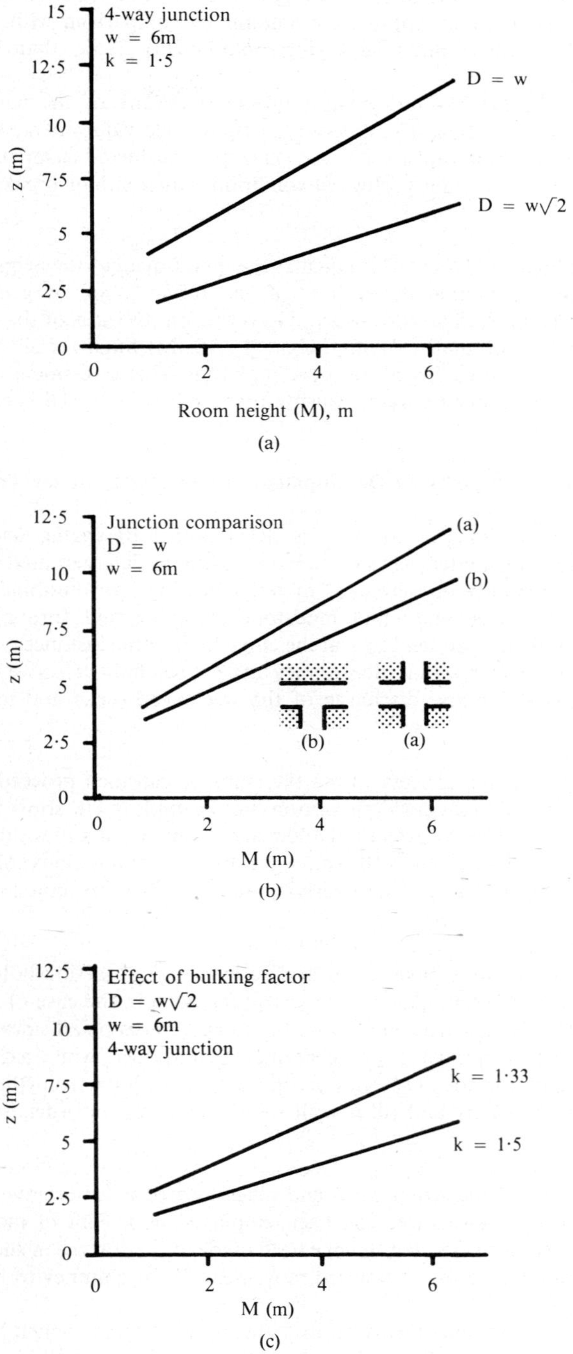

Figure 115. Influence of room height (M) on height of caving (z)

Figure 115(b) shows as would be expected that the 3-way mine room junction decreases the projected caving height by an appreciable amount by comparison with the 4-way type of junction, especially in those mines employing room heights greater than 3m.

The bulking factor (k) has the greatest influence of any of the variables and this is demonstrated in Figure 115(c). The diagram compares two values of k; for $k = 1{\cdot}5$ this is regarded as a realistic value whilst for $k = 1{\cdot}33$ this is considered as an extreme value which could be encountered in certain geological conditions where minor breakage occurs as with strong rocks.

Comparison of Figures 114 and 115 indicates that assessment of the height of caving should take into account the mining dimensions of the rooms, type of geology and bulking characteristics of the immediate roof strata. However, on the basis of the assessment carried out here it would appear that a caving height 4 - 10 (M) might occur but with increased likelihood of a height up to 7M being more probable. This assessment does not take into consideration the possibility of water causing the chimney of caved rock to flow into the mine.

Influence of Overlying Aquifer on Development of Collapse-Chimney Towards Surface

If a collapsed mine junction allows the development of caving and formation of a subsidence chimney which intersects an overlying aquifer as demonstrated in Figure 116, then there is a strong possibility of wet caved material flowing into the mine. Encountering an aquifer can lead to caved clays and mudstone softening and forming material of the consistency of mud which can tend to seal the chimney in some instances; this can promote a build-up of a head of water within the chimney and the resultant effective pressure at the seal may cause sudden release and discharge of the wet caved rocks and mud into adjoining rooms.

Consequently, the caving process within the chimney can then proceed to heights well in excess of those predicted in the previous section. For example in UK stratified ironstone room and pillar mining operations at a depth of 100m and room height and width of 6m, sink-holes have appeared at the suface after 3 - 10 years since the original junctions collapsed, but only in direct association with water from an overlying aquifer which prevented natural choking by bulking of the caved roof rocks.

Where collapsed junctions have posed a significant risk to a sink-hole appearing at the surface by virtue of intersecting an overlying aquifer, then in the case of an operating mine suitable walls have been constructed to dam the collapsed area on all sides. This preventative measure has been highly successful. In exceptional circumstances where a chimney was known to be approaching important properties at the surface, drilling from the surface has been carried out to tap the cavity and fill it with suitable material in order to arrest its upward movement.

Instrumentation involving strain wires and plugs located in a borehole over a subsidence chimney approaching the surface, has been employed as a control measure; strain wire readings using an extensometer device have been made at the surface in such situations. Such readings have allowed progress of upward movement of the chimney to be observed.

Figure 117 shows diagrammatically an instrumentation scheme which has been used over collapsed mine junctions in order to observe whether further instability is taking place which could result in a sink-hole appearing at the surface. Anchor plugs are secured at selected horizons within a borehole and attached strain wires are brought out to the surface where they

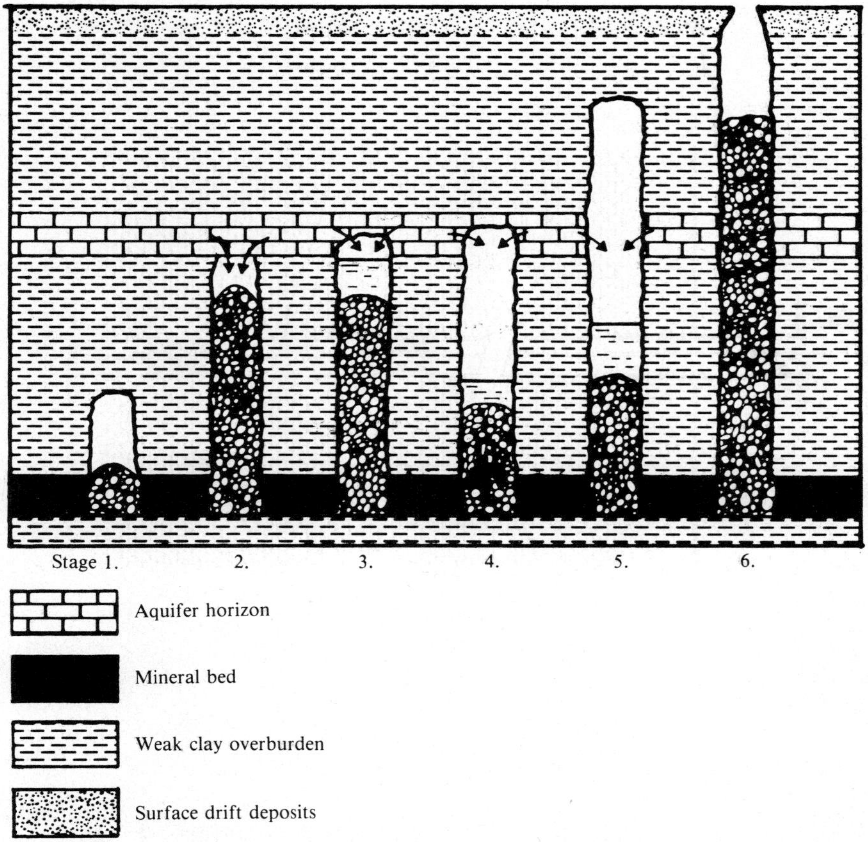

Sinkhole development stages

1. Early stage in development.
2. Chimney taps aquifer horizon and allows water to drain into cavity.
3. Build-up of water pressure by ponding can promote instability of plug of caved rock and result in flowing into mine.
4. Caved mass flows into adjoining mine rooms.
5. Chimney continues caving due to increased available space for broken rock.
6. Sinkhole can emerge at surface from considerable depths (greater than 10M).

Figure 116. Influence of chimney intersecting overlying aquifer and changing potential caving height above collapsed mine junction.

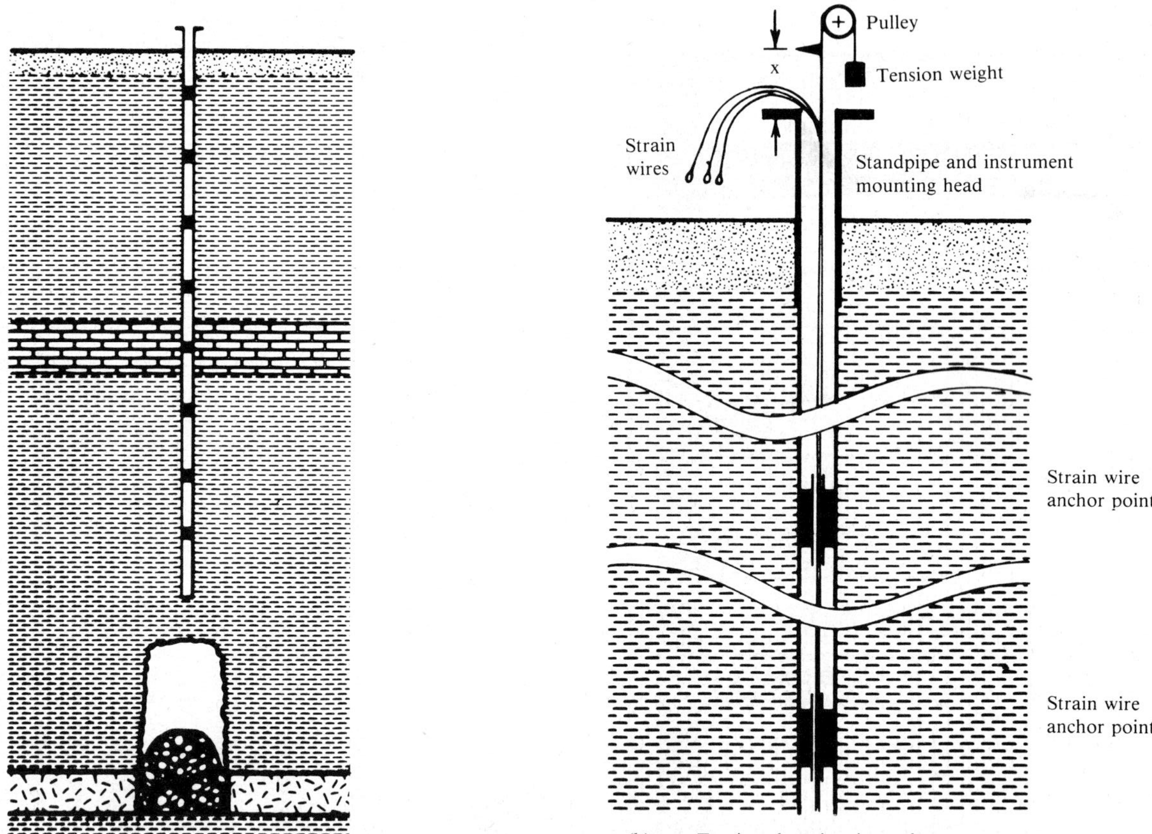

(a) Borehole location with strain wire anchorages.

(b) Tensioned strain wire and basic displacement measurement (x).

Figure 117 Scheme of strain-wire instrumentation for detecting inter-strata stability above a collapsed mine junction which could lead to sink-hole occurrence.

are tensioned prior to reading with a suitable fine measuring instrument. Roof ... instability is indicated by the strain wires moving into the borehole.

A field station established for the purpose of observing ground stability at various inter-strata positions between a collapsed junction and the surface is shown in Figure 118.

Sink-hole Occurrences Associated with Pittsburgh Coal Mine Workings

Bruhn, Magnuson and Gray (1978) have discussed data from 354 subsidence incidents over the Pittsburgh Coal Bed in Pennsylvania and the following points have been abstracted from their paper:

1 Sink-holes occurred where the overburden thickness was less than 50m, whilst the majority of holes developed at depths of less than 15m.

2 Most sink-holes had a mean diameter of less than 3m, although a few had diameters of up to 10m; their depths were generally not more than 6m below surface.

3 There was a strong tendency for sink-holes to develop along, and within close proximity to, the seam outcrop on hillsides; over 70% of the recorded sink-holes were within 150m of the outcrop.

4 Some sink-holes developed within 10 years after mining, whilst a few others are reported as occurring 100 years or so later. Around 60% of the sink-holes had occurred within 50 years after mining.

5 Sink-holes were observed to out number subsidence troughs by 30 to 1. (This is probably because the majority of the subsidence case histories used had overburden thicknesses of 25m and less.)

6 Subsidence troughs over abandoned room and pillar workings have exhibited a dish-shaped profile with diameters generally in excess of 10m and usually with not more than 1 m of subsidence at the centre.

The above points indicate that considerable uncertainty can exist in attempting to predict the occurrence of sink-holes in the absence of detailed geological and mining knowledge pertaining to particular mined-out areas. It underlines the importance of carrying out detailed historical and site investigation studies prior to surface development in such subsidence affected locations.

Subsidence over Room and Pillar Mines in Illinois

Marino and Mahar (1985) have discussed subsidence over room and pillar mines in Illinois and report that the shape and magnitude of the subsidence profiles observed were found to be strongly influenced by the thickness of soil and rock overburden, the percentage of coal extracted and the shape of the mine pillars. Their study involved room and pillar mines of relatively shallow depth, namely 30 - 70m with the thickness of extraction in the range 2 - 3m.

Figure 118 Field station of strain wire instrumentation in process of tensioning for subsequent reading.

Three principal modes of mine failure were identified as the causation of subsidence at these room and pillar mines in Illinois:

1 Roof collapses above rooms giving rise to pits and sags (subsidence holes and troughs).

2 Pillar crush giving rise to trough subsidence.

3 Pillar punching into mine floors also resulting in trough subsidence.

It would appear that combinations of pillar crushing and punching into soft floors occur frequently.

Hunt (1980) reported observations on subsidence holes and troughs and found that both these forms occurred together in the depth range of 23 - 50m. Sink-holes were not observed where the depth of mining was greater than 50m. Significant thickness of surface deposits such as soils, clays, sands and gravels tended to inhibit the formation of sink-holes.

The report of Marino and Mahar (1985) demonstrated that surface subsidence troughs of up to 300m across could develop over room and pillar coal mine workings where pillar crushing and punching into mine floors occurred. Maximum values of subsidence in such troughs was given as up to 0·6 - 1·0m. The degree of extraction (in plan as measured from mine plans) where surface subsidence troughs were observed over room and pillar workings ranged from 47 to 70% with an average of 65%. It was recognised that the ensuing subsidence over room and pillar workings was related to the degree of extraction, but it would appear that local geological conditions and strength of the mine pillars and integrity of the immediate roof and floor all played important roles in influencing the occurrence of subsidence.

Mahar and Marino (1982) have given earlier information on the nature of sink-hole characteristics as observed above room and pillar coal mine workings in St. David, Illinois. Sink-holes were observed to have diameters in the range 0·6 to 10m with depths to the top of the collapsed soil surface in the range 0·6 to 6m; with overhanging sides also occurring.

Sand Model Demonstration of Sink-hole Characteristics

Sand models have been used to demonstrate the flow behaviour of granular material through an orifice representing a draw-point associated with a mine stope. The same form of demonstration can be used for a sink-hole development under particular conditions.

Figure 119 illustrates the development of localised movement of the sand as it is drawn from an orifice at the base of the model. The model shows the following features:

1 movement of the sand particles is within a narrow vertical plane,

2 propagation of movement is confined within the narrow vertical plane and the movement process rapidly reaches the upper surface of the model,

3 continued out-flow of the sand produces a conical depression adopting the angle of repose of the sand,

4 the conical depression simply enlarges at the angle of repose of the sand by further out-flow of the sand,

5 material from upper layers rapidly finds its way to the base of the model.

Figure 120 shows the fully developed state of the sand model demonstration with the draw ellipsoid clearly evident. The size of the sand particles influences the demonstration. Additionally damp sand can induce the formation of a distinct cavity within the column of draw and show how the movements progressively decrease by collapse of the sand into the cavity.

The sand model discussed here considers the draw phenomenon and in relation to sink-hole development shows some important features regarding the relatively localised nature of the collapse process. It also shows that for further movement to occur there must be some means of removing sand from the base of the column of moving material. The fairly rapid downward movement of sand from the surface layer, shows how in some circumstances surface features such as trees can appear at the mining horizon. The sand model does not take rock caving behaviour fully into account but is a useful demonstration of the sink-hole problem.

Conclusion

Surface subsidence above room and pillar mining layouts can take two main forms, namely sink-holes and troughs. Sink-holes are mainly depth related unless they are influenced by water which can result in the chimney of caved rock and mud flowing into adjoining rooms and allowing caving to proceed to heights in excess of 7 - 10 (M). This occurrence is fairly rare,

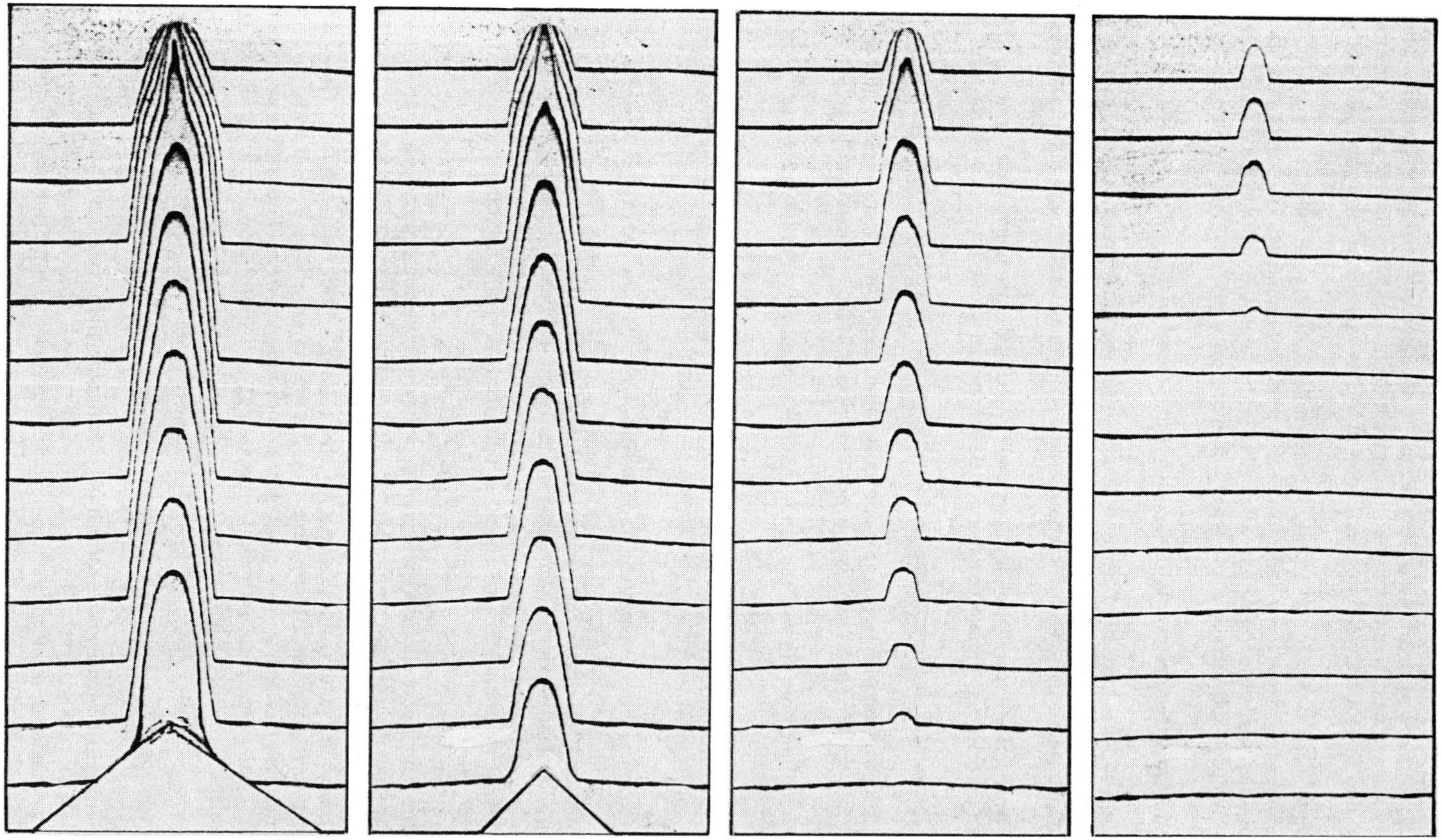

Figure 119 Sand model illustrating development of a sink-hole in unconsolidated material

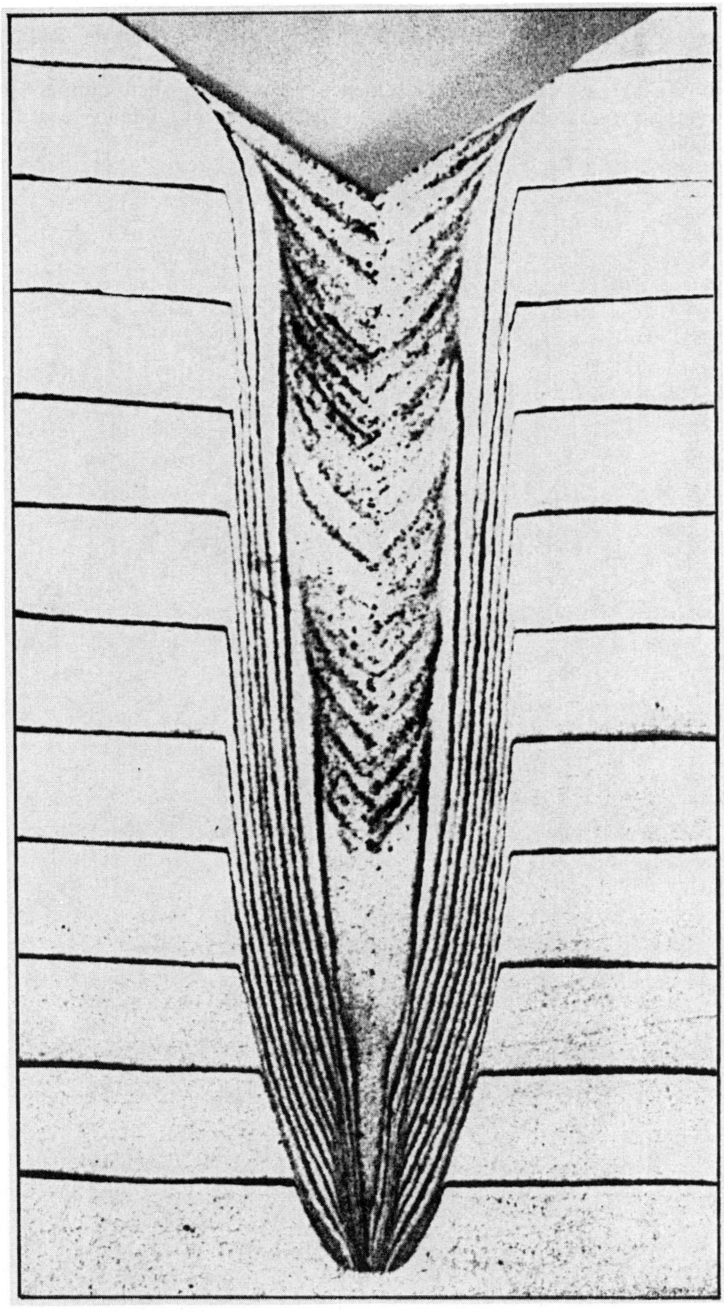

Figure 120 Sand model illustrating fully developed formation of sink-hole in unconsolidated material.

however, in abandoned mines which may be flooded and tend to discourage large-scale flows of rock from the chimney into water-filled rooms.

Local geology and natural strength of the immediate roof are important factors, but the mining dimensions and geometry of workings are of equal importance and should be considered in making assessments of subsidence risks above such mines. History of subsidence events above particular mine layouts offers important guidance regarding trends in surface stability.

CHAPTER 10

SUBSIDENCE ASSOCIATED WITH STEEPLY INCLINED SEAMS

Prediction of subsidence overlying steeply inclined seams calls for special considerations not just of angle of dip but also of depth below the surface and direction of working since each can have a major influence on the character of the resulting surface subsidence. A number of investigators have addressed the problem of predicting subsidence over steep seams and in most cases the main concern appears to have been the depiction of the position and shape of the surface subsidence depression.

This chapter examines the common forms of mining steep seams in relation to longwall extraction, and goes on to discuss the general character of the accompanying subsidence. Attention is mainly devoted to the prediction of subsidence but aspects of surface strain and displacement are also considered. Particular comments are made on the nature of ground movements and crack propagation associated with the working of steeply pitching seams.

Mining aspects of steeply inclined seams

The angle of inclination of steeply pitching seams has a major bearing on the direction of working longwall extractions. In the case of gently dipping strata of say less than 20° to the horizontal, considerations of caving may be of secondary importance to working direction. Seams inclined at 30° and more, however, introduce special problems connected with gravity which strongly influence selection of face orientation relative to dip.

Figure 121 illustrates the three principal directions of working a longwall face in a steeply inclined seam.

Along the Strike. This is the most common direction for longwall working and has the following attributes:

1. The gate roadways are driven so as to be level, or with a slight gradient to facilitate drainage; this greatly assists transport and access to the working face.

2. Roadway drivage is faster with this direction of working; machine stability and horizon control are easily accomplished in addition to simplicity in rock loading and transport from the heading; support setting is also less arduous in this direction.

3. Mineral transport down the working face is assisted by the gradient and readily controlled by suitable conveying systems.

4. Roof control is effectively achieved by supports stabilised by inter-linking and caved rock is easily controlled and retained within the waste area.

5. Mineral cutting is efficient and relatively easily controlled; the mineral cutting machine is effectively stabilised and operation is relatively straight forward.

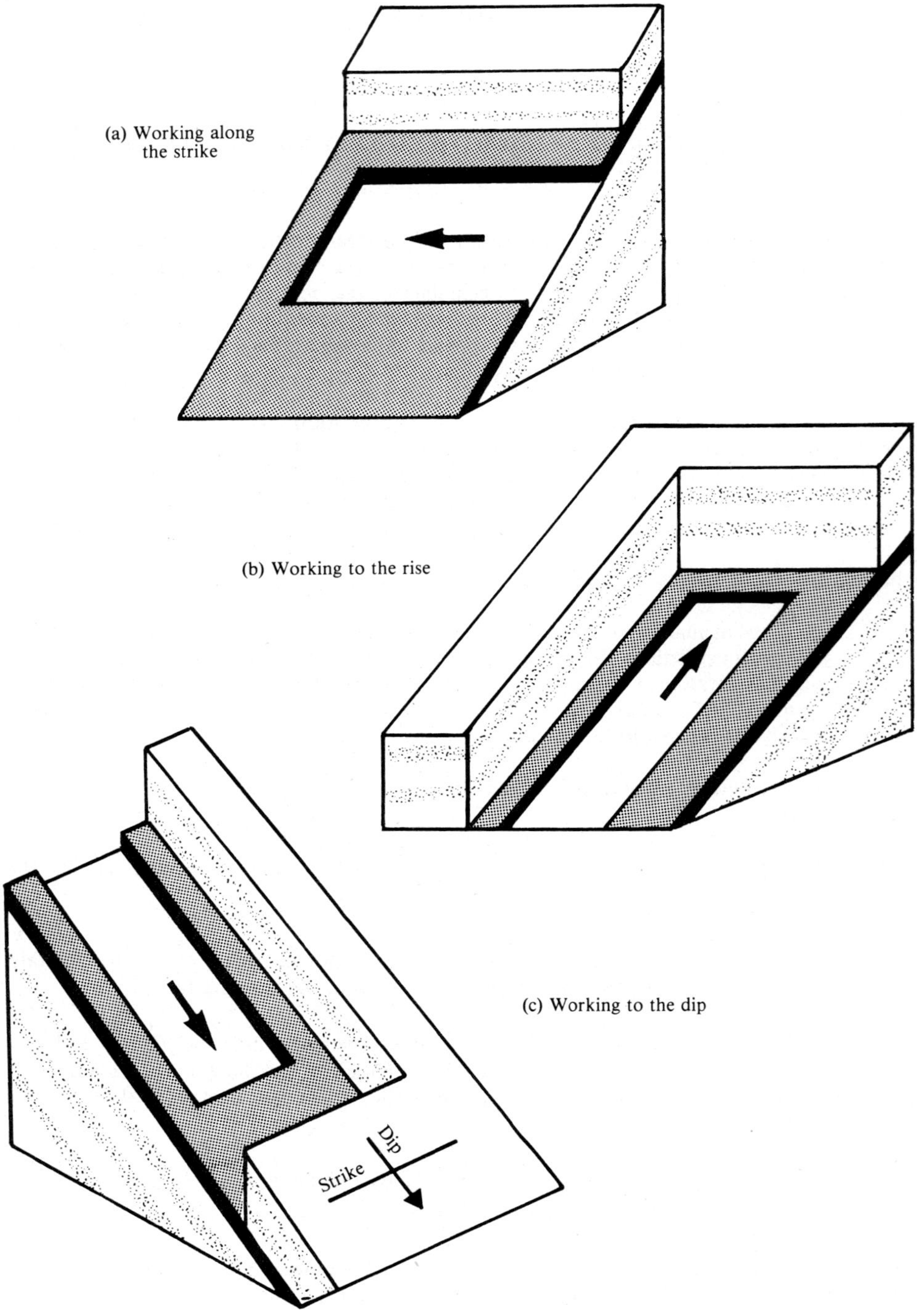

Figure 121 Principal directions of working steeply inclined seams by longwall extractions.

6. Water drains down the face and simply collects in a convenient sump in the main (lower) gate.

7. Methane dilution and removal from the face area is effectively achieved.

8. There is greater potential for higher rates of output from faces working along the strike.

To the rise. The method is less convenient than working along the strike, although it has some features which other directions of working do not possess. Its main characteristics are listed as follows:

1. Roof caves fall away from the working area; consequently safety against roof falls from the goaf affecting the working face is improved with this direction of working.

2. Coal slabbing off from the working front can hamper face operations and decrease relative safety.

3. Machine stability is more difficult to achieve as is also cut consistency.

4. Methane is more likely to collect on the face; also methane is more naturally encouraged to enter the working face area from the goaf.

5. Conveying of mineral along the face is generally less efficient.

6. Water drains naturally away from the working face.

7. The gate roadways are more difficult to construct and support.

8. The face supports may prove more difficult to advance.

To the dip. Working to the dip is more favoured than to the rise, and is frequently a natural extension of mining as the depth of working increases. This direction of working has the following features:

1. Roof control behind the working face calls for a highly effective support system to control heavy caves and prevent their encroachment on to the face area.

2. Coal production operations may encounter problems connected with conveyor loading and experience lower carrying efficiency.

3. Water can give rise to conveyor starting-up and operational problems, and may also influence roadway formation speed and efficiency.

4. Methane control is better than with working to the rise.

5. Transport in the gates is less efficient than when working along the strike.

6. Production potential is markedly less than when compared to working along the strike.

The vast majority of steeply inclined mine workings operate longwall extraction along the strike in view of its major advantages over the other two directions of working, especially when considering production potential, operational advantages and improved control over environmental factors.

Discussion has focussed here on longwall working of steep seams. Room and pillar operation is rarely applied in steep seam situations owing to technical problems associated with cutting, loading and transport on steep gradients. Longwall operation has overwhelming advantages over room and pillar working in steep seams especially regarding vastly improved production potential and overall safety during the mining operations quite apart from the more efficient extraction of reserves.

Horizon mining. The main form of layout for extracting coal seams in steeply inclined conditions by longwall working is by means of the horizon mining technique. Figure 122 illustrates the principle of horizon mining as applied to the successful extraction of several seams grouped together as shown. Access to develop the mine involves the use of horizons and sub-levels from which gate-roadways are driven for setting up longwalls which are normally operated by retreat mining as shown in Figure 122. The uppermost seam is usually extracted first in order to avoid caving and subsidence effects disturbing the other seams. Staple shafts are an integral feature of horizon mining layouts, since they greatly assist in facilitating speedy access from one level to another for both men and materials. These shafts need protection from the effects of mining.

A more detailed illustration of a longwall operating in a horizon mining layout is shown in Figure 123. The face line is not generally on the line of full dip, but slightly offset in order to allow the bottom gate to lead the upper one so that the push-over rams can assist in correcting for downhill creep of the face conveyor.

Horizon mining involves the driving of several roadways across the Coal Measures containing the group of seams. Development effort is much greater than for comparable flat situations. However, where only steeply-inclined seams exist, horizon mining is the most economical and efficient means of extracting such coal seams. Retreat mining is mainly practised.

Subsidence arising from horizon mining layouts tends to be concentrated. There is scope, however, for the phasing of the longwall extractions so that the development of ground strain at the surface is more spread out. Additionally, it is sometimes possible to plan the longwall extractions in successive seams so that there is a tendency for some strains to cancel out those of opposite sign.

Most subsidence troughs in horizon mining layouts have their major axes parallel to the strike of the seam.

General character of mining subsidence in steep seam situations

The extraction of a longwall by working along the strike of a steeply inclined seam gives rise to the general character of subsidence as depicted in Figure 124(a). Caving of the roof behind the longwall tends to shift to some extent down the gradient thereby resulting in asymmetrical development of ground movement. The fact that the top gate is at a shallower depth than the bottom gate also implies that the subsidence at the rise side of the face is likely to have a more marked disturbing effect on the surface where the seam is close to outcropping as is shown in Figure 124(a). The latter point does, however, depend upon depth below surface since in some very deep situations the effects of subsidence on the surface may be

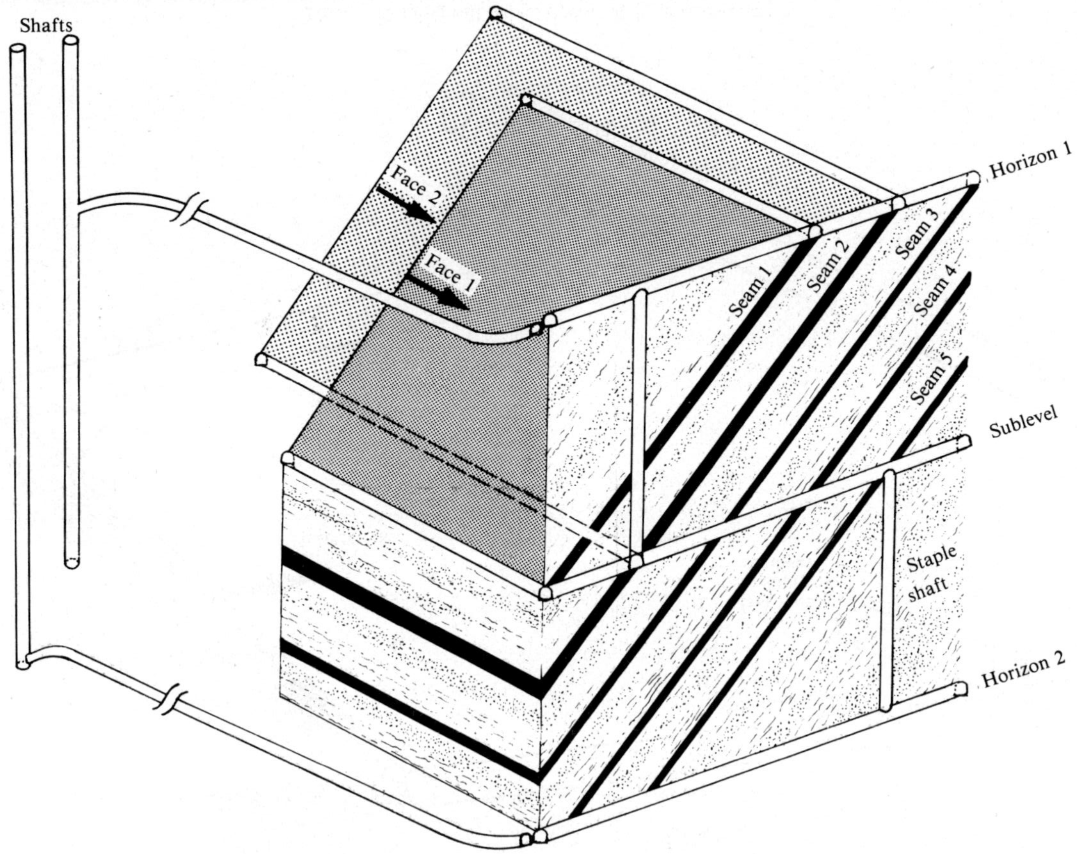

Figure 122 Illustrating the principle of horizon mining

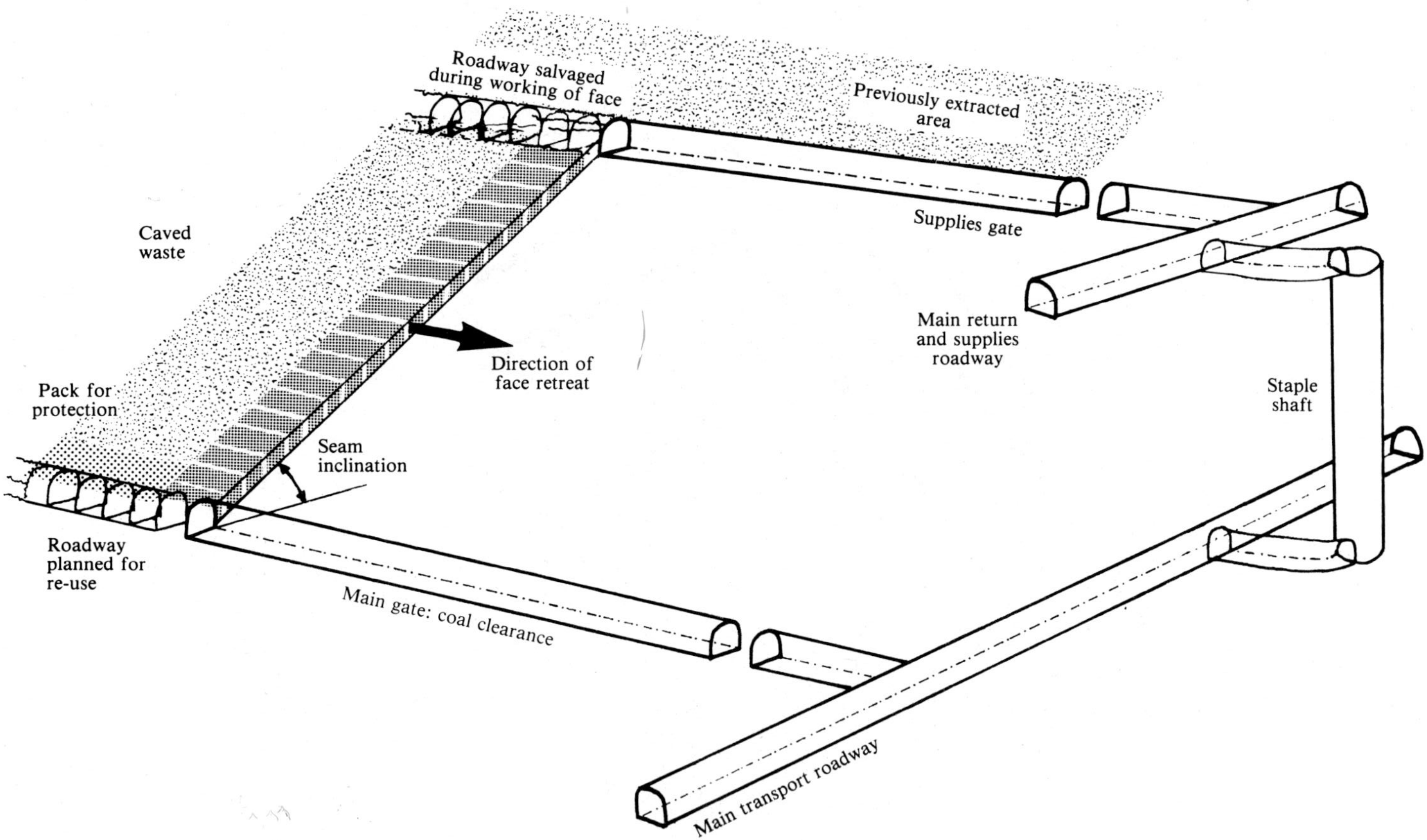

Figure 123 Longwall retreat operation in horizon mining

hardly discernible on either side of the face. The same argument applies to the gradient also, since if the angle of dip is particularly shallow, then the surface subsidence effects on both sides of the longwall may be hardly distinguishable between them. Consequently, when making any general comparisons between longwall subsidence effects on the surface in steep seam situations, due account needs to be taken of gradient and depth.

Figure 124(a) is illustrative of the typical surface subsidence features which are associated with steeply inclined seams in a very shallow depth situation. The surface subsidence profile is displaced down the gradient and the angle of influence increases on the dip side whilst decreasing on the rise side. There is a high concentration of ground strain on the rise side which affects the whole of the strata up to the surface. The surface overlying the rise side of shallow depth steeply inclined seams can experience significant fracturing effects, and can be a potential path for surface water gaining access to the mine workings.

Figure 124(b) shows the projected representation of strata movements for the subsidence condition illustrated in Figure 124(a). Strata movement migration is shown to be towards the dip side of the longwall.

Ground movement patterns indicated by physical models

Physical models employing sand and plaster materials of carefully determined strength and deformation properties have been used at the Department of Mining Engineering, University of Nottingham, to investigate caving and subsidence behaviour of longwall extractions in steep seams. The models cave and subside under their own weight as a natural phenomenon and consequently reflect the natural behaviour of the longwall mining process.

Figure 125 shows the principal fractures and form of surface subsidence indicated by physical models for longwall extractions in seams dipping at 25° and 35° respectively. The test rig used in these experiments was 3m long and 1·5m high and each model was constructed in layers parallel to the seam in order to produce a sedimentary structured representation. Figure 125 shows the models after caving and subsidence had ceased. There are similar patterns of movement indicated by these models.

In order to highlight the significance of the fractures generated during the caving and subsidence phases of each of these models, Figure 126 has been drawn to allow ease of comparison. The following features are evident with these shallow depth model situations:

1. The greatest localised sub-surface concentration of fracturing occurs on the rise side of the face in both cases, and the fractures break through to the surface in these particular cases.

2. The dip side is characterised by the occurrence of a major surface fracture of almost vertical disposition lying over the rib-side in both cases. Both these major fractures occurred in the tensile strain region of the subsidence trough. They do not appear to link up with deeper fractures generated by caving of the mined out part of the seam. These major fractures are indicative of appreciable tensile strain at the surface over the dip-side of the longwall extraction.

3. Fracturing around the dip side rib occurred in both models, although when the angle of dip was 35°, the fractures were starting to hade over the solid.

4. The subsidence trough positions are shown together with a chain line indicating

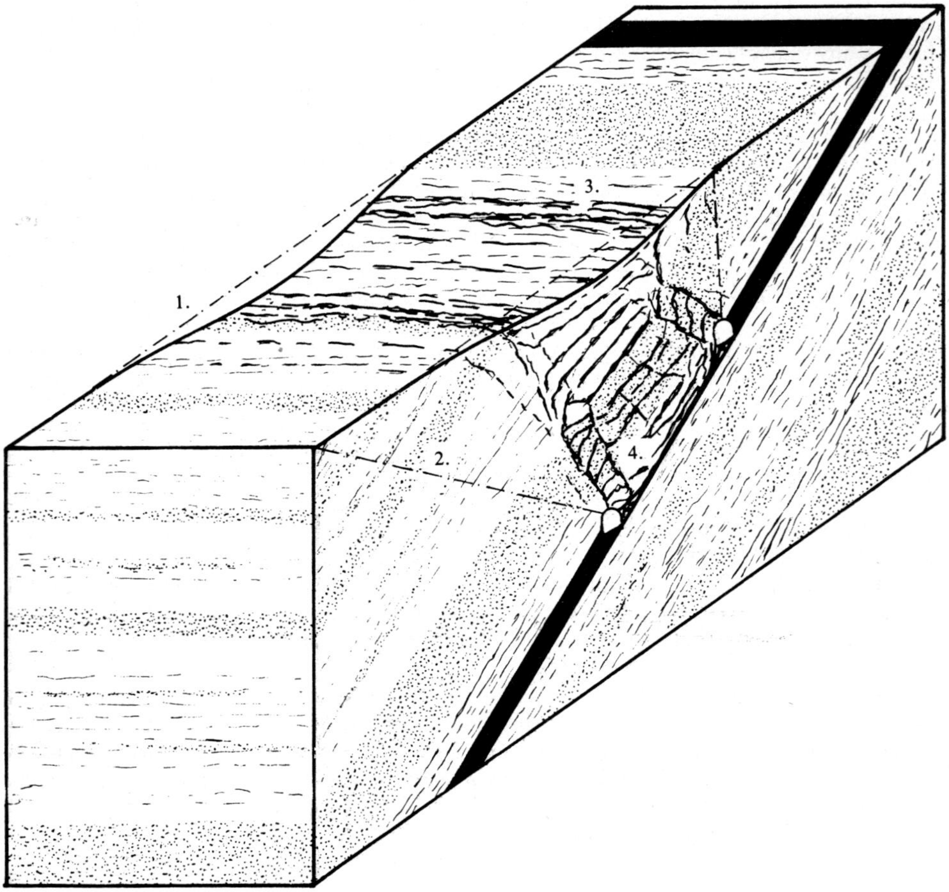

1. Surface subsidence trough displaced to the dip side.

2. Dip side experiences broadening of the area of subsidence influence especially that of tensile strain.

3. Conversely the rise side experiences more confined sub-surface subsidence effects, with greater concentration of ground strains at the surface.

4. The immediate roof beds cave and tend to slump down the gradient under the effect of gravity, thereby establishing a general subsidence pattern which propagates upwards.

Figure 124(a) General character of subsidence arising from longwall extraction in a steeply inclined seam

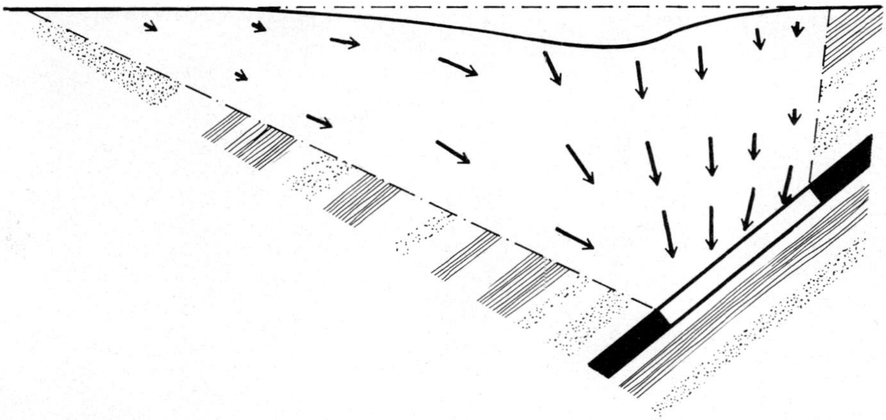

Figure 124(b) Projected strata movement pattern relating to steep seam extraction situation shown in Figure 124(a)

where the SEH graph would predict the position of the centre of the resulting depression. There is some consistency between the model subsidence trough and its SEH predicted central position for the seam dips considered here.

5. The rise side rib is clearly subjected to a degree of tearing action and results in the opening of tensile fractures.

6. There appears to be a tendency for some rotation about the dip side rib and this encourages the translatory movement at higher levels which causes the major vertical fracture seen with both models thus giving rise to an appreciable tensile strain region at the surface.

7. The central area of each face appears relatively free from open or major fractures.

Figures 127(i) and (ii) show photographic details of these two models particularly the nature of the fractures over the rib sides. Fracturing around the dip side ribs is much more confined than on the rise side.

Similar models in steeper situations indicate that there is a tendency for uplift to occur, over the rise side rib at the surface. Also, the caved roof beds are more greatly influenced by a tendency to slide down the gradient, and wedging of individual blocks becomes increasingly evident.

Rotational character of subsiding ground mass

Investigations by Degirmenci, Reddish and Whittaker (1988) involved examination of the pattern of displacements between the mining extraction and the surface for seam inclinations in the range 25° to 55° dip. Scaled models constructed of sand and plaster as shown in Figures 125, 126 and 127 were used. Measurements were made on a grid of pins inserted into the face

(b) Subsidence model of caved longwall (35°).

Figure 125 Fractures associated with longwall mining of seams dipping at 25° and 35°.

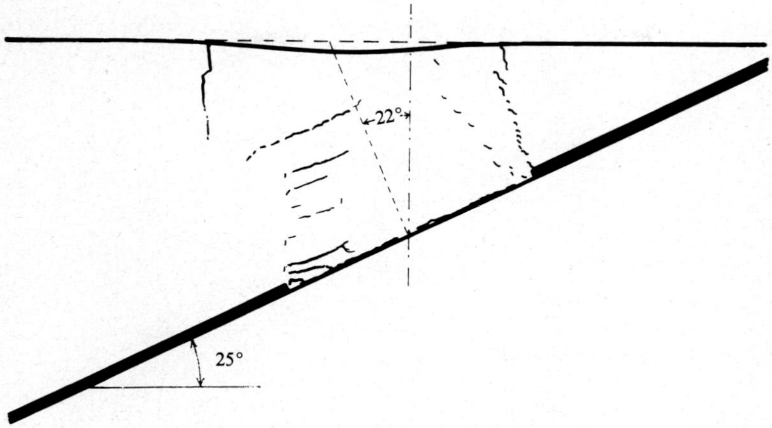

(a) Fracture development and subsidence profile for longwall extraction dipping at 25°

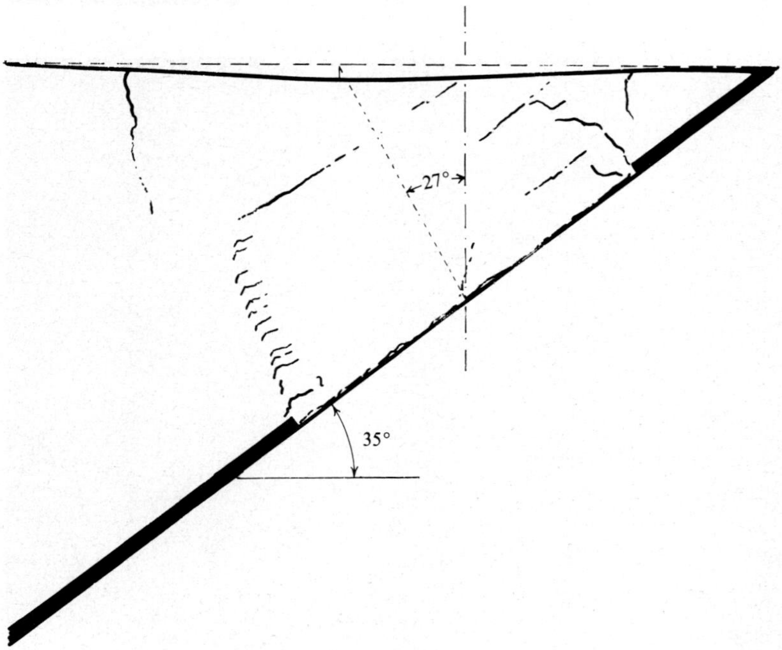

(b) Fracture development and subsidence profile for longwall extraction dipping at 35°

Figure 126 Principal fractures and subsidence profiles for longwall extractions in steep seams

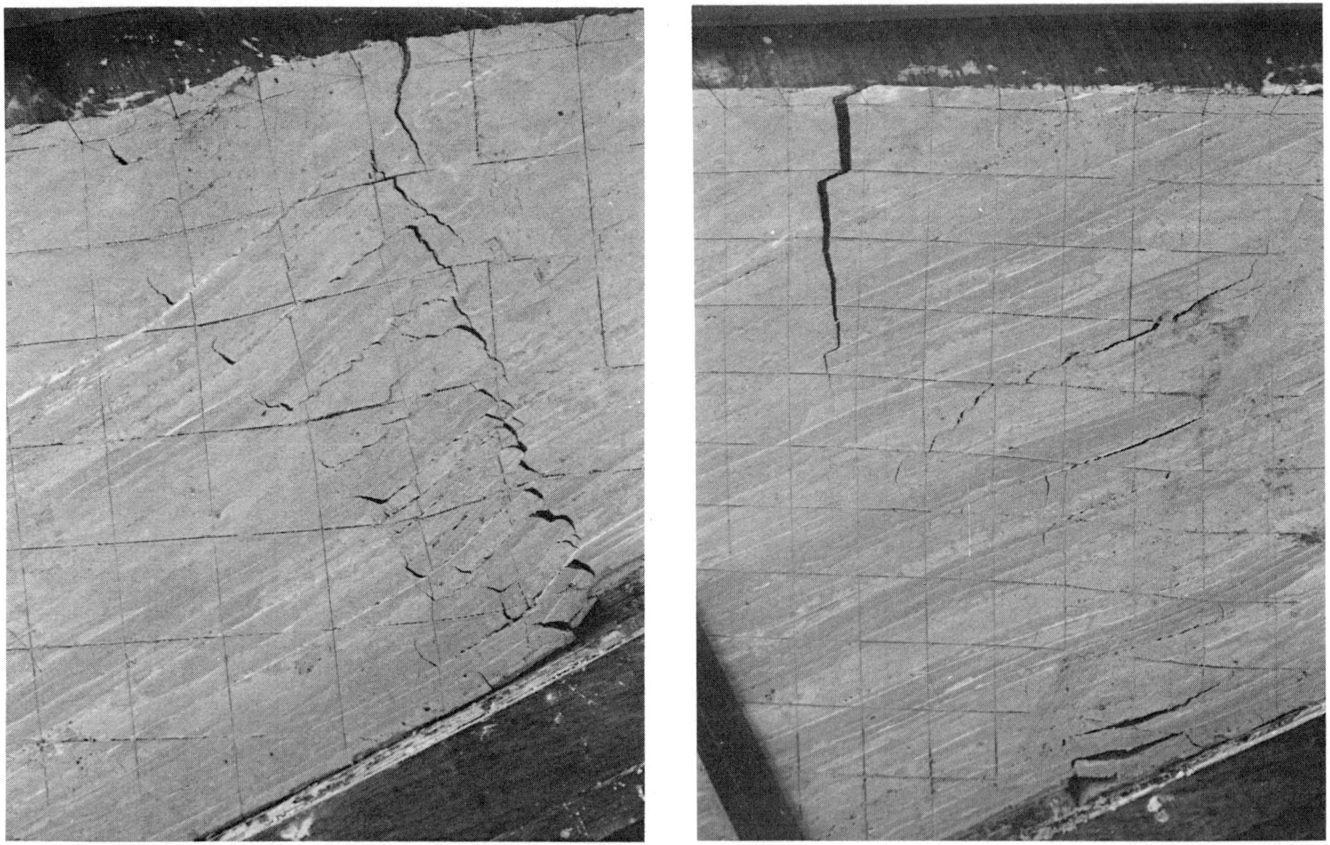

Figure 127(i) Subsidence model for longwall dipping at 25°
Right: dip side; left: rise side.

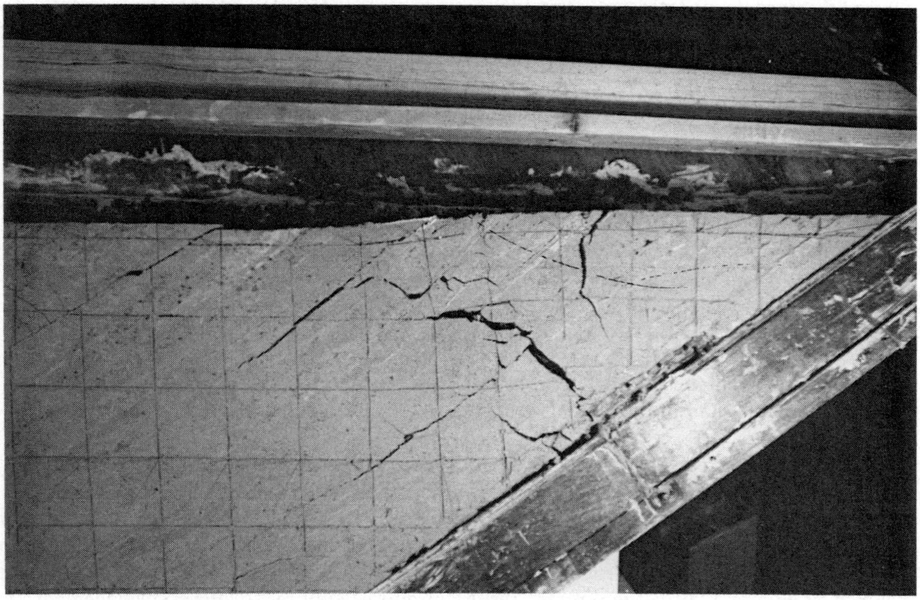

(a) Rise-side fractures

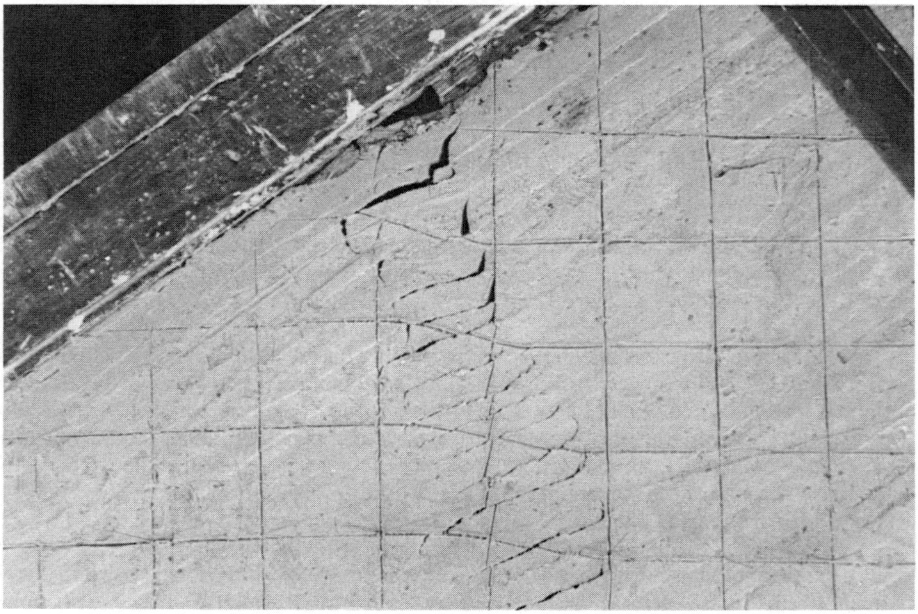

(b) Dip-side fractures

Figure 127(ii) Subsidence model for longwall dipping at 35°

of the model section in order to observe vertical and horizontal displacements. Results of ground displacement vectors for seam inclinations of 35°, 45°, 45° and 55° dip are presented in Figure 128(a). The patterns of ground movement indicated by these results demonstrate the following features:

(a) There is clear evidence of a tendency for the subsiding ground to try to rotate about a point on the dip side of the extraction. This point appears to be located over the solid rib side some distance away from the subsiding ground mass.

(b) The orientation of the vectors indicates that the surface will experience its greatest horizontal strains on the dip side of the extraction; the rise side is more remote from this point of rotation although it is recognised that the rise side may have its own point of rotation even though it is likely to be not very significant when compared to the dip side.

(c) Subsidence in steeply inclined seams is clearly demonstrated to be combined with appreciable lateral movement, with the latter increasing towards the surface.

The general pattern of ground movement shown in Figure 128(a) is summarised in Figure 128(b) which illustrates the concept of rotation during the subsidence phase. The fact that the predominance of ground movement is affected by the lower point of rotation means that this will more greatly govern the character of the ensuing surface subsidence. Consequently for equal surface subsidence on each side of the longwall extraction, as shown by the inset diagram of Figure 128(b), the dip side results in appreciably more lateral displacement and correspondingly increased tensile strain. This is borne out by the occurrence on the dip side of the significant tensile crack at the surface with both the cases illustrated in Figure 126.

A later chapter deals specifically with the mechanics of ground movement as indicated by physical models, and further comments are given on the effects of steeply inclined seams regarding crack occurrences and propagation.

Surface subsidence prediction over gently dipping seams

SEH (1965) describes a method of adjusting the position of the subsidence profile to take gradient into account for seams of inclination up to 20°. Hall and Orchard (1963) discussed the field observations which led to the adoption of this procedure. The basic principle involved is demonstrated in Figure 129(a).

Ogden and Orchard (1959-60) have reported observations made in the North Staffordshire Coalfield of western England which examined the consistency of projecting a normal from the centre of the longwall extraction to intercept the surface. Figure 129(b) shows results for various depths below surface and seam gradients. All the projected normals meet at the point of maximum subsidence P. These results confirm the validity for determining the position of maximum subsidence by simply projecting a normal from the face centre to the surface as is the case with Figure 129(a).

The maximum subsidence (S) is determined from knowledge of the longwall width/depth ratio and the extracted seam height (M) as is the case with flat seam subsidence predictions, see Chapter 6. Subsidence profile positioning is made by projecting a line at right angles to the face centre-line. Where the projected normal intercepts the surface, this represents the position of the maximum subsidence. This procedure of displacing the position of maximum subsidence towards the dip side obviously leads to bringing the limit of subsidence nearer to

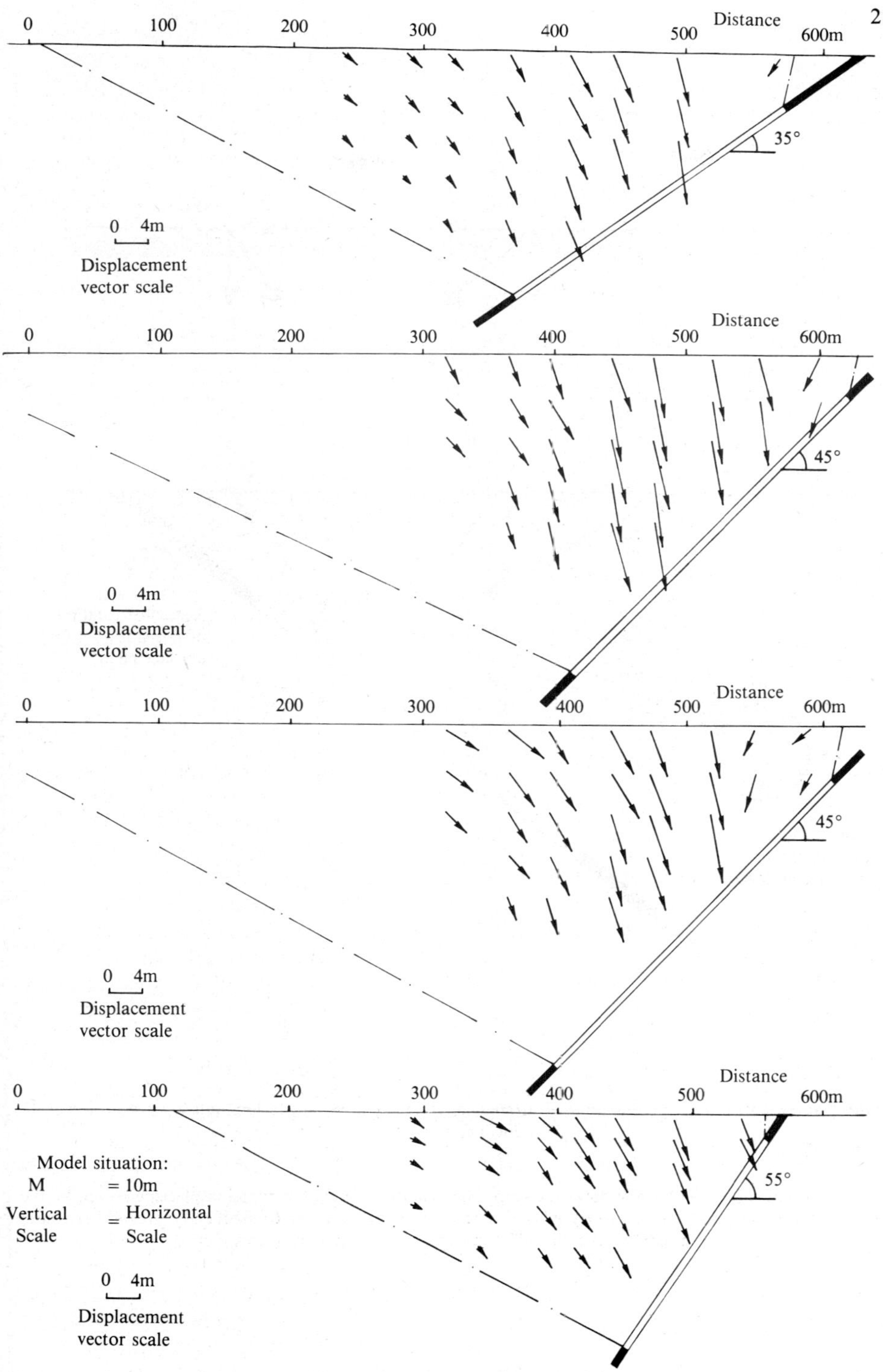

Figure 128(a) Displacement vector diagrams for longwall extractions in steeply inclined seams. Physical model results, after Degirmenci, Reddish and Whittaker (1988).

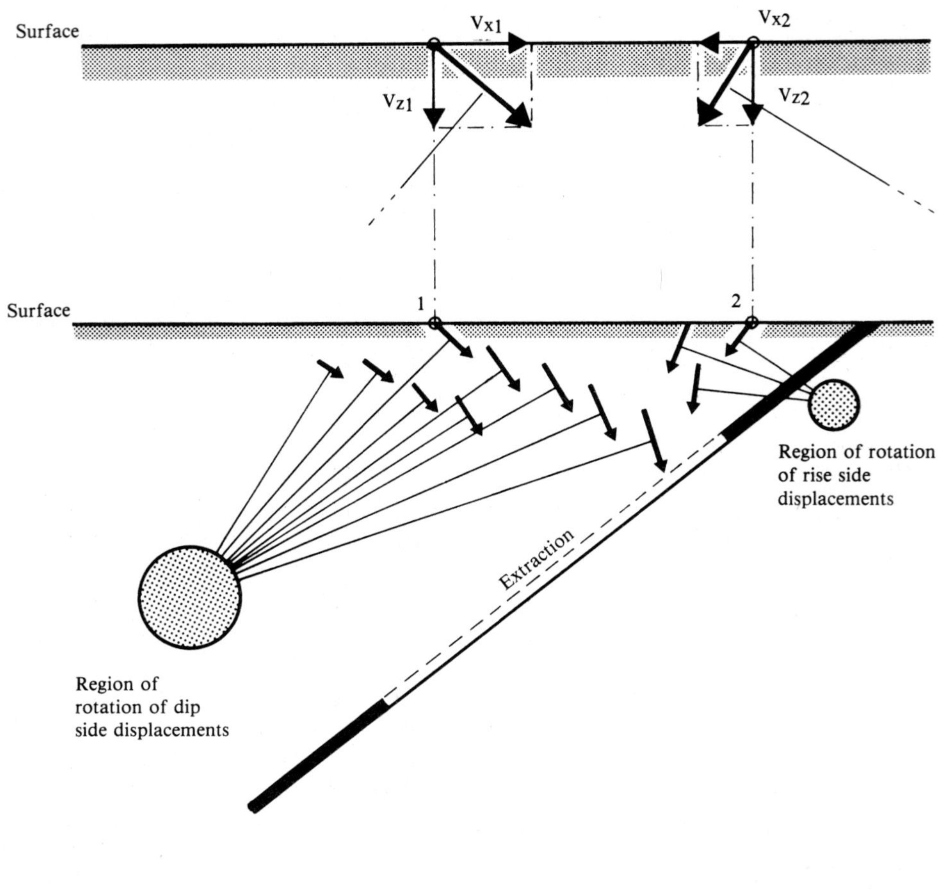

Figure 128(b) Generalised representation of ground movement pattern above longwall extraction in steeply inclined seam.

Note: Points 1 and 2 show that they experience significantly different horizontal displacement even for the same magnitude of V_Z, owing to the positions of rotation centres. Consequently, greater horizontal strain is experienced over the dip side as compared to rise side of the extraction.

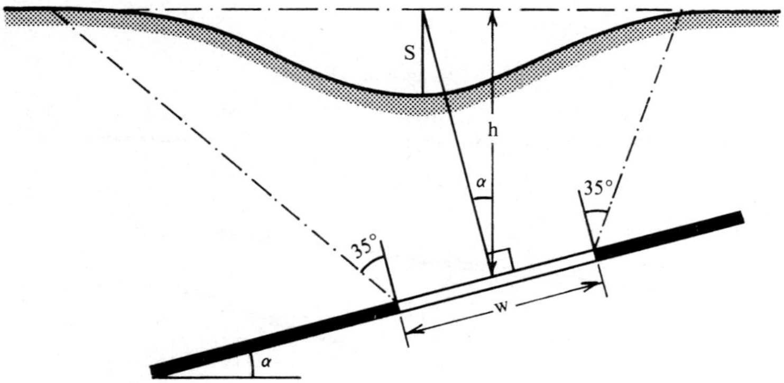

Figure 129(a) Location of subsidence profile in gently dipping seams (up to 20°)

the extraction on the rise-side whilst extending the limit further away on the dip side, even when assuming a symmetrical profile as can be virtually the case with very shallow gradients.

This form of treatment for positioning the subsidence profile is extended in the following section.

Surface subsidence profile positioning over longwall extractions in steep seams generally

A more comprehensive treatment of the surface subsidence profile positioning is put forward in SEH (1975). Figure 130 shows the method with the relevant symbols and graphical relationship between limit angle and dip of the seam. Profile determination is for a longwall advancing along the strike of the seam. The profile is positioned using the interception of the surface of three projected lines from the middle and each edge of the longwall extraction. These three projections are determined in orientation by specified angles measured from the vertical. The value of each angle is simply read off the graph presented in Figure 130.

It will be noted from the graphs showing limit angle plotted against dip of seam that beyond around 30° dip the lines are essentially extrapolated on the basis of the small amount of reliable data available on subsidence behaviour in steep seams. However, these graphs reflect experiences gained in European coalfields, notably West Germany, and reported by Rom (1964). Similar reasoning is given in a publication of Stojanowic and Pataric (1979) who suggest limit angles for steeply inclined longwall extractions.

Figure 131 presents examples of surface subsidence profiles plotted from the SEH (1975) design method given in Figure 130. These subsidence results demonstrate the predicted asymmetric profiles produced for seam angles between the horizontal and vertical positions. There is considerable broadening of the subsidence trough in the range of seam dip of 30° to

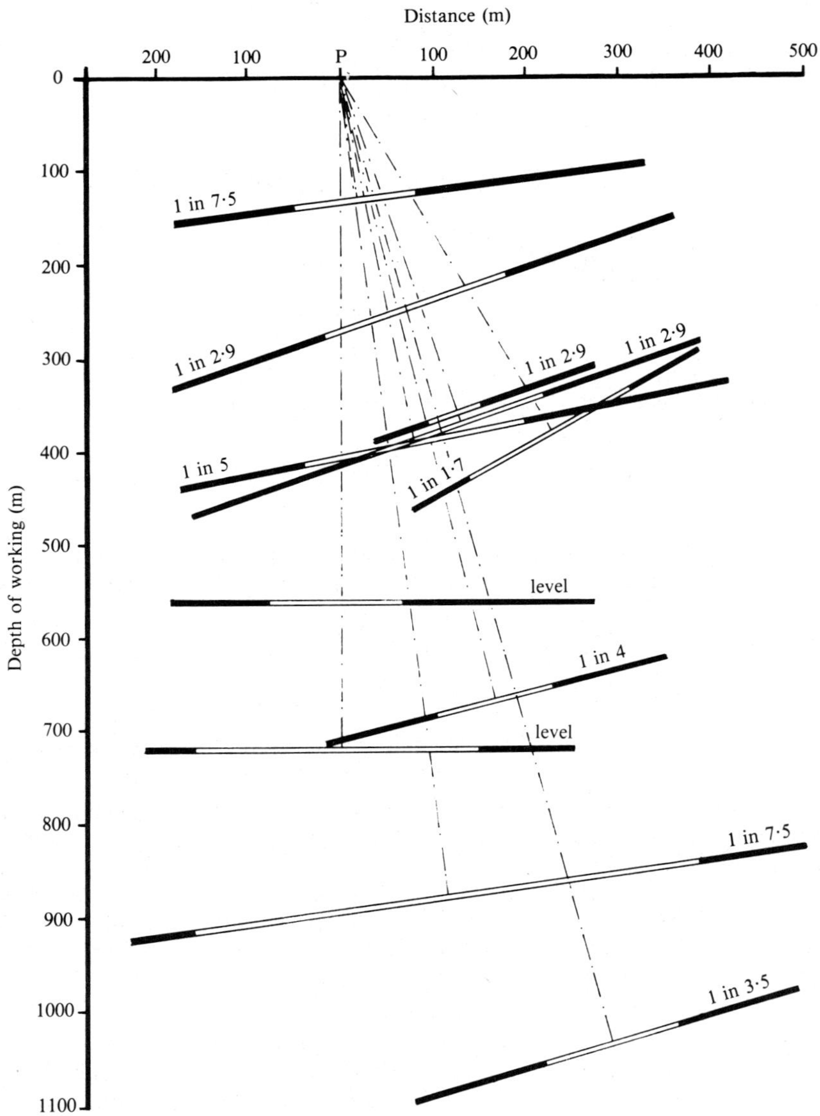

Figure 129(b) Position of longwall extractions and projected normal from face centre in relation to the point of maximum subsidence at P, after Ogden and Orchard (1959-60).

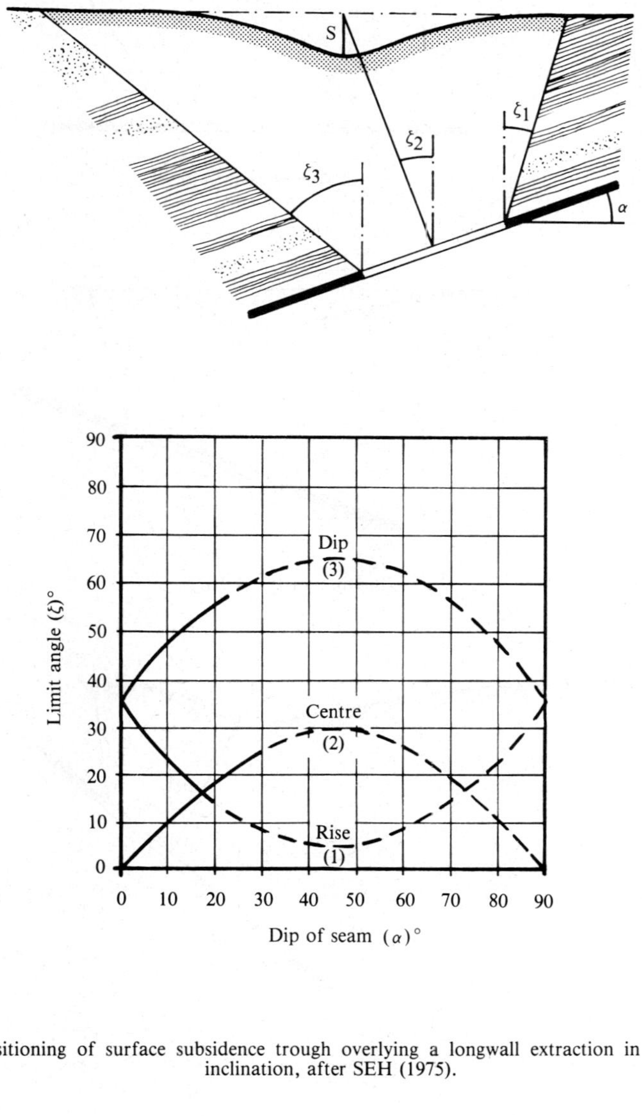

Figure 130 Positioning of surface subsidence trough overlying a longwall extraction in relation to seam inclination, after SEH (1975).

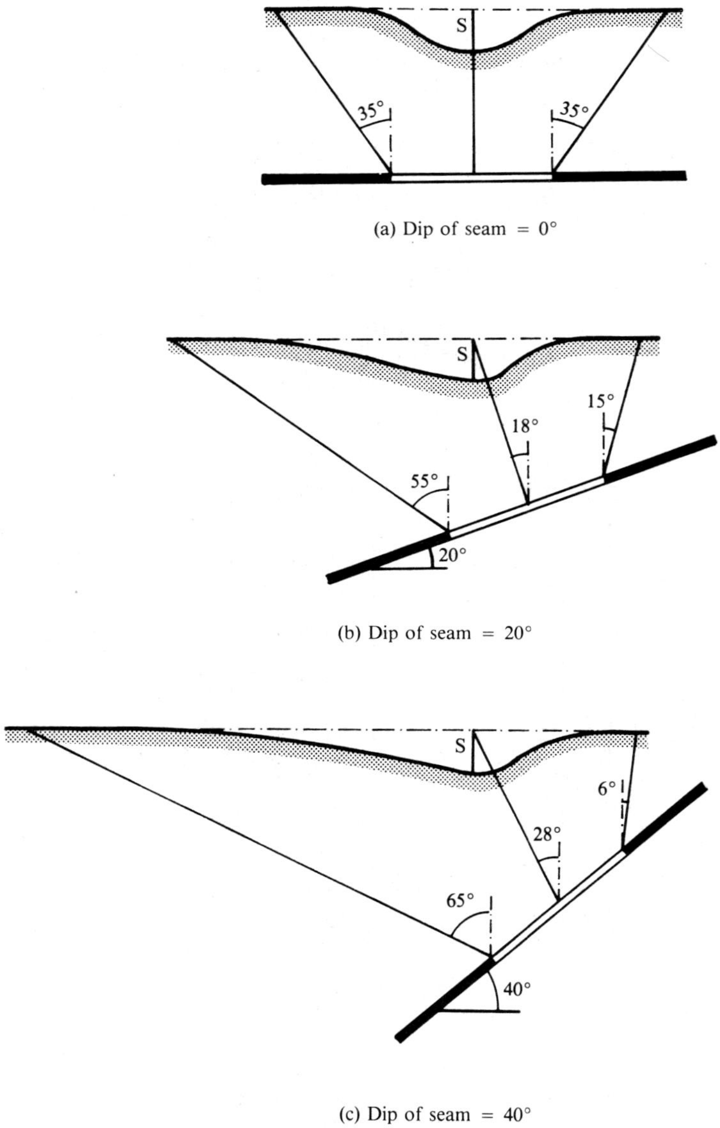

Figure 131 Surface subsidence profiles. For seam dip values of 0°, 20° and 40° according to SEH (1975).

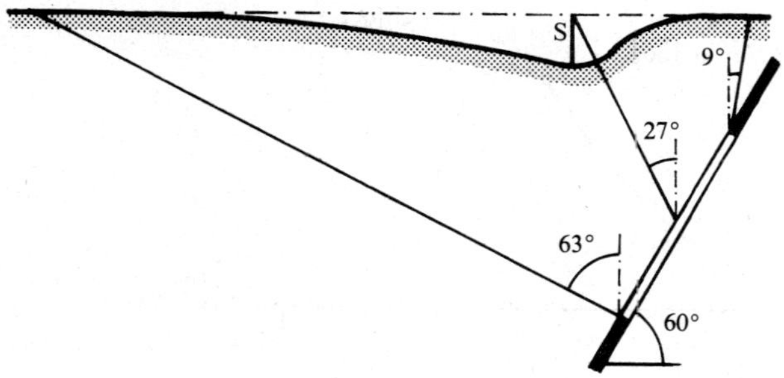

(d) Dip of seam = 60°

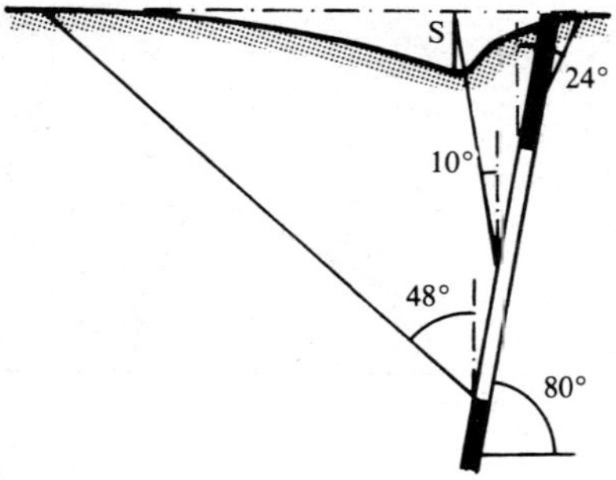

(e) Dip of seam = 80°

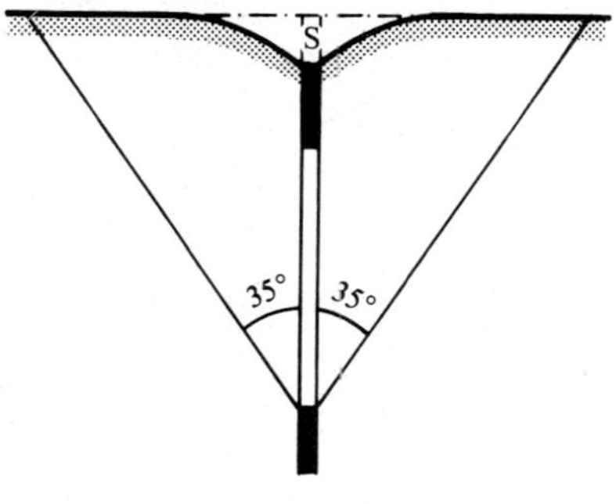

(f) Dip of seam = 90°

Figure 131 Surface subsidence profiles (continued). For seam dip values of 60°, 80° and 90° according to SEH (1975)

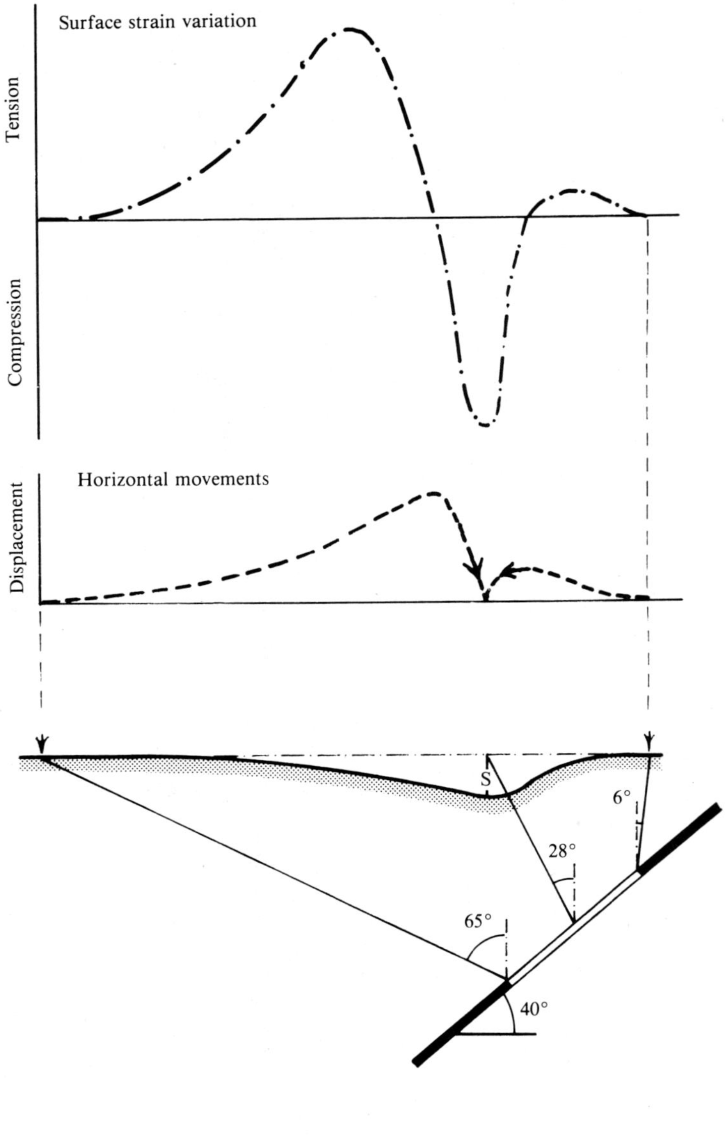

Figure 132(a) General character of surface ground strain and horizontal displacements above longwall extraction in a steeply inclined seam, based on SEH (1975) subsidence profile.

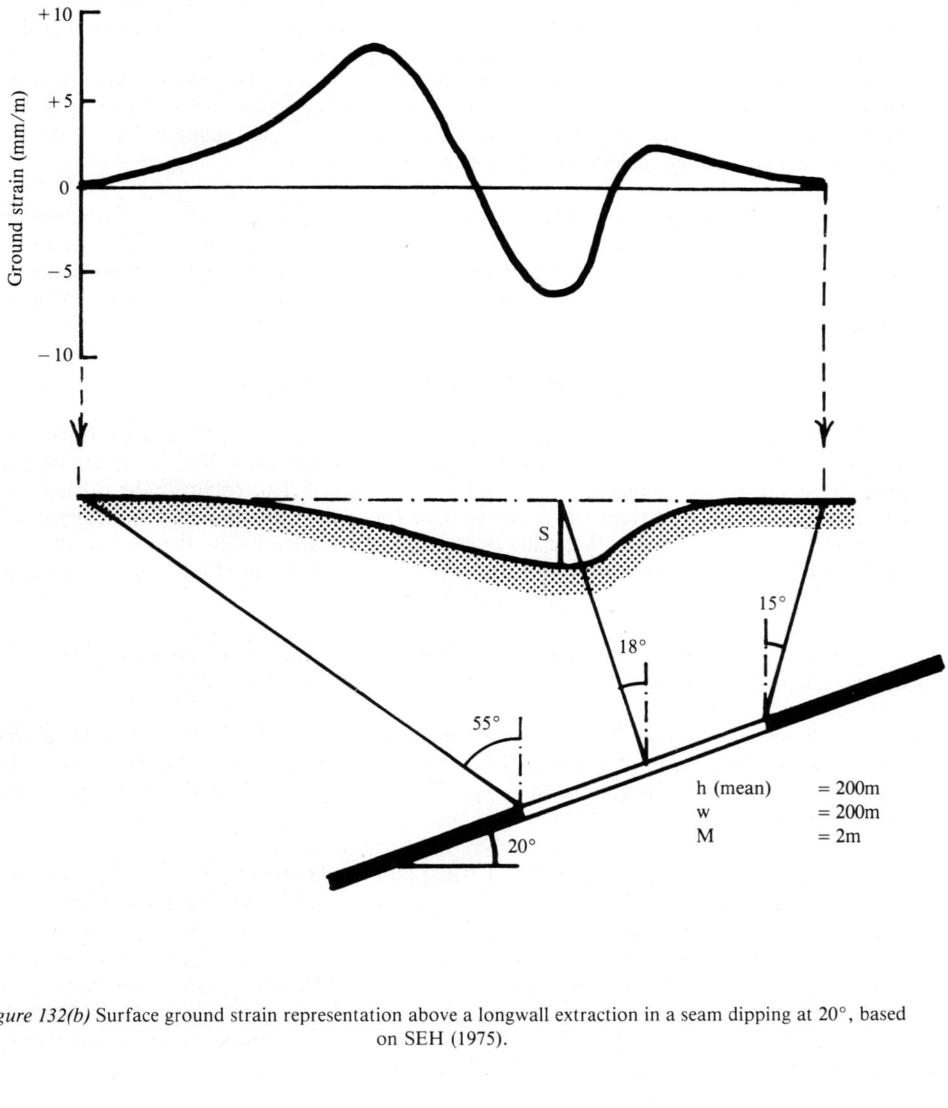

Figure 132(b) Surface ground strain representation above a longwall extraction in a seam dipping at 20°, based on SEH (1975).

70°. All the cases predict that where steeply dipping seams occur, then there is increased confinement of tensile strain on the rise side of the subsidence profile, whilst broadening of the tensile region on the dip side.

For seams in excess of 60° dip, the predicted profiles are probably of a tentative nature, and especially the vertical seam situation. In the latter case, the role played by the crown pillar in modifying subsidence behaviour is still relatively uncertain and is an area still deserving further detailed investigation. The predicted profile for a vertical seam as given in Figure 131 is probably the best available assessment to-date. The reasoning behind predicting a sharply pronounced subsidenced trough over the vertical seam, is based on the assumption that for surface subsidence to occur then inward collapse of the side walls takes place thereby attempting to fill, either completely or partially, the excavation. This process will undermine the bridging ability of the overlying strata. Consequently, any lowering of the strata on each side of the pillar will allow the pillar itself to subside. Kratzch (1983) suggests the formation of two separate subsidence troughs in the vertical seam situation.

The strength of the rocks around the seam will play a role, and the weaker their overall strength the more the likelihood of a single subsidence trough forming. In very strong rock formations, caving may not occur. However, where the crown pillar is of limited size and close to the surface, it has not been unknown for the excavation to become exposed due to pillar deterioration.

Surface subsidence strains over longwall extractions in steep seams

The general character of the surface ground strain profile over longwall extractions in dipping seams has been discussed by Rom (1964) who indicated that the effect of seam gradient was to greatly change the limit angles of subsidence. This results in the rise side limit of subsidence moving substantially closer to the edge of the extraction and as a consequence greatly reduces the length of the tensile strain zone. Correspondingly, the magnitude of the horizontal surface ground strain values are reduced on the rise side of the longwall extraction (assumed to be moving along the strike of the seam).

The dip side of the longwall extraction results in extending the zone of tension, and thereby increases the tensile strain magnitudes on this side of the working.

An important feature of the work of Rom (1964) is that on the rise side of the subsidence profile the transition point is at about 0·5(S) whilst on the dip side it occurs at about 0·7(S) thus resulting in generating greater bias towards tensile strain development on the dip side of the extraction.

The general shapes of the horizontal strain and displacement profiles are presented in Figure 132(a) for a longwall extraction in a seam dipping at 40°. The subsidence limit angles and location of the maximum subsidence (S) are in accordance with SEH (1975) and as in Figure 131(c). This representation of the horizontal strains indicates a significant difference in tensile strain mangnitudes between the rise and dip sides of the extraction. Compressive strain is indicated to be increased in relation to a comparable flat seam situation. Considering the horizontal displacement profile, this shows a relatively minor proportion over the rise side whilst a substantial amount gradually builds up on the dip side.

Empirical determination of strain magnitudes in steeply inclined longwall workings.

Recognition of the appreciable difference in tensile strain magnitudes on both sides of the longwall extraction in steeply inclined measures is given in SEH (1975) and a method of adjusting equivalent flat seam tensile strain values for use in steep seams is put forward. These values are presented here in Table 16. The data given in Table 16 are based largely on examination of measured strains from subsidence investigations in UK coalfields. As stated in SEH (1975) the proportional correction values have been rounded off to allow the sum of the dip side and rise side strains to equal twice the maximum tensile strain for a flat seam, that is 2(+E). SEH (1975) states that this method of strain adjustment is intended as a useful guide rather than as an exact design procedure. It is intended to give useful guidance in the gradient range shown in Table 16.

When using the correction procedure for steep seams, SEH (1975) states that the mean depth and plan width of the longwall extraction should be used.

An example of the correction procedure is presented in Figure 132(b) which considers the horizontal surface strains above a longwall situation depicted by Figure 131(b). In this example the mean depth = 200m, extraction width = 200m, extracted seam height = 2m, seam gradient = 20°.

The calculated data are:

Plan width of longwall	=	188m
Equivalent w/h ratio	=	0·94
Maximum subsidence (S)	=	1·6m
+E = 0·65 S/h (flat seam)	=	5·2mm/m
−E = 0·55 S/h (flat seam)	=	4·4mm/m
For 20° dip (Table 16):		
$+E_{rise}$ = 0·435 (5·2)	=	2·26mm/m
$+E_{dip}$ = 1·565 (5·2)	=	8·14mm/m
−E = 1·565 (4·4)	=	6·89mm/m

The calculation of the maximum compressive strain (−E) for the steep seam situation using flat seam strain data is not referred to by SEH (1975) when discussing the treatment of dipping seams. Subsequent analysis of surface strain as discussed later in this chapter indicates that the maximum compressive strain is likely to increase similar to the maximum tensile strain on the dip side of the longwall extraction. Consequently, for the purpose of representing the complete strain profile across the longwall shown in Figure 132(b), the value for −E as determined for the flat seam condition has been increased by a factor of 1·565 which is consistent with determination of $+E_{dip}$ for this particular gradient. The accurate positioning of the maximum tensile strain on the dip side becomes especially important in steep seam working in view of the surface damage potential of this particular zone. It is clear that the accuracy with which these strain zones can be determined depends very much on the method of predicting the position of the subsidence trough. The method of determining the subsidence limits and position of the maxmimum subsidence as advanced by Rom (1964) has been used as a basis for the UK SEH (1975) steep seam prediction model. These two methods employ relatively small differences in specification of limit angles and position of maximum subsidence. However, such small differences in angular positioning of the subsidence trough can lead to appreciable differences in predicting the location of the zones of high strain. Attention is focussed on comparing these empirical design procedures with results obtained by influence function methods discussed later in this chapter.

Steeply inclined seams result in surface subsidence troughs exhibiting features associated with the dip of the strata. There is a tendency for some sliding to occur along the bedding planes, so that natural weakness planes can open appreciably especially where the dip is 45° or more. Personal observations by the authors of steeply pitching coal seams in Spain indicate

Seam dip (degrees)	Gradient	Proportion of Normal Tensile Strain	
		Rise	Dip
33·7	1 in 1·5	0·29	1·71
26·6	1 in 2·0	0·35	1·65
21·8	1 in 2·5	0·41	1·59
18·4	1 in 3·0	0·46	1·54
15·9	1 in 3·5	0·51	1·49
14·0	1 in 4·0	0·56	1·44
12·5	1 in 4·5	0·60	1·40
11·3	1 in 5·0	0·64	1·36
10·3	1 in 5·5	0·69	1·31
9·5	1 in 6·0	0·73	1·27
8·7	1 in 6·5	0·76	1·24
8·1	1 in 7·0	0·80	1·20
7·6	1 in 7·5	0·83	1·17
7·1	1 in 8·0	0·85	1·15
6·7	1 in 8·5	0·87	1·13
6·3	1 in 9·0	0·89	1·11
6·0	1 in 9·5	0·91	1·09
5·7	1 in 10	0·92	1·08

Table 16 Table of data for adjusting equivalent flat seam tensile strain values for use in steep seams, after SEH (1975).

that large strains on the sides of a longwall working result in the opening of natural fissures and development of appreciable stepping, see Figure 133.

Influence function methods applied to mining subsidence and displacement prediction

Attention has been focussed on the prediction of surface subsidence and displacements arising from longwall mining of steep seams, using various influence function methods and the SEH (1975) design procedure, see Whittaker, Reddish and Ren (1988) on whose paper the following comments are based.

The influence function approach to flat seam surface subsidence prediction has been discussed in Chapter 4, and it is now examined in relation to steep seam conditions.

Circular influence area method (one angle method). The area of influence around a point P overlying a longwall extraction in a steep seam is assumed to be circular in shape. The radius of the circle of influence is governed by the depth below surface and the angle of draw which is of equal value in all directions with this method. The maximum subsidence is located over the centre of the extraction, whilst the two limit angles (γ_L, γ_H) are automatically adjusted by the method. Figure 134(a) illustrates the basic principle of this method.

Two angle elliptical method (two angle method). The angle of draw is considered to change in relation to the dip of the seam, and the area of influence is elliptical in shape with a corresponding displaced point of maximum influence (z). The limit angles are determined from previously established field observations. Figure 134(b) shows the basic principle with

Figure 133 Illustrating opening of natural fissures and development of surface stepping overlying longwall extraction in a seam dipping at greater than 75°

reference to three surface points $P_{1,\ 2,\ 3}$ whose two limit angles define the limits of influence of seam extractions.

Figure 135 shows the projection method adopted. The ellipse central point (0) bisects the line AB. Further subdivision of the area of influence is achieved by projection of the limit line intercepts and central point onto the level plane. The two halves are then divided as if they were a flat seam problem using the stochastic function (see Appendix 1). The line positions are projected into the seam to produce the elliptical rings as shown. This method generates its own nadir angle by virtue of the level seam assumption employed.

The nadir angle μ is given by equation (60).

$$\mu = \tan^{-1}\left(0{\cdot}5\left[1/\tan\gamma_H - 1/\tan\gamma_L\right]\right) \qquad \ldots(60)$$

Three angle elliptical method. The method is essentially an extension of the previous two angle method, except that the nadir angle (influence central angle) can be selected to achieve more realistic results. Figure 136 illustrates the basic principle in respect of projecting the magnitude of influence over that part of the seam within the limit angles γ_L and γ_H.

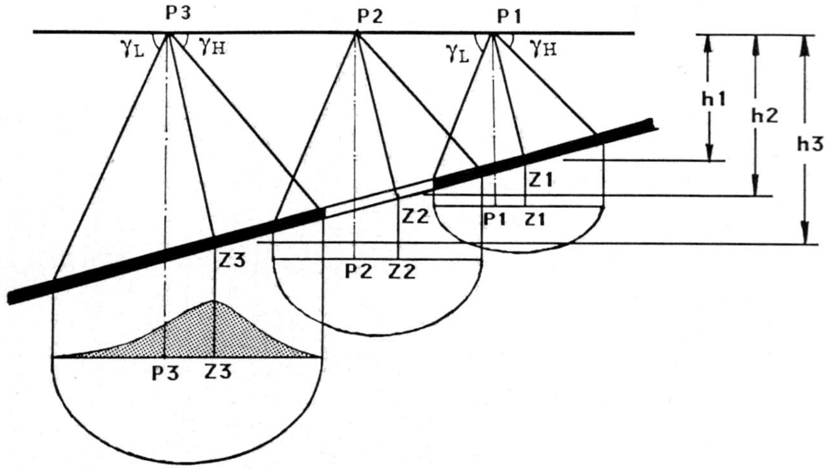

(a) Circular influence area method (one angle method)

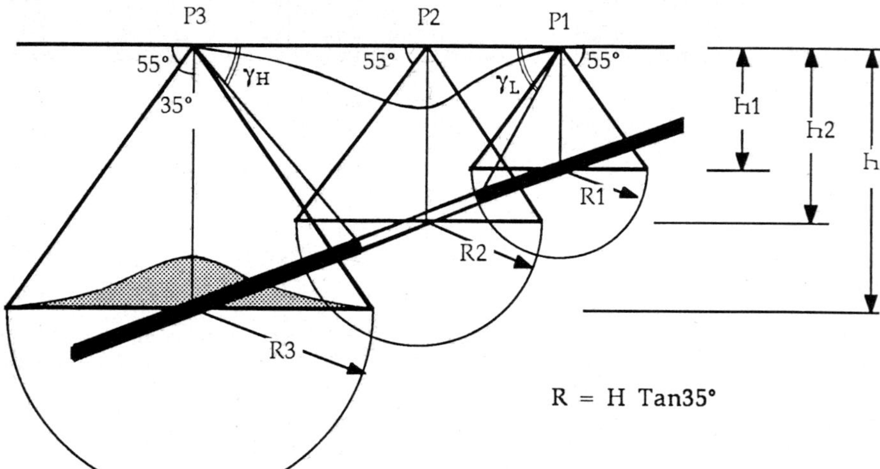

(b) Two angle elliptical method (two angle method)

Figure 134 Comparison of circular and two angle elliptical influence area methods.

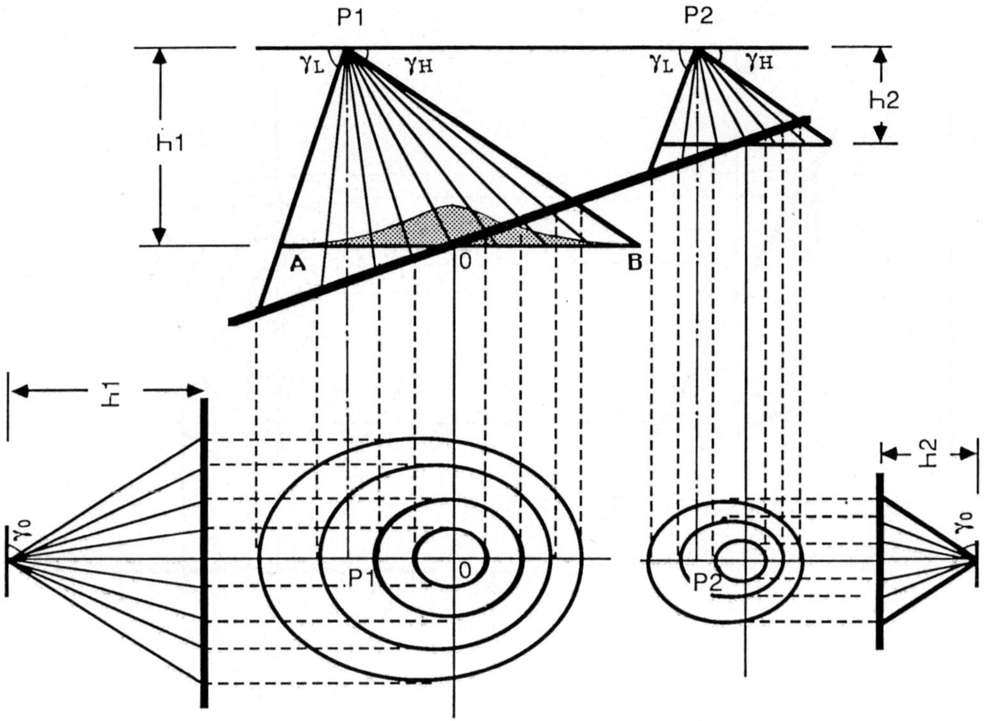

Figure 135 Illustrating method of projecting subsidence limits and ellipse positions with the two angle influence function method.

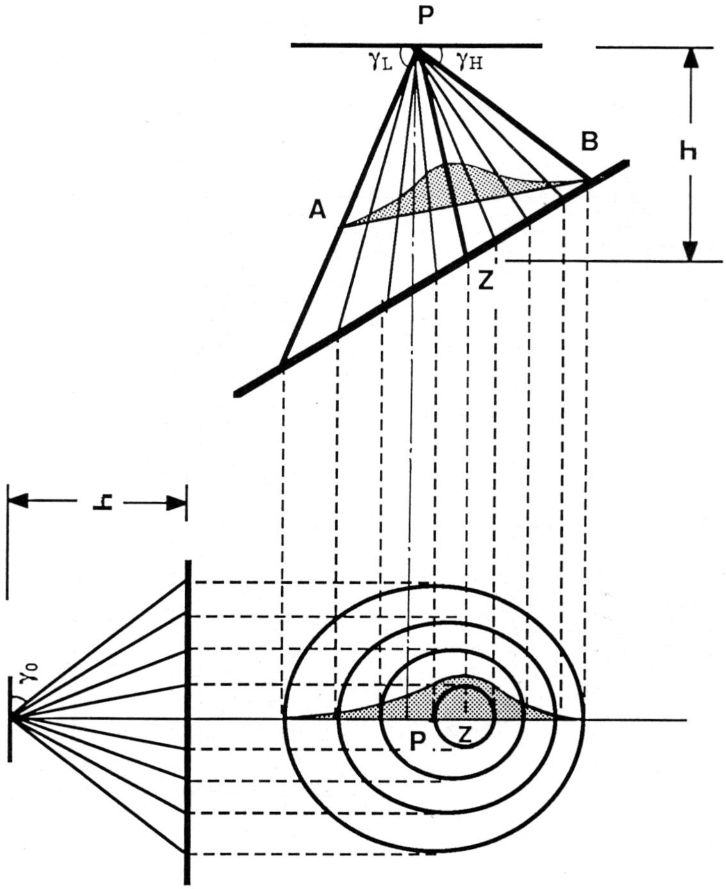

Figure 136 Illustrating the basic principle of the three angle elliptical influence area method for subsidence determination in steeply inclined seams.

The main controlling factor is the positioning of the focal point z and this is determined from knowledge of the nadir angle μ. The focal point z represents the maximum centre of influence in its attraction of the surface point P. The value of the nadir angle (μ) has been reported by Kratzsch (1983) as given by equation (61) for some mining districts, presumably West German coalfields, but he points out that in Upper Silesia and the Netherlands its value has been observed to be $0{\cdot}7\alpha$ and $0{\cdot}5\alpha$ respectively.

$$\mu = 0{\cdot}5\,(\gamma_L - \gamma_H) \qquad \ldots(61)$$

In respect of equation (61), the values of γ_L and γ_H need to be established by field observation. Kratzsch (1983) reports that for a dip angle (α) of 40° a nadir angle (μ) of 5° is recommended, and for α = 60° the corresponding value of μ = 15° whilst at α = 80° then μ = 25°. It should be recognised, however, that local conditions may influence the value of μ quite significantly.

Figure 137(a) illustrates comparison of the nadir and zenith angles, whilst Figure 137(b) shows limit angle in relation to seam dip.

Maximum subsidence determination. The SEH (1965, 1975) empirical design procedure for determining the maximum value of subsidence in flat seams has been discussed in Chapter 6. It was shown that the extraction width to depth (w/h) ratio is a major controlling parameter and the result is expressed as a function of the extracted seam height. A basic assumption made here in extending this design procedure to steep seams is that maximum subsidence will also be a function of the cosine of the angle of seam inclination so that the subsidence (S) is given by equation (62).

$$S = a\,M \cos\alpha \qquad \ldots(62)$$

where a = maximum subsidence factor for flat conditions, where a = 0·9 for UK coalfields and most European coalfields

M = extracted seam height (normal to the seam),

α = dip of the seam.

Figure 138 shows a graphical relationship between S/M and seam inclination.

Displacement calculation. The focal point method has been used to calculate displacements of surface points overlying longwall extractions in steeply inclined seams. Appendix 1 gives details of the application of this method to level seam situations. The focal point method is equally applicable to inclined seams, although an additional assumption needs to be made during the calculations. Whilst the level seam situation results in a circular shape of influence area, it appears that in the case of inclined seams the influence area is elliptical in character. The focal point theory remains the same as for a level seam, namely in that each extraction element exerts an influence on a surface point P, and so attracts P to move towards the source of attraction.

Figure 139 illustrates in section and plan the extraction of the element dA. The subsidence component is given by dV_Z and this can be determined by the influence function. Consequently, the horizontal displacement vector dV_{xy} can be calculated by equation (63).

$$dV_{xy} = dV_Z \tan\zeta \qquad \ldots(63)$$

where ζ = angle subtending the vertical and the line meeting the surface point P from the extraction element dA.

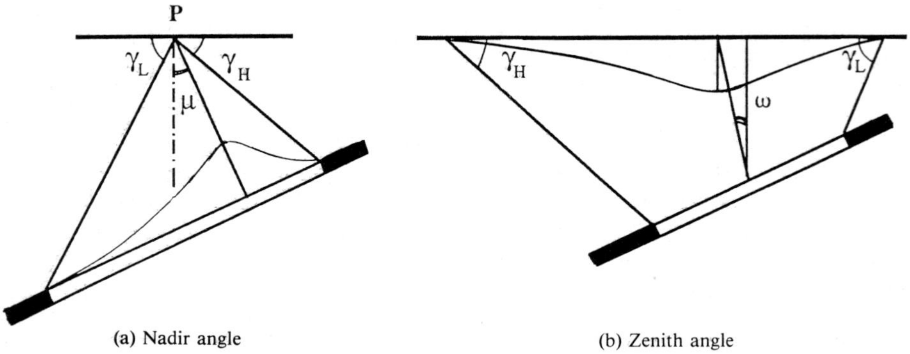

Figure 137(a) Illustrating the nadir and zenith angles as applied to steeply inclined extractions

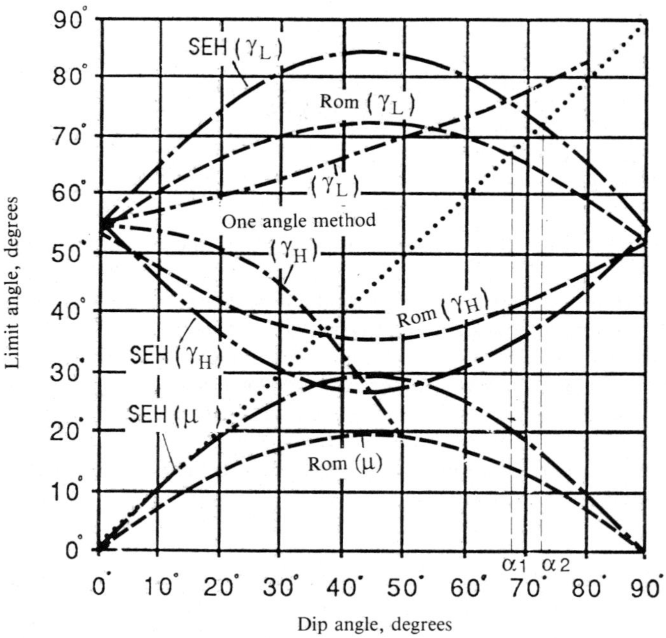

Figure 137(b) Comparison of limit angles in relation to seam dip angle.
Note: For one angle method $\mu = 0$.

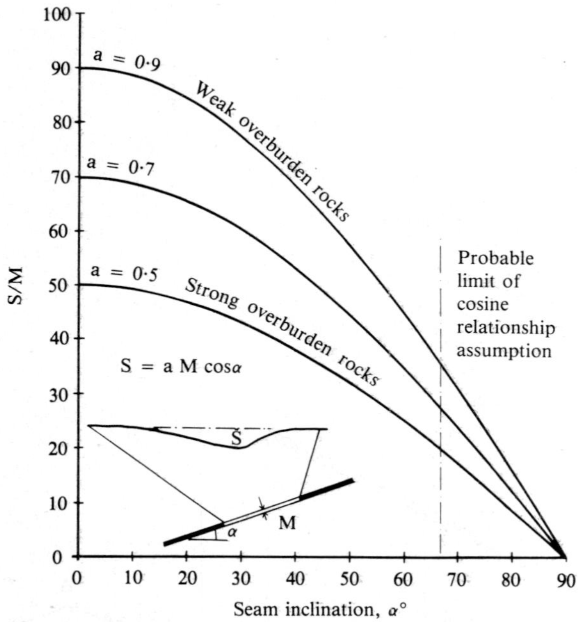

Figure 138 Effect of seam inclination on maximum subsidence.

Displacement calculation is the same as that for level seams, see Appendix 1 for details. The displacement occurring at point P is the summation of the influence of every element that has been extracted, and is expressed in general form as given by equation (64a and b).

$$V_X = \sum dV_X \qquad \text{...(64a)}$$

$$V_y = \sum dV_y \qquad \text{...(64b)}$$

Total extraction of the whole influence area will not necessarily result in a nil horizontal displacement at P owing to the non-symmetrical character of the influence area of an inclined seam. Consequently, with a super critical extraction there is an area of uniform displacement down dip over its centre, which contrasts with the zero displacement in the centre of the surface subsidence trough of a flat seam.

Examples of subsidence and displacement output. In order to compare the subsidence and displacement calculation output for inclined seams as opposed to a flat seam situation, a width of extraction of 600m was chosen, and a central depth of 350m employed. The results calculated using the three angle method are presented in Figure 140(a), (b) and (c). The broadening of the subsidence trough on the dip side of the inclined seam situations is in evidence, as is the general concentration of surface ground strains. The maximum subsidence at the surface is less with the inclined seam cases as compared to the flat seam situation; this is undoubtedly due to the effective reduction in plan width of extraction in relation to depth even though the rise side of the inclined extractions is appreciably closer to the surface. The strains are significantly changed in the inclined seam situations as compared to the flat situation, this being particularly apparent when $\alpha = 50°$.

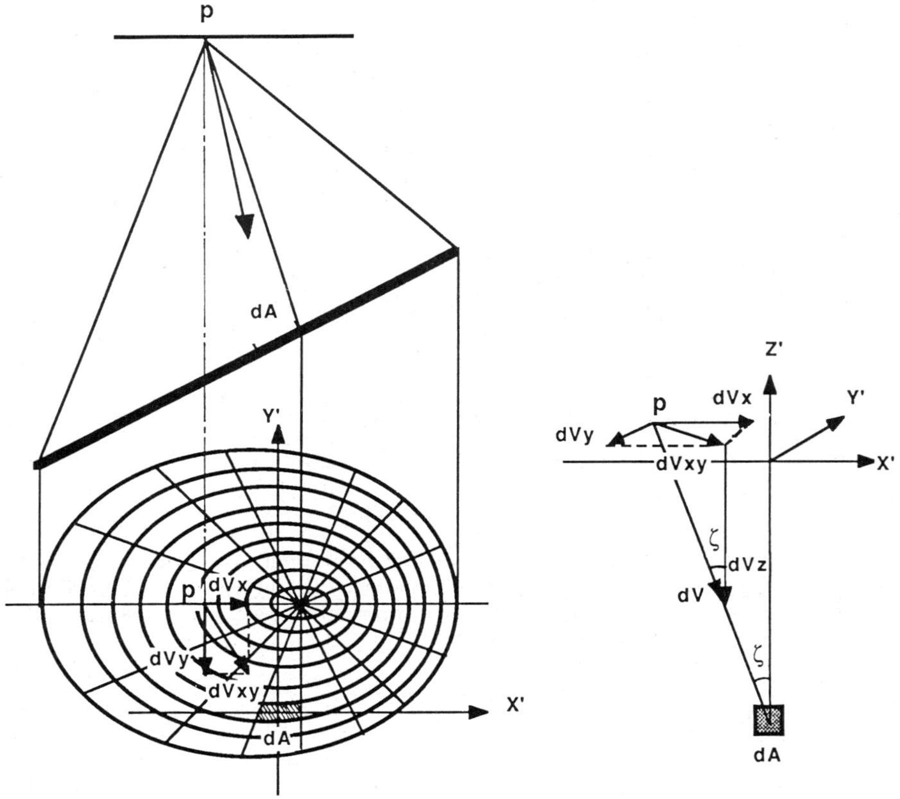

(a) Extraction of element dA in section and plan

(b) Vector representation

Figure 139 Illustrating the method of displacement calculation.

Table 17 gives a summary of maximum principal strains shown in the strain output diagrams of Figures 140(a), (b) and (c). The extracted seam height was 3m in all cases; the flat seam situation maximum subsidence was 2·7m for this super critical width of extraction.

Seam inclination	Maximum principal strains					Figure No. (text)
	Compression		Tension			
	Transverse	Corner	Rise	Dip	Corner	
	mm/m	mm/m	mm/m	mm/m	mm/m	
0	−1·3	−2·1	+1·75	+1·75	+1·55	140(a)
25°	−2·0	−2·9	+0·72	+1·63	+1·75	140(b)
50°	−2·0	−3·65	+0·55	+1·1	+1·6	140(c)

Table 17 Summary of maximum principal strains for flat, 25° dip and 50° dip longwall super critical extractions, as determined from displacements by influence function method.

It is shown in Table 17 that for the longwall extraction configuration used changes in strain are experienced as follows:

1. Compressive strain is increased with increasing dip for both the flank (transverse section of the longwall) and corner regions of the extraction. The latter experienced a 74% increase, over the flat seam condition, when the dip was 50°, whilst in the transverse direction the increase was 54%.

2. Tensile ground strains at the surface are indicated to decrease with increasing dip for a comparable flat seam situation as employed here. Both the rise side and dip side maximum principal strains decreased, although in differing proportions. For the 25° dip condition the rise and dip maximum principal tensile strains when reduced to the same proportions as given in Table 16, that is the SEH (1975) proportioning to allow their sum to equal 2, gives 0·61 (rise): 1·39 (dip). This ratio indicates the difference between the maximum tensile strain to be less than would be anticipated by SEH (1975), although it is recognised that this is about at the limit of reliability of SEH (1975) regarding availability of steep seam subsidence data.

3. The corner tensile strains increase over the flat seam situation, for both the angles of dip considered. Tensile strain magnitudes in the corner region of the longwall extraction are not generally taken into consideration in routine subsidence strain assessments. These results suggest that the corner regions should be taken into consideration.

Considering the results in relation to Figures 140(a), (b) and (c), due attention needs to be drawn to the size of the zones over which these principal strains manifest themselves at the surface. As the gradient increases the tensile strain zone on the rise side diminishes significantly. Conversely, the tensile zone on the dip side increases significantly. This is consistent with the reported findings of Rom (1964) and Kratzsch (1983).

The principal strain configurations presented in Figures 140(a), (b) and (c) show that in the corner region of the subsidence trough, it is fairly common to have principal strains compressive in one direction and tensile at right angles to it at several points.

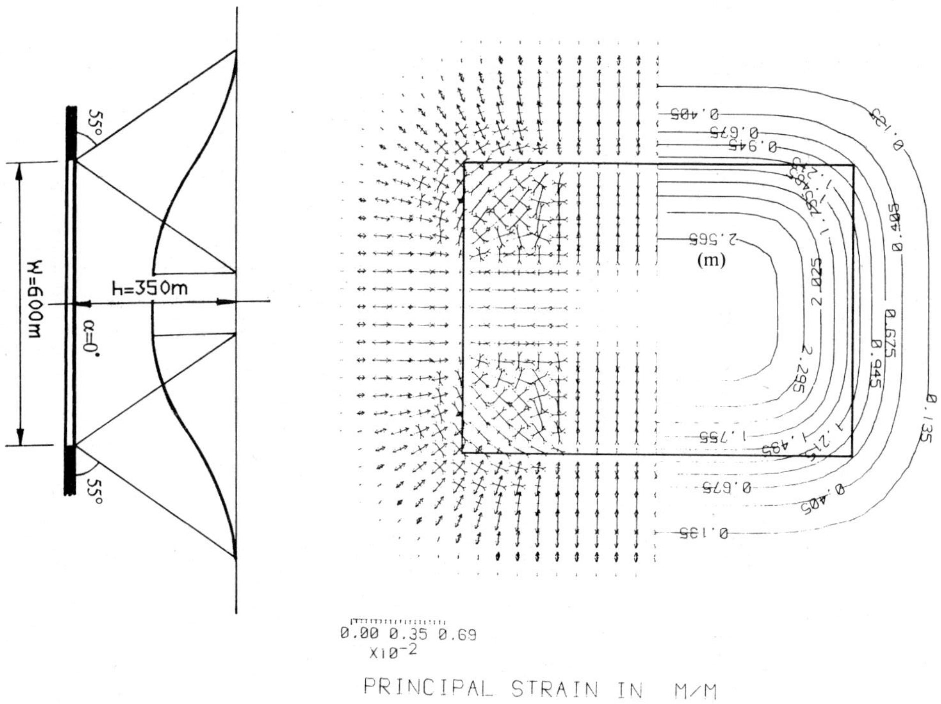

Figure 140(a) Subsidence and principal strain comparison: super critical extraction, $\alpha = 0°$.

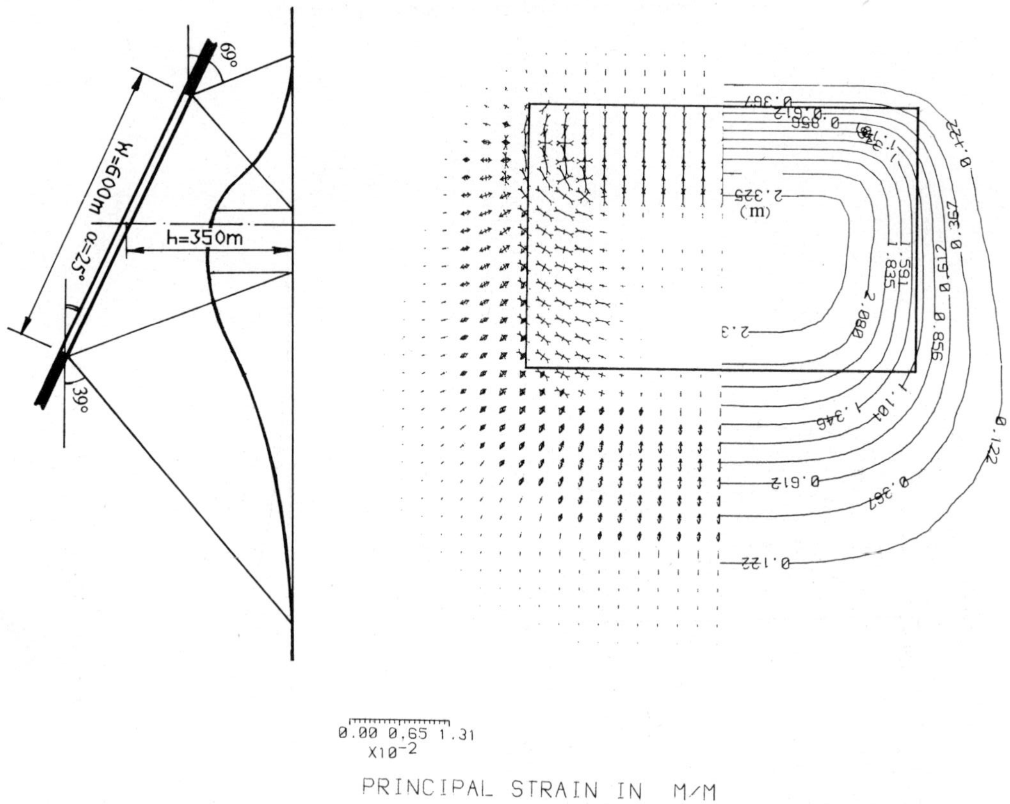

Figure 140(b) Subsidence and principal strain comparison: super critical extraction, $\alpha = 25°$

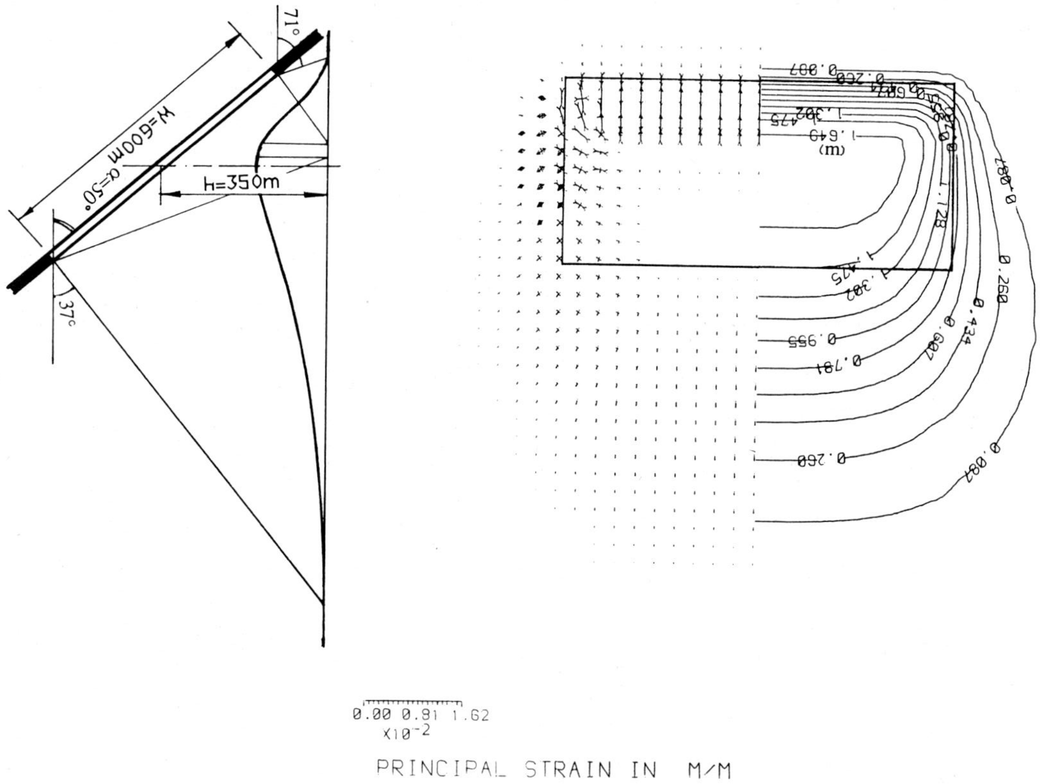

Figure 140(c) Subsidence and principal strain comparison: super critical extraction, $\alpha = 50°$.

Case history comparisons using different prediction methods. Three case histories have been used to allow comparison of the prediction methods discussed here. The seam inclinations considered are $\alpha = 13{\cdot}4°$, $\alpha = 43°$ and $\alpha = 62°$; the first case history is from the UK western coalfield whilst the other two are from the northern coalfield of China.

Figure 141(a) and (b) is a case history where the seam is gently dipping, $\alpha = 13{\cdot}4°$, and illustrate that under these conditions the SEH method produces the closest comparison with the measured curves for both subsidence and displacement. The other prediction methods produce the correct shape of curves, but with a tendency to over predict the maximum values of subsidence and displacement.

Figure 142(a) and (b) present the results for a case history where $\alpha = 43°$. It is evident that the three angle prediction (influence function) method gives the closest comparison as regards subsidence profile position and magnitude. The SEH method is tending to displace the subsidence trough too far on the dip side. Considering the displacement results, there appears to be a rise side effect which the prediction methods do not indicate. The three angle method gives the more satisfactory result for this case study comparison.

Figure 143(a) and (b) examine a case history comparison where $\alpha = 62°$. The mining situation was complicated by the presence of the pillars, but it does show that both the SEH and three angle influence function methods were predicting results which were fairly consistent with the measured values.

Case history examples from UK coalfields

Figures 144 to 148 present subsidence and strain profiles for observations made in UK coalfields above longwall extractions in seams pitching at different gradients.

Although a fault intercepted the seam on the left-hand side of the longwall of Figure 144 it was judged to be essentially of a localised nature and not sufficiently significant to greatly influence the surface subsidence results. The subsidence and strain profiles display the onset of the effect of gradient, namely a slight displacement in the position of the subsidence trough down the gradient, an increase in compressive strain, and a difference in magnitude between the maximum tensile strain values on the rise and dip sides of the longwall.

Longwall panel No. 1 of Figure 145 displays very similar characteristics to those of Figure 144. However, there is the additional effect of extracting panel No. 2 and thereafter panel No. 3 so that the combined result of all extractions can be seen. The effect of increasing the width of the extraction has not greatly changed the tensile strains on the rise side of the extracted area, nor the magnitude of the compressive strains. The tensile strain magnitudes on the dip side of the longwall are greater than on the rise side, but the increase in extracted width does not appear to have greatly influenced their magnitude. The shape of the subsidence trough is consistent with that which would be expected for gradient effects.

Figure 146 shows a case example for a gradient of 19° dip again with a displaced subsidence trough. The depth of mining is fairly shallow, with h_{mean} being only 46m. The maximum tensile strain on the dip side is substantially greater than on the rise side, ie. about four times, and is considered to be exceptional by comparison to anticipated results. However, the general pattern of subsidence behaviour is fairly consistent with the previous examples.

The case example presented in Figure 147 demonstrates the character of subsidence arising from longwall mining in a seam dipping at 21°. The subsidence trough is displaced

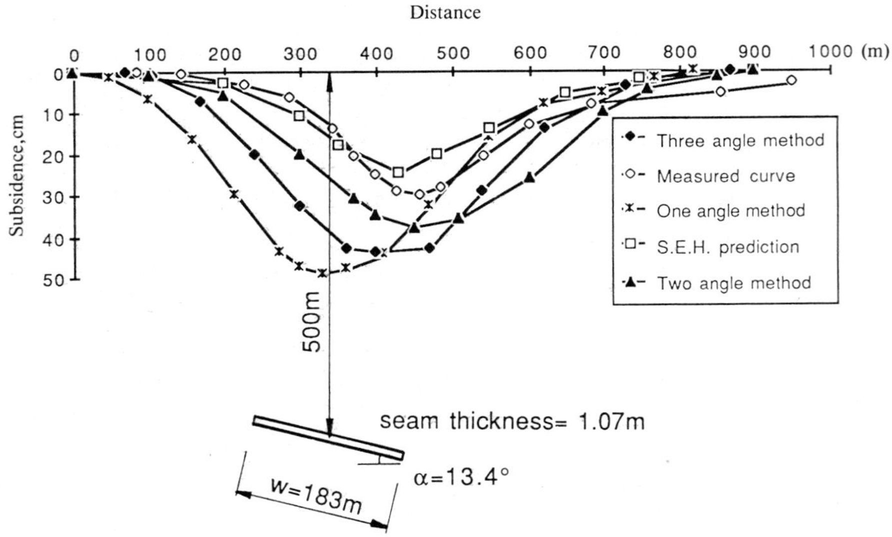

Figure 141(a) Case history comparison, from UK western coalfield; α = 13·4°. Comparison of subsidence profiles.

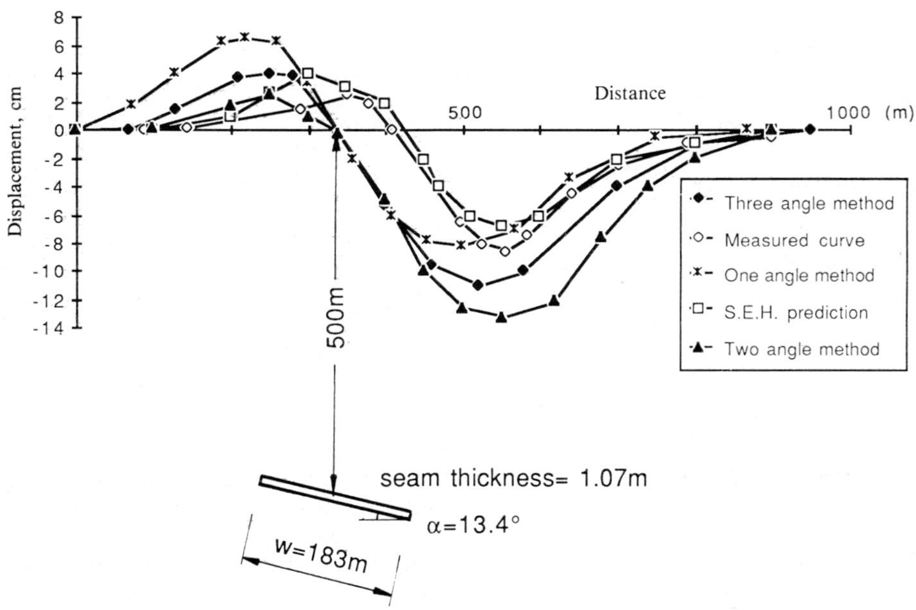

Figure 141 (b) Case history comparison, from UK western coalfield; α = 13·4°. Comparison of displacement profiles.

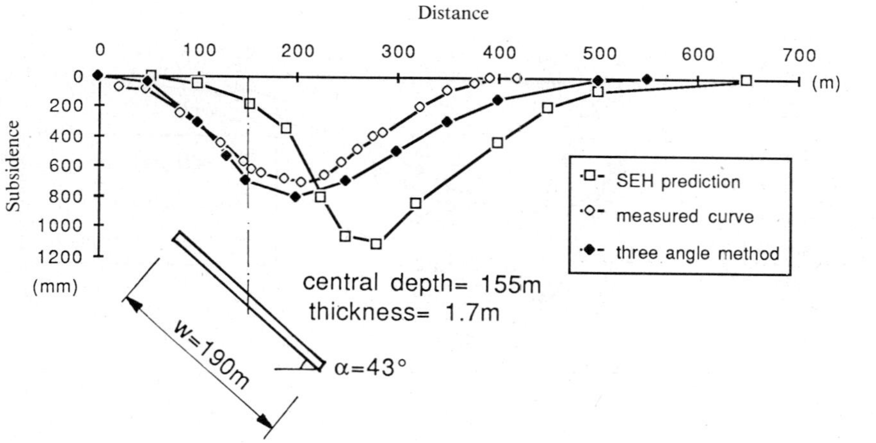

Figure 142(a) Comparison between influence function method, SEH normal prediction and the measured profile in respect of subsidence: α = 43°. Northern coalfield, China.

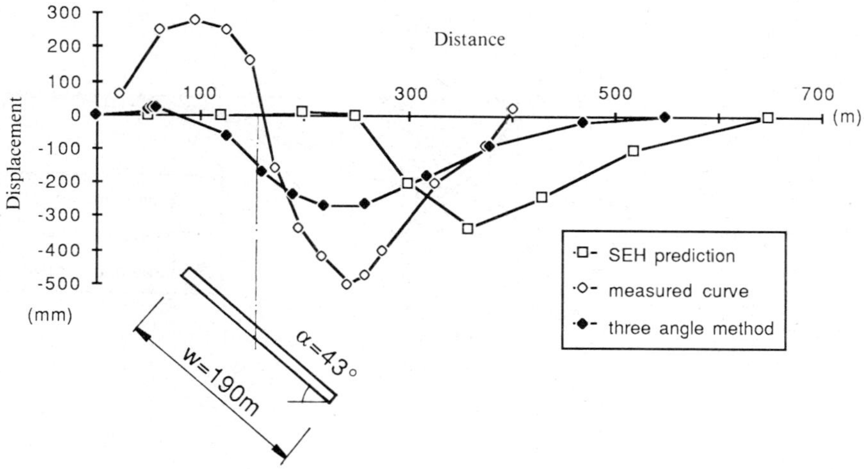

Figure 142(b) Comparison of predicted curves with the measured displacement profile: α = 43°. Northern coalfield, China.

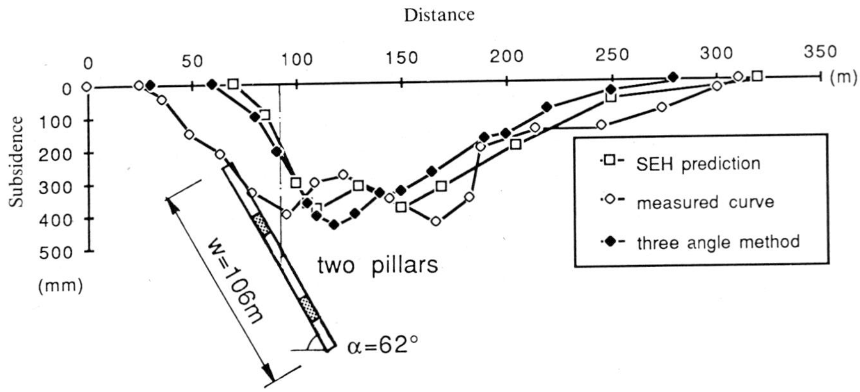

Figure 143(a) Comparison between influence function methods, SEH normal prediction and measured profile in respect of subsidence: $\alpha = 62°$. Northern coalfield, China.

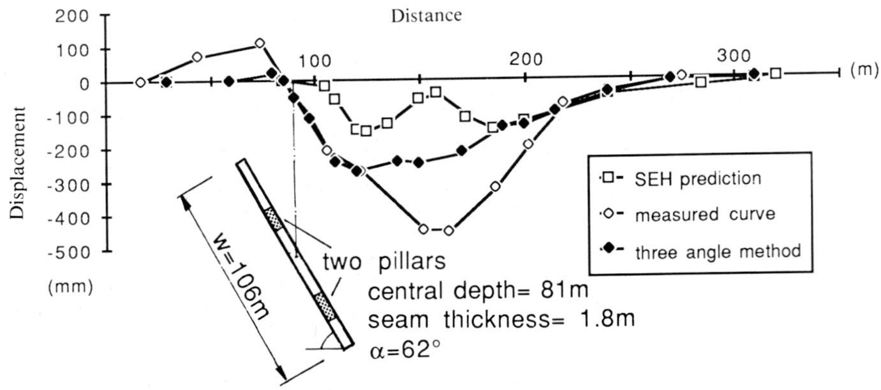

Figure 143(b) Comparison of predicted curves with the measured displacement profile: $\alpha = 62°$. Northern coalfield, China.

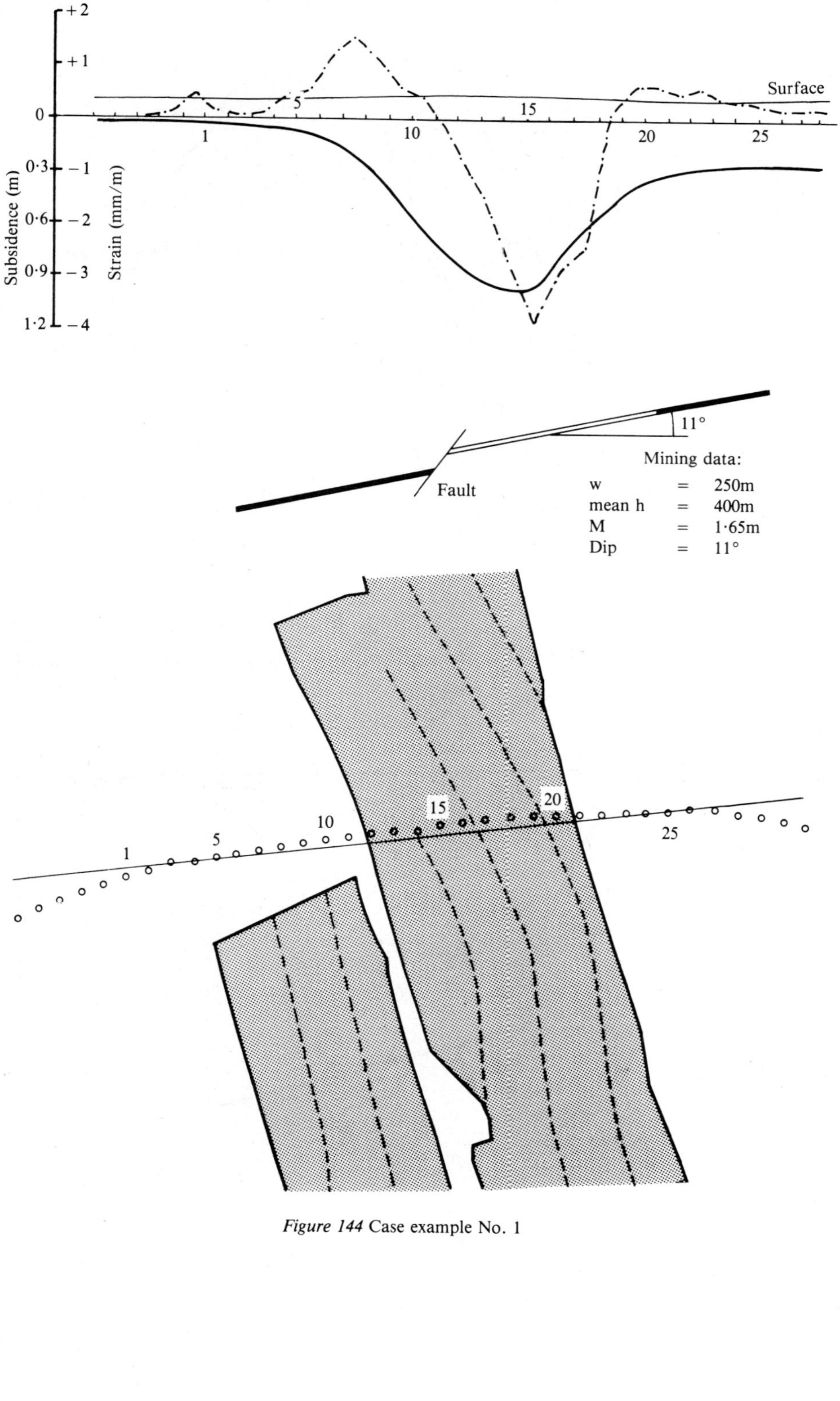

Figure 144 Case example No. 1

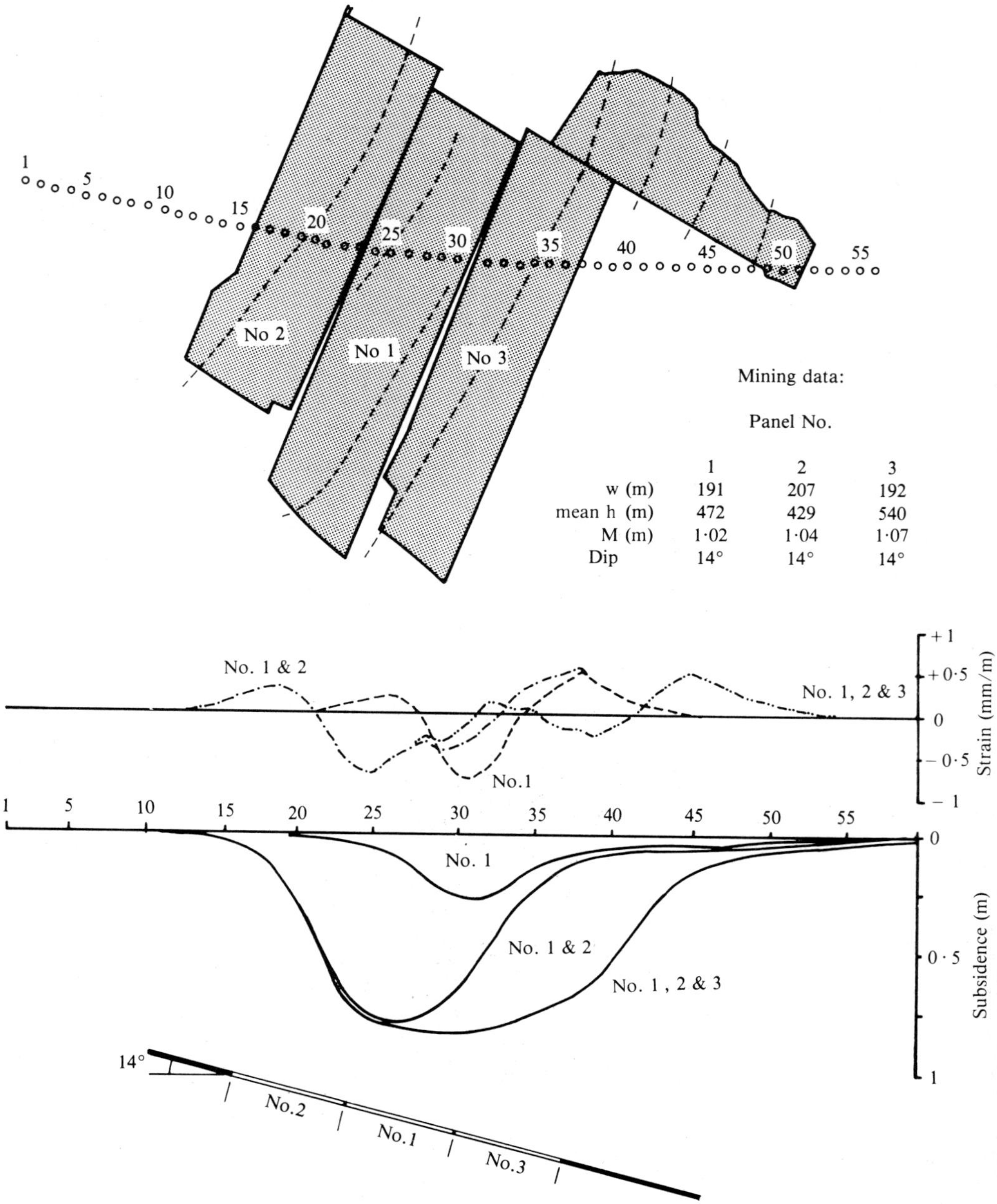

Mining data:

Panel No.

	1	2	3
w (m)	191	207	192
mean h (m)	472	429	540
M (m)	1·02	1·04	1·07
Dip	14°	14°	14°

Figure 145 Case example No. 2

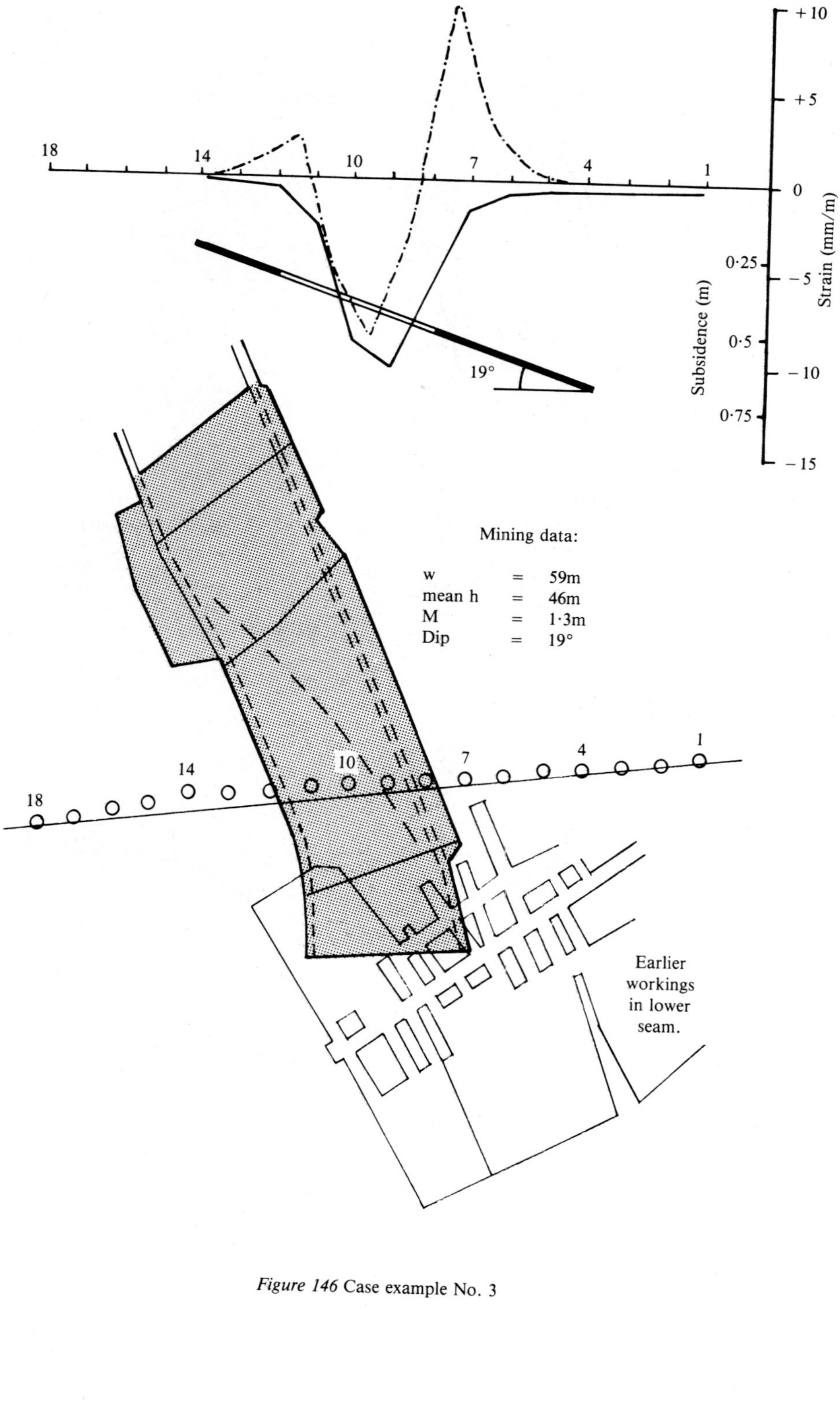

Figure 146 Case example No. 3

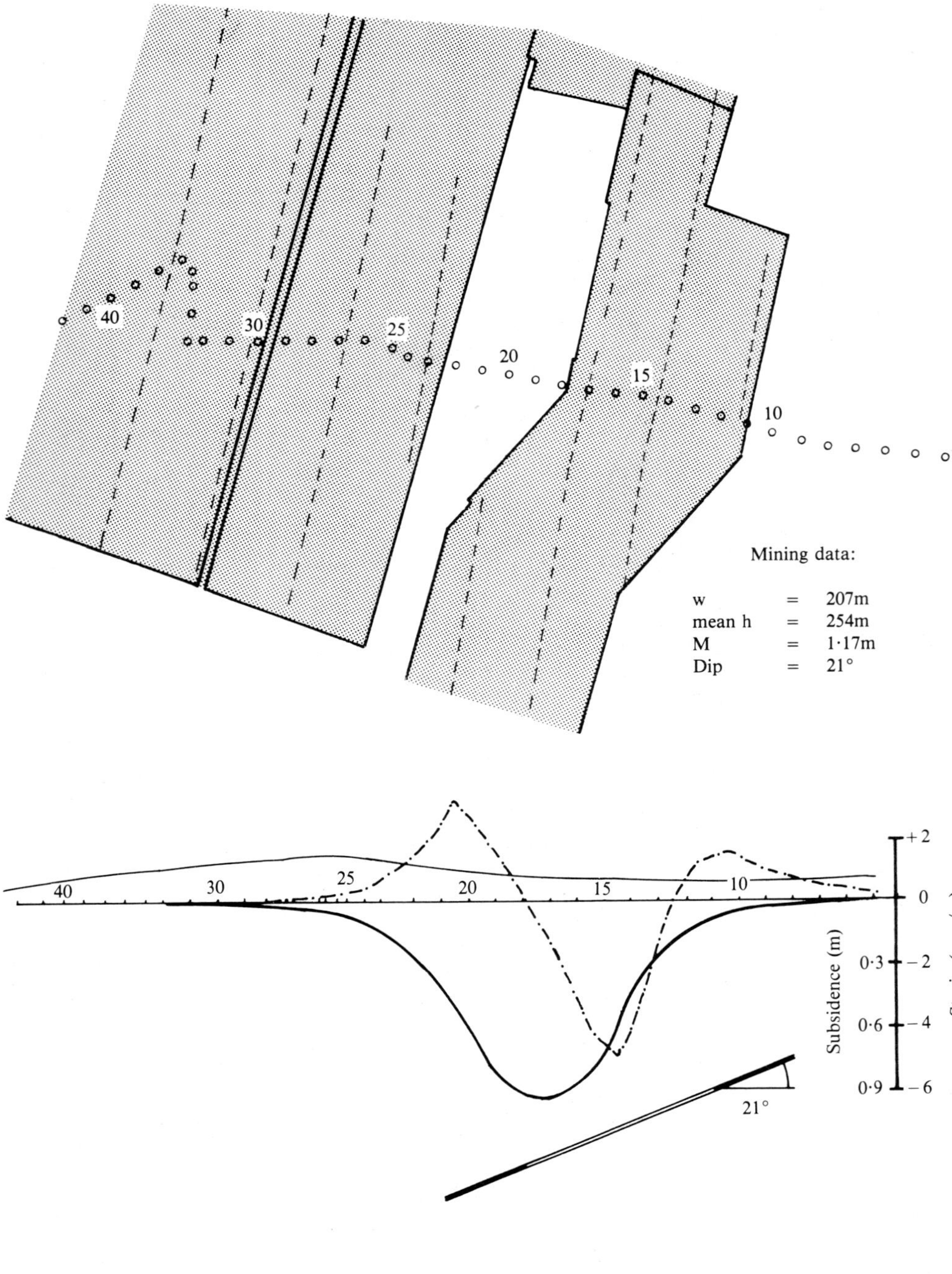

Figure 147 Case example No. 4

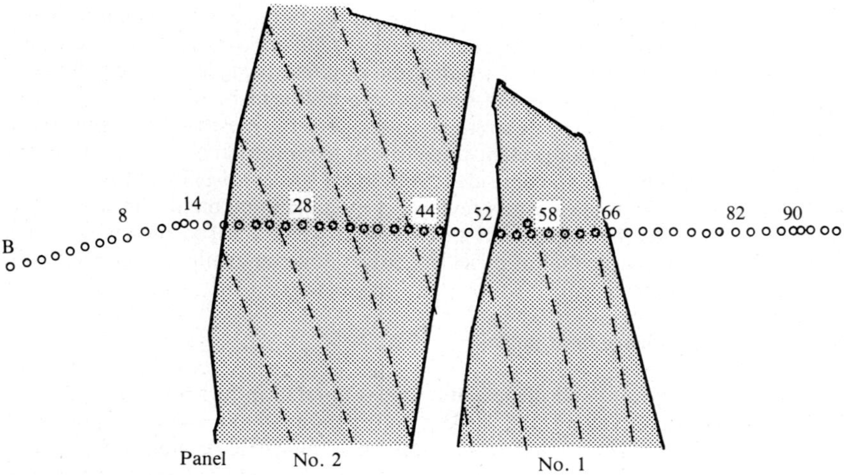

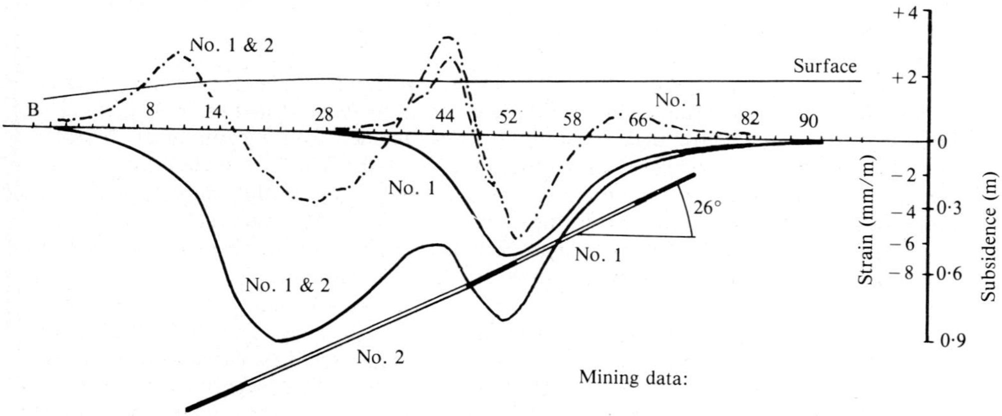

Mining data:

	Panel No. 1	Panel No. 2
w (m)	128	241
mean h (m)	147	241
M (m)	1·3	1·3
Dip	26°	26°

Figure 148 Case example No. 5

down the dip as expected. The maximum tensile strain values on the rise and dip sides again differ in magnitude, but not to the same extent as Figure 146 nor Table 16 for such a gradient.

Figure 148 shows the effect of extracting two faces in succession on a gradient of 26°. Following extraction of panel No. 1, the subsidence trough dispacement and difference between maximum tensile strain values for the two sides of the extraction are similar to those discussed previously in this section. The compressive strain appears to have increased significantly. The extraction of panel No. 2 resulted in a rib pillar being left between the two workings and consequently its effect on the subsidence profile can be seen particularly in displacing the hump down dip. The tensile strain increase due to the presence of the pillar was not significantly more than had been present prior to extraction of panel No. 2. However, after extraction of the second panel the dip side tensile strain was consistent with that of the first panel. The fact that the rise side tensile strain was greater than that on the dip side in this situation was entirely due to the presence of the rib pillar which masked the effects. The general pattern of subsidence behaviour of panel No. 2 does however exhibit consistency with the other cases presented here.

Further discussion on subsidence limit angles

Zilavy (1968) has considered subsidence limit angles in relation to seam gradient for a longwall extraction operating along the strike, and Figure 149(a) shows a graphical representation of his results. The curved relationships of limit angle on both sides of the longwall show a similar general character to that of SEH (1975) although with decreased magnitudes of change in limit angle. These results appear to relate to experiences in European coalfields. Zilavy's subsidence limit values tend to indicate that at fairly low magnitudes of seam gradient, there is not such an appreciable change in the subsidence limits, but beyond the 60° dip value the limit angles experience rapid change back to the flat seam value as verticality is approached. The limit angles for the level seam condition are much less than those generally reported for West European coalfields.

Subsidence limit angle data for seams pitching at different gradients have been collated by Ren, Reddish and Whittaker (1988) and presented in the same format as SEH (1975). The limit angle data and centre angle values have been obtained from measurements taken in different countries. The results are presented in Figure 149(b). The limit angle values for dip-side and rise-side situations are plotted separately, however, as are the centre angle data in order to ease comparison and assist clarification. The SEH (1975) and Rom (1964) limit angle relationships in respect of seam gradient are included for comparison. There are a number of features exhibited by Figure 149(b) and these are listed as follows:

1. The three suggested relationships show some marked differences. The SEH (1975) curves were based on limited data at relatively shallow gradients, and these curves appear to have been extrapolated to be consistent with Rom (1964) especially concerning general shape. SEH (1975) probably represents a highly conservative assessment of the likely limits of subsidence in that the curves encompass almost all the data values plotted in Figure 149(b). The curves of Ren, Reddish and Whittaker (1988) are judged to represent the mean relationship of the data points plotted, and may be regarded as the anticipated relationship.

2. Due consideration should be taken of the fact that the data points plotted were probably obtained by surveying to differing orders of subsidence measurement accuracy. Additionally natural ground movement effects will have exhibited different orders of magnitude from country to country. Quite apart from these aspects however, it is considered that the mean curves indicate important trends.

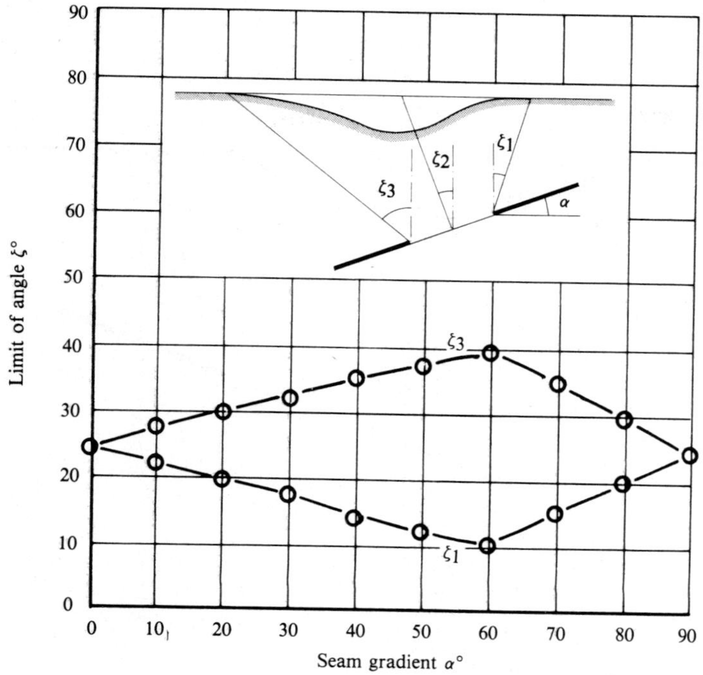

Figure 149(a) Subsidence limit angles in relation to seam inclination according to Zilavy (1968)

3. For seam gradients up to around 20° dip, the subsidence limit curves are all very similar as regards the dip-side situation. Beyond 25° dip there is a marked departure involving all three curves. No data points support the highly conservative curve of SEH (1975) beyond 25-30° dip. Even the less conservative curve of Rom (1964) does no correspond with these data points for dips greater than 25-30°.

4. The rise-side subsidence limit angle values show appreciable scatter of data, although the mean curve of Ren et al appears to be consistent with the relationship suggested by Rom.

5. The centre angle data which represent the position of maximum subsidence, show that there is general agreement between all three curves up to around 20° dip. Beyond this value of dip, however, the SEH (1975) relationship appears to greatly exaggerate the shift in position of the subsidence trough. There is consistency between Rom and Ren et al regarding the predicted shift in maximum subsidence for steeply inclined seam conditions.

Considering the dip-side limit angle curves, it appears from the subsidence data measurements that the gradient does no have such a marked effect as previously anticipated for seam gradients greater that 25-30°. The authors would expect that at gradients approaching the vertical condition, small changes in gradient would have a much less effect as compared to similar gradient changes close to the flat condition when considering such effects on the limits of subsidence.

Subsidence terminology:

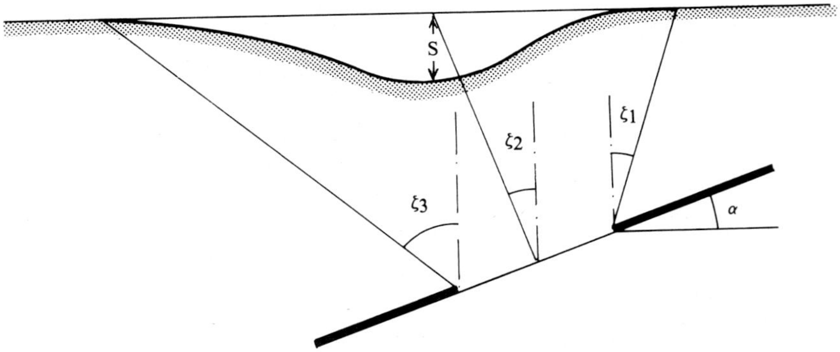

Subsidence data:

No	Source
1	Western Coalfield, UK: British Coal
2	Western Coalfield, UK: British Coal
3	Western Coalfield, UK: British Coal
4	Western Coalfield, UK: British Coal
5	Western Coalfield, UK: British Coal
6	Western Coalfield, UK: British Coal
7	Northern Coalfield, China: Peking Coal Mining Research Institute
8	Western Coalfield, UK: British Coal
9	Japanese Coalfield: Hiramatsu et al (1979)
10	Mid-West Coalfield, China: CIMT (1982)
11	Mid-West Coalfield, China: CIMT (1982)
12	Mid-West Coalfield, China: CIMT (1982)
13	Central Coalfield, China: CIMT (1982)
14	Northern Coalfield, China: CIMT (1982)
15	Ostrava-Karvina Coalfield, Czechoslovakia: Zilavy (1968)
16	West European Coalfields: Stojanovic & Pataric (1979)
17	West European Coalfields: Stojanovic & Pataric (1979)
18	Austrian Tertiary Coal Basin: Perz (1957)

Figure 149(b) Subsidence observations in relation to limit angle and seam gradient.

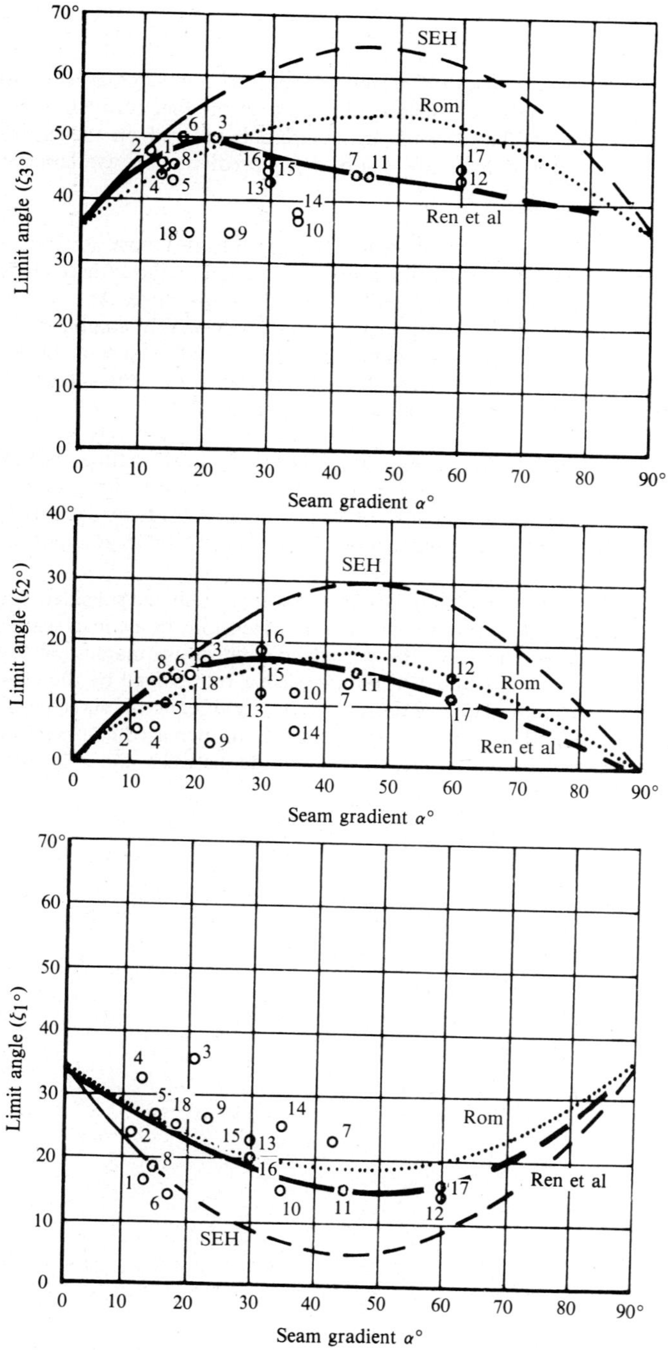

Figure 149(b) (continued). Subsidence observations in relation to limit angle and seam gradient.
Note: Subsidence data according to Ren, Reddish and Whittaker (1988).

Concluding remarks

Subsidence associated with inclined seams produces a non-symmetrical profile. The position of the maximum subsidence is displaced from over the face centre line, to generally the dip side of the extraction. The largest strains are encountered on the dip side of the extraction. A number of different methods of subsidence prediction are available for inclined seam conditions.

The SEH (1975) empirical design procedure is probably the most accurate for gently dipping (α <20°) seam situations, whilst the influence function method (three angle method) has indicated fairly close comparison in the α = 45° inclination range. At α = 60° both the SEH (1975) and influence function method gave fairly reasonable comparisons with the measured values. For seam inclinations greater than α = 60°, there is little published information available, so that the effect of near vertical seams on surface subsidence behaviour is still largely based on extrapolation or hypothesis.

The general impression formed is that the SEH (1975) steep seam subsidence profile positioning method could be improved for gradients around 45° dip and greater. The treatment of subsidence limit angles and position of maximum subsidence appears to be closer to that suggested by Rom (1964) for longwall extractions in this seam gradient range.

An important conclusion drawn from the comparison of predicted subsidence and strains for a flat seam with comparable extraction and depth dimensions in inclined seams, was that with increasing inclination of the seam the subsidence and strain magnitudes appreciably changed. This latter feature is attributed to the increasing role played by the effective plan width of extraction in relation to depth in decreasing the significance of subsidence effects at the surface on the rise side although it gives rise generally to increased compression over the extraction, and an appreciable difference in magnitude between the rise and dip side maximum tensile strain values.

CHAPTER 11

SUBSIDENCE ASPECTS OF SLOPING GROUND SURFACES

The discussion on subsidence behaviour of the surface is now extended to sloping ground conditions. In earlier chapters, attention has been focussed on the assumption that the surface overlying the influence of subsidence arising from underground excavation and subsequent collapse is essentially flat or that whatever slope exists is of no significance. Substantial underground mining operations take place in hilly and mountainous terrains in several countries. Subsidence needs to be carefully considered in such circumstances for a number of reasons.

Firstly, there is the changing depth of mining, with seams frequently exposed at the surface. Secondly, depending upon the nature of the immediate surface rocks, the effect of mining subsidence can result in surface cracking which may have implications on natural drainage and slope stability. Thirdly, the effect of the sloping ground can induce additional subsidence effects not generally encountered with the flat surface condition, particularly the tendency for increased displacements down the slope.

There are several other aspects to be taken into account and these will be discussed in relation to the more common methods of underground mining.

Room and Pillar Mining

Room and pillar mining in hilly country frequently encounters situations where the seams outcrop into the sloping surface thus giving rise to rapid changes in depth of cover. Where the depth of cover is less than 10 - 30m, there is a significant likelihood of the overburden being heavily weathered. This results in roof control within the room becoming more difficult. There is increased risk of collapses within the rooms and at junctions giving rise to sink-hole and surface depression problems.

Figure 150 illustrates the general subsidence problem arising with room and pillar mines close to the outcrop of coal seams in hilly country. Pillar failure close to the outcrop is comparatively rare owing to the depth of cover being small, unless the pillars have been deliberately robbed or influenced by adjacent de-pillaring operations. In view of the overburden being weakened by weathering effects, sink-holes and conical depressions frequently cluster together over the room and pillar workings close to the outcrop of the seam. The depth of the weathering effect depends upon the climatic conditions. Heavy rainfall conditions favour increased depth of weathering even where sandstones overlie coal seams. The Godavari Coalfield of India located near to Hyderabad, lies within the tropics and is subjected to heavy monsoon rains. This has given rise to sandstones being weathered to depths frequently down to 30m with increased risk of roof collapses at junctions and in some cases accompanying inflows of sand into the mine. The form of surface subsidence is that of a conical depression immediately above the collapsed junction. Such surface depressions can result in water gaining access to the mine, especially from local water courses or aquifer horizons.

The overburden depicted in Figure 150 may consist of mudstones either with or without

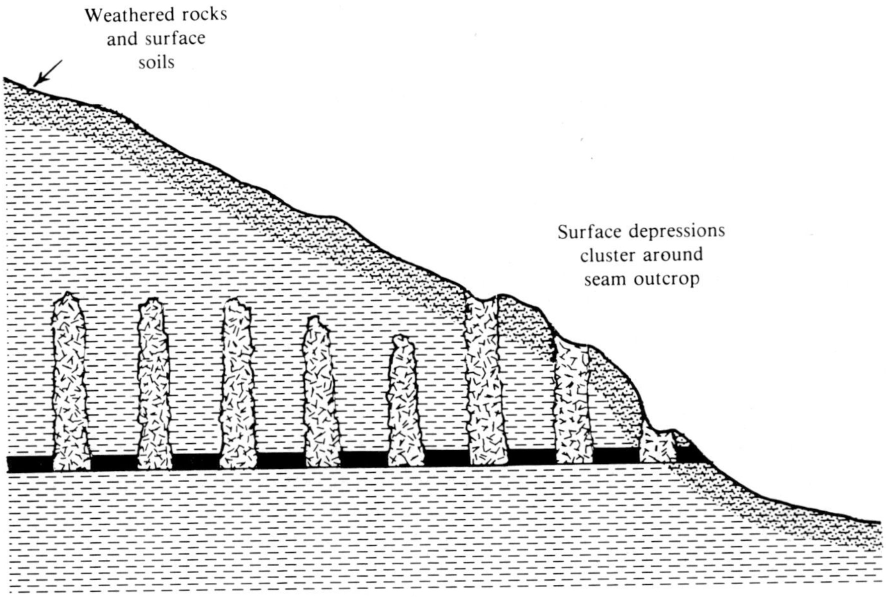

Figure 150 Surface subsidence above room and pillar workings in hilly country

sandstones or other rock types. The mudstones degrade during weathering and can result in impervious clay bands tending to discourage surface waters from penetrating deeply into the bed rocks. Their weak overall strength, however, allows ready collapse of junctions close to the outcrop, and the frequent occurrence of sink-holes and depressions appearing at the surface. The sink-holes may seal themselves in time by virtue of the breakdown of mudstones to form a clay plug.

The collapse process is of a fairly confined nature. Characteristically the beds immediately above the failed junction being subjected to gravitational effects thereby tend to displace mainly vertically without the collapse chimney being biased by the presence of the slope. This pattern of movement occurs even though the weaker surface materials are encountered on the dip-side of the slope as the sink-hole is about to emerge at the surface. Naturally, the dip-side of the freshly exposed sink-hole has an increased tendency to collapse initially, but this is soon followed by slumping from the other sides.

The main influence of a sloping surface overlying room and pillar workings appears to be that of decreased depth near the outcrop giving rise to increased likelihood of sink-holes and conical depressions appearing at the surface. The presence of this form of surface subsidence generally changes the pattern of drainage near the outcrop, with the sink-holes acting as drainage points. Lines of sink-holes along the contours of the slope above the outcrop of the worked seam give a slope weakening effect by virtue of the perforated character of the overburden. Pillar extraction at the outcrop of the seam can result in slumping of the overburden.

Longwall mining: shallow and super-critical extractions

Extraction of coal seams by longwall mining from the outcrop of a seam on a hillside generally gives rise to the form of subsidence displayed in Figure 151(a). The extraction sequence illustrated is essentially that of undermining the toe of the slope and thereby inducing a state of instability which results in the overburden collapse process shown. As undermining progresses the slope forming the overburden fails due to shearing followed by slumping down the gradient. The frequency of shearing and size of the sheared blocks depends upon the degree of competence of the overburden. Water has a major weakening effect in such situations. Further mining into the hillside is likely to experience less overburden shear steps appearing through to the surface, with increased development of a more gradual lowering of the overlying beds. Caving behind the longwall induces roof fractures hading over the goaf area. These fractures are likely to intercept the slope instability induced shear planes at shallow depths of cover, but as mining progresses further into the hillside their likelihood of interception becomes less. The surface can be characterised by large open cracks which in mountainous terrain generally call for remedial measures where they appear close to surface water courses. Flexible overlapping sheets secured to the bed of mountain streams has proved effective in northern Yugoslavia in preventing excessive inflows of water gaining access to coal mine workings.

Although longwall extraction from the outcrop is practised in some countries as shown in Figure 151(a), the more common approach is to leave a pillar of coal at the outcrop and commence working in ground less affected by weathering. This is likely to avoid the initial slumping of the overburden around the outcrop. Figure 151(b) shows the subsidence initiation and trough development in respect of longwall extraction where a coal pillar is left at the outcrop. There is likely to be significant surface crack development immediately over the extraction side of the pillar left at the outcrop owing to the concentration of tensile strain. Stepping may form also around this locality. The general pattern of subsidence trough development, however, is that of high initial ground strains, and as the depth of cover increases the leading tensile strain progressively decreases. The surface over the advancing face line experiences a slight increase in gradient with the possibility of tensile cracks occurring and opening ahead of the face but their magnitude and significance will be dependent on the extracted seam height and depth below the surface. Where a longwall extraction travels at right angles to that shown in Figure 151(b), namely parallel to the outcrop then a transverse subsidence trough can be expected to develop. This will be similar to that depicted in the third phase of Figure 151(b).

Prediction of subsidence trough profile: SEH (1975) method

The main approach to predicting the subsidence trough profile where a sloping ground surface exists, is to use a graphical projection procedure. Figure 152 shows the empirical method given by SEH (1975) which is based on subsidence knowledge established in UK coalfields. Basically, the panel w/h ratio is used to deduce the maximum subsidence (S) in relation to the extraction thickness (M) as with a level surface situation, the depth (h) being the mean value. This method has increased validity when applied to sub-critical extractions. The smaller the w/h ratio the greater is its accuracy.

As shown in Figure 152 the subsidence values along the plotted profile for the assumed flat surface, are then projected proportionately on to the actual inclined surface. The limits of subsidence are assumed to apply to sloping ground surfaces from a level extraction horizon, as is the case with a flat surface. An angle of draw of 35° is employed as suggested by SEH (1975), although in other situations a different value for the subsidence limit may be employed, see for example Abel and Gentry (1978) who applied this method to interpretation

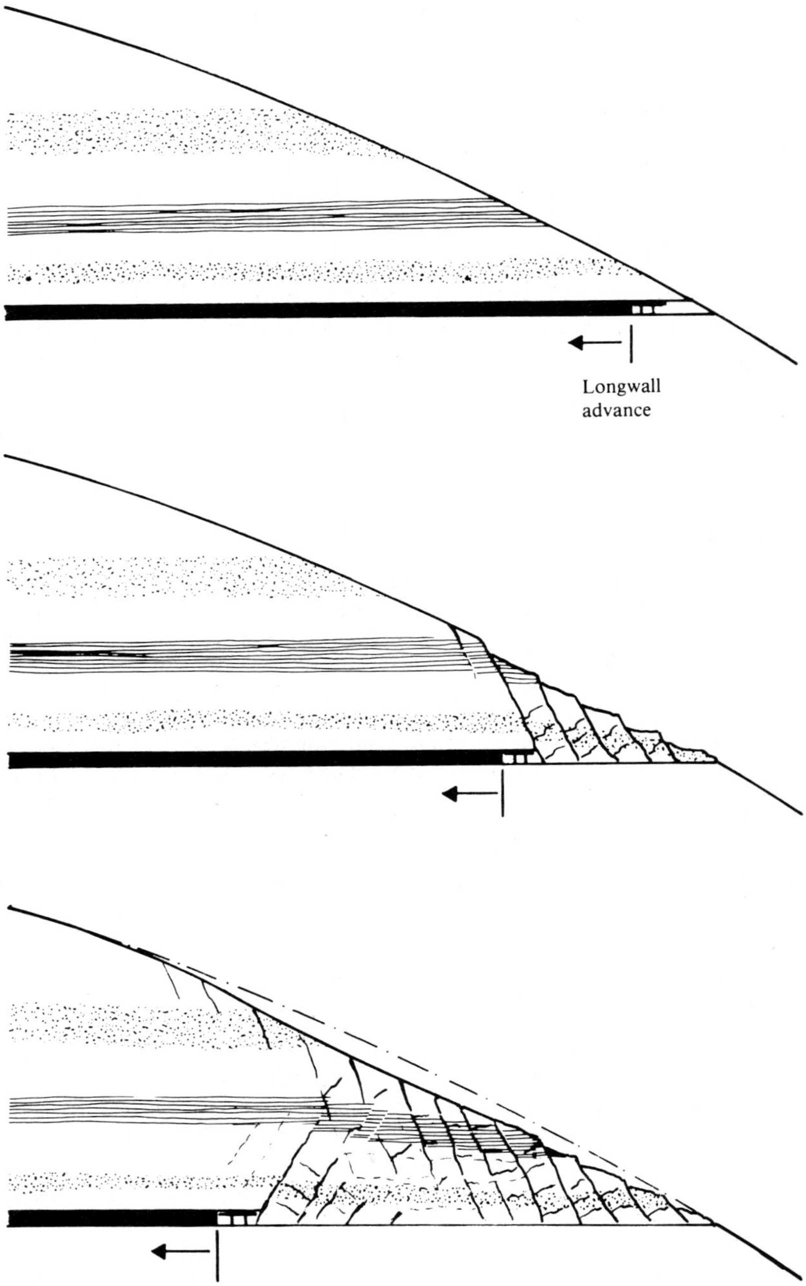

Figure 151(a) Illustrating initiation of subsidence and progression to trough development with increasing depth of cover for longwall working from the outcrop.

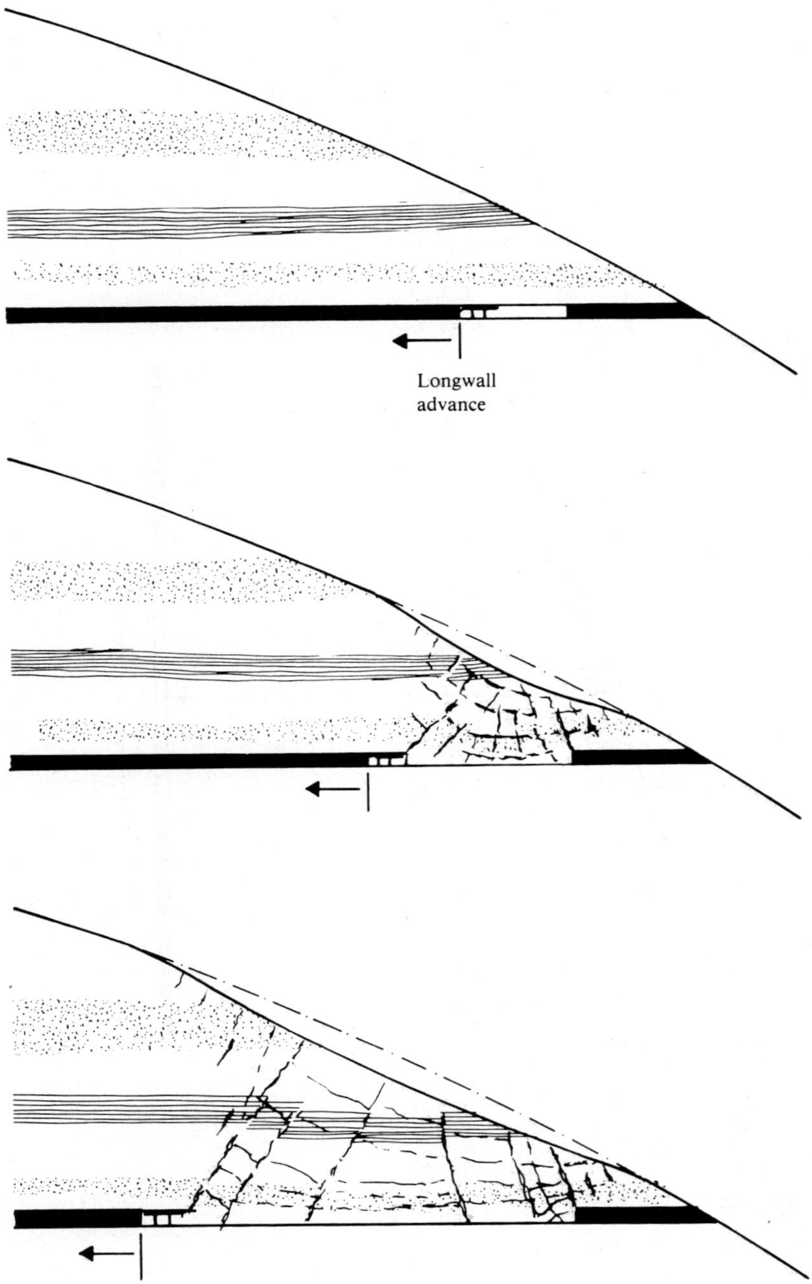

Figure 151(b) Subsidence initiation and trough development as longwall depth of cover increases after leaving coal pillar at outcrop.

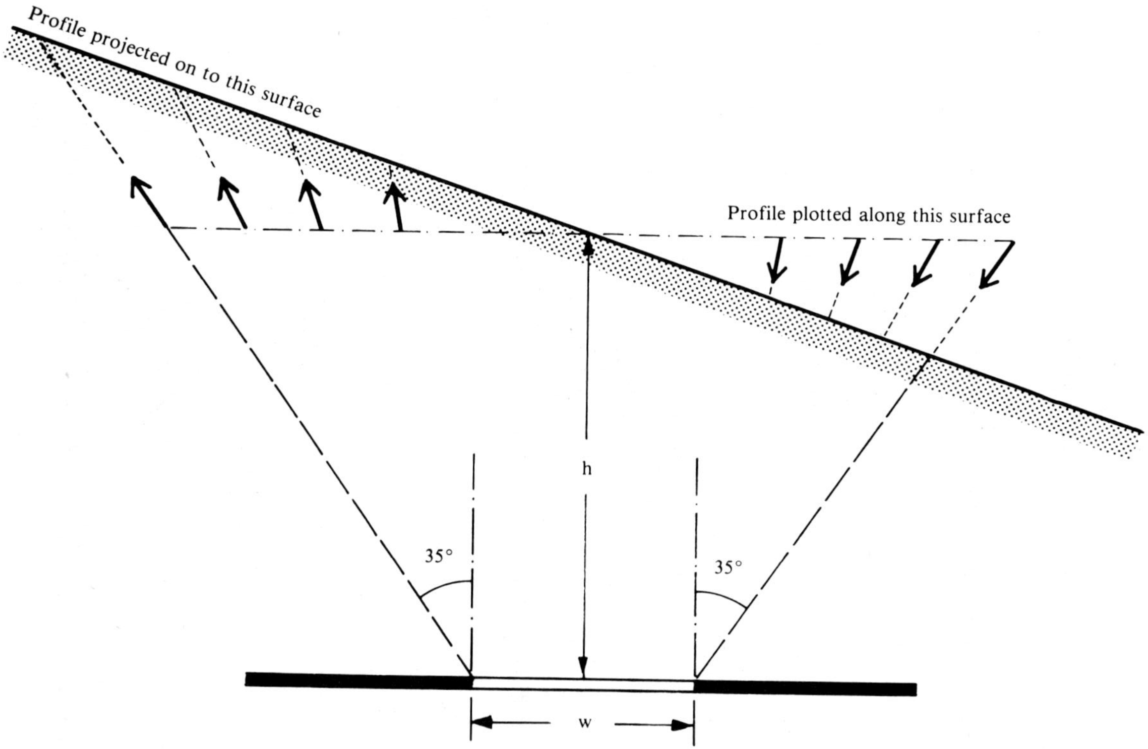

Figure 152 Method of projecting subsidence profile on to sloping ground surface, after SEH (1975)

of subsidence results at York Canyon Mine, Raton, New Mexico. In effect, this method of subsidence profile prediction results in a significant increase in the extent of subsidence influence on the rise side with a decrease in the extent on the dip side of the sloping ground.

Examples of the type of subsidence profile predicted for increasing surface ground slope are presented in Figures 153 and 154. The flat surface profile result is included in Figure 153 for comparison.

Sloping ground surface subsidence, strain and displacement profiles: influence function method

The sloping ground situation has been examined by the stochastic model influence function one-angle method in order to appreciate the patterns of predicted surface ground strain and displacement. Surface slope angles of 15°, 30° and 45° to the horizontal have been used. A longwall extraction width (w) of 200m, a mean depth (h) of 400m and extracted seam height (M) of 2m have been employed.

The basis of strain treatment is illustrated in Figure 155. The strain in the x-direction is expressed as the ratio of the differential displacements (V_{x2} - V_{x1}) and the plan length of (L_x) of the bay (L_o) between observation points P_1 and P_2, as given by equation (65).

$$e_x = (V_{x2} - V_{x1}) / L_x \qquad (65)$$

The displacement (V_x) for each point is expressed in terms of its absolute movement in the x-direction.

Results from the influence function method are presented in Figures 156, 157 and 158 for the respective surface ground slope values of 15°, 30° and 45°. Considering the results collectively, the following observations are made:

1. The subsidence profile is similar to that predicted by SEH (1975) and presented here in Figure 153. The magnitude of the maximum subsidence is, however, predicted to be 20 -25% greater with the influence function method as compared to that given by SEH (1975).
2. The magnitude of the compressive strain ($-E$) progressively increases with steeper values of surface ground slope.
3. The tensile strain zone on the up-slope side of the extraction obviously increases in extent as the slope angle becomes larger, but the magnitude of $+E_{\text{up-slope}}$ decreases.
4. Conversely the magnitude of $+E_{\text{down-slope}}$ increases as the surface slope steepens, but the extent of tensile strain zone decreases.
5. There is an appreciable difference between the magnitudes of the maximum tensile strain values for the up-slope and down-slope situations, and this difference increases for steeper ground slope conditions.
6. The ground displacement profiles all indicate significant down-slope values on the up-slope side of the extraction.

The main effect of increasing the surface ground slope angle is to greatly increase the zone of tension of the up-slope side of the extraction. There is also a steepening of the slope locally. There would appear to be conditions which favour increased movement down the slope owing to the significant zone of tensile ground strain created on the up-slope side. The down-slope side of the subsidence trough experiences increased magnitude of tensile ground

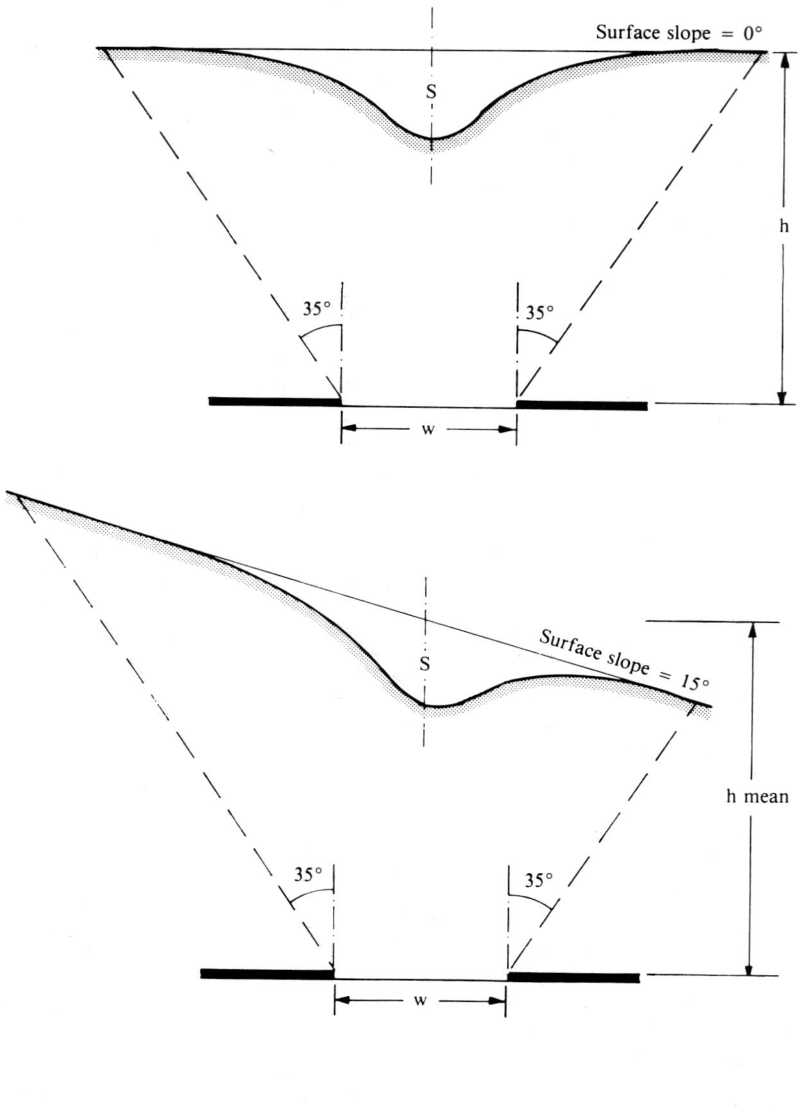

Figure 153 Comparison of subsidence profiles for flat and 15° sloping surfaces by SEH (1975) method.

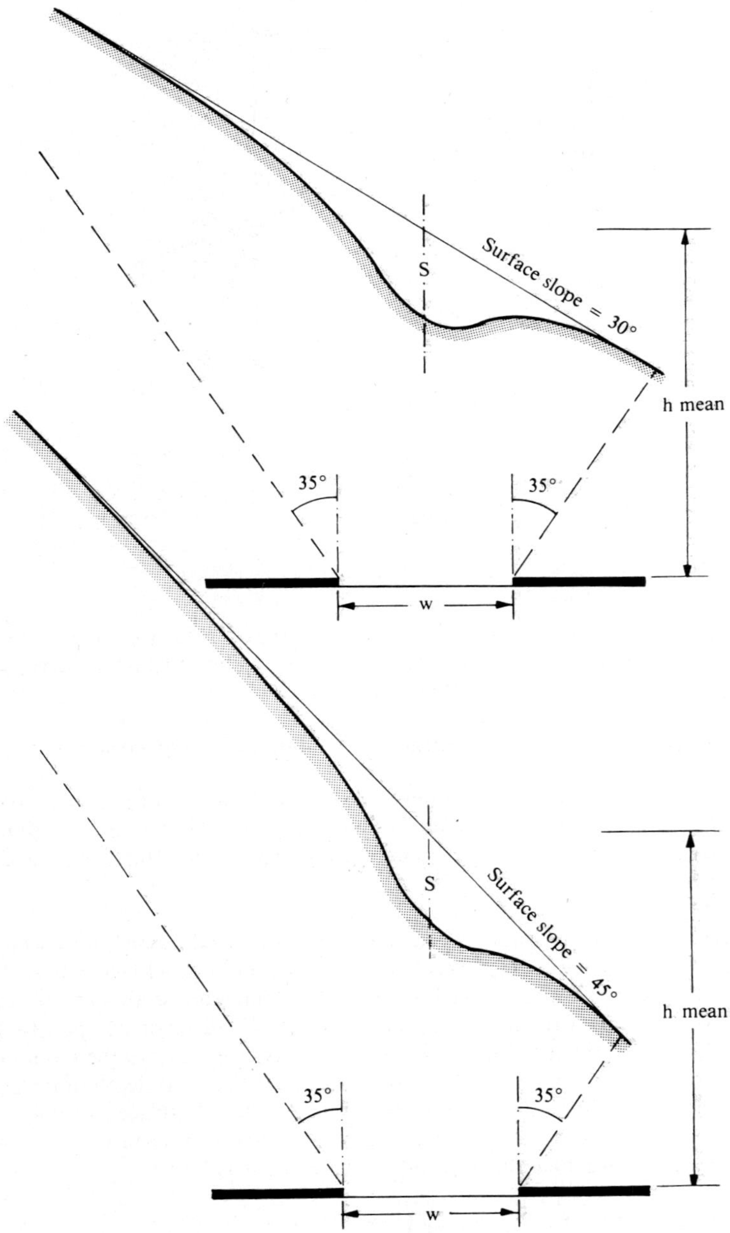

Figure 154 Comparison of subsidence profiles for 30° and 45° sloping surfaces by SEH (1975) method.

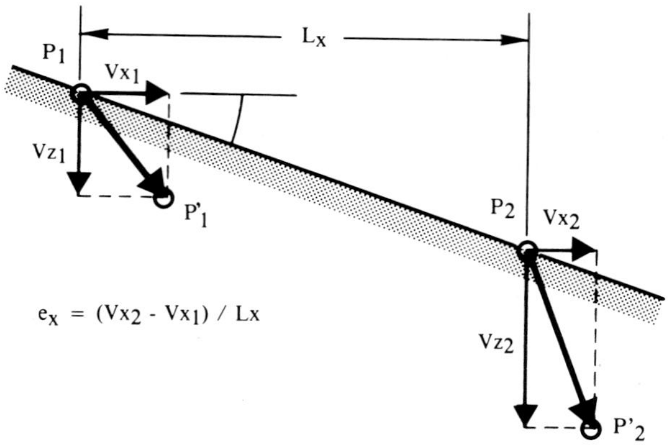

Figure 155 Strain and displacements of points on sloping ground surface.

strain effects according to this form of prediction. The zone of influence is less however, and the rate of change of induced slope due to subsidence is tending to favour increased slope stability with accompanying reduced ground slope angle.

Observed behaviour of sloping ground surfaces in relation to mining subsidence

Several investigations have been performed in different countries of the effects of mining subsidence on the surface, but comparatively few have been devoted to those situations where appreciable sloping ground surfaces exist. Some of these more important investigations are referred to here.

Subsidence observations in Canada. A coal seam 15m thick and dipping into a mountain-side which itself had a slope of 35-40° was being extracted on a panel layout basis by high-pressure hydraulic monitors which allowed removal of the full seam section at one pass. This site near Sparwood, British Columbia, Canada was studied and reported upon by Fisekci, Chrzanowski, Das and Larocque (1981). The coal seam outcropped into the mountain-side. Subsidence observations were made over an extraction area where the depth of cover ranged from 50m to 300m (at hill top). These authors reported that the subsidence movements ceased within one year. The subsidence effect was mainly characterised by surface cracks and steps with the main effects predominantly vertically over the extracted area.

Subsidence observations and findings in the USA. The report of Siriwardane and Amanat (1984) discusses subsidence prediction with reference to undulating/hilly terrain by means of a non-linear finite element method. Difficulties were experienced by these authors in selecting an appropriate value of reduced elastic modulus to give satisfactory comparison between measured and predicted values of maximum subsidence. An appropriate reduction factor was obtained from a statistical model. The Drucker-Prager yield criterion was employed in the analysis. The form of the results obtained was that of indicating the principal plastic and

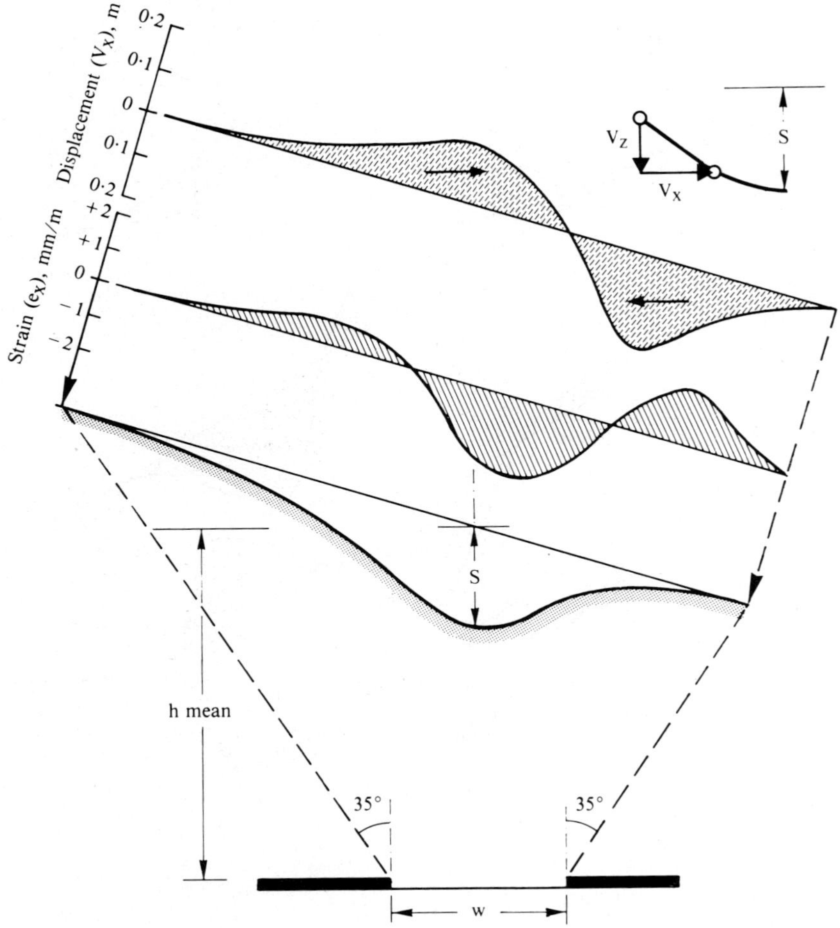

Mining Data

w	=	200m
M	=	2m
h mean	=	400m
Surface slope	=	15°
S	=	1·15m

Figure 156 Subsidence, strain and displacement above longwall extraction with a surface ground slope of 15°

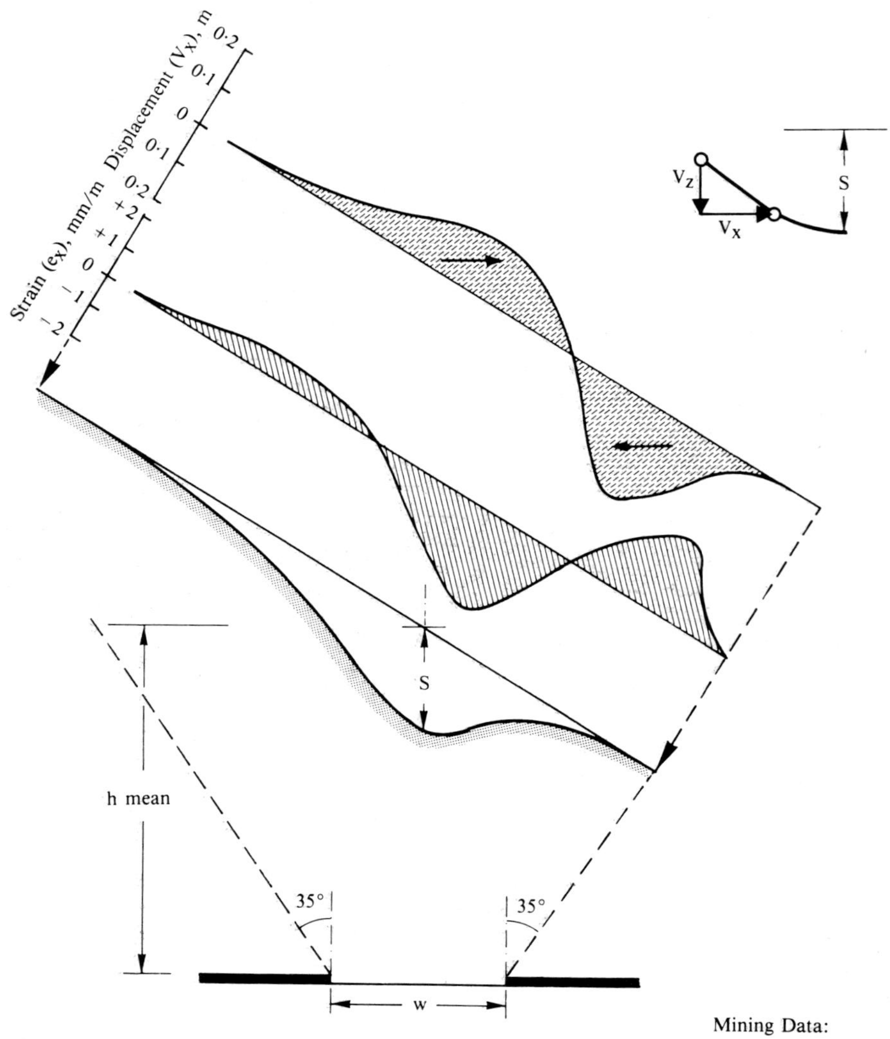

Mining Data:

w	=	200m
M	=	2m
h mean	=	400m
Surface slope	=	30°
S	=	1·165m

Figure 157 Subsidence, strain and displacement above longwall extraction with a surface ground slope of 30°.

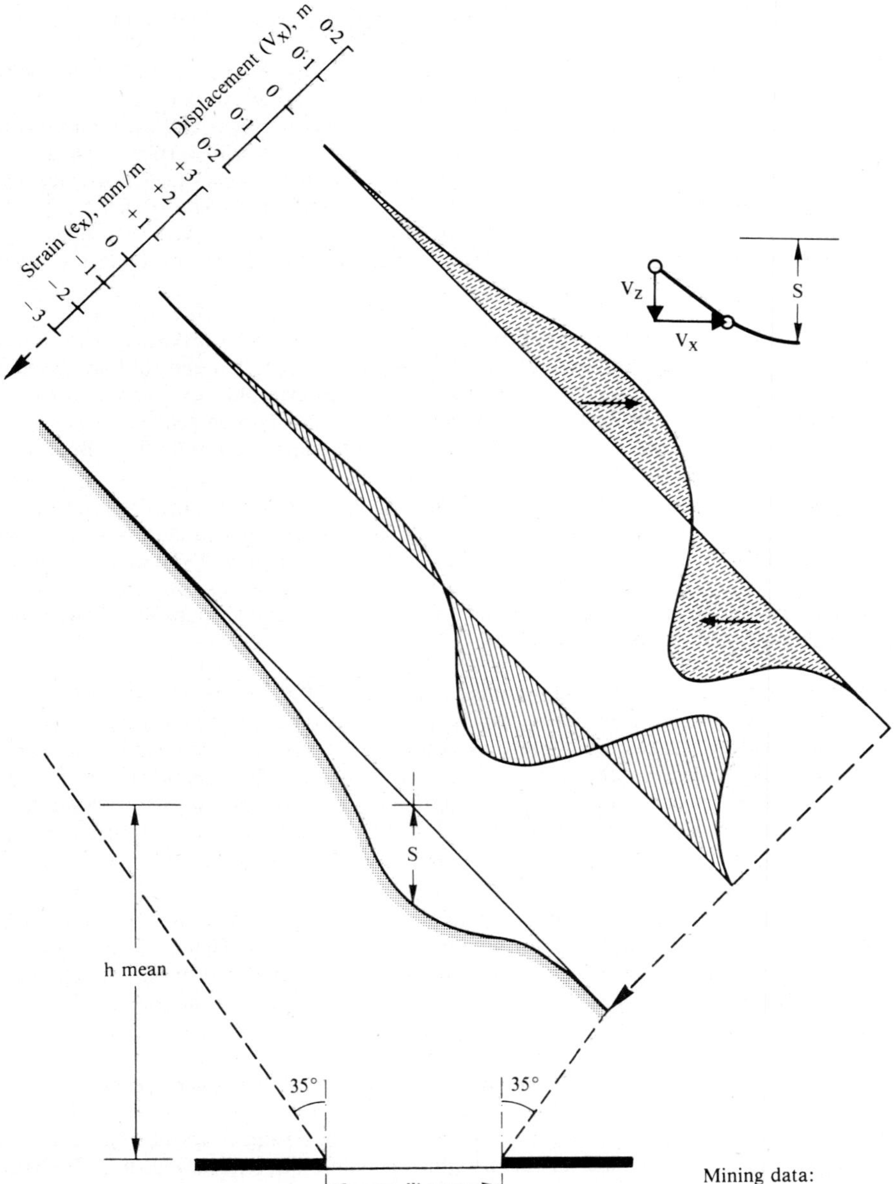

Mining data:

w	=	200m
M	=	2m
h mean	=	400m
Surface slope	=	45°
S	=	1·19m

Figure 158 Subsidence, strain and displacement above longwall extraction with a surface ground slope of 45°.

tensile zones, particularly over the hill-side which in this case examined was the panel rib side. The results appear to be of a rather generalised nature without sufficient refinement to indicate the likely pattern of crack development on the sloping ground surface.

Some interesting observations have been reported by Allgaier (1982) who examined surface subsidence in hilly terrain overlying longwall workings at Book Cliffs, Utah. The overburden consisted of mainly sandstones with intervening shales and carbonaceous units and coal seams. The surface terrain was referred to as rugged with vertical cliffs and steep unstable slopes; deep and narrow canyons were also featured on this site. The extracted seam height was 1·8m whilst the panel depth varied, for example 340-460m (Panel 3), 230-320m (Panel 4). The results presented by Allgaier indicated that the subsidence troughs did not appear to be greatly extended by the steep surface ground slopes overlying the longwall extractions.

The publication of Gentry, Stewart and King (1981) describes subsidence observations in New Mexico in a similar situation to that of Book Cliffs, Utah. The general finding reported was that increased subsidence occurred in the topographically higher areas with correspondingly less subsidence in topographic lows. Overburden tended to be attracted towards the topographic lows as was also indicated by Allgaier's results from Book Cliffs.

Conroy and Gyarmaty (1982) show subsidence results observed in the Eastern Coal Basin which clearly demonstrate the trend for horizontal displacements to be down the slope on overlying hilly topography. The topographic relief can be as high as 137m in this Coal Basin. These authors report that in some extremely steep localities up to 1·2m of horizontal displacement has been recorded which may mask the horizontal strains anticipated, in this case, to be of the order of 1 to 5mm/m.

Subsidence observations and findings in the UK. Franks and Geddes (1984) have researched into mining subsidence effects on sloping ground surfaces. A finite element numerical modelling study was made of mining subsidence situations corresponding to geographical locations typical of the South Wales Coalfield valleys. The authors reported that the sloping ground surface appeared to have a greater influence on horizontal movements and ground strains than on vertical movements and tilts. Their finding appears to be supported by Figures 156, 157 and 158.

Some interesting observations were made by Forrester and Whittaker (1976) on colliery spoil heaps which were subjected to mining subsidence effects. The spoil heap slopes and crests were investigated regarding their response to subsidence and it was concluded that excess settlements were observed at the crest coupled with heave at the toe. There was clear indication of increased displacement down the slope.

Summary of potential ground slope instability arising from mining subsidence

When considering potential ground slope instability arising from the effects of mining subsidence, there are a number of factors which should be examined and these are listed as follows:

1. The significantly broadened zone on the up-slope side encourages the opening of existing weakness planes favourably orientated relative to the direction of longwall extraction. The opening of existing weakness planes such as fissures, joints and fault planes can lead to decreasing the structural integrity of the slope.
2. A slope weakened by tensile strain effects in the form of crack development, becomes

vulnerable to the access of water and possible resulting decrease in natural stability.

3. Steep faces and exposed rock scars can be subjected to increased toppling effects. Small changes in tilt can have an appreciable influence on the natural stability of cliffs and overhangs.
4. There is a resultant movement down the slope which can accentuate any natural creep taking place.
5. In mountainous terrain, major rock structures can experience stability changes. Narrow canyons and scree areas may prove to be vulnerable to changes in tilt.

Concluding remarks

Mining subsidence effects on sloping ground surfaces can result in extensive zones of tensile strain developing on the up-slope side of the mined-out areas, in addition to localised steepening of the ground slope. There is an accompanying down-slope movement trend which can be of special significance where slopes exhibit natural creep. The increased zone of tensile strain can be of importance in encouraging the opening of natural and induced fissures, cracks and other weakness features which can lead to aggravation of stability by entry of water.

Where mining is taking place in regions having appreciable sloping ground surfaces, due consideration should be given to the effect of mining subsidence giving rise to any possible changes in the natural stability of such slopes.

CHAPTER 12

ROCK-SALT AND POTASH MINING SUBSIDENCE ASPECTS

Rock-salt and potash deposits give rise to surface subsidence problems under particular conditions. Such minerals are frequently found in association with marls which generally inhibit infiltration by water, although water does gain access to parts of such deposits particularly through geological disturbances of which faults represent the main form of potential path. The formation of solution cavities by fresh water action coming into contact with evaporite deposits is recognised as a form of natural subsidence when such cavities collapse and/or undermine the supporting ability of the overburden and give rise to surface depressions or holes. Extraction of rock-salt can be by solution mining in the form of brine pumping or by room and pillar operation or other types of positive mechanical extraction. There is a wealth of knowledge on rock-salt mining in Cheshire, U.K. This is particularly the case with the resulting surface subsidence effects which arose from certain forms of mining which were practised in the past. The principal features of the Cheshire rock-salt subsidence experiences are reviewed here.

Attention is also focussed on the subsidence behaviour of the overburden when mining rock-salt and potash deposits. Observations made in different countries are discussed and general patterns of movement are compared.

North American experiences of subsidence overlying rock-salt formations

Ege (1984) has discussed the mechanisms of surface subsidence resulting from solution extraction of salt and has drawn attention to the fact that subsidence causation arises from partial or total collapse of underground cavities. He highlights the importance of the strength properties of the overlying rocks especially regarding their ability to sustain support of cavities created by dissolution of rock-salt. Void migration to the surface is suggested as one mechanism of such subsidence, whilst subsurface erosion of susceptible overlying beds in association with the rock-salt can also result in void formation with surface subsidence potential.

Ege discusses mainly rock-salt subsidence aspects within North America, giving particular attention to the middle Devonian Prairie Formation extending from North Dakota and Montana through Saskatchewan and Alberta to the Northwest Territories. Natural subsidence has been taking place on a geological time scale, and voids have given rise to subsidence collapse features at the surface. Crater Lake, 244m diameter and 6m deep is quoted as a typical subsidence feature of this type.

Solution mining of rock-salt is reviewed by Ege and he comments upon the importance with one particular method of employing a pad of oil, or entrapped air, to control the upward progression of the dissolution process. This allows the solution cavity to concentrate in the lower part of the rock-salt deposit so as to promote increased stability against failure of the roof of the chamber. The geometry of the solution cavity is better controlled with this method.

The risk of sink-hole occurrences resulting from solution mining of rock-salt in central

Kansas has been discussed by Walters (1977). A particular example given by him occurred in October 1974 at brine extraction operations in Hutchinson, Kansas, when a sink-hole of 90m diameter developed over a 3-day period and resulted in railroad tracks being suspended over the subsidence hole. The rock-salt bed involved is some 105m thickness and lies around 120m below the surface. There are unconsolidated deposits of sands, gravels and soils overlying the rock-salt and associated Permian shale. Salt has been produced at this locality since 1888 and many cavities exist.

Sink-hole occurrences over brine pumping operations has been a common form of surface subsidence in North American salt field extraction operations. A 90m diameter and 15-18m sink-hole appeared over a 12 hour period at Barton County, Kansas during 1959 which appeared to be due to a salt-water disposal well which allowed unsaturated brine to come into contact with a salt face at a depth of approximately 300m below the surface, Walters (1977).

Solution mining has been carried out in the past at Grand Saline, Texas, where the major Grand Saline salt dome exists. A 15m deep sink-hole occurred within the city of Grand Saline during 1976, Grand Saline Sun (1976).

The Grosse Ile brine field extraction operations at Detroit, Michigan have experienced sink-hole subsidence events. The Grosse Ile island is located in the Detroit River and has dolomites and sandstones overlain by unconsolidated deposits. The main salt formation starts at about 340m below the surface and extends to a depth of around 400m. The island is extensively covered with brine wells. Sink-holes appeared during 1971 which had diameters of 60m and 135m respectively, Landes and Piper (1972). These subsidence occurrences demonstrate that brine pumping operations within rock-salt deposits of appreciable depth can still result in sink-holes appearing at the surface providing the conditions favour their development.

Brine-field subsidence occurrences have been reported by Terzaghi (1970) in connection with operations at the adjacent rock-salt deposits of Windsor, Ontario. Brine pumping gives rise to a small amount of general lowering of the surface as part of the operation. Sink-holes are comparatively rare occurrences in such operations.

Introduction to the Cheshire rock-salt subsidence occurrences

Exploitation of rock-salt in Cheshire, U.K., has given rise to a history of surface subsidence which has been discussed by Calvert (1915). These evaporite deposits were formed during the Triassic period and occur in the Keuper marls. The salt beds are flat and up to 40m in thickness. Around the Marston and Winsford localities, the rock-salt occurs as two main beds, being referred to as the Top Bed and Bottom Bed and each being typically 20-30m thickness. The intervening marl bed is generally 8-10m thick. The Bottom Bed is around 100m deep at Northwich and 150m deep at Winsford. Surface drift deposits generally cover the Triassic beds in the rock salt region.

Mining of rock salt in Cheshire has been carried out for about three centuries. The first mining operations were concerned with exploiting the Top Bed. The late seventeenth and early to mid-eighteenth century mining operations consisted mainly of sinking a shaft to the Top Bed and working radially from the shaft leaving relatively small pillars of rock salt to support the immediate marl roof. The consequences of employing the marl as a roof over these early workings soon became apparent as the mined-out regions readily collapsed and caused the appearance of subsidence features at the surface notably conical depressions and subsidence holes. Underground exploitation was later concentrated in the lower section of the Bottom Bed, which was the purest rock salt in addition to providing scope for stable roof conditions during working.

Figure 159 shows a diagrammatic representation of the early and later approaches to mining the Cheshire rock salt, together with the surface instability effects which resulted from working the upper part of the Top Bed.

The old workings in the Top Bed later experienced the breaking in of fresh water thereby giving rise to brine reservoirs which provided another phase to exploitation of these salt deposits by brine pumping which led to the main cause of the surface subsidence occurrences in the Cheshire salt mining field.

The period 1750 to 1800 witnessed appreciable surface subsidence events in the form of "mine fall-ins" or "rock-pit holes" as they were generally known in the region immediately to the north of the town of Northwich. These subsidences were as a result of mining the upper section of the Top Bed, and were essentially funnel-shaped depressions.

Ashton's Salt Mine collapse and resulting subsidence

Calvert draws attention to Ashton's Mine which was working the Top Bed as early as 1789, and records of that time show the depth of the shaft as 38m, and the thickness of the extraction as 10·7m with support pillars of 20·6m^2 plan area. Records show that the proportion of support by the mine pillars to that excavated to be 1:9. This suggests that the support pillars were probably 5 x 5yd (4·63 x 4·63m) supporting a nominal area of 15 x 15yd (13·9 x 13·9m) and resulting in 89% extraction. The 1789 records refer to 32 pillars left for support of an area of 7200yd^2 (6183m^2). Calvert reports that Ashton's Mine collapsed during 1838 and constituted a major surface subsidence occurrence which engulfed the winding engine and surface structure of the mine with accompanying loss of life. At the time of the mine collapse the Top Bed had been worked and the Bottom Bed had been developed and was in the process of being working. Records indicate that prior to the mine collapse, the pillars in the Top Bed were in a state of deterioration and this appeared to influence the decision to discontinue working the upper bed. It would appear that working the Bottom Bed contributed to accelerating the strength deterioration of the support pillars in the Top Bed although time dependent behaviour must have been a factor also. It is also recorded that at the time of the collapse of Ashton's Mine, water inflow into the workings was regarded as very little and did not influence pillar strength deterioration; pillar stress and time-dependent behaviour were the principal factors involved.

Figure 160 shows a reconstruction of the subsidence collapse of Ashton's Mine.

Early features of the Cheshire rock-salt mines

The early rock-salt mines of Cheshire employed a single shaft for both downcast and upcast ventilation. Such mines exploited the Top Bed. It was common for the shaft to stop short of the rock-salt and access be gained to the salt via a side drift so that the main shaft could facilitate water entrapment in a sump. Although the Top Bed mine workings did not generally extend greater than around 90m from the shaft, some of the workings eventually connected and thus allowed independent downcast and upcast shafts to exist. The room and pillar mine workings usually involved taking a working height of 9-11m with pillars of around 5yd (4·6m) square to provide support of the overburden. The mines in the Top Bed were frequently affected by roof instability and subsequently water related problems. The

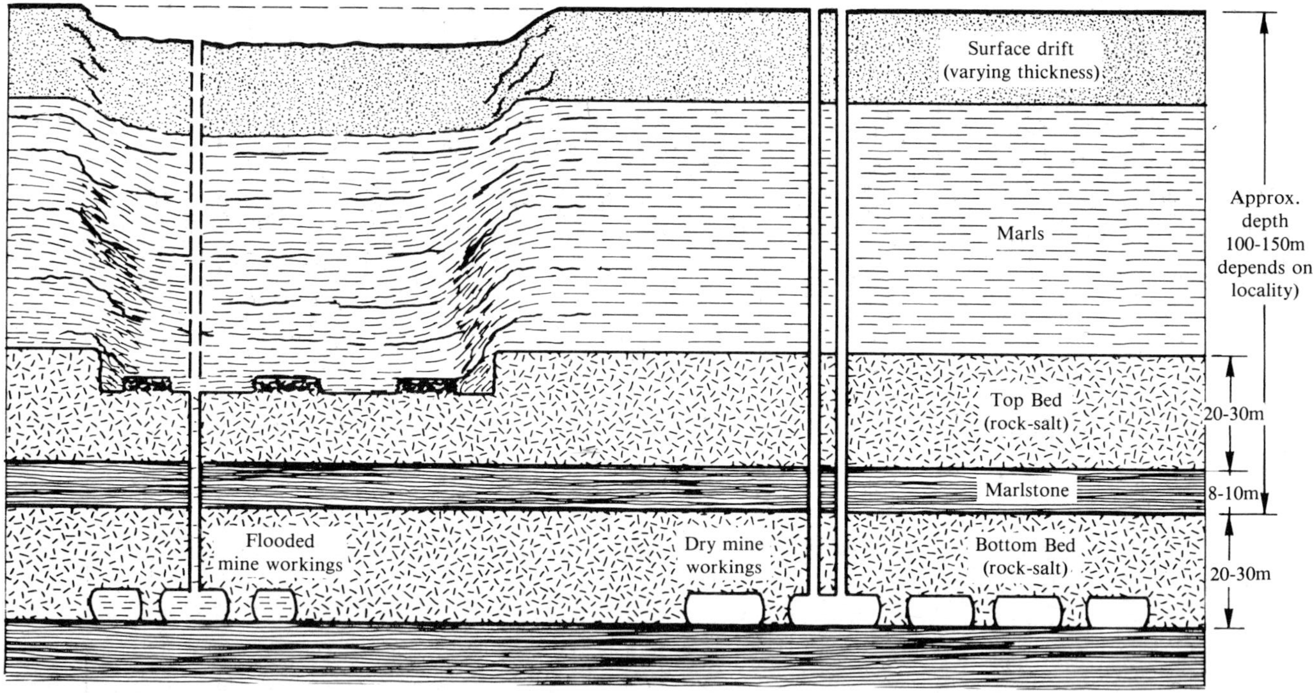

Figure 159 Mining stability features of working the Cheshire rock-salt by early and later methods.

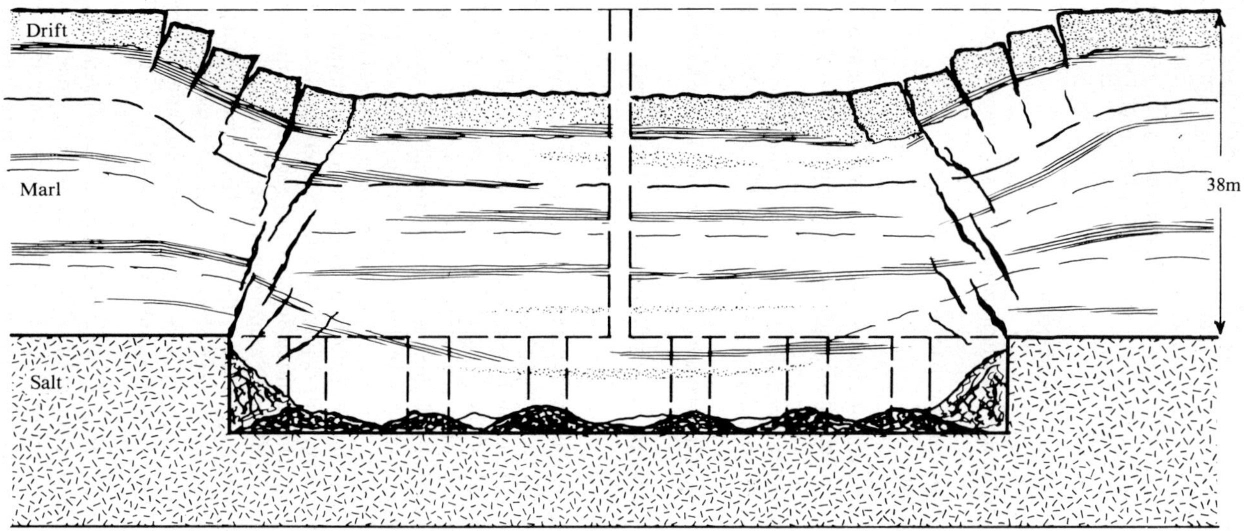

Figure 160 Re-construction of the Ashton Mine collapse which engulfed the shaft-top buildings. Note: The mine pillars before collapse are shown by the broken lines and these were 10·7m high.

thickness of rock-salt left at roof level was generally insufficient to promote longterm stability of the span between the support pillars.

Mining of the Bottom Bed was generally based on two shaft operations placed 10-14m apart, and frequently a third shaft for the drainage and pumping of groundwater. The latter shaft was of limited depth and sufficient to act as a natural drainage feature to prevent water gaining access to the mine. Preventing the rock-salt dissolving by the presence of fresh water was a particular problem in the access shafts which called for special protective measures in the lining employed.

Character of the Cheshire rock-salt subsidences

Calvert (1915) has reported that the Cheshire rock-salt subsidences occurred in two main forms:

1. Funnel-shaped holes of varying diameter and depth.
2. Trough-shaped hollows.

The first category of subsidence occurrence is explained as being a direct consequence of a mine collapse within the Top Bed salt deposit. Calvert subdivides the trough-shaped hollows as follows:

a) Deep troughs exhibiting stepped sides; the overlying surface is very much disturbed.

b) Shallow troughs with no stepping of the ground at the sides or any appreciable disturbance of the surface.

c) Very shallow and almost indiscernible hollows extending over a significant area.

Figure 161 presents a summary of the principal forms of subsidence encountered around the Cheshire salt mining field.

Natural subsidence by erosion of the upper interface of the Top Bed by the action of groundwater tends to give rise to widespread shallow troughs in the Cheshire salt field. Hull (1881) gave an account of surface subsidence occurrences in the salt districts of Cheshire and remarked upon the dissolving of the salt beds as a natural process of underground water networks; subsequent discharge of such waters in the form of brine springs dates back to pre-mankind times. He recorded that such natural springs were still to be seen in the Cheshire salt districts even though large quantities of brine were being extracted from underground locations by pumping. Hull associated the Cheshire meres (natural lakes) with natural subsidence resulting from dissolution of the salt. He discussed the natural process of water percolating through the overlying drift deposits and marls overlying the rock-salt bed which formed an aquiclude horizon. The resulting dissolving of rock-salt continued until the water contained 21 to 25·5% salt in solution. Pumping such brine to the surface created a cycle of events which ultimately resulted in surface subsidence. Hull reported that Northwich had subsided at least by 20m within the memory of man. He surmised that the Top Bed of the rock salt had almost been dissolved away in places. He also remarked upon the constant building up of the banks supporting roads, canals, railways and culverts.

Those subsidences of the deep trough category exhibit a tendency for fairly sudden occurrence. The first appearance is that of sudden lowering of a circular ground mass of a few metres diameter. After occurrence of such a depression, the circle increases in size accompanied by deepening of the hole. The observed diameter of such holes has been around 8m and 4-5m deep. Conversely, an adjacent hole exhibited a diameter of some 90m. A common depth of such holes was observed to be 12-15m, which in time filled with water forming small lakes. It was common to see several such lakes due to subsidence clustered together.

The lakes formed in the Northwich and Winsford localities due to salt mining operations cover many hundreds of acres in extent; their local name is that of "flashes" or "rock pit flashes" and they have depths as great as 15m or so. A mining subsidence lake of greater than 1km in length was not unknown even in the late nineteenth century in these localities.

Williams (1897) has referred to mining subsidence engulfing the River Weaver at Witton Brook which resulted in the entire river flowing into mine workings through a subsidence hole which suddenly appeared. The subsidence lakes have been known to suddenly drain into abandoned underground workings enlarged by brine pumping operations.

It is recorded according to Calvert that when a major sudden subsidence was taking place the surface subsidence lakes and ponds experienced violent bubbling, and in some cases geyser effects, as expelled air was finding its way out from the collapsing mine workings. This phenomenon took place over a period lasting a number of hours. A major subsidence event of this type occurred in 1880 on the outskirts of Northwich at Dunkirk over extensively honeycombed abandoned salt-mine workings and resulted in the Wincham Brook being swallowed through a vortex in the subsidence hole; the subsidence crater was some 90m in diameter, and about 4m deep. A chimney stack of about 27m height leaned and collapsed during this subsidence occurrence. This subsidence appeared to trigger further lowering of the surface and disturbance to neighbouring subsidence lakes which experienced a sudden lowering of their water levels. Subsidence effects eventually extended to a diameter of over 150m, circumscribed by extensive crack development and steps of around 0·6m. Some ten thousand tons of water flowed into the underground mine workings.

Some of the subsidence lakes periodically drained into the ground below, followed by subsequent natural restoration of the lake surface. It appears that the puddling action of the marls/clays within the subsidence hole plays a role in temporarily arresting the downward flow of water to the subterranean salt cavities and collapsed mine workings. The brine pumping action generated the potential for more voids to form within the next bed and consequently the conditions favouring the release of the puddled plug within the existing subsidence hole. Calvert points out that in the Marston district the overburden appeared to remain suspended over solution cavities within the salt for some considerable time; subsequent collapses involved tens of thousands of tonnes of rock and overburden which resulted in creating large subsidence holes which soon filled with water. An example of such a subsidence hole is that which occurred following brine extraction at Marston and reported by Calvert to have a diameter of around 60m and maximum depth of some 12m which contained water to a depth of 3m as observed shortly after its appearance. The volume of the subsidence hole was given as 11 900m^3 by Calvert. The subsidence hole eventually filled with water. At Northwich, the rock-salt is fairly shallow with overlying weak marls and drift deposits. Consequently, the underground cavities formed did not appear to have the opportunity to enlarge excessively before the collapse process commenced; the subsidence holes rarely exceeded 5-7m in diameter and were usually of a conical shape.

Principal effects of the Cheshire rock-salt subsidences

Calvert draws attention to the following particular examples of subsidence as associated with the salt mining and brine extraction operations in Cheshire:

1. The main effect has been that of extensive flooding especially along the course of the Witton to the north of the town of Northwich.

2. Railways through the mining area have experienced significant effects of subsidence, with the need for repeated raising of tracks. A local line needed its embankment raised by as much as 9m over a period of 40 years.

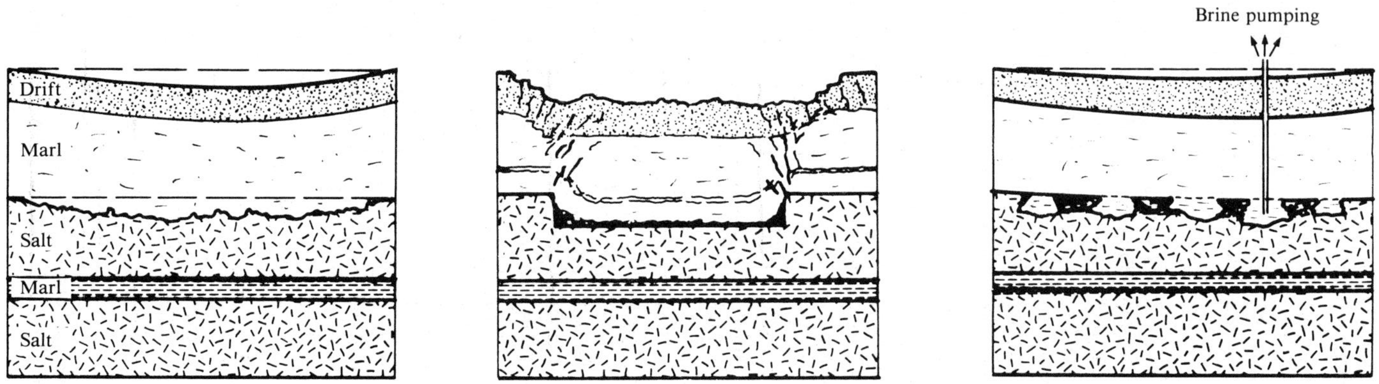

Natural subsidence:

1. Predominant sub-surface erosional behaviour.

2. Top Bed mainly affected by subterranean drainage patterns,

3. These subsidence effects in Cheshire have tended to be gradual and widespread and generally occur over long periods of time.

Deep troughs:

1. Have usually arisen due to collapse of mine workings in the Top Bed owing to over-extraction and inadequate pillar support.

2. Under-designed pillars may fail in time due to long-term effects of high stresses and/or erosion due to water ingress.

Shallow troughs:

1. These have been usually associated with excessive and uncontrolled brine pumping.

2. Mine pillars have gradually decreased in size owing to the dissolution process by the action of fresh water flowing through previously mined rock-salt beds.

3. The subsidence tends to be widespread and usually occurs gradually.

Figure 161(a) Some principal forms of subsidence associated with the mining of Cheshire rock-salt

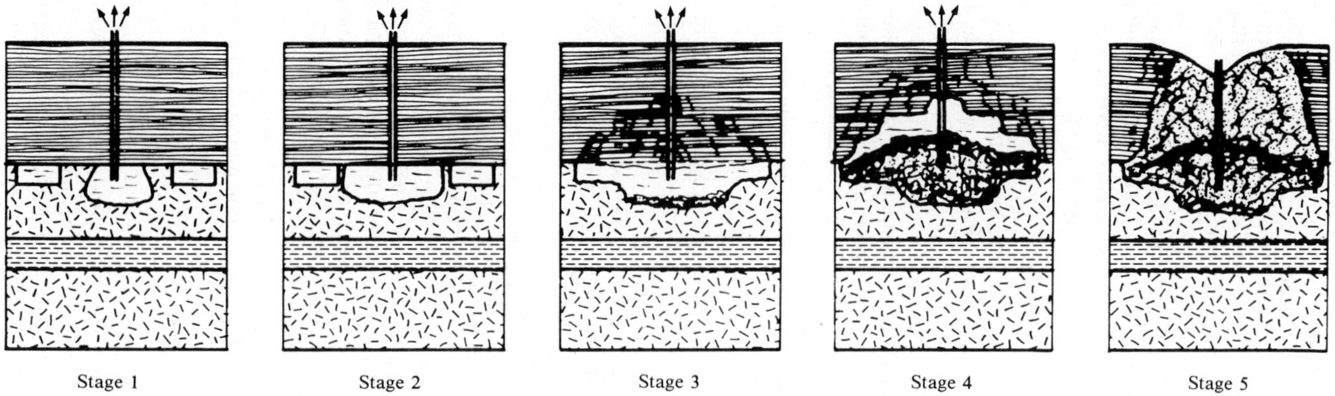

Figure 161(b) Collapse process initiated by brine pumping leading to a conical depression occurring at the surface

Notes on stages:

1 & 2. Enlargement of old mine workings by brine extraction.

3. Roof deterioration due to large underground expanse created by dissolution of rock-salt pillars.

4. Early stages of major underground collapse of rocks above brine extraction horizon.

5. Upward progression of collapse process reaches surface and creates subsidence depression.

3. Canals and river navigation have suffered subsidence effects with the need for bridge and embankment modification work. Some parts traversed by canals have experienced almost 30m of subsidence over about 100 years.

4. The period around 1880 was a time when subsidence effects around Northwich and Winsford reached màgnitudes which raised national concern and resulted in the lodging with Parliament of "The Cheshire Salt Districts Compensation Bill". There was appreciable damage to and destruction of property, and formation of large lakes in subsidence depressions. There was damage to water-mains, sewers and gas-pipes on a continuous basis; bridges sank and needed restoration and raising. Agriculture lands experienced significant cracking and subsidence. Churches were damaged, some beyond repair. Some properties required raising and rebuilding up to three times in a period of only eleven years.

5. There was evidence of depressions and uneven surfaces along streets and roads which needed continuous repair and maintenance. Roads subsided by as much as 10m.

6. A striking feature appears to have been the tilt of surface structures and properties. The leaning of chimney stacks and buildings overhanging the street; many broken panes of window glass, extensively cracked walls both internally and externally of houses were in strong evidence. There were difficulties in maintaining floors in a level condition. Doors were frequently becoming jammed and distorted. The use of tie bars and timber shoring to safeguard damaged buildings was commonly applied.

Figures 162, 163 and 164 illustrate subsidence effects at the surface arising from rock-salt mining in Cheshire.

The major subsidence occurrences resulting from mining the rock salt and the associated surface effects were reduced by controlled brine pumping.

Subsidence arising from brine pumping at Tuzla, Yugoslavia

Jeremic (1964 and 1976) has examined the nature of the subsidence problems occurring with rock-salt mining. He identifies the difficulties in controlling formation of cavities by solution mining, coupled with the rheological behaviour of such rock types results in creating uncertainties in predicting anticipated subsidence and consequently concludes that forecasts are often made by analogy with known occurrences. Subsidence arising from solution mining of rock-salt under the city of Tuzla, Yugoslavia is used to demonstrate the type of problems which can arise with this form of relatively uncontrolled mining. The subsidence described by Jeremic in respect of Tuzla appears to be mainly in the form of a large surface depression. Damage to surface structures by tensile strain effects and tilting appears to be a common feature.

Stability of rock-salt mining operations

Potts, Potts and Szeki (1972) have described field studies and laboratory investigations relating to underground mining of rock-salt in Cheshire at the Meadowbank Mine, Winsford. The rock-salt occurs in two main beds separated by 9 to 12m of Marl; the upper bed is 30-51m thick and lies at about 60m below the surface, whilst the bottom bed is around 25m thick. The working level discussed by Potts *et al* mined rock-salt at an extraction height of 7·5m near the base of the lower bed; the mine floor in the area concerned was 177m below surface. These

(a) House tilting

(b) House subjected to compression

Figure 162 Houses affected by subsidence in the Cheshire rock-salt mining field

(a) Road crossing subsidence zone

(b) Road passing through edge of subsidence zone

Figure 163 Roads affected by subsidence in the Cheshire rock-salt mining field

Figure 164 Cheshire flashes — widespread subsidence depression in the Cheshire rock-salt mining field

authors discussed stability investigations in a part of the mine where the square pillars were 19·5m wide and 7·5m high giving a pillar width/height ratio of 2·6; the pillars were laid out in straight lines, that is employing 4-way room intersections with 39·5m between centres. The plan percentage extraction was 75%. The district of the mine studied was a 6-heading development leaving successive rows of pillars with five in each row. This study as reported by Potts *et al* demonstrated that a well-designed mining layout in the Cheshire rock-salt can achieve a high degree of extraction but still allow a stable mining situation to be achieved. The pillar configuration and dimensions used were shown by deformation measurements taken within the mine to achieve a high order of stability.

Hedley (1972) reported on investigations into roof stability at a Canadian salt mine where a 24m thick bed of rock-salt lying at a depth below surface of 535m was being extracted by the room and pillar method of mining. The room width was initially 18m and square pillars of 64m side length were used, giving around 40% extraction. The mining layout dimensions were later revised to give a room width of 14m and square pillars of 45m side length. The room height was 12m, which allowed 6-9m of rock-salt roof beam and 3m in the floor to be left. Roof slabbing was encountered when a room width of 18m was used. No surface subsidence is referred to. The pillars are particularly stable.

Subsidence effects resulting from mining of Louisiana salt domes

Miller and Stoakes (1985) have pointed out the different subsidence mechanism involved with mining salt domes as opposed to bedded structures. The massive nature of the salt domes overlain by soft sediments give them a unique setting. The Five Island salt mines of Louisiana on the Gulf Coast have employed room and pillar methods of mining. The mining depth is generally in the range 350-450m below surface. The degree of extraction is in the range 45% to 75%. Room heights are 6-9m; pillar dimensions are 21 x 30m to 73 x 73m; room widths are in the range 18 to 30m generally although one mine did use a room width of 48m. Having established the initial development layout, the excavations are then benched downwards, with benches of up to 30m. Where multi-level mining layouts are employed, it has been the practice to superimpose the workings and leave sill pillars between the mining levels; the sill pillars have varied in thickness between 38m and 122m.

The overlying rocks deform with time as a consequence of the time-dependent behaviour of the rock-salt dome mine workings. Mine shafts have moved laterally. Miller and Stoakes draw attention to the analysis and prediction of surface subsidence being complicated by the salt's rheological behaviour. There is a time delay between the mining phase and the resulting surface subsidence. These authors appear to concur with Kratzsch (1983) that subsidence over evaporite room and pillar old mine workings is slowed down but not necessarily complete. The continued slight subsidence effects appear to be due to a secondary creep phase of pillar deformation. This will of course be dependent upon the original size of the pillars and depth below surface coupled with the degree of extraction carried out.

Miller and Stoakes suggest the following reduction and preventative measures in respect of surface subsidence:

1. Little or uneconomical extraction of the salt.
2. Use of controlled room and pillar mining with barrier pillars.
3. Appreciation of stress relief methods.
4. Use of controlled multi-level mining.

These authors put forward the following remedial measures for combating subsidence:

1. The use of backfilling with waste material.
2. Flooding of the mine.

They also conclude that mining methods can be successfully designed to help control the resulting subsidence of the surface, and that measurements should be employed to ascertain the degree of control being attained.

Subsidence and ground control aspects of mining potash

Potash commonly occurs as a bedded structure which lends itself to extraction by several forms of room and pillar mining or by some form of longwall mining operation.

The report of the seminar on Canadian potash mining and ground control reviewed the mining methods in common use world-wide, CANMET (1985). Room and pillar mining without depillaring accounts for 51% of the total potash mines; the mining method with caving and backfill respectively represent 7% and 4% of the total. Shortwall and longwall operation accounts for 13% of the total mines, whilst sublevel stoping represents 15%, cut and fill and block caving 2% each and lastly solution mining which is only 4% of the total.

Potash mining operations take place in wide ranging depth conditions, the deepest being Saskatchewan in Canada with 960-1500m, although other potash mining districts operate in fairly deep situations, for example Italy 800-900m, France 850-1000m, Utah, USA 800-1000m, Whitby UK 1150m and New Brunswick, Canada 500-1000m. Wide ranging depth conditions also exist in West Germany 350-1200m, East Germany 350-1100m, Spain 400-850m, USSR 200-900m and New Mexico, USA 220-550m.

This Seminar identified subsidence and ground control as important areas for investigation.

Ground control behaviour of potash mining in Spain

Oyanguren (1972 a) has described an investigation into the ground movement and control aspects of extracting two potash beds in close proximity at the Esparza Mine in Spain. A bed of carnallite around 2·5m thick overlies a 2m thick bed of sylvanite, there being an intervening thickness of salt of around 1m. Around 70m of salt overlies the carnallite. Figure 165 shows Oyanguren's ground deformation results obtained behind a longwall extraction. At around 180-200m behind the face line, the ground appears to have almost stabilised. The results were obtained from anchor points located in boreholes drilled from the surface.

Figures 166 (a) and (b) show further results reported by Oyanguren. These show the vertical component of strata pressure acting in the vicinity of a longwall extraction in potash, the first corresponding to pressure distribution normal to the face, whilst the second is transverse to the direction of working. These results show that the front abutment strata pressure and flank abutment strata pressure zones are confined to a very narrow zone and occur close to the extraction edge. This latter distance is indicated to be about 5m in both sets of results. The depth of mining in these case study results is 400m. Figure 166 (a) shows that the goaf pressure becomes stabilised at around 350-450m behind the face line. Oyanguren reported that the results gave no indication of the presence of any rear pressure abutment effects. Figure 166 (b) shows a transverse section at a significant distance (greater than the depth) from the face line. The results show that pressure build-up in the goaf begins to occur at about 50m from the ribside and takes around another 100m or so before attaining coverload pressure.

In the discussion on his paper Oyanguren (1972 b) reports that surface subsidence immediately above the longwall face was only some 3% of the extracted seam height (M) and only 15% of M at 50m behind the face. The subsurface beds exhibited delayed subsidence behaviour owing to their plastic deformational character.

Salas (1979) has further considered the potash subsidence data used by Oyanguren (1972 a) obtained at a mine in north west Spain. A longwall extraction layout with an excavation height (M) of 2m of sylvanite was studied. Table 18 gives the basic subsidence data relating to this longwall.

Type of extraction	Longwall
Goaf treatment	Caving
Subsidence classification	Super-critical
Extraction thickness	2m
Depth below surface	200m
Transverse profile subsidence limits	25-28° (to vertical)
Value of maximum subsidence	0·67 (M)

Table 18 Subsidence data relating to longwall extraction of potash in Spain, based on information reported by Salas (1979).

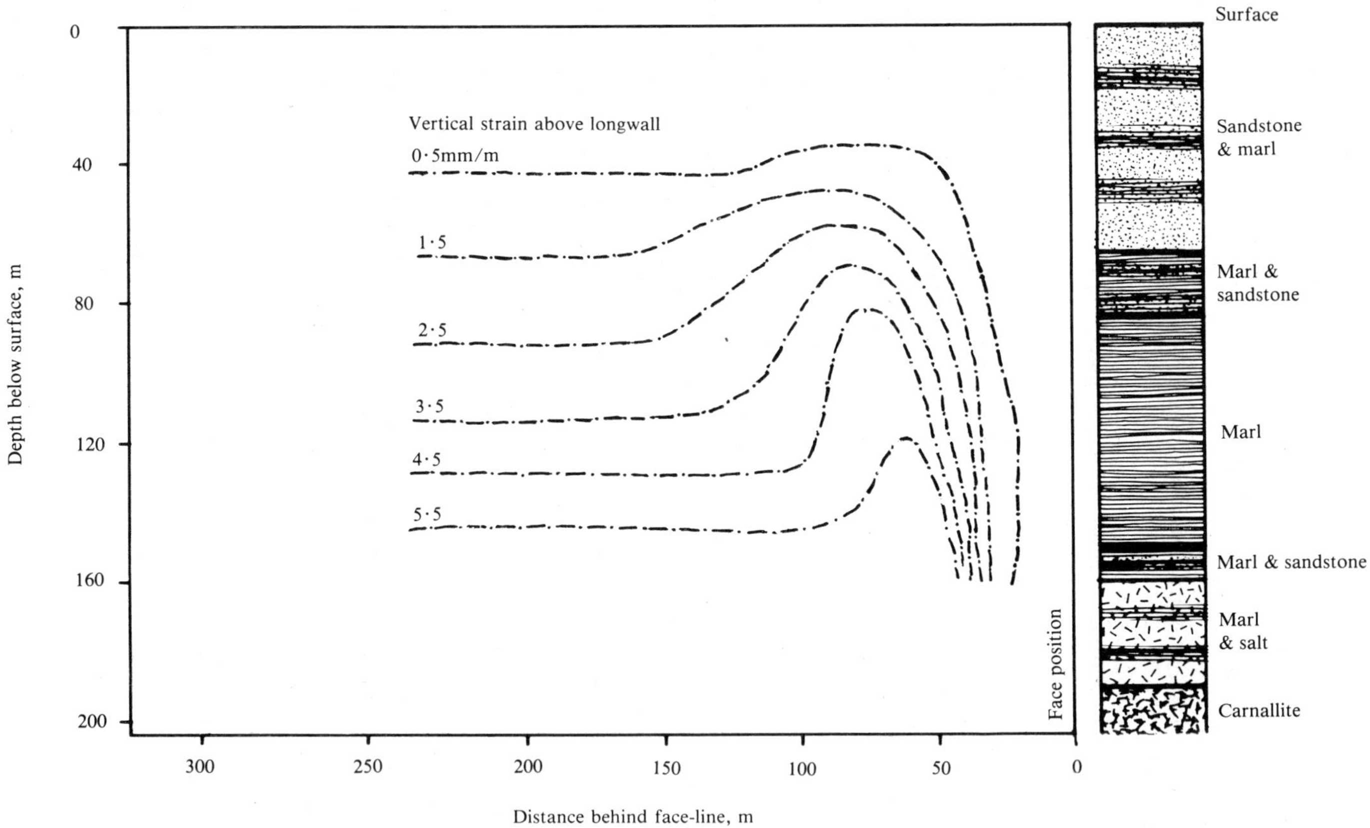

Figure 165 Sub-surface vertical ground strain behind longwall, after Oyanguren (1972)

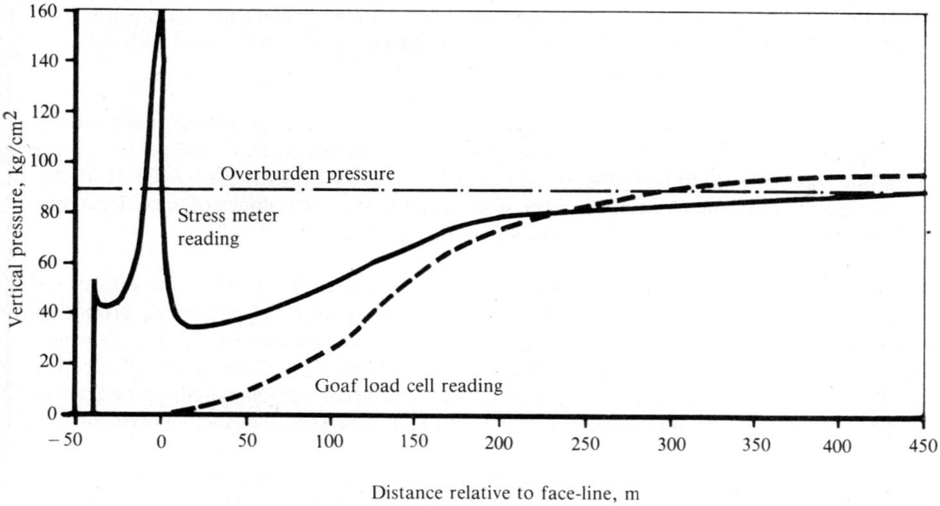

Figure 166(a) Pressure distribution normal to the face at depth below surface of 400m, after Oyanguren (1972)

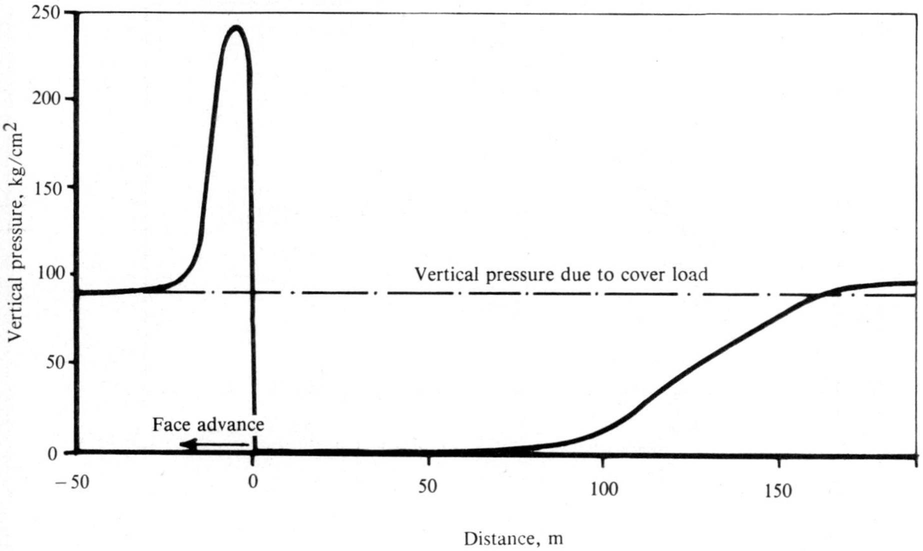

Figure 166(b) Strata pressure distribution on a line longitudinal to the direction of longwall face advance for a depth below surface of 400m. After Oyanguren (1972).

Figure 167 shows subsidence results reported by Salas (1979) in respect of the development of subsidence in relation to face advance. The U.K. SEH (1975) curve is also included for comparison at this depth of mining. There is fairly close agreement exhibited between the potash mining results and the U.K. SEH (1975) predicted relationship. The overburden, as shown in Figure 165, consists mainly of sandstones and marls with salt and marl close to the carnallite horizon.

Forster (1967) also interpreted surface subsidence data from observations made above longwall extractions in potash mines in north-west Spain. He reported that subsidence above the face-line was observed to be 1·7% (M) compared to 14% (M) for British coalfields. It was observed that the movements occurred very slowly during the early stages of subsidence but then developed rapidly along the lines of the classical profile.

Figure 168 shows the general layout of the longwalls and the positions of the subsidence observation stations. The point made by Forster regarding the slow response of the ground in the development of subsidence at the surface is clearly demonstrated in Figure 169 which shows the partially developed dynamic (longitudinal) subsidence line after 123m advance for longwall JP - 45. Conversely, Figure 170 shows the subsidence development curve for Station 10 above longwall JP - 45 and illustrates minor subsidence effects over the solid rib-side.

A transverse subsidence profile is presented in Figure 171 which supports the conclusion of Forster on the development of subsidence at the edges of the excavated area. These results show that the maximum subsidence was 0·6 (M).

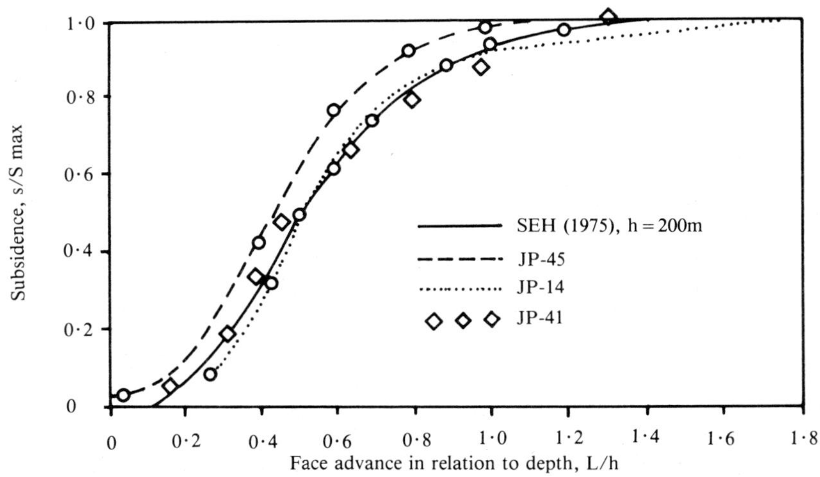

Figure 167 Comparison of subsidence development curves for potash extractions with SEH (1975), after Salas (1979)

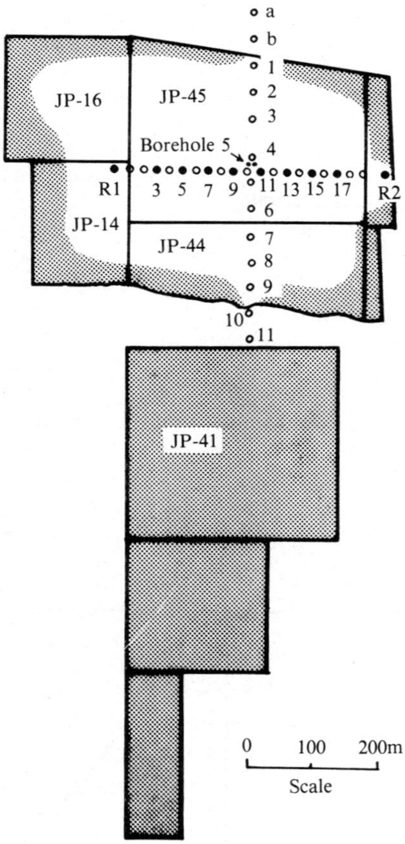

Figure 168 Location of surface observation stations and Borehole 5 in relation to Face JP-45, after Forster (1967)

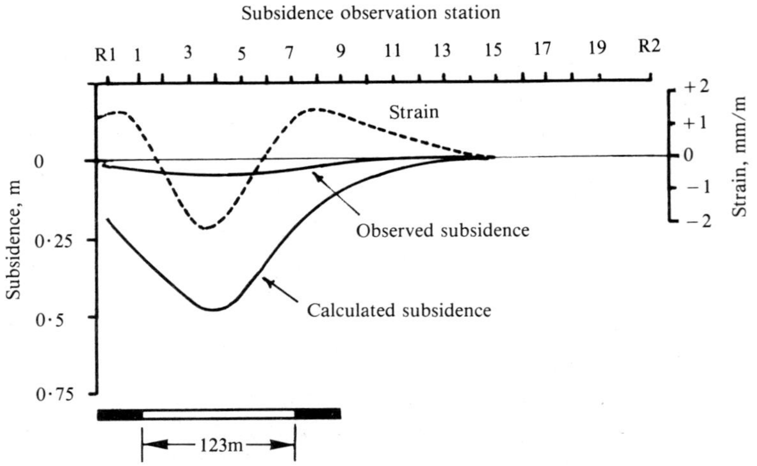

Figure 169 Longitudinal (dynamic) subsidence profile after 123m of advance at JP-45

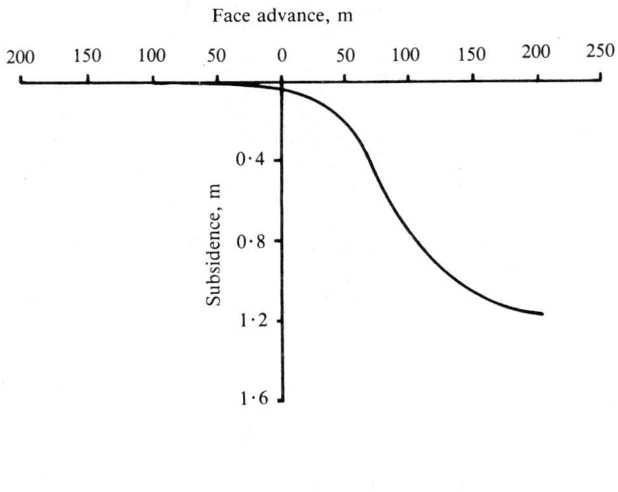

Figure 170 Subsidence development curve observed at Station 10 at JP-45

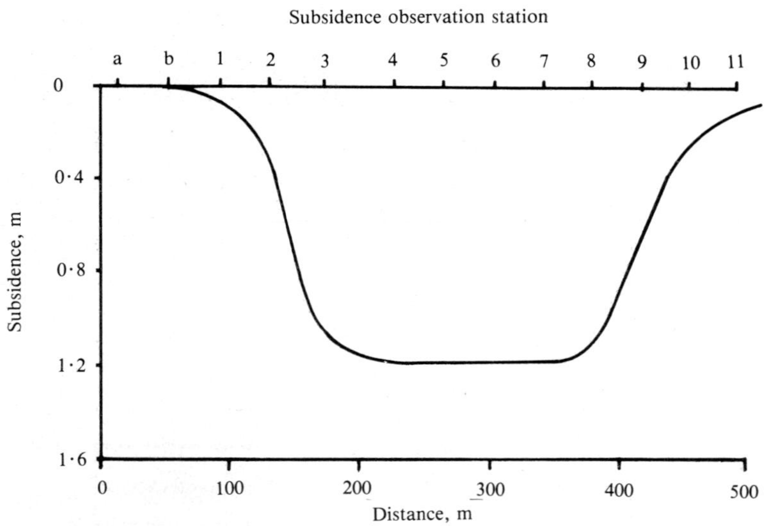

Figure 171 Transverse subsidence profile for JP-45, after Forster (1967)

An important part of Forster's investigations was concerned with examining the nature of sub-surface development of ground movement as observed during the longwall caving process and the resulting subsidence phase. Figure 172 shows Forster's results for Borehole 5 in respect of the absolute movement of borehole anchor points and the surface in relation to face advance. There is uniformity in the progressive development of subsidence as shown by these results taken from a borehole located approximately within the central area of the extraction. The main bed separation phase is indicated to occur after the transition point is encountered and during development of the final part of the subsidence curve. The vertical displacement contours have been interpreted by Forster as shown in Figure 173 (a) whilst Figure 173 (b) shows his interpretation of the "strain contours" which appear to be only in relation to the tensile strain condition and in the vertical direction only. No compressive strain values are shown but they most certainly exist particularly in the region ahead of the face-line.

Ground control and subsidence aspects arising from Canadian potash mining

Bawden and Mottahed (1986) have reviewed subsidence experiences resulting from potash mining in Saskatchewan and examined the potential application of subsidence prediction techniques to these mining conditions. Subsidence data were collated at a number of Saskatchewan potash mines. The main finding was the difficulty in accommodating time-dependent behaviour of the potash and associated rocks into existing subsidence prediction models. Another difficulty encountered was that of incorporating a complicated mine geometry into the subsidence prediction procedure. The authors expressed the view that influence function methods coupled with finite element methods appeared to offer the greatest promise as suitable potential techniques for the Saskatchewan potash mining situations.

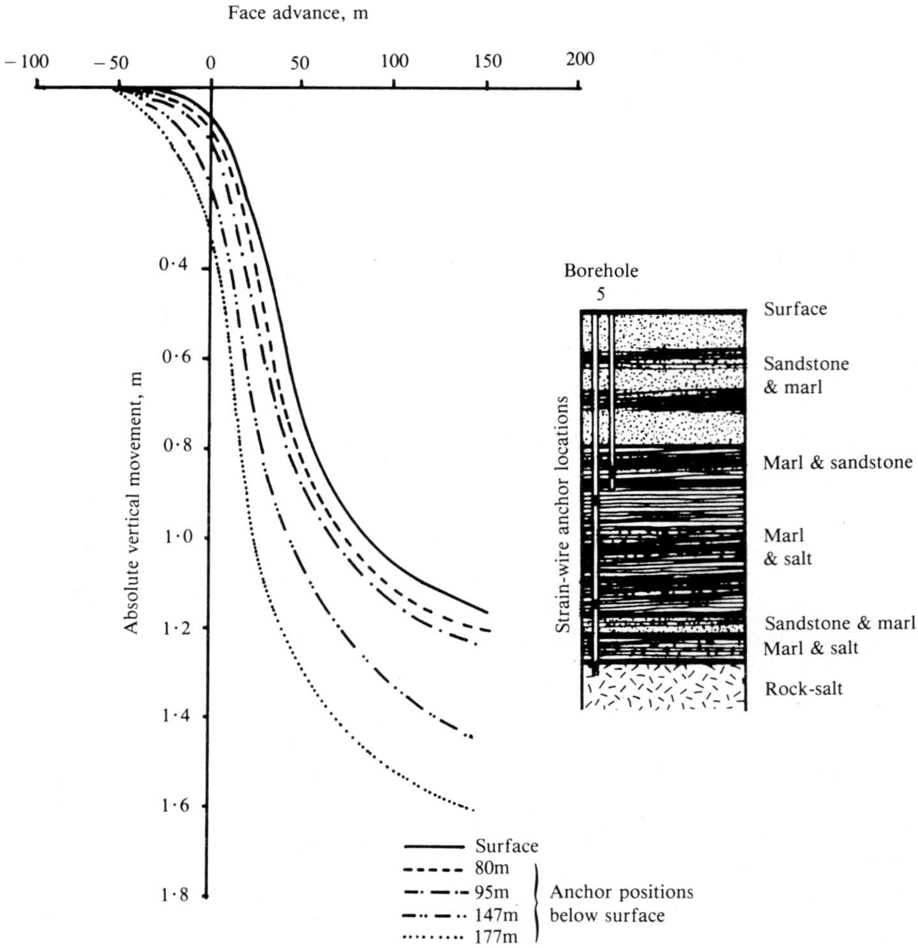

Figure 172 Development of surface subsidence and sub-surface movements in relation to face advance, after Forster (1967)

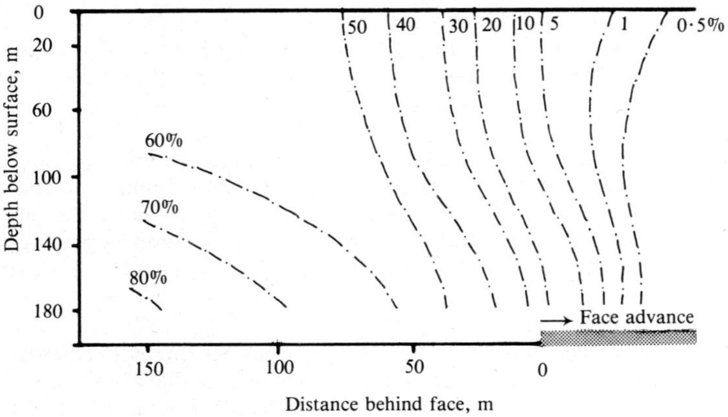

(a) Contours of vertical displacement for Borehole 5

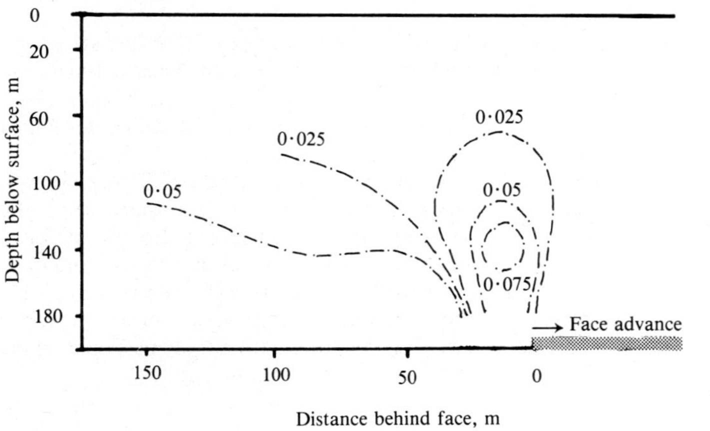

(b) Contours of strain, Borehole 5

Figure 173 Development of vertical displacement and strain contour representation behind potash longwall mining extraction, after Forster (1967)

The principal mining method of extracting potash in the Saskatchewan mines is that of continuous mining a bed 2·4-5m thickness at a depth of around 1000m. The degree of extraction was initially about 25%, but this was increased to 35-40% following improvements in mining efficiency. Within an individual working district it is common to attain +50% extraction. The two main forms of continuous mining employed are (a) pillar panel mining, and (b) chevron mining.

Bawden and Mottahed present sets of field observations which show values of maximum subsidence of 0·4m which took four years to develop over the partial extraction workings. Subsidence troughs of 4,000m width are also reported.

The subsidence observed above Saskatchewan potash mines exhibits two stages in its development. There is fairly rapid convergence immediately after mining and this usually takes 100-200 days. The second stage appears to be due to the creep of the pillars and surrounding rocks. These authors concluded that the zone area method employing suitable functions to take the initial and constant subsidence rate into account, gave a reasonable degree of prediction.

Mraz (1985) reported that subsidence profiles over some of the Saskatchewan potash mining areas exhibit a double troughing phenomenon. He reports that the main subsidence trough is of the anticipated shape and areal extent. However, Mraz points out that there is a shallow and uniform subsidence of the surface surrounding the main trough in some situations. He suggests that this extended subsidence effect is probably due to the nature of the stratigraphical column between the surface and the mining horizon.

Alsace, France potash mining surface subsidence aspects

Partial extraction layouts taking longwalls of 50-80m width and leaving intervening support pillars of around 70m have been employed with an extraction height of 2·1m, in a depth below surface range of 460-510m. The general characteristics of the resulting subsidence observed are listed as follows:

1. The maximum observed subsidence with such partial extraction layouts was 160mm.
2. The intervening pillars did not result in any significant hump effects developing in the surface subsidence profile.
3. Subsidence appeared slow to develop initially, then the main trough developed fairly rapidly after 50% S max had been attained.
4. For the depth range mined, there was a noticeable time-dependent feature associated with the exponential decay in the gradual decrease of surface subsidence over the entire trough. Even towards the extremities of the subsidence trough minor settling movements were in evidence and these appeared to be due to the time-dependent characteristics of the potash and surrounding rocks. Such effects were only of the order of a few millimetres, but they could be still measured over a two-year period.
5. The subsidence troughs exhibited classical features without any particular evidence of double troughing effects as reported to have occurred in Canada.

Longwall faces of up to 500m width and taking an extraction height of up to 3m have also been employed. Such a mining layout reduces the likelihood of support pillars tending to give rise to significant time-dependent subsidence effects especially after completion of the mining operations.

Details on Alsace potash strength properties, pillar behaviour and general subsidence aspects have been discussed by McClain (1963) and Potts (1964).

Concluding remarks

The mining operations involved in the extraction of rock-salt can result in surface subsidence of different forms. The dissolution properties of rock-salt not only serve to provide a convenient means of extraction, namely that of solution mining, but can present special problems to surface stability unless the operations are effectively controlled.

Solution mining of rock-salt can give rise to sink-holes and major surface collapses occurring over such operations. The type of overburden has a strong influence but the main factor is the size of the cavity, or cavities, created by dissolution. It is frequently to what extent such cavities are allowed to continue collapsing towards the surface, since the operation of solution mining is self-perpetuating in creating underground voids thus giving the potential for an established collapse zone to progress nearer to the surface. Additionally, this feature also occurs with natural subsidence phenomena where a fresh water course gains contact with the rock-salt body or deposit. Solution mining control is generally difficult where the brine pumping holes create cavities which link up, in view of not knowing exactly from which part of the rock-salt mass the greatest dissolution is taking place. Strong beds of rock overlying rock-salt deposits at appreciable depth can tolerate greater solution mining effects than shallow deposits with weak overburden rocks.

Instrumentation and monitoring using strain-wires anchored within the subsurface over a known solution cavity can help to observe whether such a cavity has stabilised or if it is still collapsing and approaching the surface. Subsidence levelling on a systematic basis over a network of surface observation stations overlying a solution mining affected area of concern can provide a valuable form of monitoring service especially in detecting the onset of surface subsidence. Aerial photography, and general observation from a helicopter can also be of considerable assistance where the area involved is widespread or the nature of the surface makes other forms of routine observation difficult if not impossible to achieve.

Shallow subsidence depressions as occur with rock-salt mining usually develop gradually and over an appreciable time-span. Their main effects are to increase the potential for localised flooding of the surface but this depends upon the nature of the surface before mining and also on the general climatic conditions.

Systematic underground mining by mechanical cutting and removal of rock-salt offers the most effective form of control of ground movement. The mining extraction dimensions and size of support pillars can be selected to either minimise or eliminate any subsequent likely surface subsidence which may arise from the underground operations. The excellent natural supporting ability often encountered with the roof beds immediately overlying rock-salt formations, can be utilised to good effect in promoting long-term stability of both the mine and the surface by suitable mining design which takes due account of the time-dependent characteristics of the rocks.

Longwall mining of potash deposits can result in surface subsidence features similar to those encountered with coal mining operations providing there are similarities in the nature of the overburden. Where the overburden has significant thicknesses of rock-salt or other strongly time-dependent behaviour rock types, then the subsidence is likely to develop very gradually in a delayed form by comparison to classical subsidence development curves. During the later stages of subsidence development, however, field observations show that the ground movement responds more quickly to catch up with that anticipated with classical curves.

Deep mining operations extracting rock-salt and potash of the order of 1000m which involve leaving support pillars tend to exhibit time-dependent subsidence effects at the surface. Such effects appear to be spread over the entire width of the mining subsidence trough.

Old room and pillar mines in rock-salt and potash can continue to give a creep-controlled form of subsidence, many years after mining has taken place if the conditions of extraction percentage, pillar dimensions and mining depth allow the rock stresses to be of sufficient magnitude as to produce measurable time-dependent deformation effects.

CHAPTER 13

THE TIME FACTOR IN MINING SUBSIDENCE

The chapters dealing with calculation of subsidence resulting from longwall mining operations present results on the basis that the duration of subsidence has elapsed. The operation of underground mining extraction can involve leaving the overburden supported by pillars which may or may not give rise to surface instability many years afterwards. Conversely the underground mining operation frequently involves the deliberate caving of the immediate roof beds. The subsequent lowering of the overburden as a controlled operation allows the subsidence phase to be determined in duration and importantly the time of likely cessation. The lowering of the surface creates a disturbing effect to the general state of equilibrium. A certain amount of time is generally required in order to restore the surface to its normal state.

Various investigators have been interested in seeking quantitative information on the time taken between cessation of mineral working and completion of subsidence at the surface, and Table 19 has been compiled from Lane and Roberts (1929) who reviewed work up to that time published by different authors.

Case No	Mining situation	Time to complete subsidence	Source
1	Shallow ironstone mine working Deep ironstone mine working	Few months 7 to 9 years	Hedley (1923)
2	For workings less than 90m deep	2 to 3 years	Kay (1898-99)
3	For workings less than 180m deep	2 years	Hamilton (1923)
4	For workings at a depth of 206m	1 year	Dixon (1885-86)
5	For workings at a depth of 278m	5 years	Fletcher (1866-67)
6	Depth of working was 329m	2·8 years	Lloyd (1918-19)
7	For workings less than 463m deep	Up to 5 years	Whitelock (1920)

Table 19 Time to completion of mining subsidence quoted by early investigators, compiled from data given by Lane and Roberts (1929).

Table 19 shows some widely differing quoted periods of time. The ironstone mine cases may have referred to pillared workings, and some of the others may have employed differing degrees of accuracy in subsidence observation.

Room and pillar mine workings can experience surface subsidence effects some 100 years or so after completion of the mining operations, see Chapter 9 and Kratzsch (1983). The reasons for surface subsidence occurring above room and pillar mine workings are given in Chapter 9.

Attention is focussed here on the significance of time involved with subsidence connected with longwall operations.

Study of time duration for ground movements to transmit to the surface

Corden (1964), carried out field studies in the Yorkshire Coalfield, UK and employed instrumentation to detect surface movements caused by undermining. An automatic measuring and recording apparatus was used on the surface to allow correlation with underground mining operations. Two longwall situations were studied, for Site 1: $h = 165m$, $M = 0{\cdot}81m$, $w = 167m$; and for Site 2: $h = 104m$, $M = 0{\cdot}9m$, $w = 222m$. Results presented by Corden tend to show that the recordings of vertical movement at the surface displayed a stepped characteristic, with the vertical part of the graph corresponding with the coal cutting phase of the mining operations on the longwall face. The following general conclusions were made by Corden in respect of these shallow conditions:

1. The period of rapid subsidence development at the surface could be clearly identified in the mining component parts of the underground working cycle, using an instrument of high sensitivity. The effects of holiday periods and weekend stoppages were clearly correlated in the subsidence development curve.

2. For the shallow conditions investigated the highly sensitive apparatus used allowed the possibility of each coal cutting shift to be identified.

3. The observed surface response to mining was that of (i) an "immediate" effect which appears shortly after the start of the cutting operation. This immediate effect lasted for a similar period to that taken by the cutting phase of the mining operation. Additionally there was (ii) a more protracted subsidence phase which was slower to develop. The immediate effect was absent at weekends if no cutting operation was carried out, whilst the secondary subsidence effect was considerably reduced.

4. After the face had advanced beyond the critical subsidence distance (for an angle of draw of 35° to the vertical), some further ground movement was still measured. This was undoubtedly what has later been referred to as residual subsidence effects.

5. The general physical structure and mechanical behaviour of the overburden and constituent beds results in only part of the total effect being transmitted immediately to the surface. Bed separations occur. The subsequent stress relaxation accompanied by later bed closure forms part of the equilibrium restoration process. The result of this is to produce a smoothing effect of the subsidence development curve.

Supporting discussion on this work and related aspects is given in the publication by Corden and King (1965).

Interruptions in the subsidence development curve: tilt observations in the field

Pasamehmetoglu (1972) carried out field observations of ground tilt at the surface overlying a longwall extraction of average depth 91m, extracted seam height 1·57m and extraction width 200m which was advancing along the strike of a seam dipping at 14°. An electrolevel was used to observe subsidence tilt changes as the longwall approached specially designed field stations located across its path. The technique has been described by Whittaker and Pasamehmetoglu (1981).

Figure 174 shows a set of tilt observations for a field instrumentation station located over

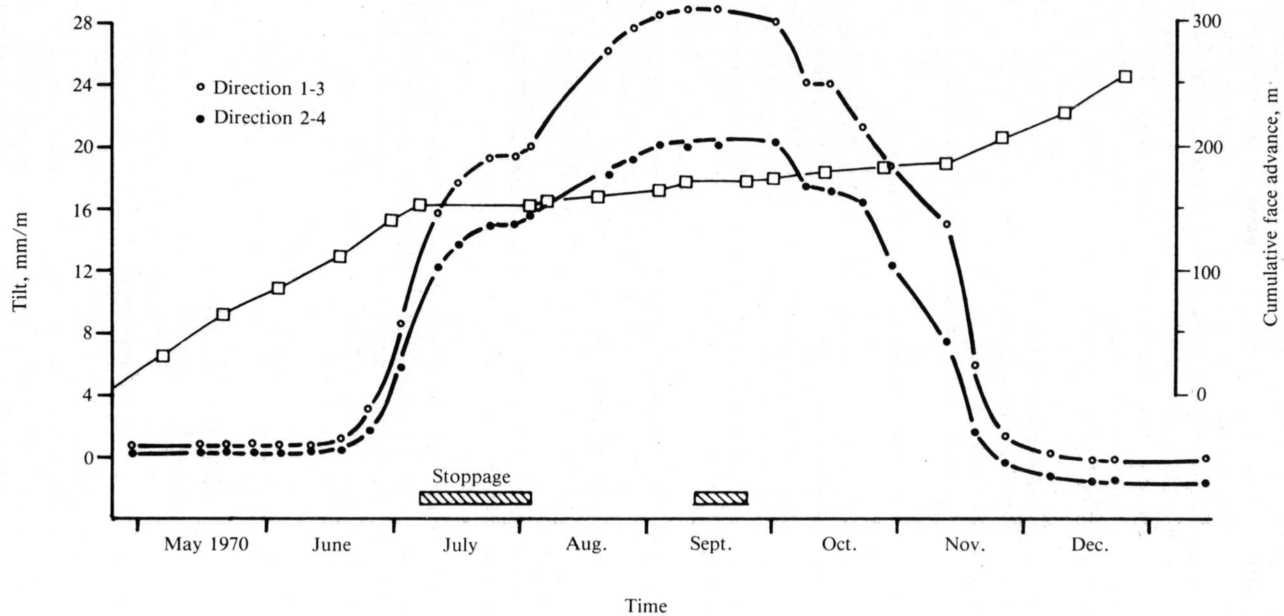

Figure 174 Tilt changes observed by electrolevel at field station NU5 in relation to time taken for longwall to completely undermine the observation point. After Pasamehmetoglu (1972).

the centre of the longwall, as reported by Pasamehmetoglu. The longwall face concerned experienced a fairly uniform rate of advance during the first part of the graph shown and the subsidence phase had begun to affect the measuring station when a stoppage period occurred of almost one month duration. The tilt graph shows that there was a marked time-dependence effect in that ground tilt continued to increase even though the face had stopped; the tilt graph demonstrates a form of exponential decay during the face stoppage period. After the face started up again its rate of advance was substantially less than during the period up to July and this continued into November when the rate was almost restored to its planned value. The period August to November experienced a further shorter stoppage. The variation in ground tilt reflects the rate of change of face advance. This longwall encountered difficult face conditions during this period and thus accounted for the change in rate of advance.

An interesting feature of Pasamehmetoglu's work is shown in Figure 175(a), (b) and (c) which presents details of observed changes in ground tilt at three different field stations in relation to a major stoppage of the longwall. An exponential decay characteristic is evident in all three cases which were located within the fully subsided section of the excavated area.

The following conclusions were drawn by Pasamehmetoglu:

1. The time-factor is distinguished from the development factor because the former has no relationship with the mining operations and the development of subsidence with time. Time-dependent subsidence continues during periods of face stoppage. The fact that the overburden rocks possess a measure of time-dependent behaviour, then it is to be expected that time-dependent deformations will take place during all stages of face advance.

2. Mining subsidence due to longwall operation falls under three broad categories:
   (i) closure due to elastic deformation and is time-independent,
   (ii) the inelastic (time-dependent) deformations, and
   (iii) closure arising from fracture and collapse under gravitational force of the overlying rocks.
   The three generally act collectively to produce a lowering of the overburden and is referred to as subsidence.

3. Measurable time-dependent behaviour of ground movement was observed during periods of longwall face stoppage. The tilt-time curve during a temporary stoppage as observed on the subsidence development curve was found to be represented by an exponential function of the form given by equation (66).

$$y = A \pm B e^{-Ct} \tag{66}$$

where y = tilt (mm/m)
A and B = constants (mm/m)
C = constant (dimension of inverse time)
t = time (days)

The initial state of ground movement at the particular point in time of the face stoppage influences the value of the constants A, B and C. The constants also depend upon the parameters: depth, extracted seam thickness, face width, position of surface point in relation to face position, and rate of change of tilt. The plus or minus sign applies to the relevant stage of tilt of the surface point at the time of face stoppage.

4. A period of 20-30 days elapsed before the ground tilt decayed sufficiently for no further change to be measurable in the case of the relatively shallow conditions investigated.

5. The field observations of tilt by electrolevel showed that on the first day of working after a long stoppage of the face, ground movements had an almost immediate effect on the field station at the surface.

6. The effect of stoppage of a longwall face over the Christmas holiday period at the Eastwood Hall site resulted in a clear step in the rate of change of tilt with time, with the ground appearing to respond similar to relaxation of a Burger's model. The extracted seam height was 1·67m, depth below surface was 260m and the face width was also 260m.

Time-dependence in mining subsidence

Orchard and Allen (1974) have examined the time duration for development of mining subsidence and pointed out that the old concept that subsidence takes years to develop, independently of the speed of extraction needed thorough examination. Their findings were that it can take years to develop where it takes that amount of time for one, two or even three longwalls to extract the critical area of a particular surface point. This of course depends upon depth below surface and the rate of advance.

Of importance to the significance of the time factor is the nature of the time-dependent properties of the rocks undergoing subsidence. Orchard and Allen draw attention to what they term as the true time factor in mining subsidence as causing subsidence to continue during temporary stationary periods such as at weekends etc. Additionally, after the longwall face finishes, there is a small amount of additional subsidence, and this is referred to as the element of residual subsidence.

A case history illustrating time-subsidence effects is presented in Figure 176(a) and (b). This shows a longwall of dimensions: $M = 0\cdot74$m, $h = 72$m and $w = 195$m which was operated by cyclic mining of cutting, filling and preparation and then repeated. The results presented in Figure 176(b) show detailed relative subsidence values and clearly demonstrate that the effects of the cutting operation have been transmitted virtually immediately to the surface. It is seen that at the weekend forming the first part of Figure 176(b), subsidence had virtually stopped. During the Monday cutting shift subsidence occurred at the rate of around 0·6mm/hour which subsequently decreased to 0·33mm/hour on the following two shifts. The subsequent working days of that week experienced subsidence rates of 1·1mm/hour and 0·5mm/hour for the cutting and intervening shifts respectively. Figure 176(b) indicates that during the cutting shift the surface did not just merely subside, but appeared to experience vertical fluctuations resembling a minor shock wave pattern effect.

Residual subsidence, location and amount

The time-dependent element is related to the active element of subsidence according to the work of Orchard and Allen (1974), and consequently residual subsidence varies in magnitude as to the position of the surface point relative to the longwall face.

A case study referred to by Orchard and Allen is presented in Figure 177(a) which shows that residual subsidence was at its greatest around the transition point, namely that part of the subsidence curve experiencing the greatest activity at the time the longwall finished. This illustration also includes residual subsidence shown as separate curves for the periods stated;

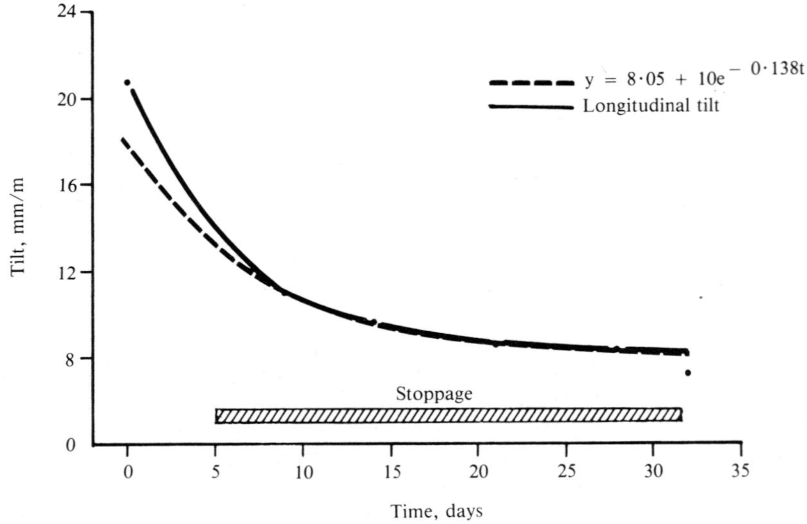

(a) Detail of tilt change with time for Station NU3

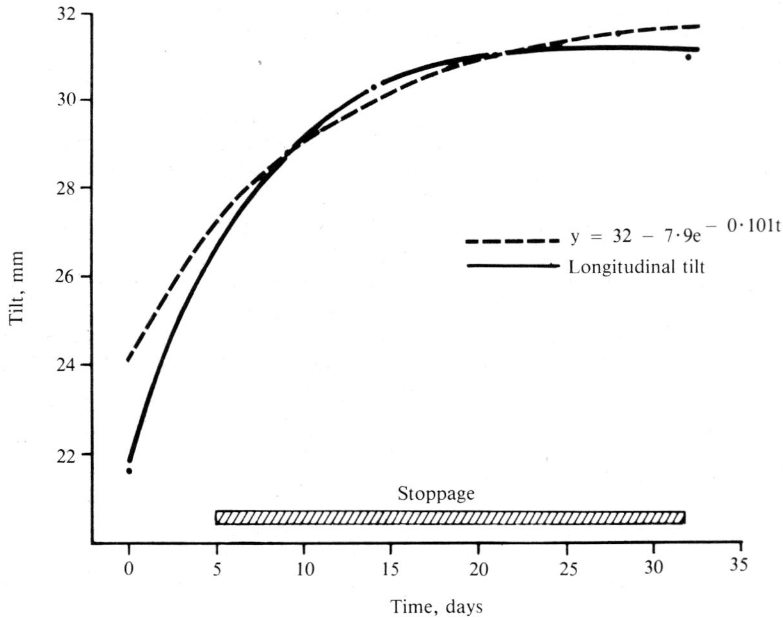

(b) Detail of tilt change with time for Station NU4

Figure 175 Detail of tilt changes in relation to longwall face stoppage. After Pasamehmetoglu (1972).

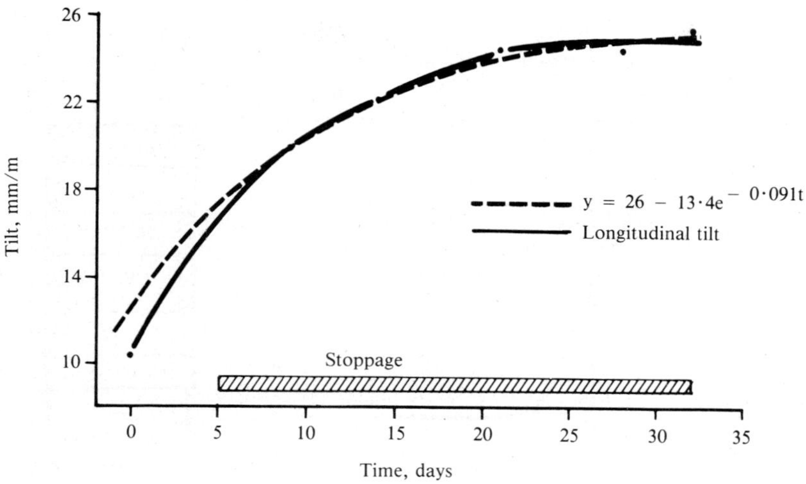

(c) Detail of tilt change with time for Station NU5

Figure 175 (continued) Detail of tilt changes in relation to longwall face stoppage. After Pasamehmetoglu (1972).

the estimated curve allows for the one day after cessation of mining as subsidence was not measured until the following day. Residual subsidence is shown here to be at a maximum of 7% of the total subsidence.

Orchard and Allen report that the maximum residual subsidence has often been observed to occur at around the half-subsidence point on the subsidence profile and represents the point subsiding the fastest when the face has stopped. Residual subsidence decreases backwards and forwards from the point of half-subsidence namely the transition point.

Additional residual subsidence data are shown plotted on Figure 177(b) and supports the comments made above regarding the location of the maximum value and the decrease in significance towards the edges of the zone of subsidence. Maximum residual subsidence values of 6·8 to 9·8% of the total subsidence (S max) are indicated by Figure 177(b). The residual subsidence effects are seen to occur with different rock types.

The amount of residual subsidence observed is commonly in the range of 5-10% and often less for UK coalfields.

Delayed residual subsidence

Orchard and Allen (1974) have reported two cases of delayed residual subsidence effects.

The first case was a longwall in the High Main Seam at Peterlee, Durham Coalfield, UK where the Coal Measures were overlain by: surface drift deposits, 9-15m; dolomitic limestone, 122m; marl slate, 0·6m; and basal Permian sand, 30m. In this situation three seams had been previously worked at a greater depth and at least about 40 years earlier. The High Main

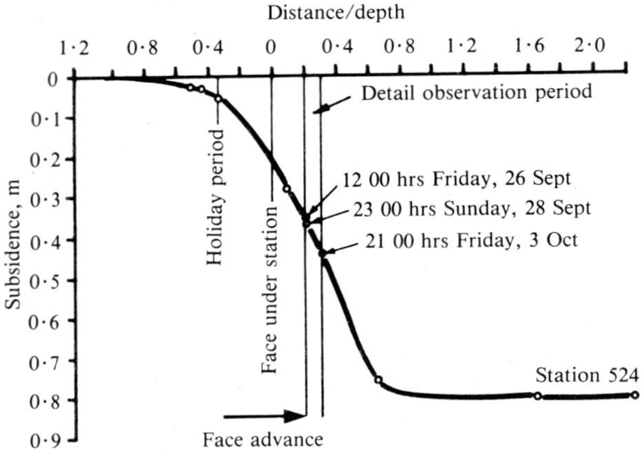

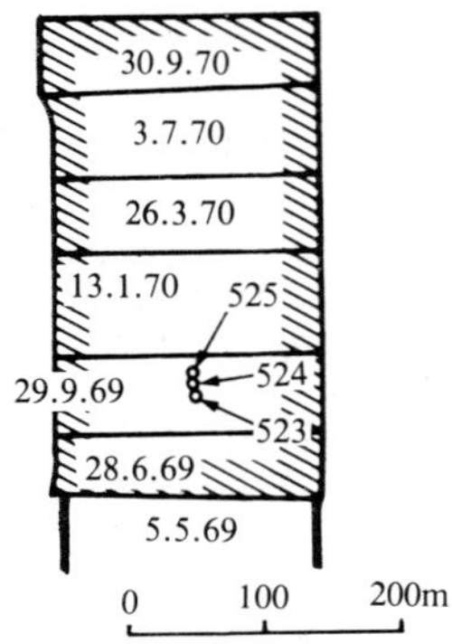

(i) Subsidence development curve and plan showing location of observation stations

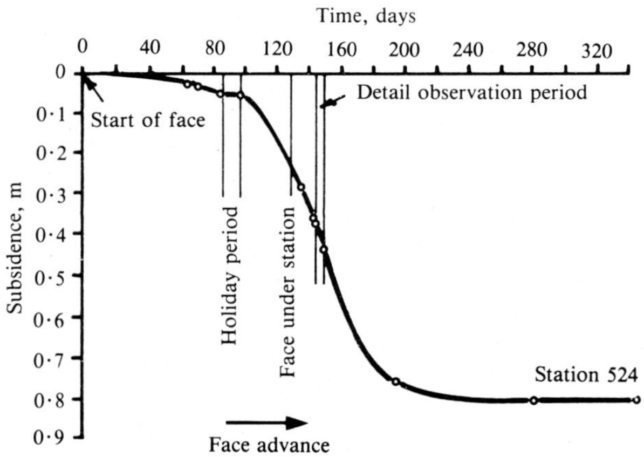

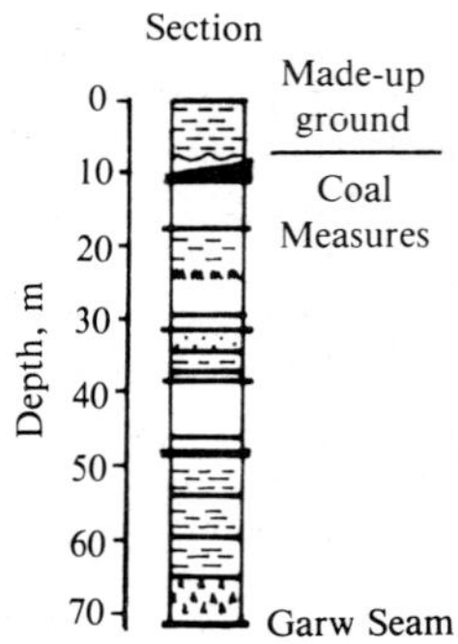

(ii) Time-subsidence curve and geological section

Figure 176 (a) Subsidence development in relation to distance and time, of Garw Seam, Blaenavon Colliery, South Wales longwall extraction — case study. After Orchard and Allen (1974).

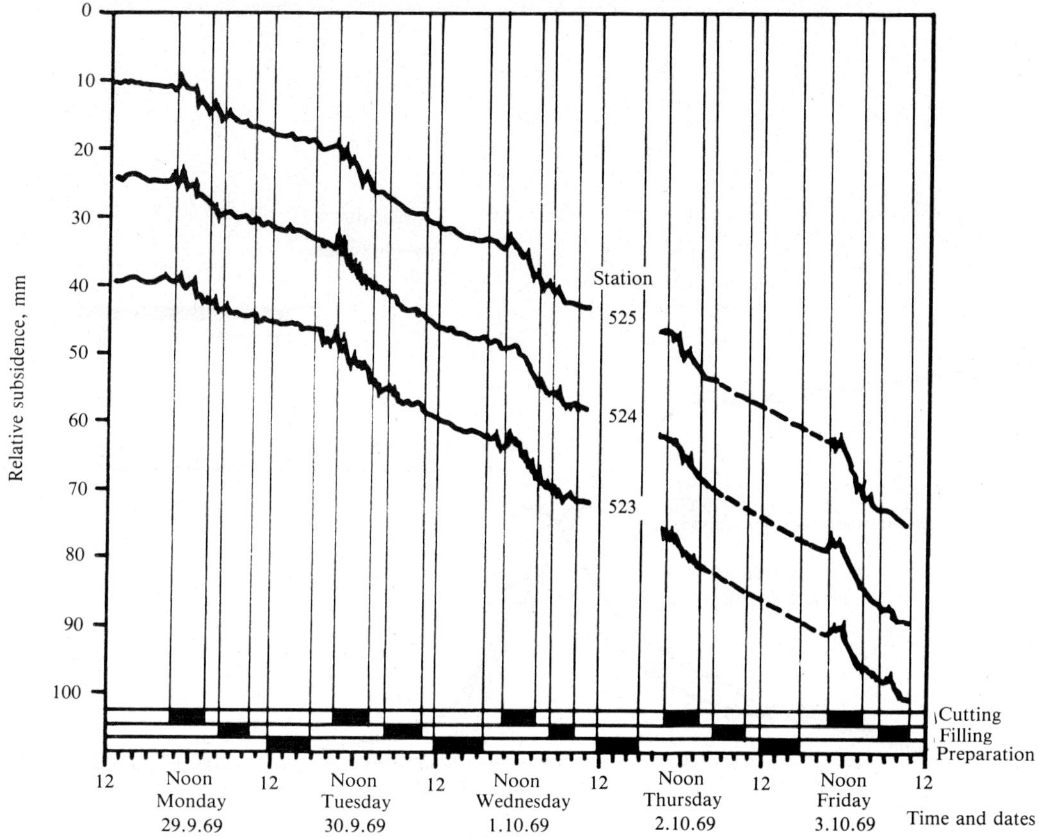

Figure 176 (b) Detailed observations of relative subsidence recorded during the cutting, filling and preparation shifts of longwall operation in the Garw Seam, Blaenavon Colliery, South Wales. After Orchard and Allen (1974).

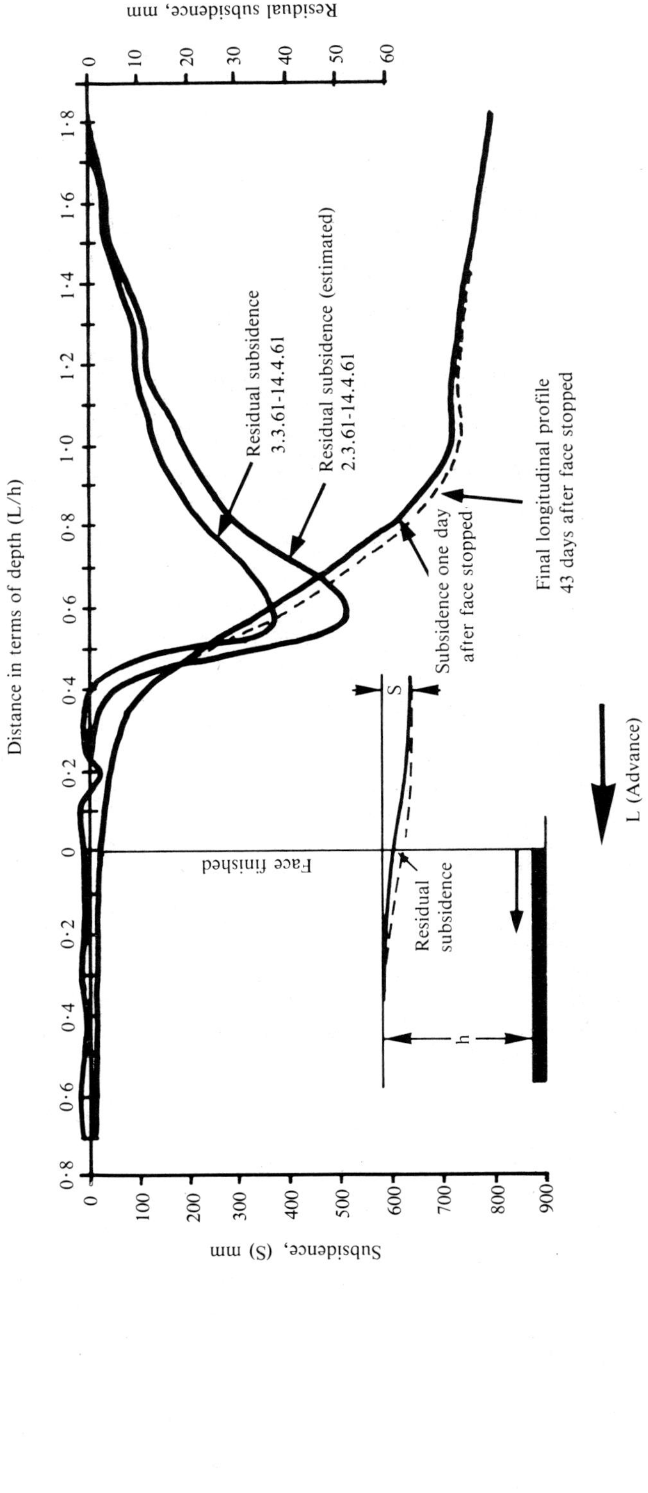

Figure 177(a) Residual subsidence profile, Shirland Colliery. After Orchard and Allen (1974).

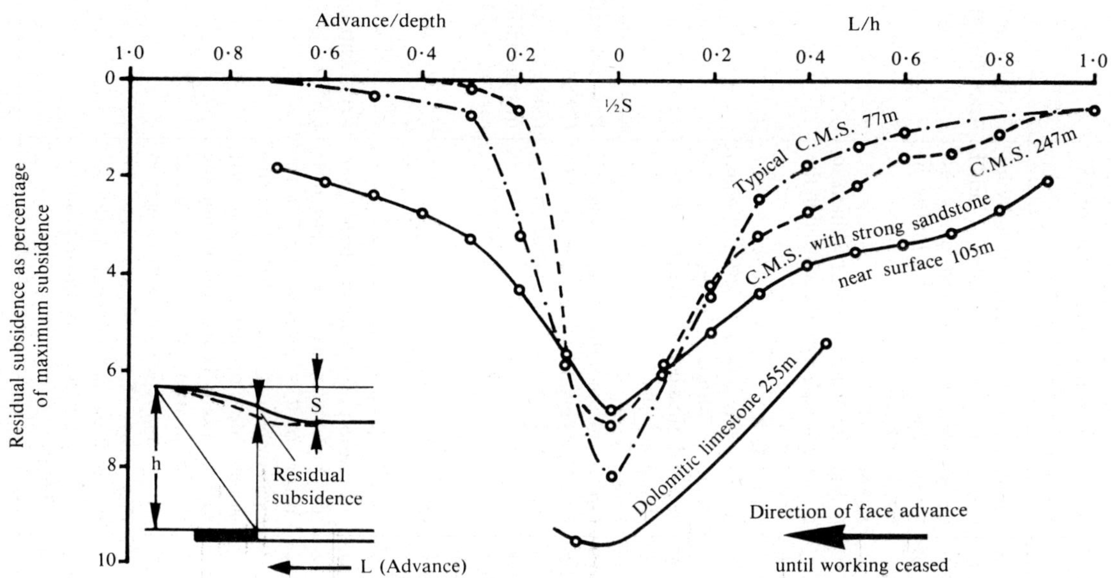

Figure 177(b) Comparison of residual subsidence profiles. After Orchard and Allen (1974).

No.	Mining situation	Time to complete subsidence	Source
1.	UK Coalfields; based on detailed examination of subsidence measurements made throughout the UK in respect of longwall workings.	A few months for a residual of 5% subsidence to occur. In some cases residual subsidence of 5-6% has taken 12 months or more to occur. The author regarded the older estimates of 5, 7 or even 10 years for the ground to become stable as very wide of the mark; such estimates were made when a limit angle of 18° was commonly accepted.	Orchard (1956-57)
2.	UK Coalfields; various longwall mining conditions giving consideration to examination of many subsidence records.	Main effect of mining is virtually instantaneous with usually a small delay for residual subsidence to finish. The time can be longer in some isolated cases: Example 1: 6 years for 94mm (9% of S max) to occur although 89mm occurred in the first 4 years. Example 2: 3·7 years for 33mm (6·8% of S max) to occur although about half the residual subsidence occurred in the first 3 months.	Orchard and Allen (1974)
3.	East Midland Coalfield, UK. Investigation of subsidence and use of highly sensitive electrolevel to detect changes in tilt. Depth of mining was 260m.	Up to 4 months after longwall travelled beyond critical area of subsidence of observation station (T.I.)	Pasamehmetoglu (1972) (Authors' analysis of original data carried out 1988)
4.	West German Coalfields	A duration of 5 years has been given for the subsidence effect with time factors of 75, 15, 5, 3 and 2% for each respective year. However, subsidence development owing to the surface point being still within the critical area appears to be included with time-dependent subsidence in this case. Consequently rate of longwall extraction would influence these time factors for the depth considered.	Flaschentrager (1957)

Table 20 Time to completion of mining subsidence from longwall operations based on current knowledge.

5.	West German Coalfields	Subsidence complete 6 months to 5 years after halting of operations (or presumably advanced beyond the critical area of the surface observation point).	Kratzsch (1983)
6.	Southern Coalfield, NSW, Australia. Longwall detail: h = 226m, w = 189m, L = 259m, M = 1·9m. Overburden predominantly sandstones with some interbedded shales. Total caving practised. Face started 23-6-65 and finished 22-12-65. Radius of area of influence = 158m.	Subsidence was complete after 200 days. This period started at the time the face passed under the surface observation point and consequently subsidence due to face position is included with time dependent subsidence. Taking account of the radius of area of influence, this gives a residual subsidence period of about 3 months.	Kapp (1973)

Table 20 (continued) Time to completion of mining subsidence from longwall operations based on current knowledge.

longwall stopped in 1967. Residual subsidence of 94mm (9% of the total) was observed to occur over a period of six years, with 50mm occurring in the first year and only 5mm in the last two years. The dolomitic limestone was considered by Orchard and Allen to be the cause of this protracted duration of residual subsidence, who also point out that less total subsidence is encountered over the dolomite than over Coal Measures in comparable conditions. The present authors would add, however, that the 5mm of residual subsidence reported as occurring during the last two years could be accounted for by natural ground fluctuations due to climatic effects or ground water level variations.

The second case referred to is also from the Durham Coalfield, UK where a Harvey Seam longwall at 105m depth was overlain by 23m of strong sandstone. After the longwall finished, 16mm of residual subsidence occurred in the following three months; the next 5 months experienced no further subsidence. Then subsidence continued and resulted in a further 17mm over the following period of three years. The total residual subsidence was 6·8% of S max. It would appear that the shallow conditions and nature of the overlying rocks played a role in this case.

Orchard and Allen (1974) conclude their paper by stating that the main effect of mining is virtually instantaneous which is accompanied by a small delay for the residual subsidence to finish and may take longer under certain conditions especially where stronger strata overlie the Coal Measures.

Duration of subsidence after longwall face finishes

Table 19, discussed earlier in this chapter, presented published data on the time to complete mining subsidence as referred to by Lane and Roberts (1929). Knowledge of the factors governing the development of subsidence has obviously greatly increased since the time of that publication which can now be viewed in the light of more recent findings. Whetton (1957) discussed the time factor and pointed out that it was connected with the area of underground excavation. He also drew attention to the commonly considered view that subsidence ceases shortly after cessation of mining in the case of shallow workings, but he remarked that the depth of the seam will influence the time for cessation of subsidence. Whetton identified four particular contributory factors which he suggested influenced the continuance of subsidence as a post-mining effect:

1. The existence of a relatively soft coal and/or floor may encourage yielding in the presence of abutment pressures along extraction edges, and is probably influenced by depth of working.

2. Gradual closure is more likely in zones of incomplete convergence, which would appear to be influenced by mining depth and physical properties of the strata.

3. The existence of bed separations over an extraction edge provides the potential for gradual closure of such voids.

4. Changes in the physical support condition of the goaf area can induce further yield and subsequent lowering of the overlying strata.

More recent discussion on the time factor has been published by Shadbolt, Whittaker and Forrester (1973) who state that the period of ground instability for a given surface point due to undermining is proportional to both the extent and depth below surface of the underground workings. Subsidence occurs during the period when the mineral is being worked within its area of influence. These authors remark that movement usually finishes soon after the workings have moved out of the area of influence, but acknowledge that the time for cessation of subsidence to occur as a post-mining effect varies according to conditions. They list the following factors as contributing to the time duration for mining subsidence to cease in respect of a surface point:

(a) Distance from the working face.

(b) The area already worked out in that seam.

(c) The composition of the rock.

(d) The thickness of the overburden.

(e) The rate of advance of the working face.

(f) The method of coal working.

(g) The quality of the goaf support.

(h) The presence of certain geological features, e.g. limestone cavities and old shallow workings can lead to extended post-mining surface deformation effects.

The NCB (1963) publication on subsidence engineering considered the residual subsidence element to be about 5% and this occurred during a relatively short period after the main subsidence phase. This document points out that the time for subsidence to occur is essentially dependent on the time taken by a longwall face to work through the critical area, and this is governed by depth and angle of draw and rate of extraction.

Table 20 presents the findings of various investigators in respect of the time to completion of mining subsidence due to longwall operations, and is essentially a summary of current knowledge. The main published works on the time factor aspects of mining subsidence indicate that subsidence development is directly associated with the rate of underground extraction within the area of influence of the surface point P. The residual effects of subsidence according to UK experiences appear to have a duration of up to 12 months or so in the majority of cases. There are a few isolated cases where residual subsidence exceeds this length of time and can be of the order of 4 to 6 years. An important feature, however, is the fact that residual subsidence is at its greatest at the time the longwall halts and the trend is for such subsidence to gradually decay in an exponential form.

Kratzsch (1983) reports a range for residual subsidence of 6 months to 5 years. It is probable that the West German experiences of residual subsidence are similar in character to those observed in the UK.

Concluding remarks

Longwall operation appears to have a virtually instantaneous response regarding the transmission of mining extraction effects to the surface according to mining subsidence studies in relatively shallow conditions in UK coalfields.

The magnitude of the residual subsidence appears to be of the general order of 5-10% of the maximum subsidence, and it is likely to be frequently less than this amount.

Residual subsidence is at its maximum immediately after the longwall face halts. There is a gradual decay in the magnitude of this value of subsidence and appears to be of exponential form.

It is over the edges of the longwall extraction that the greatest residual subsidence effects are observed, and such effects gradually decrease to zero as the subsidence limits are approached.

UK experiences indicate that the duration of residual subsidence is likely to be up to 12 months or so after the mining operations halt. In many situations it is likely to be much less, namely up to 3-4 months only. A few isolated situations can result in the residual subsidence extending to as much as 4 to 6 years.

CHAPTER 14

INFLUENCE OF GEOLOGICAL FACTORS ON THE CHARACTER OF MINING SUBSIDENCE

Shadbolt (1972) has commented upon the importance of soil mechanics, rock mechanics, geology and hydrogeology — collectively referred to as geotechnical factors — in mining subsidence engineering assessments and draws attention to the fact that some sites are, or potentially are, naturally unstable without any influence from mining subsidence. He subdivides related investigatory work into three principal parts:

The mining factor(s). These are essentially the basic movements arising from the coal extraction process and depend on extracted seam height, depth below surface and width of working etc.
The site factor(s). This includes the nature of the prevailing geotechnical conditions in addition to the effect of any previous mining either underground or at the surface.
The structure factor(s). These involve shape, size, design of foundations and superstructure, materials used, age, standard of maintenance and repair, as well as the purpose for which the structure is intended to be used.

Shadbolt (1972) argues that many permutations of conditions commonly arise with various combinations of all three of these parts, but adds that mining far outweighs the influence of the site factor whilst most buildings react in a predictable manner.

The mining factor has already been discussed in earlier chapters, whilst the structure factor is covered in a separate later chapter of this book.

This chapter is concerned with what Shadbolt refers to as the site factor, which is principally governed by geological influences and these involve surface features as well as the structural characteristics, composition and physical deformational behaviour of the overburden rocks. These aspects are now examined in relation to different conditions and as separate facets of the geological and geotechnical influences on the response of the surface to mining subsidence.

Shadbolt and Mabe (1968) researched mining subsidence in the East Midlands Coalfield of the UK. They have reported that in their estimate of ground movement induced by undermining some 10% can be attributable to the site factor, although they suggest that a change of 10% in the magnitude of maximum strain makes little difference to any structural damage so caused. Additionally, these authors report that around 25% of all the mining subsidence cases examined by them contained a measure of abnormal movement. This means that there was an observed deviation from the predicted behaviour using the standard subsidence prediction procedure. Geotechnical factors are important and need to be taken into due consideration. The geological setting of a subsidence site should be always taken into account.

Seasonal and climatic influences on variations in ground level

The main effect of moisture changes on clay deposits at the surface is that of creating shrinkage and expansion behaviour. This frequently produces localised lowering of the surface and differential horizontal movements with the potential for causing cracks in buildings to open and close periodically.

Seasonal shrinkage. SEH (1965) refers to measurements of vertical movements by the Building Research Station carried out at their Garston station. Plates were buried at depths below surface of 0·3m, 0·6m, 0·9m and 1·2m respectively. Vertical movements were recorded of 30mm at 0·3m below the surface whilst only 5mm was recorded at 1·2m depth. These measurements clearly demonstrate the susceptibility of the surface to vertical movement fluctuations when clay is present and is subjected to natural seasonal shrinkage. The substantially increased protection of lower positions relative to the surface indicates the importance of securing observation stations to an adequate depth. This also demonstrates the importance of maintaining consistently deep foundations on surface properties, especially when extensions are made to existing buildings. SEH (1965) draws attention to the problem of drying shrinkage being considerably increased owing to the close proximity of trees of the fast-growing variety such as poplar, elm and willow. Paved areas shelter clay against normal climatic drying but tree root action tends to offset this by the marked seasonal drying effect of the roots.

Tree root effects. An interesting example of the drying effect of tree roots is given in Figure 178 *(a)* which according to SEH (1965) represents results from a row of survey stations located alongside a large hedge. The results show that most of the survey stations experienced subsidence during, or subsequent to, a dry spell. The maximum subsidence experienced as a result of this climatic effect was around 20mm. Figure 178 *(a)* shows that the survey stations rose again following an increase in the rainfall. An interesting feature illustrated here is the effect of locating Station 3 near to a large tree. During the drought period, this station subsided by more than 76mm. However, much of this localised subsidence was regained following the end of the drought. The tarmac footpath protected Station 10 from the drought effects, although this station experienced a slight rise during the peak of the dry period. Subsidence observation stations located in the Millstone Grit off the coalfield experienced no subsidence variations during this period of climatic changes according to Figure 178 *(a)*.

Clay shrinkage effects. SEH (1965) explains that damage caused to buildings by clay shrinkage is a result of differential vertical settlement. The clay underneath the building is substantially protected against drying effects by virtue of the presence of the building itself. However, clay close to the external walls will not be so protected and the drying effect will result in their settlement. As a consequence the outer walls tend to pivot on the lower structure and thereby tilt outwards. The swelling of clays in winter cause such cracks to partially close. The next dry season results in further shrinkage with already fractured portions of a building tending to experience a little more settlement. A vicarage in Coalville, Leicestershire is referred to by SEH (1965) where surveying levellings were taken to investigate damage to the building; the results recorded demonstrated that the ground surface experienced a rise of 32mm between October 1959 and February 1960 and was attributed to natural seasonal effects.

The effect of rainfall changes on survey stations located in clay is clearly demonstrated in Figure 178 *(b)*. The clay soil resulted in fluctuations in level by up to 25mm as a result of rainfall variations coupled with seasonal changes which will have greatly influenced the drying character of the immediate surface clay soils.

Seasonal and climatic effects. Shadbolt and Mabe (1968) and Shadbolt (1972) reported on aspects of ground movement in the East Midlands Coalfield in the UK. Significant variations in ground level of up to ±100mm with accompanying variations in ground strain of up to ±2mm/m were reported as being due to seasonal and climatic effects as observed at survey

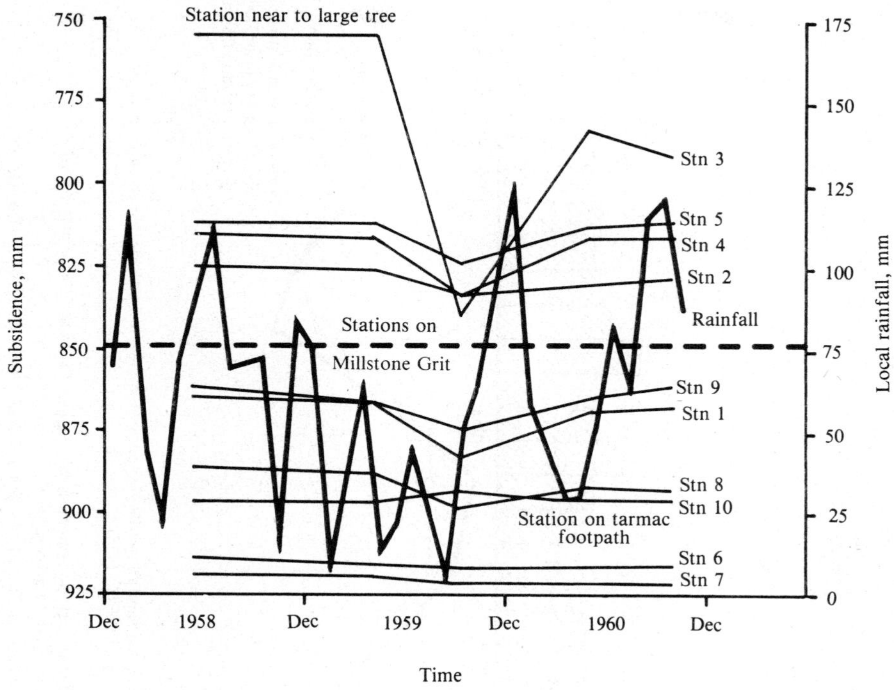

Figure 178(a) Effect of rainfall variations and climatic change on the level of survey stations (1 to 10) immediately alongside a hedge and tree. Note: also shown is the effect of the same climatic changes on survey stations located on Millstone Grit and off the Coalfield. After SEH (1965).

stations located on Triassic Lower Mottled Sandstone, a fine-grained soft sandstone mottled with yellow and white patches. They comment that pumping stations in the sandstone aquifer may have contributed to changes in hydrostatic support. They also confirm that a control line on Millstone Grit has remained stable for over 20 years irrespective of climatic effects.

An interesting example of effects attributed to fluctuations in water content of surface deposits is also referred to by Shadbolt and Mabe (1968). They report that damage was sustained by dwelling houses over full depth distance from underground workings. The houses tilted in a direction away from the mine workings. The subsequent investigation discovered the site to be a former pond and trial holes proved lacrustine deposits to a depth of 1·5m. These authors reported the damage to be due to settlement of strip foundations on organic silts whose intrinsically low bearing capacity appeared to be further weakened by water content fluctuations as a consequence of climatic conditions.

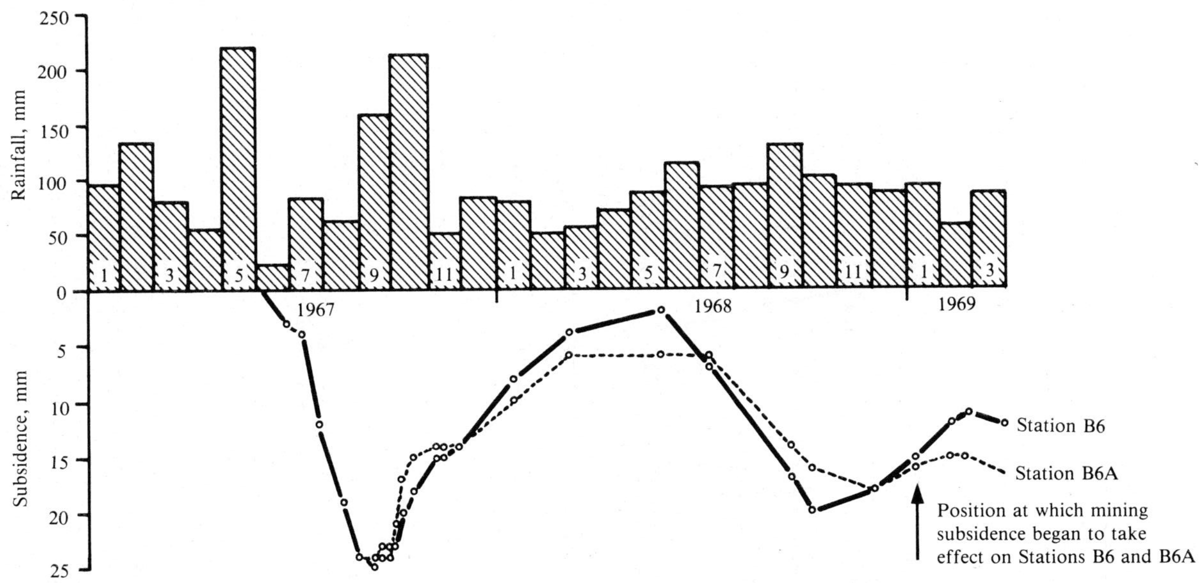

Figure 178(b) Illustrating the effect of changes in rainfall on the response of survey stations located in clay. After SEH (1975).

Notes: 1. Station B6 comprised a 16 × 760mm steel rod cemented into the ground with a 300 × 460mm deep concrete cap to ground level.

2. Station B6A comprised a 16mm steel rod set in a 300 × 1500mm deep block of concrete. This station was completely surrounded by clay.

Ground movements associated with undermining the flood plain of the River Trent

River flood plain areas experience natural fluctuations in ground water level and this can have an appreciable influence on ground behaviour. However, the influence of mining in such situations has been studied and the findings are discussed here.

Shadbolt (1972) has discussed observations of ground movement associated with mining under the flood plain of the River Trent. These observations showed that the subsidence limits were increased although the magnitude of subsidence was decreased when compared to the profile predicted by SEH (1965). Shadbolt attributed the difference in profile shapes to be due to the presence at the surface of the gravels and alluviums in this area.

Figure 179 shows the basic results referred to by Shadbolt (1972) and clearly shows the more gradual spreading out of subsidence due to the effect of the flood-plain deposits at the surface. The rib pillar is indicated to have a less marked effect on the surface subsidence profile. It would appear that the observed subsidence profile suggests a tendency for the unconsolidated and undoubtedly wet deposits to exhibit a flow characteristic towards the centre of the trough. The increased angle of draw from 35° to 45° is consistent with observations in The Netherlands where similar deposits have existed at the surface where mining subsidence effects have been measured, Grond (1953), Drent (1957) and Grond (1957).

General effects of near-surface rocks

An appreciation of the general aspects of the influence of near-surface rocks in relation to mining subsidence has been given by Shadbolt (1972). He reported on experiences in the East Midlands Coalfield in the UK where sandstones, limestones, transitional siltstones and mudstones are the common rock types present at the surface although they are frequently covered with varying thicknesses of drift deposits. Within these rock types, differing deformational behavioural features have been observed. Their geological histories have no doubt played a role here, particularly the varying ways they will have reacted to the stresses from tectonic movements resulting in folding, faulting, jointing and localised stress variations. The rocks at the surface will have been also affected by erosion and mention is made here of river valley development and geomorphological processes as important features in relation to subsidence behaviour.

Bed-rock behaviour. Shadbolt (1972) argues that in the majority of cases the near-surface rocks behave predictably when affected by mining subsidence. He goes on to suggest that for the radical minority to be discovered, then all sites on an active coalfield should be investigated. According to Shadbolt some of the most severe damage is caused by the ground fracturing under the action of tensile strain. Naturally the strength of the rocks will play a role here as well as the factors of lithology, stress intensity, jointing patterns and extent of development, faults and their distribution, and the degrees of weathering. Fissures are usually a feature of brittle rocks but they can occur along lines of maximum tensile strain if the strain is sufficiently severe as to exceed the natural strength of the rock. Shadbolt recognises that the strain itself may be feeble but it could coincide with rocks already in a state of prestressed tension due to either natural or from mining causes with the latter arising say from workings in the same or other seams. The pre-existence of natural jointing, fissuring or faulting offer potential lines of weakness which can cause concentration of mining subsidence strains and irregularities in a subsidence profile. Cavities may also pre-exist near to the surface either in the form of open joints covered by drift deposits or as enlarged subterranean water course channels or solution cavities. An important aspect mentioned by Shadbolt is that mining may have caused or influenced fissures in the rock head, which may not immediately penetrate the soil to the surface.

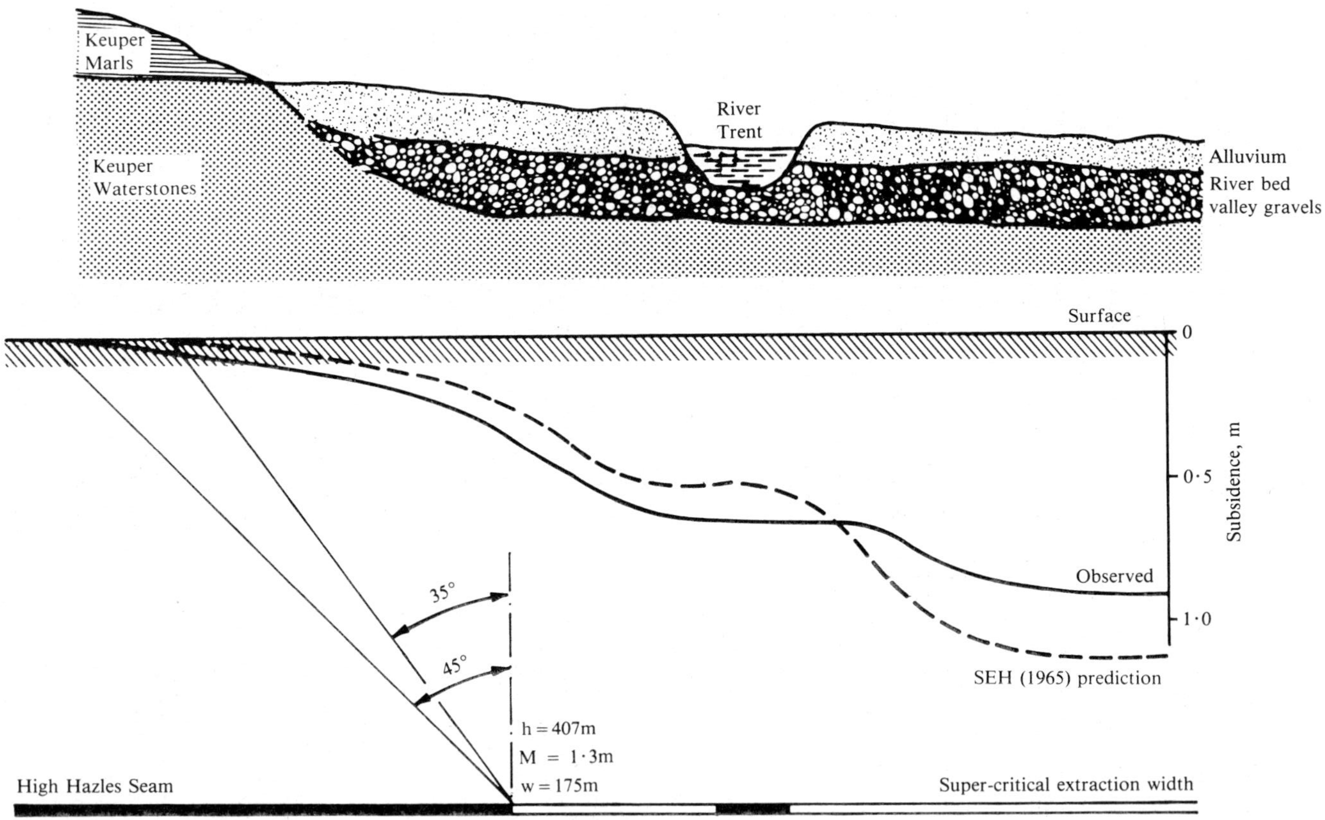

Figure 179 Showing the effect of undermining unconsolidated river flood deposits, Trent Valley, Nottinghamshire Coalfield, UK. Results as discussed by Shadbolt (1972).

Joint pattern effects. The East Midlands Coalfield has a dominant pattern of jointing in the Coal Measures and the Permo-Trias as observed at the surface. The main joint trend is north-west to south-east with a secondary system trending approximately at right angles. Shadbolt suggests that the main faulting pattern is usually parallel to the basic joint system. Since longwall workings usually work between and parallel to main faults in the East Midlands Coalfield, then the ribsides of the extracted panels often have similar bearings to the fault and joint orientation patterns.

Joint patterns when subjected to tensile forces associated with a subsidence trough tend to exhibit a rotational effect of the nature illustrated in Figure 180. Joint opening is dependent to a large extent on the ground curvature occurring around the tensile strain regions of the subsidence profile. Compressive strain regions do not tend to be vulnerable to joint effects although some relative sliding in the vertical plane of joint sets is a possibility if the strain magnitudes are particularly high.

Some geologically influenced subsidence occurrences. Some observed subsidence deformation characteristics have been discussed by Shadbolt, Whittaker and Forrester (1973) and the main features are illustrated in Figure 181. These characteristics have a major bearing on the geological setting.

Figure 181 *(a)* illustrates block shear failure as occasionally occurs with near-surface longwall extractions. Shear failure plays a significant role in the caving process of roof beds and overlying strata behind a longwall extraction. The shape of the block encourages freedom of movement in the vertical plane. Very shallow workings have experienced sudden surface subsidence of this type although it is fairly infrequent. The ground surface if suddenly lowered by this mechanism tends to give a much reduced width of stepped subsidence trough when compared to the extraction width causing the collapse. Such collapses tend to be time-dependent, and deterioration in strength of the cover rocks is a governing factor. Development of the major plane of shear failure can be of a progressive nature or may occur rapidly under some conditions.

Progressive shear block failure as a function of longwall face advance is illustrated in Figure 181 *(b)*. Shallow conditions can result in surface subsidence occurring as a result of the sliding of blocks of strata. Block tilting does not appear to be a significant factor in these conditions. A stepped surface subsidence profile is common with this form deformation process.

A further form of surface subsidence which is substantially geologically controlled is shown in Figure 181 *(c)* and this involves the presence of a strong and competent bed occurring at the surface and is underlain by weaker and more flexible strata. This form of geological setting can encourage the formation of bed separation between the strong upper bed and the weaker underlying rock. Bed separation of this type tends to exhibit inconsistent development as it depends upon the overall integrity of the upper bed; the presence of natural weakness planes can play an important role in development. The bed separation shown is obviously occurring with an advancing longwall face so that here the separations would be of a temporary nature and thus travel with the face line. However, some semi-permanent or apparently permanent bed separations can occur over the ribsides of the developing longwall.

The opening of fissures in limestone capping of weaker rocks and clays is shown in Figure 181 *(d)*. Valley bulging encourages the limestone fissures and joints to open. This is a natural process of weathering and can lead to appreciable localised subsidence effects particularly the opening of fissures which can result in instability of surface structures.

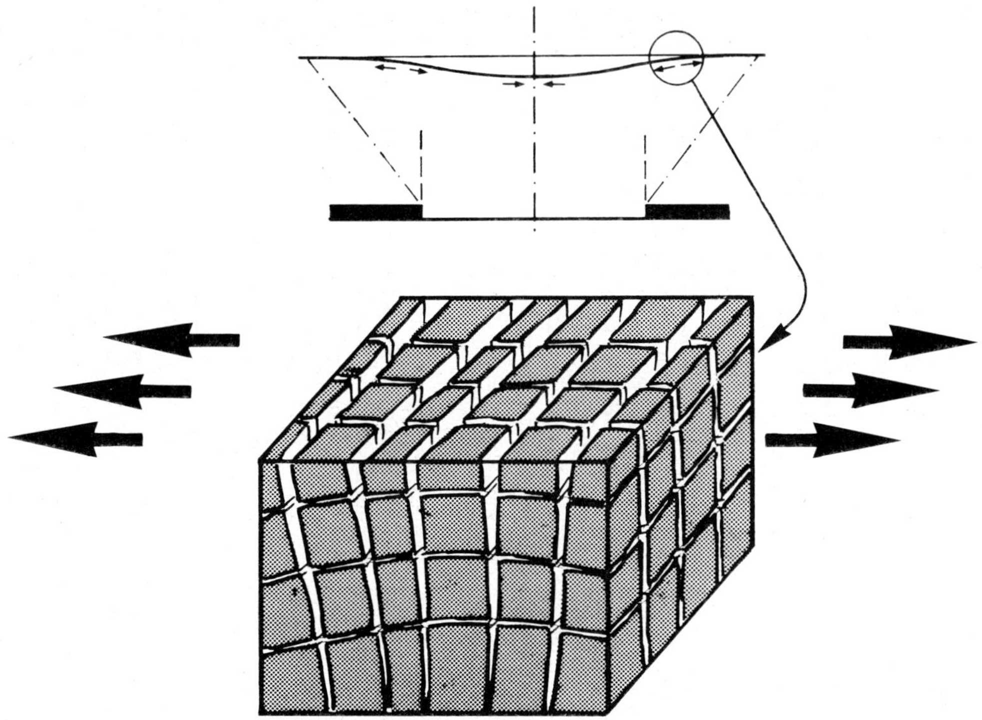

Figure 180 Illustrating jointed rock structure response to tensile forces at surface of subsidence trough

(a) Block shear failure.

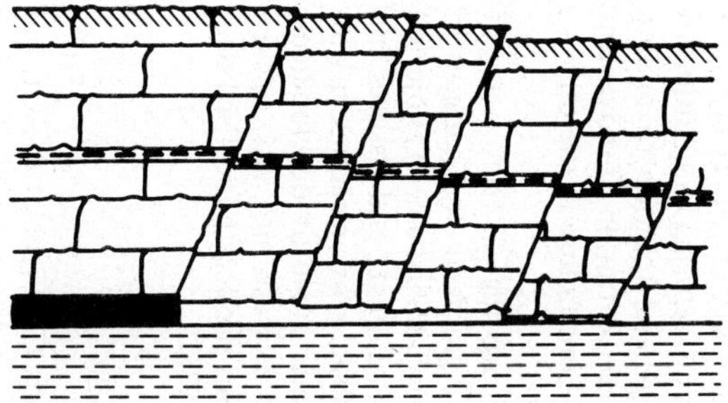

(b) Inter-movement of blocks on parallel shear planes.

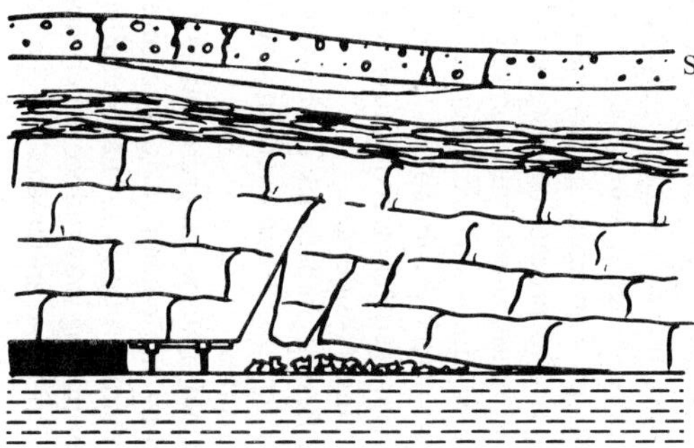

(c) Surface crack development and near-surface bed separation.

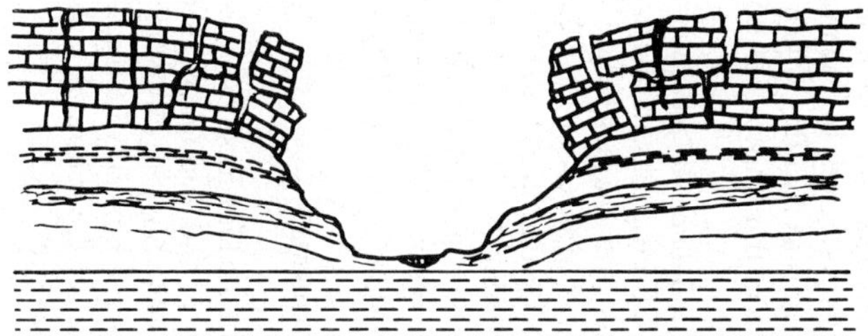

(d) Opening of surface fissures by natural weathering processes and valley bulging.

Figure 181 Subsidence occurrences influenced by geological features associated with lithology and shallow depth. Based on Shadbolt, Whittaker and Forrester (1973).

Forrester and Whittaker (1976) have suggested that the jointing pattern observed in the East Midlands Coalfield at the surface and tending to give multi-jointing in some sandstones, often results in transmission of mining movements in an orthodox form. Conversely these authors remark that limestones can exhibit susceptibility to block fracturing in addition to being subjected in varying degrees to chemical solution. The general reaction to mining subsidence takes such forms as fissuring which tends to promote concentration of movement in one extreme to apparent transmission of severe movement without failure in the other.

Hydrogeology and its influence on localised subsidence. Hydrogeological factors are of importance in mining subsidence as the latter can alter drainage gradients and patterns in addition to ground water flow characteristics. Near-surface rocks which have been subjected to appreciable subsidence strains and have experienced opening of joints and fissures can result in increased or changed ground water flow behaviour. Localised erosion can occur with cavities forming which may be of subsidence potential at a later stage of development. Conversely, there may be a natural tendency for loose material to wash into such opened joints or fissures and result in a filling action or in some situations produce a localised dam effect to the water drainage pattern, Forrester and Whittaker (1976).

Abnormal surface movements

A detailed study of surface movements in the East Midlands Coalfield of the UK has been reported by Forrester and Whittaker (1976). They also include a discussion on abnormal surface movement and point out that this is regarded as inconsistency with the predicted result and where it occurs this can normally be traced to the ground structure relating to the site in question. These authors list the following eight principal sources of abnormal movement as associated with a surface site:

1. The concentration of ground movement at a fault intercepting the surface can produce a surface step.
2. The concentration of ground strain at a joint plane can result in the appearance of a surface fissure.
3. The occurrence of bed separation between the outcropping rock and interbedded weaker rocks can result in promoting surface fissures.
4. The process of valley bulging, where weak rocks underlying strong cover rocks result in the former slowly squeezing into a valley by surcharge and natural weathering action, can give rise to promoting instability of the overlying rock. If the overlying stronger rock formation is naturally jointed, then the constituent blocks may camber and thus allow creation of surface fissures. This natural process of weathering can be aggravated by the effects of mining.
5. Naturally occurring geotechnical features such as areas prone to landslips may be disturbed by mining subsidence.
6. Existing surface fissures in bedrock can be disturbed by mining subsidence especially if opened. This can lead to fill and overlying loose material becoming quickly eroded and reveal surface fissures.
7. The presence of old shallow underground and surface mine workings which may have become masked by subsequent surface features may be disturbed by later mining.
8. The effect of previous mining may have resulted in creating significant residual ground strains, and later mining may induce superimposition of additional strains to produce effects in excess of those anticipated by the later phase of mining subsidence.

Figure 182 shows the localised effect of the concentration of mining subsidence strain at a pre-existing weakness plane in Bunter sandstone bedrock and thus resulting in the creation of a surface fissure of the type referred to under 6 in the list above.

Figure 182 Illustrating concentration of mining subsidence strain at a pre-existing weakness plane and resulting in a surface fissure, Nottinghamshire Coalfield, UK.

Observations on the influence of surface geology on the character of mining subsidence

Detailed field studies on surface geology influences on the character of mining subsidence have been reported by Whittaker and Breeds (1977) who examined subsidence observations made in the East Midlands Coalfield of the UK. The main geological features of the coalfield are the three main rock types which occur at the surface, namely Triassic sandstones (predominantly coarse-grained to medium-grained), Permian limestones and Carboniferous Coal Measures comprising of mudstones, siltstones, sandstones and several coal seams. The subsidence study involved examination of some forty separate observation line investigations covering a region of about 1000km^2. There are parts of the coalfield where only Coal Measures are present at the surface whilst other concealed parts have Triassic and Permian rocks present at the surface. The purpose of the study reported by Whittaker and Breeds was to compare the general characteristics of mining subsidence as observed in areas having these three main rock types. This coalfield has been worked continuously for over 100 years; the coal seams outcrop to the west and gently dip (about 4°) towards the east. There are varying thicknesses of superficial formations, mainly glacial deposits. The Triassic is up to 400m thick in parts of the coalfield with the Bunter Pebble Beds/Mottled Sandstones tending to predominate. The Permian is up to 130m thick and consists mainly of Magnesian Limestones although marls and breccia are also present. The Permian lies unconformably on the Upper Carboniferous which is up to 900m thick and consists of the Upper, Middle and Lower Coal Measures respectively. The Triassic sandstones and Permian limestones are noted for their brittle joint controlled character whilst the Coal Measures exhibit greater plasticity on a massive scale.

The subsidence data referred to here were obtained by precise levels and linear measurements with steel tapes along survey observation lines set out in the path of an advancing longwall face. The usual layout involved positioning around 15 subsidence stations at regular intervals along each observation line. This interval was generally 0·1 or 0·05 (depth). The research procedure adopted by Whittaker and Breeds involved plotting the basic field data and obtaining a mean subsidence profile rather than using the traditional procedure of drawing straight lines between the measured values for each of the stations. The maximum subsidence was determined to an accuracy of at least ±2·5mm.

The mining subsidence case studies used are summarised in Table 21 which gives details of the longwall extraction dimensions, depth of working below the surface and the maximum subsidence. The cases cover a fairly wide range of depth namely 198m to 833m.

Subsidence aspects. Figure 183 *(a)* shows a graphical presentation of the maximum subsidence values compared with the UK national curve. These results demonstrate increased statistical scatter for w/h being greater than around 0·5 for the cases considered. There is evidence of a tendency for less subsidence to occur with some of the Triassic sandstones and Permian limestones cases when compared to the national curve, and this may be explained by such rocks being influenced by block movement. Some such blocks may not return exactly to their original positions after completion of the subsidence development phase. Considering the rock types generally, there does not appear to be any marked difference in the subsidence results for w/h being less than around 1·0.

Figure 183 *(b)* shows a comparison of the subsidence results for over the ribside in relation to the w/h ratio. There is a fairly even distribution of values for the Coal Measures and Permian cases. The Triassic sandstone cases show the most statistical scatter which suggests surface block behaviour rather than the more fairly uniform deformational behaviour of pliable rocks. The consistency of the Permian data points in relation to the national curve suggests that the discontinuities are not significantly dictating the subsidence behaviour of the Permian as much as with the Triassic sandstones.

Positional aspects of ribside subsidence. Figure 184 shows relationships for the positions of maximum tensile strain (+E), the maximum slope (G) and the transition point (½S) relative to the ribside for various values of the longwall width to depth ratio. Figure 184 *(a)* shows the Permian limestone and the majority of the Coal Measures to be in generally close agreement with the national curve. The Triassic sandstone cases tend to have their maximum tensile strain further away from the extraction edge, that is in a region of less constraint to block movement although this point appears to only apply for w/h > 0·4.

Figure 184 *(b)* indicates that surface rock type does not have any significant influence on the position of G (the maximum ground slope) in relation to the ribside. A similar conclusion is drawn in respect of the position of half maximum subsidence (½S) in relation to the ribside, the results of which are shown in Figure 184 *(c)*. These results give additional evidence to the position of maximum ground slope corresponding with the position of half maximum subsidence (½S).

S/h coefficient data analyses. Table 22 presents derived S/h coefficient data established from back analysis of the mining subsidence data of the case studies given in Table 21. The procedure used has been that applied in the UK to analysing subsidence data for the purpose of constructing convenient design charts for predicting strains and slopes associated with longwall subsidence profiles. These data values have then been plotted to compare with the national curves.

Case No	w (m)	h (m)	w/h	M (m)	S (m)	S/M (%)
Coal Measures Case Studies						
1	132	350	0·38	0·94	0·297	32
2	143	349	0·41	1·02	0·370	36
3	169	353	0·48	0·94	0·421	45
4	218	354	0·62	0·94	0·625	67
5	149	354	0·42	1·00	0·435	43
6	215	220	0·98	0·91	0·695	76
7	209	305	0·69	1·55	1·245	80
8	247	424	0·58	1·70	1·270	75
9	150	311	0·48	1·37	0·772	56
10	200	273	0·73	1·68	1·328	79
11	198	408	0·48	1·57	1·180	75
12	66	279	0·24	1·55	0·131	8
13	177	222	0·79	1·00	0·729	73
14	225	351	0·64	1·53	0·870	57
15	228	363	0·62	0·97	0·886	92
16	211	341	0·61	1·22	0·841	70
17	178	497	0·35	1·80	0·558	31
Triassic Bunter Sandstone Case Studies						
18	245	235	1·04	1·07	0·755	71
19	274	240	1·14	1·07	0·870	82
20	201	833	0·24	1·39	0·220	16
21	290	601	0·48	1·24	0·642	52
22	201	602	0·33	1·32	0·264	20
23	341	227	1·49	0·95	0·800	85
24	229	223	1·03	1·05	0·929	88
25	252	252	1·00	0·94	0·907	96
26	329	268	1·23	0·97	0·938	97
27	311	285	1·09	1·16	0·942	81
28	553	448	1·24	1·30	0·985	76
29	268	400	0·67	1·26	0·900	71
30	220	314	0·70	1·07	0·840	79
31	220	316	0·70	1·07	0·870	82
Permian Limestone Case Studies						
32	281	241	1·16	1·30	0·922	71
33	210	261	0·80	1·22	0·924	76
34	238	279	0·86	1·22	0·958	79
35	216	292	0·74	1·22	1·060	87
36	234	198	1·18	1·12	0·830	74
37	216	273	0·79	0·89	0·780	88
38	185	401	0·46	1·24	0·833	67
39	570	279	2·05	0·87	0·790	90

TABLE 21 Mining Subsidence Data from Field Case Studies

Note w = longwall face width; h = depth below surface; M = extracted seam height; S = maximum surface subsidence. Source: Whittaker and Breeds (1977).

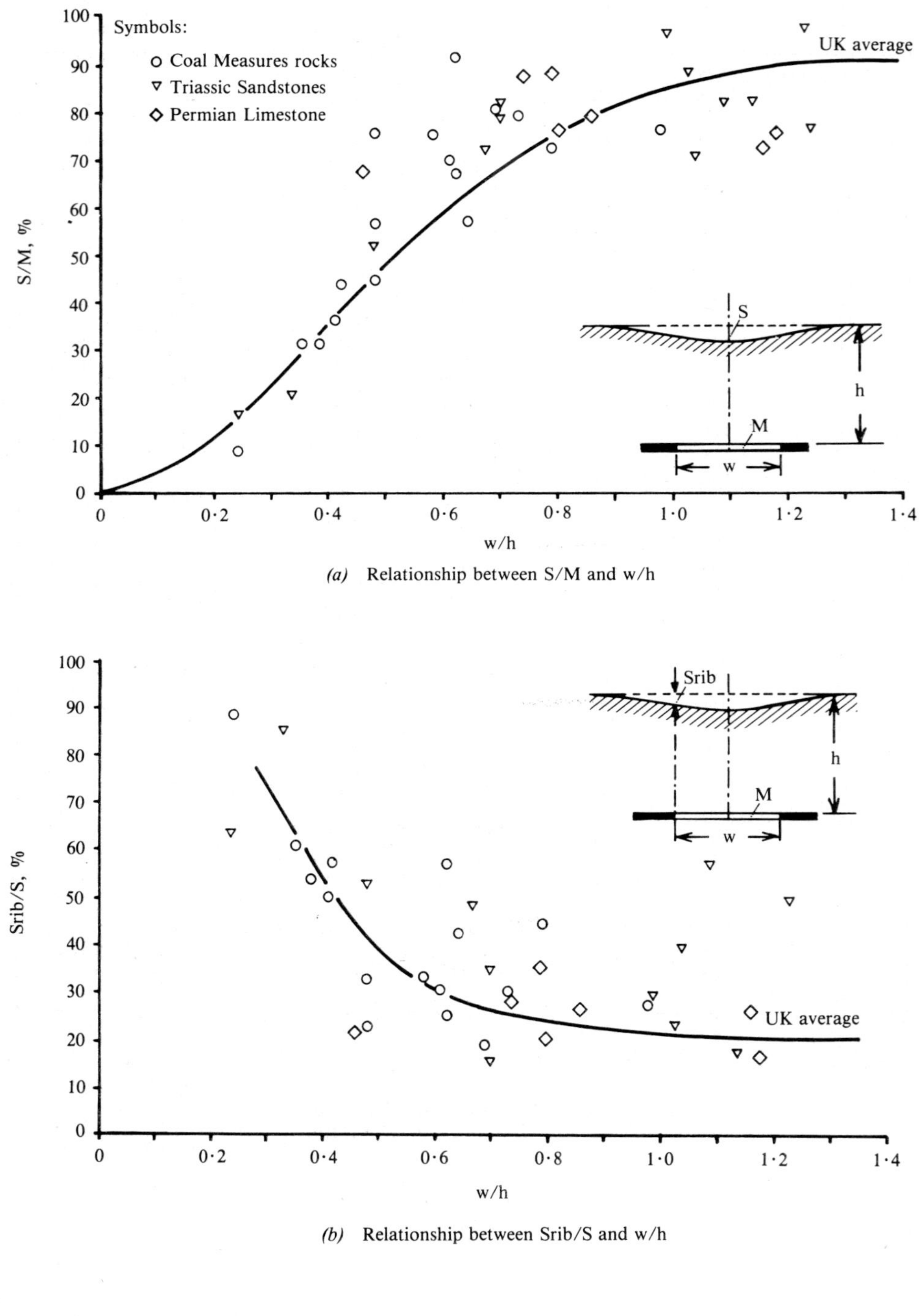

Figure 183 Subsidence data values compared with UK national curve. After Whittaker and Breeds (1977).

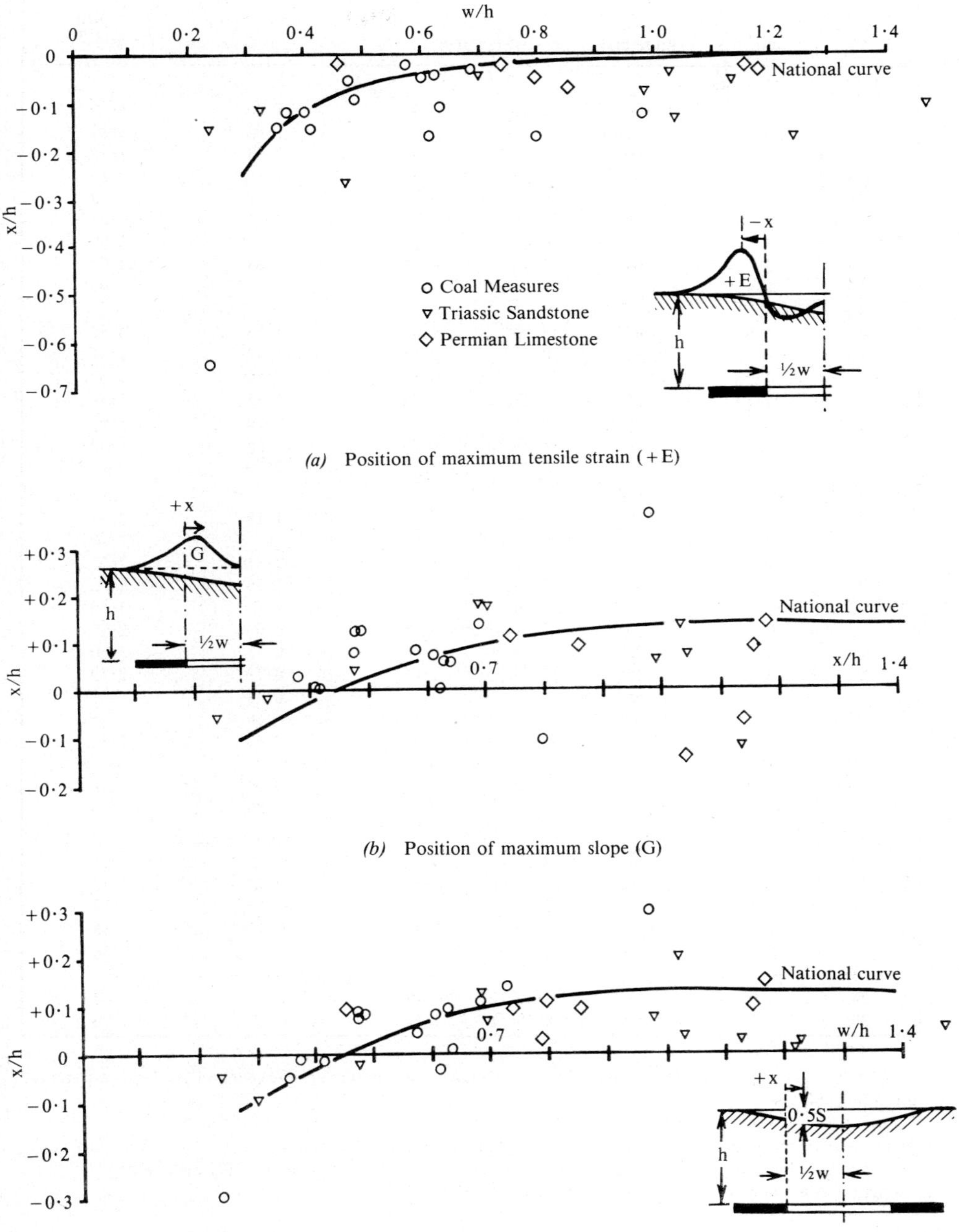

(a) Position of maximum tensile strain (+E)

(b) Position of maximum slope (G)

(c) Position of transition point (½S)

Figure 184 Comparison of the positions of maximum tension (+E), maximum slope (G) and transition point (½S) in relation to the ribside for various w/h ratios.

Case No.	+E (mm/m)	−E (mm/m)	G (mm/m)	+E ÷ S/h (× 10^{-3})	−E ÷ S/h (× 10^{-3})	G ÷ S/h (× 10^{-3})
Coal Measures Case Studies						
1	0·78	1·44	2·63	0·92	1·70	3·10
2	0·75	2·31	3·94	0·62	1·92	3·27
3	1·15	2·90	4·44	0·96	2·43	3·73
4	1·16	2·05	4·52	0·66	1·16	2·56
5	1·05	2·30	3·95	0·85	1·87	3·22
6	1·00	4·20	8·75	0·32	1·33	2·76
7	3·50	—	20·80*	0·86	—	5·09
8	2·18	5·03	11·90	0·73	1·68	3·97
9	—	—	11·16	—	—	4·49
10	—	—	—	—	—	—
11	3·08	5·45	12·80	1·06	1·88	4·42
12	0·27	0·91	0·85	0·58	1·94	1·82
13	1·06	1·75	8·26	0·32	0·53	2·52
14	1·10	2·85	8·47	0·44	1·15	3·42
15	0·73	2·20	8·81	0·30	0·90	3·62
16	1·83	4·10	8·79	0·74	1·65	3·53
17	1·20	3·13	3·27	1·067	2·79	2·91
Trias Bunter Sandstone Case Studies						
18	2·60	3·80	7·50	0·81	1·18	2·33
19	3·65	—	9·81	1·00	—	2·70
20	0·30	0·73	0·71	1·15	2·81	2·73
21	1·56	1·89	3·23	1·46	1·76	3·02
22	0·35	—	1·60	0·80	—	3·65
23	2·50	2·24	7·31	0·85	0·63	2·07
24	1·94	2·28	12·56	0·46	0·55	3·00
25	3·65	2·15	10·05	1·01	0·59	2·79
26	—	1·56	—	—	0·44	—
27	—	2·00	—	—	0·66	—
28	0·44	0·91	2·90	0·19	0·41	1·31
29	—	—	—	—	—	—
30	2·20	2·55	9·00	0·82	0·95	3·36
31	2·20	2·20	9·60	0·79	0·79	3·48
Permian Limestone Case Studies						
32	1·75	1·75	10·05	0·46	0·46	2·63
33	2·12	1·48	11·70	0·60	0·42	3·30
34	1·20	1·35	9·60	0·35	0·39	2·78
35	1·25	1·22	11·50	0·34	0·34	3·16
36	2·70	—	10·70	0·65	—	2·56
37	—	—	—	—	—	—
38	1·02	1·75	7·20	0·49	0·84	3·47
39	3·30	1·67	6·70	1·16	0·59	2·36

TABLE 22 Derived S/h coefficient data from back analysis of Mining Subsidence data from field case studies given in Table 21

Note +E = maximum tensile strain; −E = maximum compressive strain; G = maximum ground slope; S = maximum subsidence; h = depth; * = abnormally high value; — signifies data not available. Source: Whittaker and Breeds (1977).

Figure 185 *(a)* shows the relationship between the coefficient for maximum tensile strain (+E) in relation to the w/h ratio. There appears to be fairly close agreement between curve No. 3 (all cases) and the national curve. The curve for the Triassic sandstones results is consistently higher than the national curve indicating that with this rock type there is a greater likelihood of experiencing tensile strain concentrations and thereby gives increased magnitudes of maximum tensile strain. This latter feature is in general evidence across the range of w/h ratios, namely 0·2 to 1·2, employed in this study. The Coal Measures results tended to give lower values of maximum tensile strain for w/h ratios greater than around 0·6.

The −E coefficient of S/h has been plotted against w/h and is presented in Figure 185 *(b)*. The fitted curve (No. 4) for all cases is close to the national curve. Whittaker and Breeds (1977) concluded that in their interpretation rock type did not appear to greatly influence the value of the −E coefficient and that compressional strain effects can be more reliably predicted.

Figure 186 shows the coefficient of maximum slope (G) plotted against w/h. The results suggest that for the range of rock types present at the surface as used in this study, there is no significant influence made by rock type to the value of the maximum slope.

Polynominal curve fitting was carried out on the subsidence data plotted in Figures 185 and 186. Statistical data and corresponding coefficients of determination and correlation are given in Table 23.

Although the data discussed here apply to the overall subsidence profiles, due regard needs to be taken of possible localised effects as can occur with a fissure or fault intercepting the surface.

Significance of geological structure

The publication ICE (1977) draws attention to the importance of geological structure in ground control aspects of subsidence. In particular the nature of the overlying strata, faulting systems and fractures and breaks in the strata are identified as special factors. Faults, however, are regarded as having more serious effects than other features in producing displacements at the surface. ICE (1977) suggests that variations in the types of strata between the extracted seam horizon and the surface do not effect the maximum amount of subsidence. This is most certainly not the case when considering the presence of massive sandstones overlying coal seams as experienced in Australia and North America, to mention only two. The influence of lithology of the cover rocks will be considered separately. ICE (1977) does recognise, however, the importance of certain rock types such as Magnesian Limestone, which may exhibit significant fracturing and intense differential displacements with the potential to cause serious damage at the surface.

Effects of faults on surface subsidence behaviour

King, Whittaker and Shadbolt (1974) have pointed out that the main influence of faults is that of giving rise to irregularities in mining subsidence with the potential for surface steps to occur, although these authors draw attention to the depth and dimensions of the extraction influencing the probability of such irregular behaviour becoming discernible in the first place.

Kratzsch (1983) on page 172 of his book presents a case study which demonstrates that a fault produces a marked localised subsidence profile irregularity. The study shows that a well-defined step in the subsidence profile became permanently established. There is a further important feature demonstrated, however, and that is there were high early movements

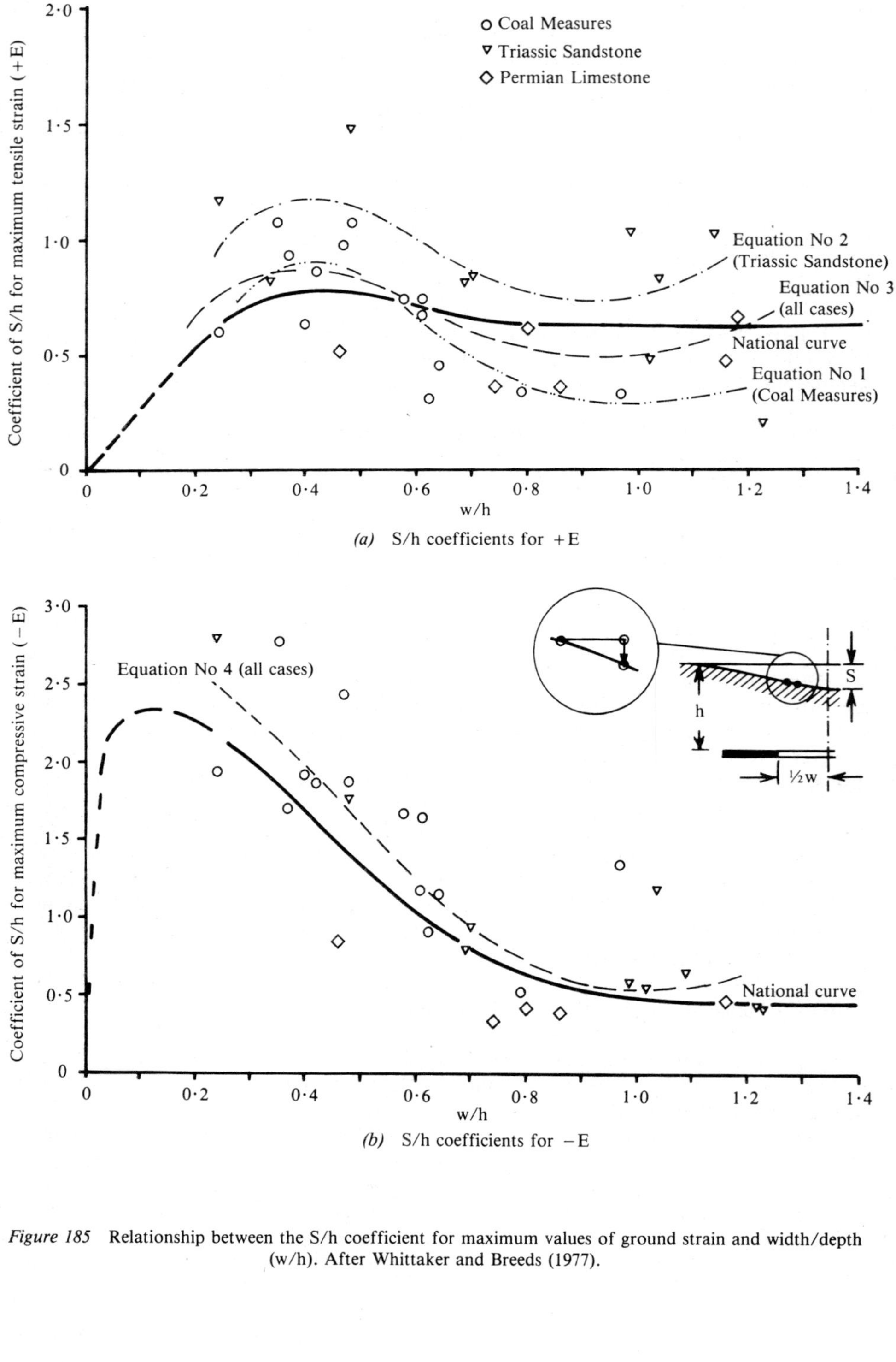

Figure 185 Relationship between the S/h coefficient for maximum values of ground strain and width/depth (w/h). After Whittaker and Breeds (1977).

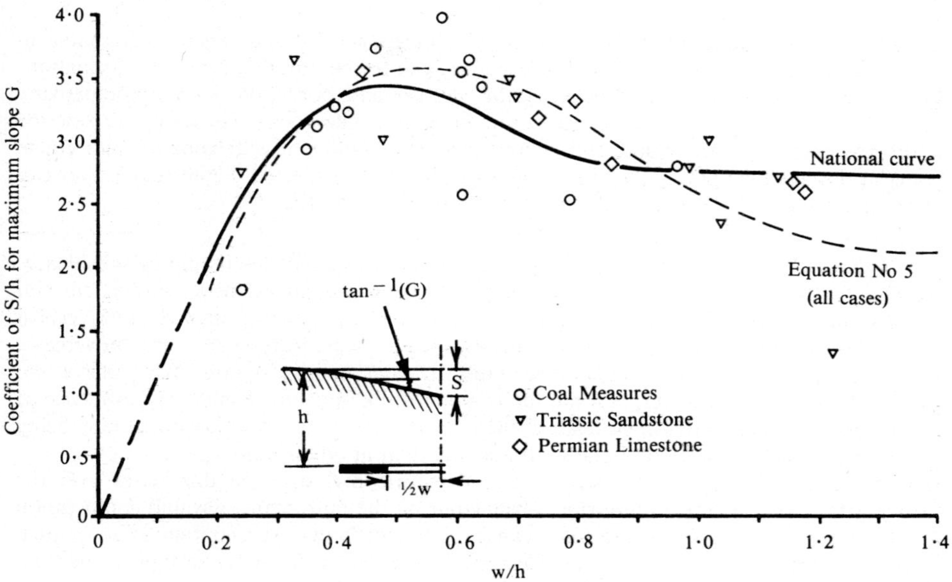

Figure 186 Relationship between the S/h coefficient for maximum slope (G) and width/depth (w/h). After Whittaker and Breeds (1977).

Coefficient (S/h)	+E	+E	+E	−E	G
Rock Type	Coal Measures	Trias Sandstone	All cases	All cases	All cases
Curve	1	2	3	4	5
Degree coefficients					
0	−2·18	−0·88	−0·30	−2·54	−1·35
1 (w/h)	20·35	13·78	8·18	2·94	23·30
2 $(w/h)^2$	−45·34	−30·74	−18·55	−17·08	−36·31
3 $(w/h)^3$	39·32	26·08	15·45	17·32	21·19
4 $(w/h)^4$	−11·83	−7·47	−4·27	−5·19	−4·19
Coefficient of determination	0·57	0·29	0·21	0·72	0·67
Coefficient of correlation	0·75	0·65	0·45	0·85	0·90

TABLE 23 Statistical equations for S/h coefficient curves

Note +E = maximum tensile ground strain; −E = maximum compressive ground strain; G = maximum tilt; S = maximum subsidence; h = depth below surface; w = width of longwall extraction. Source: Whittaker and Breeds (1977) observed at the position of the fault indicating its sensitivity to initial ground movements. It did not, however, exhibit the same behaviour after the surface step had become established.

General character of faults. The term 'fault' is fairly general and can be represented in several forms, but is essentially a well-defined break in the continuity of the rock formation. Faults can occur in any plane between the vertical and horizontal with accompanying displacements which may be of the order of millimetres or kilometres. The term may describe faults possessing a clean smooth shear plane or one containing a wide zone of fault gouge material. Depending upon the nature of the fault, it is to be expected therefore that they can have different effects on the development of a subsidence profile.

Figure 187 shows a range of fault types which can be associated with mining subsidence.
Normal faults. This type of fault mainly results from ground extension associated with tectonic movements. Normal faults are considered to occur under a state of high vertical stress and relatively low horizontal stress thereby creating a net increase in lateral movement. It is the most commonly encountered fault in coal bearing strata, particularly the Carboniferous Coal Measures. Normal faults are most likely to move under the influence of mining because being a tensional feature, the two surfaces of the fault are potentially being pulled apart, and hence friction plays a lesser role than in other fault types.
Reverse faults. Reverse faults occur where the horizontal stress predominates over the vertical stress with lateral compressional behaviour on the fault plane. Ground compression due to tectonic movements is the main cause of reverse faults. The configuration of such faults encourages increased frictional effect across the two surfaces of the fault plane, and thus decreases its sensitivity to movement due to mining subsidence.
Thrust and lag faults. These fault types are generally found in areas of severe folding and distortion of the strata. They frequently have high angles of hade.
Strike (wrench) faults. This type of fault generally occurs when a block of strata is subjected to high horizontal shear forces. Movement is predominantly horizontal. It is less common in coalfield situations than other fault types. These faults can extend for hundreds of kilometres and frequently affect regional tectonic movements.

Effect of fault on subsidence development curve

An example of the effect of a fault on the shape of a subsidence development curve is shown in Figure 188. This case study was observed at the Riddings Drift Mine in the Yorkshire Coalfield, UK where the Shafton Seam was being worked at a depth of 185m below surface. The positions of the subsidence observation stations relative to the longwall face are shown in Figure 188 *(a)*. The subsidence development curves clearly show the nature of the irregularity within the profile in the proximity of where the fault intercepts the surface. The position around the fault indicates that it was sensitive to earlier detection of subsidence effects than at other positions along the observation line. High localised strains occur in the proximity of the fault at the surface.

Localised strain effect of fault at surface

When a geological fault intercepts the surface a step can occur in the subsidence profile. This can result in causing severe damage to surface structures which might straddle the fault-affected zone. An example of a fault causing a high degree of localised damage to a 5m high brick wall is shown in Figure 189. A distinct line of damage is in evidence as the subsidence effect creates a shearing action through the wall. The local surface step caused by the fault and as observed during the subsidence development phase was 0·3m. This is typical of the effect of a fault on surface structures. Walls can experience buckling and significant localised

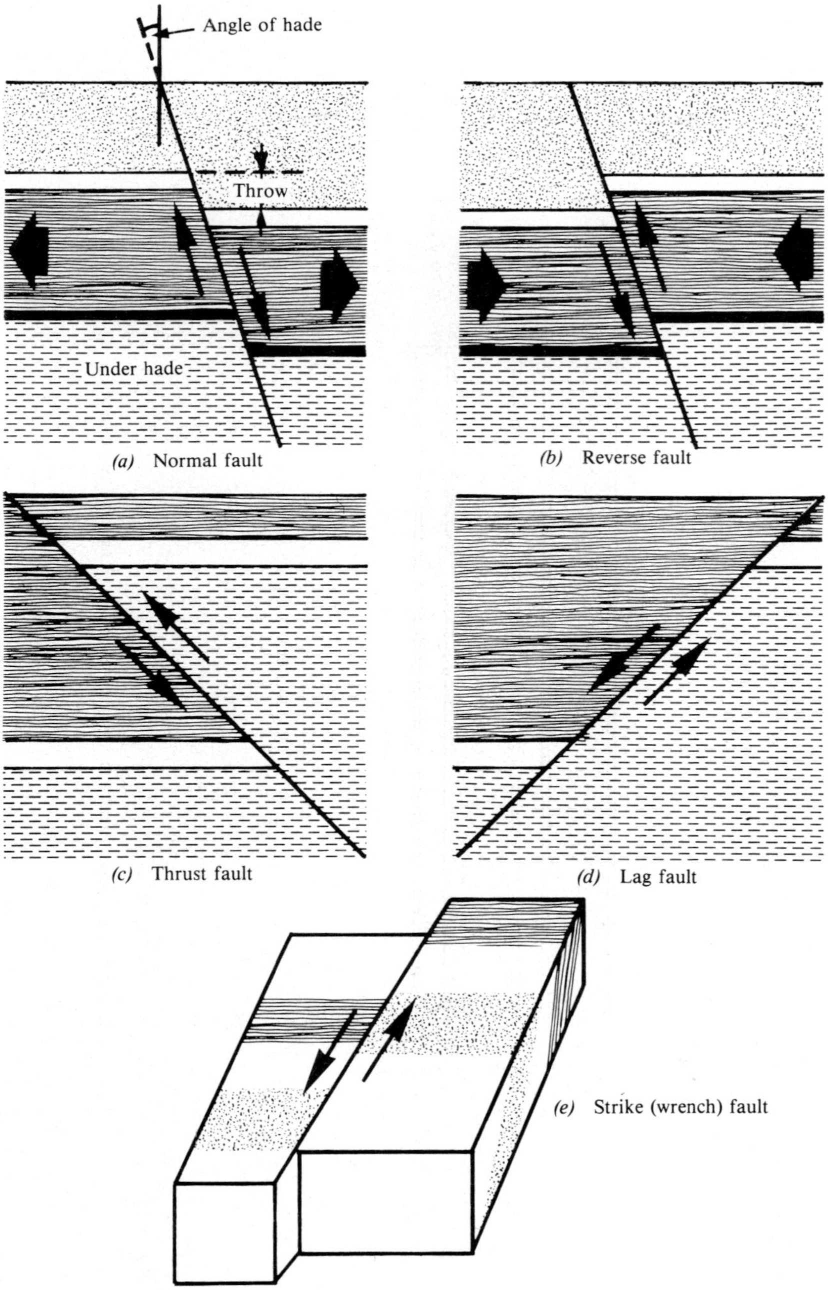

Figure 187 Some common fault types

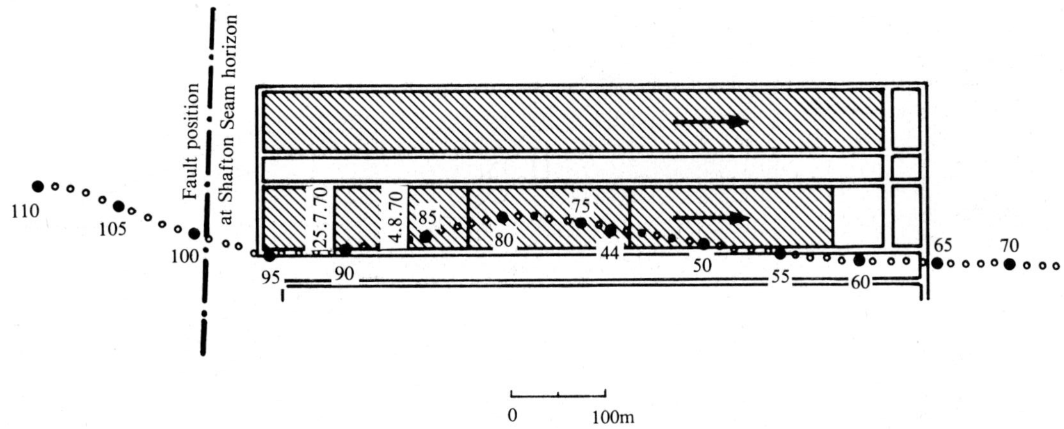

(a) Location of subsidence observation stations relative to longwall face position

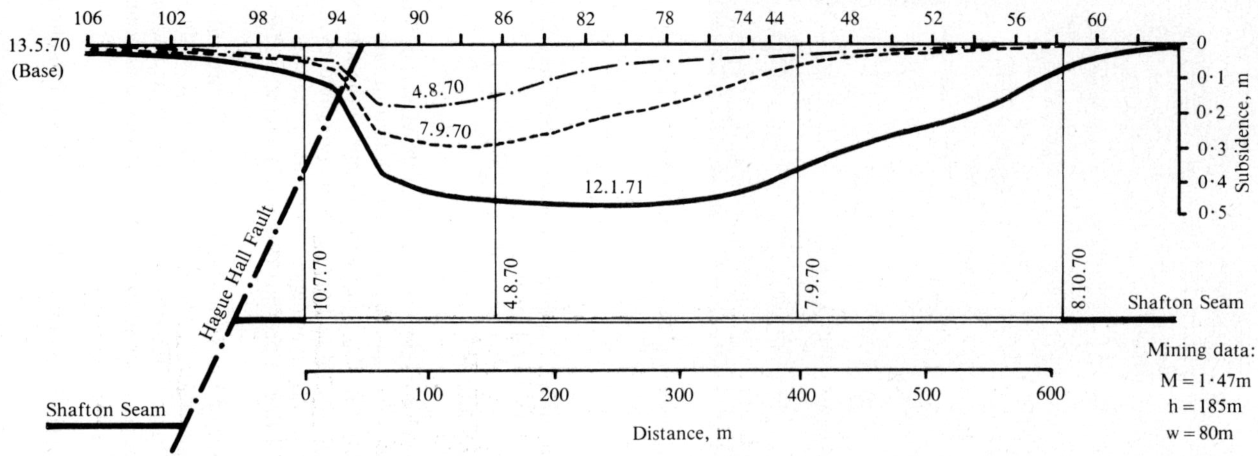

(b) Subsidence development curves

Figure 188 Illustrating the effect of a fault on the subsidence development profile. Based on Arcamone (1980).

Figure 189 Illustrating effect of fault intercepting the surface under a 5m high brick wall subjected to mining subsidence.

separations between brick layers. Surface buildings would suffer similar effects in addition to severe localised distortion of window and door frames.

Study of effect of faulting on subsidence

Lee (1965, 1966) has reported his findings of a study of the effects of faulting on mining subsidence. Figure 190 shows a typical case history used by him to rationalise the general effects. He paid particular attention to the occurrence of steps in the subsidence profiles of the longwall extractions studied. The photographic results presented by Lee suggests that compressional strain effects are in strong evidence at several of the sites where faults have affected the subsidence profile. Table 24 presents details of the case studies used by Lee.

An interesting representation of Lee's findings is given in Figure 191. This shows the position of the fault-produced step at the surface relative to the edge of the mined-out area. The majority of the steps occur close to the edge of the mine working and indeed lie mainly over the extracted area. No steps are reported to have been observed as occurring outside of the subsidence limit angle of 35°.

Particular comments of Lee (1966) on the study of the effect of faulting on subsidence are listed as follows:

1. Four case studies taken from the Lancashire Coalfield showed that there was a 50% reduction in subsidence at the fault by comparison with standard subsidence prediction procedures, and that the angle of draw was not affected. Marr (1961) gives further discussion on these case histories.

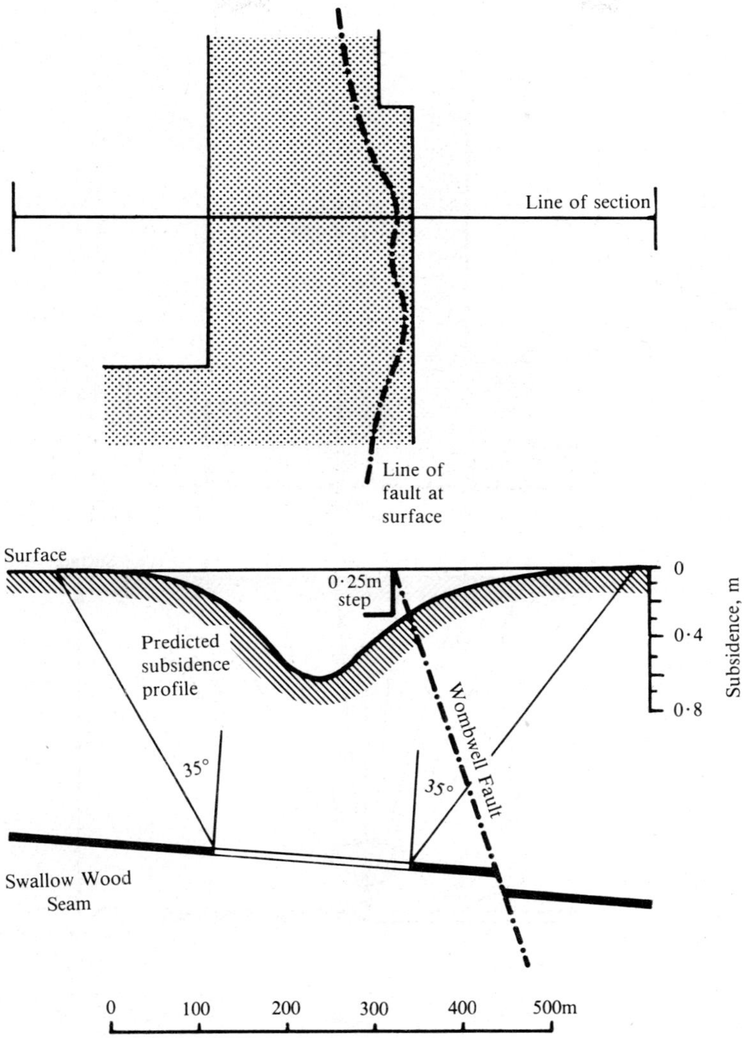

Figure 190 Case history showing occurrence of step at surface due to Wombwell Fault. After Lee (1966).

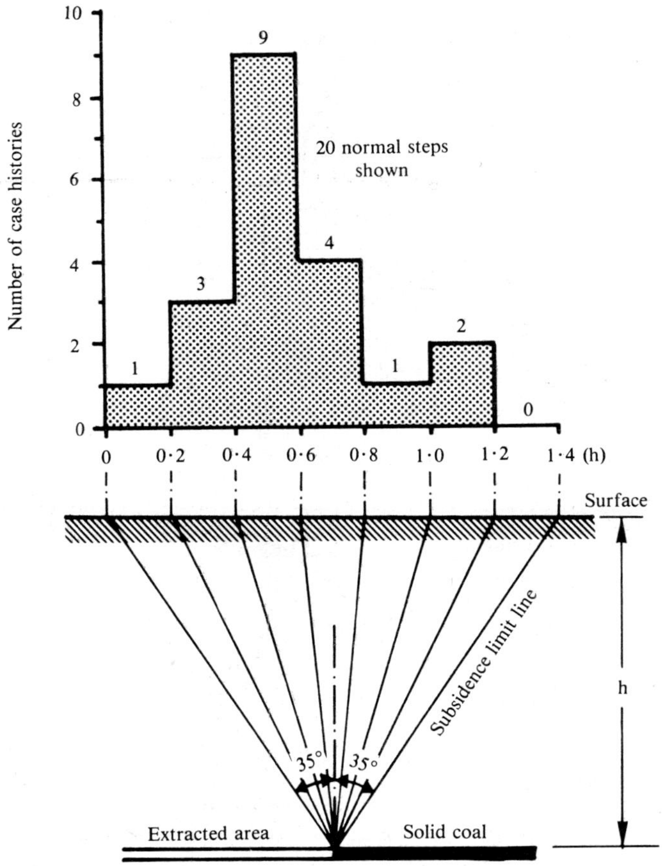

Figure 191 Distribution in subsidence profile of surface steps due to faulting in relation to position from the edge of the extraction. After Lee (1966).

Reference number of section		Seam depth (m)	Fault throw (m)	Direction of displacement related to the last working	Size of step (cm)	Predicted subsidence at step position (cm)	Maximum predicted subsidence for that working (cm)
1		134	11	towards	10	30	52·5
2		157	83	towards	20	30	62·5
3		394	5	away from	22·5	5	82·5
4		324	46	away from	37·5	85	92·5
5		204	69	towards	22·5	52·5	75
6		194	5	away from	37·5	2·5	45
7		194	5	towards	15	2·5	45
8		398	7	towards	15	72·5	77·5
9		370	9	towards	5	2·5	92·5
10		194	28	towards	22·5	20	95
11		204	9	towards	15	15	65
12		204	9	towards	10	30	65
13		185	9	towards	22·5	30	100
14		315	18	towards	25	37·5	60
15		111	46	towards	45	150	150
16		296	9	towards	22·5	12·5	12·5
17		704	83	towards	45	140	150
18		241	14	towards	37·5	117·5	125
19		555	46	towards	15	15	17·5
26	Lancashire examples	667	278	towards	5	10	15
27	Lancashire examples	1074	741	towards	7·5	15	22·5
28	Lancashire examples	481	426	towards	45	90	125
29	Lancashire examples	518	28	towards	15	32·5	42·5

TABLE 24 Details of the section examples, after Lee (1966)

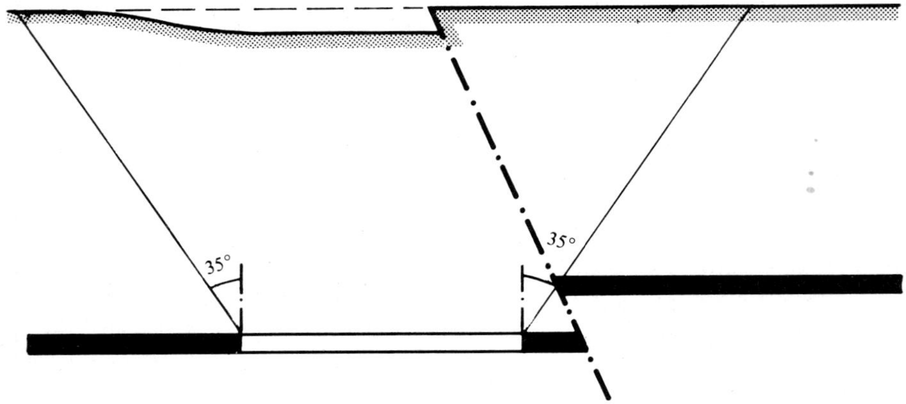

(a) High chance of stepping.

Notes:
1. Normal fault (low friction).
2. Fault close to ribside.
3. Mining limited to one side of the fault.
4. Workings under the hade.

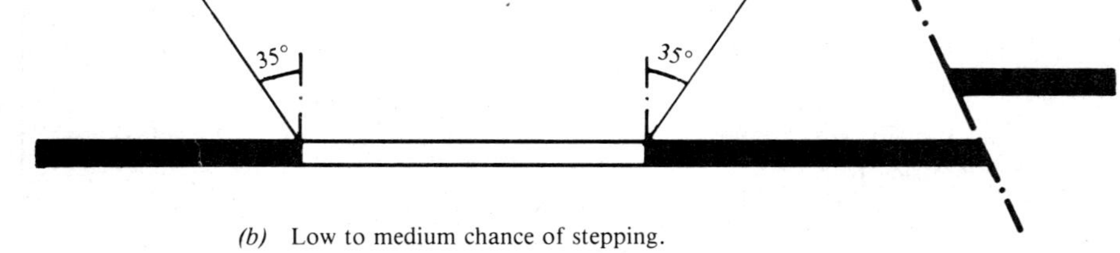

(b) Low to medium chance of stepping.

Notes:
1. Normal fault.
2. Fault near limit line.
3. Mining limited to one side of the fault.
4. Workings under the hade.

Figure 192 Mining and fault situations influencing occurrence of step in subsidence profile.

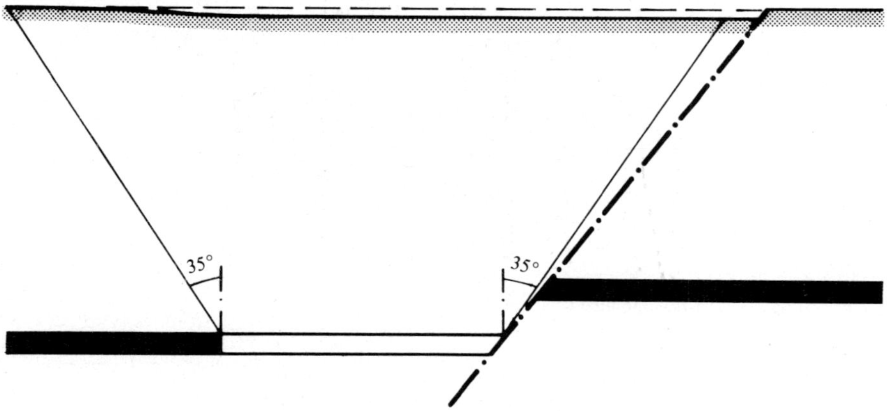

(c) Rare but possible chance of stepping.

Notes:
1. Normal fault (high hade angle).
2. Fault beyond normal limit line.
3. Mining limited to one side of the fault.
4. Workings over the hade.

Figure 192 (continued) Mining and fault situations influencing occurrence of step in subsidence profile.

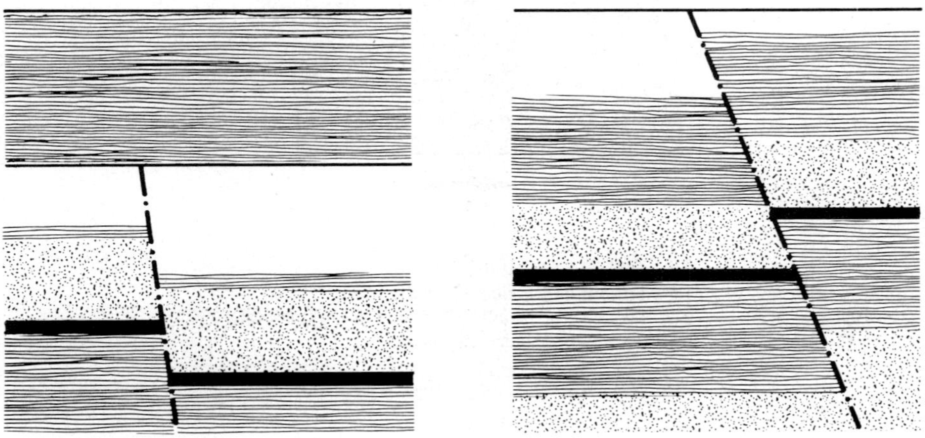

(a) Incropping fault. Effect on surface of fault presence usually negligible, although it depends upon cover thickness.

(b) Outcropping fault. Fault presence likely to be of significance in step occurrence during subsidence, depending upon the extraction position relative to the fault.

Figure 193 Comparison of incropping and outcropping fault situations in relation to effect on subsidence profile.

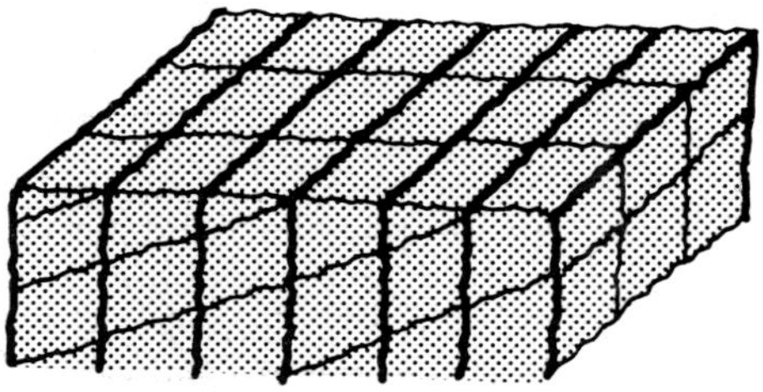

(a) Jointed structures

1. Joints frequently well-developed in one direction and often poorly developed in others.
2. These fractures show no signs of movement.
3. Joints appear to be related to other forms of tectonic movement such as faults and folds; joint density increases in proximity of these latter features.

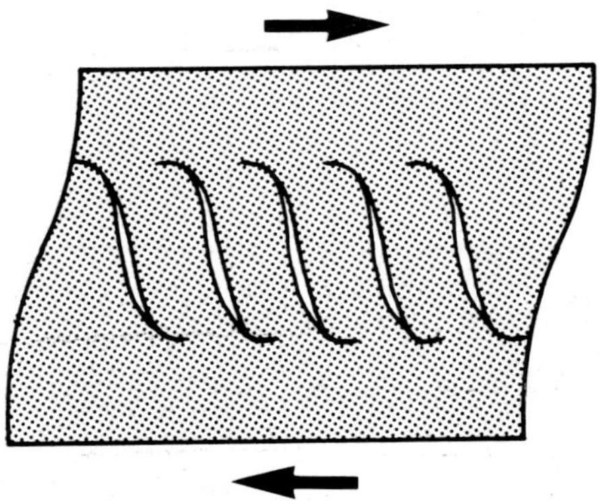

(b) Tension gashes (fractures)

1. Formed under states of tensile stress.
2. En echelon arrays of tension gashes indicate deformation under shear stress.

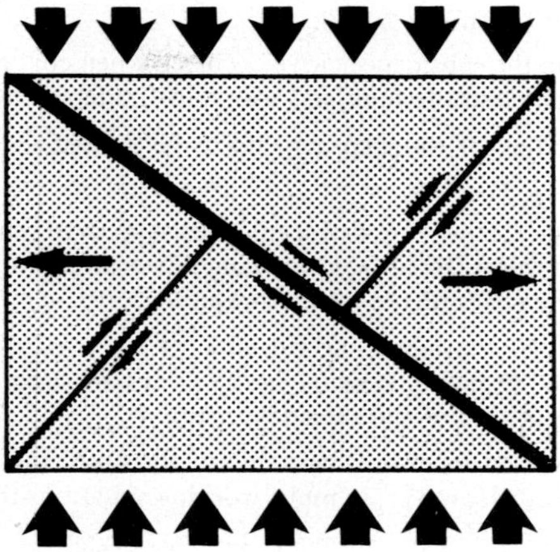

(c) Conjugate fractures

1. Predominance of stress in one direction of sufficient magnitude can cause shear failure.
2. Major shear plane commonly encounters accompanying displacements with striation of failure plane.
3. Secondary shear plane develops at right angles to main plane as part of the failure movement process.

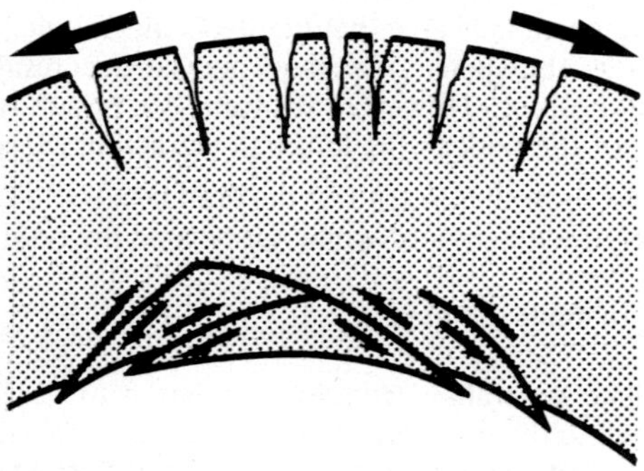

(d) Tension and shear fractures due to tectonic folding.

1. Essentially directional controlled.
2. Intensity and magnitude of fractures related to curvature of fold.

Figure 194 Some common examples of fracture developments in rock structures. Based on Blès and Feuga (1986).

2. Subsidence profiles examined from the Dutch State Mines also indicated similarities in the stepped character produced by fault interception. However, on occasions more than one step occurred.
3. Most of the steps in the subsidence curve occurred when coal extraction was complete under the fault plane and the fault was around 0·2 (depth) over the extracted area.
4. The length of the line of the step appears to be limited to approximately the area worked.
5. The step magnitude over the extracted area appeared consistent for particular workings although this differed from working to working. The steps decreased rapidly in magnitude at the edges of the mine workings.
6. The results indicated that the first working within a particular area can cause a significant step to develop. Subsequent workings in other seams can re-activate movement at the position of the subsidence step.
7. Where coal is extracted from only one side of the fault, the subsidence limit can be substantially reduced. However, in multi-working situations the normal angle of draw of 35° is likely to prevail.
8. The likelihood of movement being experienced at the fault does not appear to be influenced by depth but mainly on the longwall face position.
9. The most significant steps were observed to occur when coal was extracted underneath the hade of the fault. Such steps are usually down towards to the current extraction, but on occasions steps are encountered in association with old workings which exist in the fault region.
10. Excessively high strains accompany movements around the fault region at the surface. The differential movement is mainly in the vertical plane.
11. Predicting the magnitude of differential movement on faults appears to be only possible with workings on the upthrow side and beneath the hade of the fault. The field evidence examined indicated that the step's magnitude is likely to be of the order of a third of the anticipated maximum subsidence resulting from the longwall extraction at that position. However, this value can be substantially influenced by the relative position of the fault to the longwall extraction causing the movement.
12. Field observations show that the step at the fault is likely to form over the entire width of the extraction although not extending far beyond the edge of the mining extraction.

Generalised likelihood of fault stepping

Figure 192 presents three generalised conditions relating to likelihood of fault stepping. Proximity of the longwall to the fault and the fault hade are important factors. Additionally, the frictional properties of the fault plane and the nature of the loss of constraint of the strata associated with the fault are equally important factors. Faults encountered underground do not usually extend to the surface. It is those faults which outcrop that are more likely to cause irregular surface movements. Figure 193 illustrates incropping and outcropping fault situations.

Hellewell (1988) has given a review of the influence of faulting on ground movement due to coal mining operations and makes special reference to UK and European experiences. He draws attention to the findings of Marr (1964) who reported that the angle of draw is not affected by the presence of a fault. Also faults outcropping at the surface are frequently associated with step development which once formed further coal extraction causes the step magnitude to increase and concentrate strain.

Surface fissures associated with mining subsidence

Attention has already been drawn earlier in this chapter to the observations of Shadbolt (1972) who reported that fissures usually occur in brittle rocks and along lines of maximum tensile strain if it is sufficiently intense as to induce failure or cause opening of pre-existing lines of weakness within the rock structure at the surface.

Fissures are mainly of natural origin with occurrences linked with tectonic movements. They are encountered in surface bed rock conditions above mining areas and consequently can pose special problems in association with subsidence.

Common examples of fracture developments in rock structures

A comprehensive discussion on fracture developments in rocks has been given by Blès and Feuga (1986). Figure 194 presents a summary of common fracture developments. Essentially, fractures in rocks can range from symmetrical sets of joints to tension controlled gashes (fractures). Folding curvature can produce both tension and shear fractures. Conjugate fractures appear to mainly result from different combinations of vertical and horizontal stress components and displacement along the resulting shear planes is a common feature.

Rock type influences the nature of fracture development, particularly the size of constituent blocks formed between fractures although the degree of curvature of folding will also be a major factor influencing the intensity and distribution of the resulting fractures. Stress intensity and degree of constraint are major factors also.

Some naturally formed fractures such as tension gashes and fractures occurring on the tension surfaces of tectonically folded strata will be of the open variety. In the main, however, natural fractures are generally of the closed variety and this is probably due to the main stress environment being compressive in nature.

Fissure occurrence in near-surface rocks

Although some open tension fractures can occur as a natural feature, closed fractures generally predominate. The term fissure is essentially by definition that of an opening initially resulting from some fracture process but being also subjected to widening at some stage.

Bed rocks are commonly covered with more recent deposits of soils, clays, sands etc. UK coalfields have glacial deposits present at the surface. On exposure, the bed rocks exhibit marked weathering effects, particularly degradation of weaker rock types. Sandstones and limestones show significant localised weathering along major joint planes. In the UK coalfields the bed rocks have been subjected to the surface denudation effects of successive periods of glaciation. The joints between blocks have consequently been exposed to a wide range of weathering effects. The result has been to create fissure development mainly through natural processes. Such effects obviously decrease in intensity with increasing depth below the surface, although subterranean water courses can give rise to widening of joints and formation of cavities at appreciable depths. The result is for bed rocks of the sandstone and limestone types to exhibit a marked block-controlled deformation characteristic when subjected to differential subsidence. Movement will tend to concentrate along lines of least resistance, with strain effects being greatest at the vertical lines of fissure development. The thickness of cover of surface soils will be of special significance, since a thin layer can result in rapid occurrence of a fissure line at the surface owing to soil collapsing into a fissure system which may have been opened by subsidence. If a fissure system has been disturbed by mining subsidence, subsequent rainfall and downward percolation of surface waters can give rise to later fissure opening at the surface owing to the washing down action of fine materials into a deeper fissure system. Plugging of the fissure system with suitable fill may be required in order to arrest the action of the fissure in allowing further surface soils washing down. This type of surface instability does not necessarily require mining subsidence to trigger its occurrence. Indeed changes in climatic and drainage conditions can cause such fissures to produce opening manifestations which affect the surface.

Figure 195 shows some typical natural fissures as observed in different rock types. Fissures subjected to tensional strains are of particular importance to mining subsidence.

Consequently, the surface overlying mining areas may already contain natural weaknesses within the bed rock and these may have an influence on the development and character of the subsidence resulting from mining operations.

King, Whittaker and Shadbolt (1974) report that the behaviour of sandstones to mining subsidence effects tends to be mainly influenced by joint pattern and structure. Differential slipping between constituent blocks occurs as a result of lateral surface strain. Bunter sandstones sometimes fissure but their natural jointing patterns tend to predominate and such discontinuities prove difficult to detect owing to the presence of surface drift deposits. Magnesian Limestones exhibit brittle properties and variable bedding in addition to possessing well-developed systems of jointing which have a regional trend. Limestone fissures may have been subjected to weathering effects and have widened and subsequently become infilled with clays and other loose materials. Consequently irregular behaviour is likely when affected by subsidence with the strain tending to concentrate along the fissure.

Where lateral constraint is reduced or even removed, then opening of joints is likely. This can arise naturally due to the processes of weathering. It can also occur in man-made excavations for example railway cuttings and quarries. Joint opening of course also occurs in connection with mineral extraction operations.

Figure 196 shows a case history of subsidence and surface ground strain in the vicinity of a natural fissure. A significant strain anomaly occurs in a position of very little subsidence. This demonstrates the effect of vertical discontinuities in encouraging the concentration of surface strain. This latter feature forms the basis for protection of surface structures by means of trenching around the foundations and is discussed in a later chapter.

Field observations on occurrence of surface fissures with mining subsidence

Detailed studies of the occurrence of surface fissures in association with mining subsidence have been carried out in the Midlands' Coalfields of the UK. The purpose of the study was to examine the character of the appearance of fissures in mining subsidence and in particular to establish quantitative information on their occurrence. The subsidence data have been analysed in relation to location of fissure development, type of bed rock present at the surface and magnitude of tensile strain. All the subsidence data relate to longwall mining extractions in a depth range of 50-830m below the surface.

Fissure location. Figure 197 shows distribution diagrams for the observed location of surface fissures associated with longwall mining subsidence. Three generalised rock type conditions are considered, namely Coal Measures, Triassic Sandstone (mainly Bunter series), and Permian Limestone (mainly Magnesian) being present and forming the bed rock at the surface.

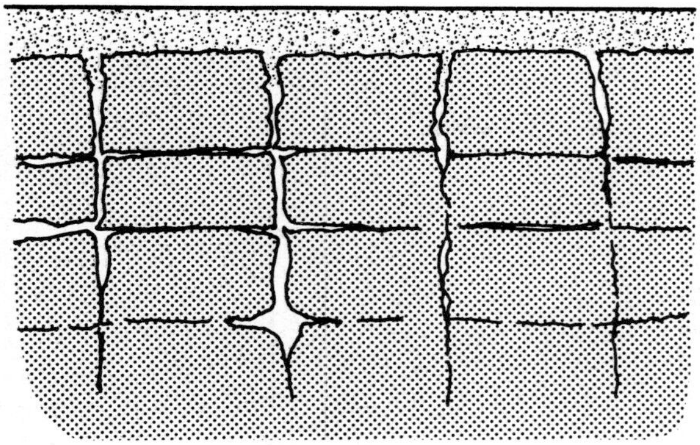

(a) Fissure development in sandstones.

1. Surface denudation and successive periods of glaciation and weathering results in joints in bedrock having been widened.
2. Water courses and sub-surface drainage can result in localised widening of jointed structures.
3. Fissure development is mainly a near-surface development.

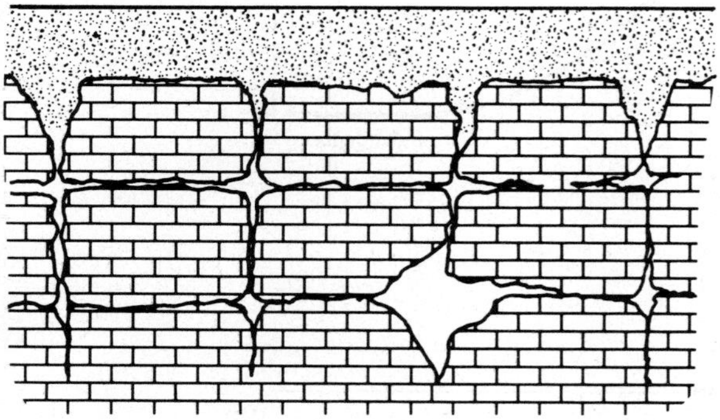

(b) Fissure development in limestones.

1. Limestones affected by denudation cycles during glaciation and subsequent natural weathering are significantly prone to fissure development.
2. Solubility of limestone results in additional widening of joints especially in connection with water courses; sub-surface erosional features are also generally present.
3. Fissure development is more prominent within immediate bedrock.

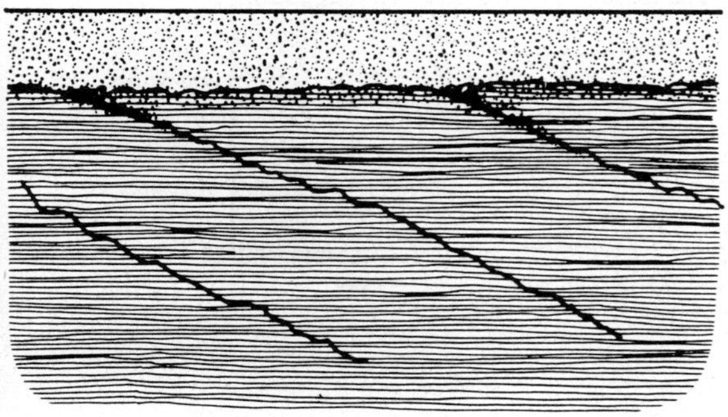

(c) Fissure development in mudstones/shales (Coal Measures).

1. Significant degradation of bedrock in near-surface situations.
2. Mudstones inhibit fissure development owing to slaking properties and general degradation behaviour.
3. Fissures can occur if well-jointed sandstones exist at, or near, the surface. See *(a)* above.

Figure 195 Fissure development in bedrock due to natural weathering effects.

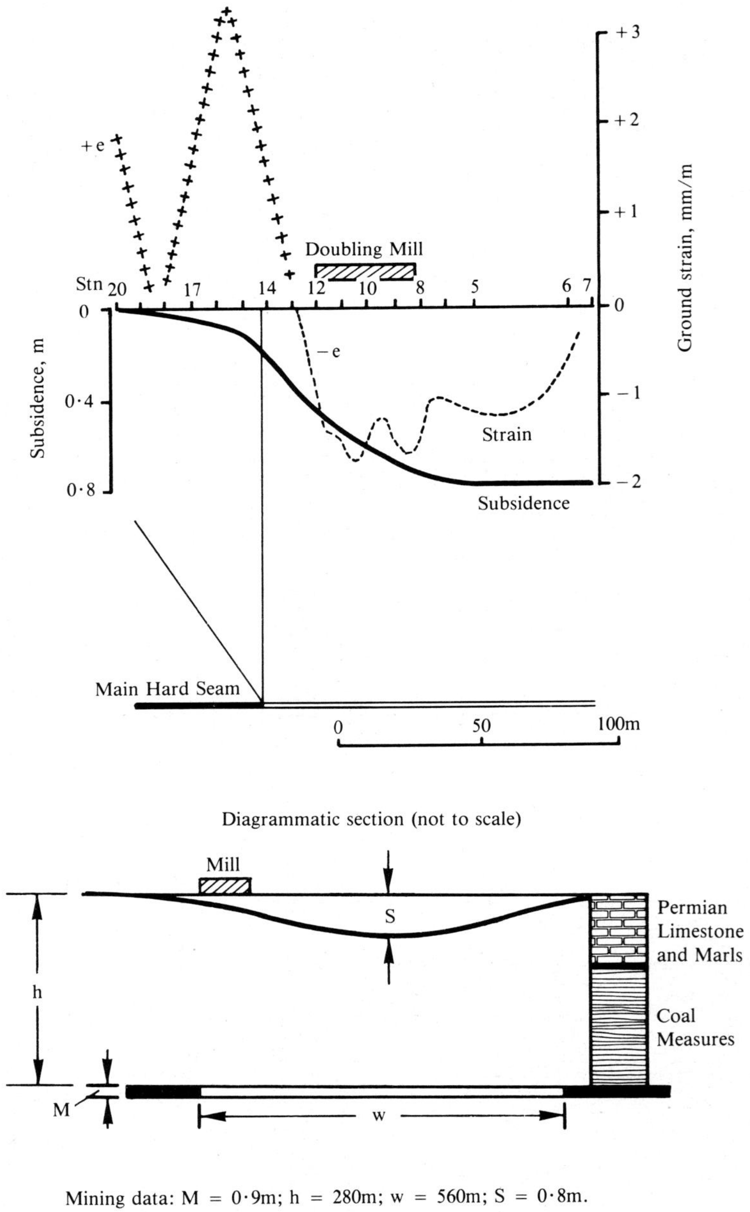

Figure 196 Irregularity in ground strain profile produced by geological discontinuity. After King, Whittaker and Shadbolt (1974).

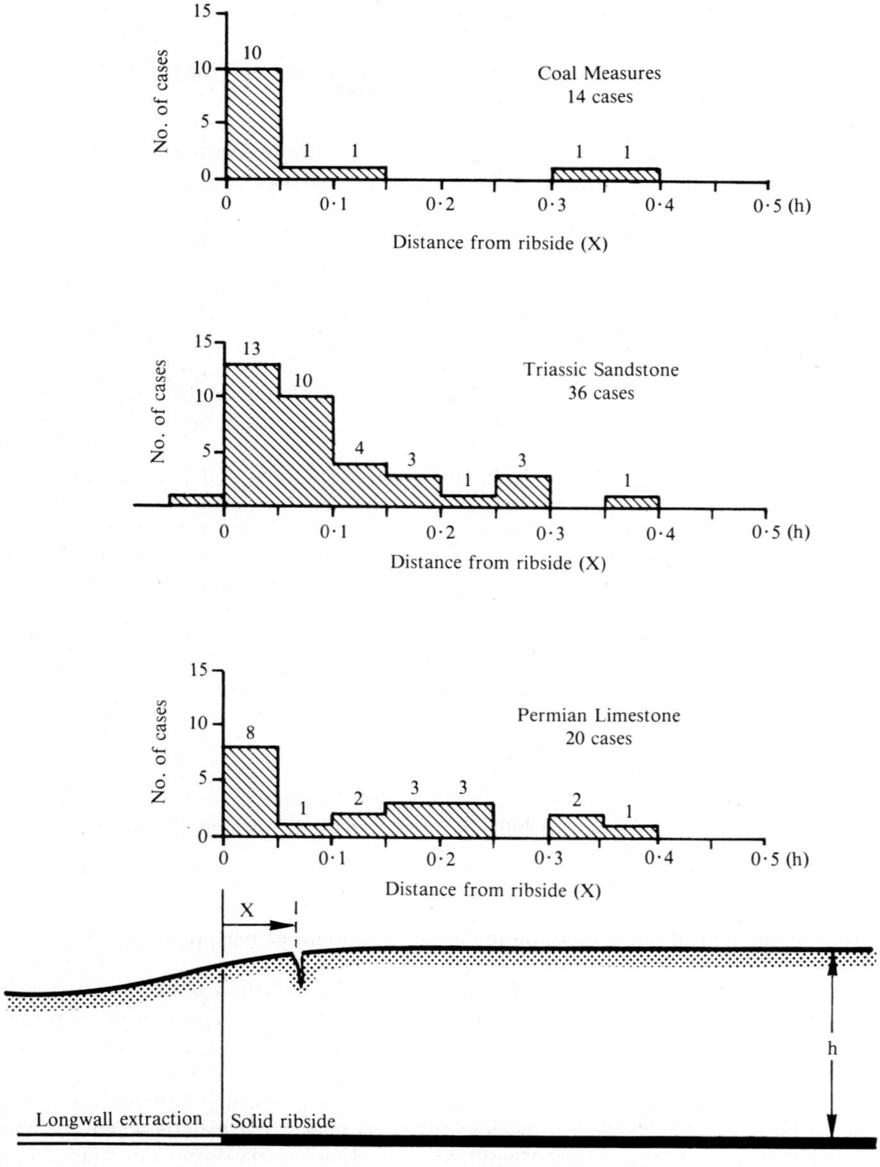

Figure 197 Observed location of surface fissures associated with longwall extractions in Midlands Coalfields, UK

The fissures mainly occurred near the position of maximum tensile strain with the Coal Measures cases; a few occurred further over the ribside.

Considering the Triassic Sandstone cases most occurred within 0·1(h) of the ribside; the others were spread out further over the ribside.

There was a wider spread of occurrence of fissures with the Permian Limestone cases although a high concentration was observed close to the extraction edge.

In all cases no fissures occurred beyond 0·4(h) from the ribside.

Tensile strain at position of fissure. Figure 198 presents distribution diagrams of tensile strain values at the surface position where the fissures were observed. The tensile strains were calculated as being due to the extraction conditions and mining dimensions.

The Coal Measures cases show that no fissures occurred for tensile strain values less than +4mm/m. The fissures occurred over a wide range of tensile strain, up to +20mm/m.

The Triassic Sandstone and Permian Limestone cases show that fissures occurred at much lower orders of tensile strain by comparison to the Coal Measures. Indeed a substantial number of fissures occurred in both the Sandstone and Limestone in the range 0-2mm/m tensile strain. It would appear that these rock types are much more sensitive to the effects of tensile strain in that they will exhibit fissures at relatively low magnitudes of tension.

General features observed with fissure occurrence. The subsidence data employed in this study of surface fissure development indicated the following general features additionally to those referred to above concerning position and magnitude of tensile strain:

1. The surface fissures appeared at the time of mine working.
2. The fissures were generally singular with widths up to 0·46m although the widths tended to be in the range 0·1 to 0·3m.
3. Some of the fissures appeared as multiple stepped features.
4. A number of fissures exceeded 200m in length as a single opening; the longest recorded was 283m.
5. The fissures ran parallel to the edge of the mining extraction.

Influence of strength of cover rocks on the resulting amount of maximum subsidence

The strength of the cover rocks overlying a longwall extraction influence whether the immediate roof will cave. If the longwall is sufficiently wide so as to encourage the immediate roof to fail and cave, the upper beds may still possess high strength and be massive in character so that the caving effects may rapidly dissipate thus reducing the amount of vertical displacement transmitted to the surface. European coalfields frequently experience maximum subsidence of the order of 90% of the extracted seam height, particularly where the cover rocks consist mainly of typical Coal Measures strata. However, increasing amounts of more massive strata such as limestones and sandstones tend to reduce the maximum amount of subsidence experienced above longwall extractions.

Examples of different cover rock conditions are presented in Figure 199 which clearly show how subsidence arising from longwall mining differs in respect of the case history data used for various countries. The examples from India, Australia and North America have substantially stronger cover rocks, in addition to being shallower, than the European examples shown. The data for India and Australia show that the strata do not subside at low values of w/h. The information was collated by Shadbolt (1987).

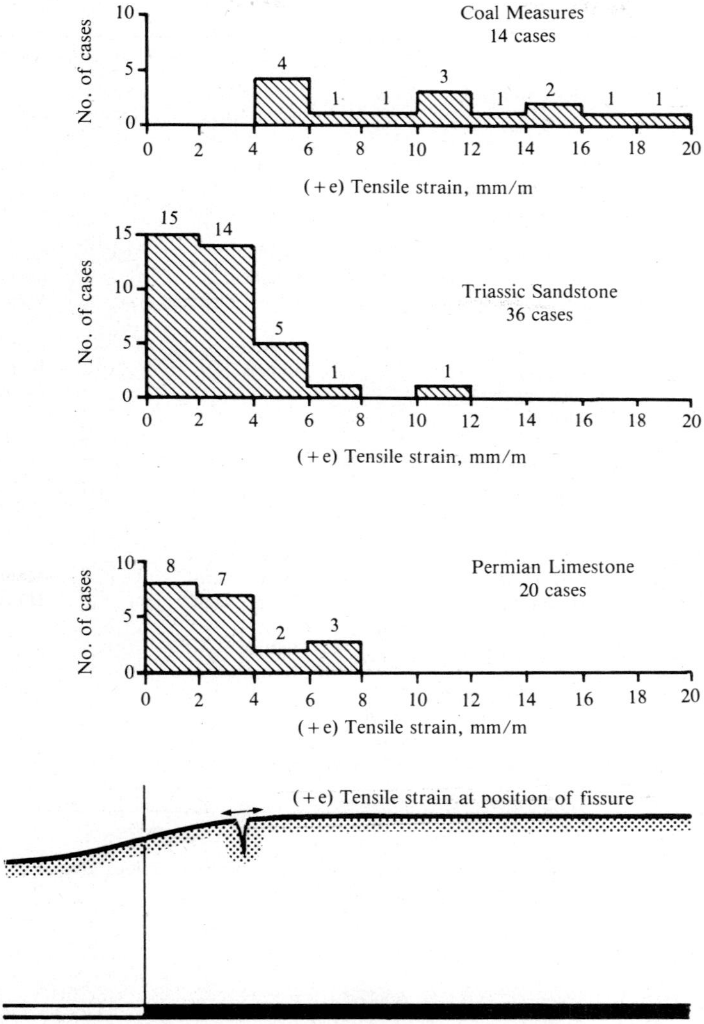

Figure 198 Magnitude of tensile ground strain at the position of surface fissure occurrence associated with longwall extractions in Midlands Coalfields, UK.

Angle of draw	Ground conditions	Location	References
45°	Deep alluvial sediments overlying rock overburden	Limburg Coalfield, The Netherlands	Pöttgens (1978)
41°	Deep alluvial sediments overlying rock overburden	Trent River Valley, United Kingdom	Hazine (1977)
35°	Rock overburden predominantly composed of shales and siltstones	Midlands Coalfield, United Kingdom	SEH (1975) Orchard (1957) Wardell (1969)
23°-29°	Rock overburden predominantly composed of shales and siltstone	Benton, Illinois	Nieto (1979) Wade and Conroy (1977)
*26°	Rock overburden contains thick, strong sandstone beds	Parts of the European Coalfields	Wardell (1969)
25°	Rock overburden contains thick, strong sandstone	Scranton and Wilkes-Barre, Pennsylvania	Montz (1930) Howell (1979)
16°	Rock overburden contains thick, strong sandstone beds	Raton, New Mexico	Gentry and Abel (1978)

*Angle of draw determined from value of critical width to depth ratio.

TABLE 25 Summary of angles of draw associated with coal mines in various ground conditions, after O'Rourke and Turner (1979)

A further example of the influence of strong cover rock conditions is presented in Figure 200. The authors of the report presenting this information, O'Rourke and Turner (1979), draw attention to the large proportions of thick sandstone contributing to reducing subsidence by comparison to the SEH (1975) prediction.

Influence of ground conditions and strength of cover rocks on the limits of subsidence

The work of O'Rourke and Turner (1979) discusses cover rock conditions in relation to the angle of draw. Table 25 shows the data assembled by these authors to demonstrate that increasing strength of cover rocks reduces the limits of subsidence.

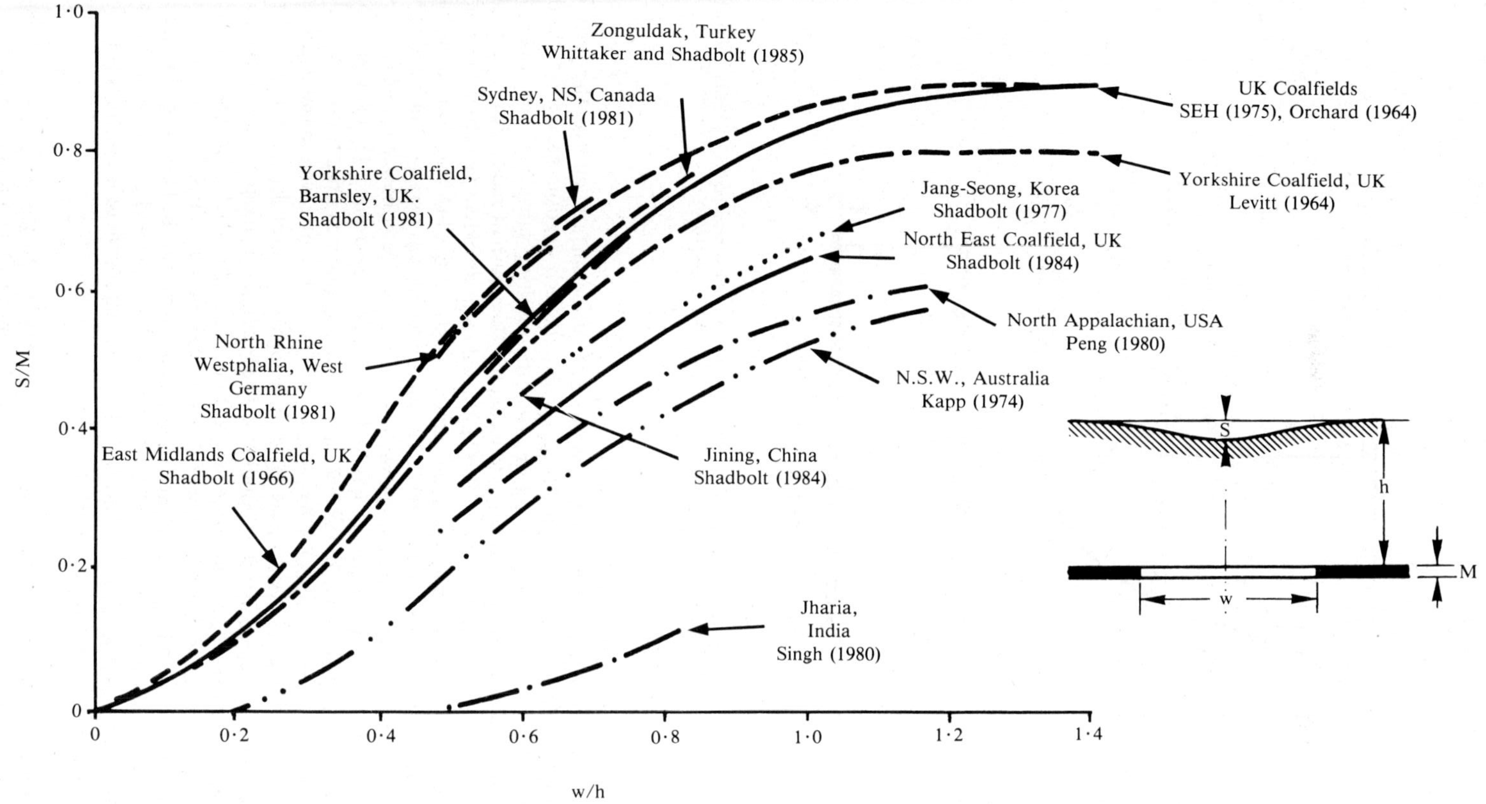

Figure 199 Relationships between S/M and w/h for caved longwall extractions in various coalfields. By courtesy of Dr. C. H. Shadbolt (1988).

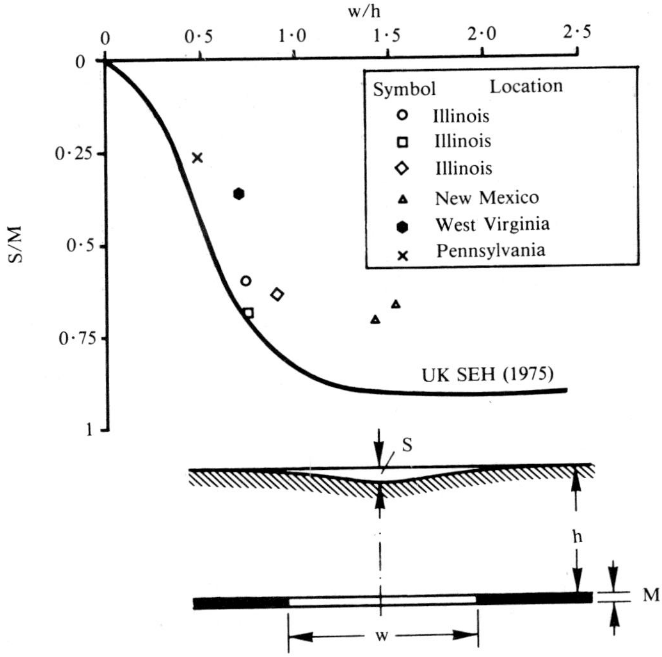

Figure 200 Subsidence related to US longwall extraction dimensions and compared to UK SEH (1975) for caved goaf conditions. After O'Rourke and Turner (1979).

Where unconsolidated and saturated deposits exist at the surface, an irregular form of subsidence behaviour is likely to occur. In particular, if such deposits gain access to the underground mine in the form of an inrush, then the resulting subsidence depression, or crater, is likely to be widespread in relation to its depth. The inrush of saturated peat deposits into the Knockshinnoch Castle Colliery, Scotland resulted in an extensive subsidence crater even though the material gained access to the mine through a single heading close to the base of the unconsolidated deposit, Bryan (1951). The mine heading was rising at 1 in 2 and holed through to a deposit of clay, sand and gravel located in a surface hollow/channel erosional feature within the coal bearing strata. The hollow/channel was around 13·4m deep. Lake mud lay on top of the basal sand and gravels, and above the mud was a deposit of peat which had a maximum thickness of about 3·6m. The top 0·6m of the latter was consolidated as it was above the field drainage system level. The remainder of the peat deposit was considered to be in a saturated state and in a fluid condition. The heading which tapped the base of these unconsolidated deposits experienced water inflow from the roof for about a week immediately prior to the inrush. A period of heavy rainfall caused a three-fold increase in the inflow of water into the mine and appeared to trigger instability at the heading end which led to collapse of the roof up to the base of the bed of boulder clay and gravel, mud and peat which flowed into the heading. Apparently the peat continued to flow for some time eventually filling up kilometres of underground roadways.

The dramatic rescue within 2 days of 116 men whose escape was cut off is described in detail by Bryan and this occupies a major place in the annals of coal mining history.

The subsidence crater was about 80m width, even though the depth to the base of the deposit was only around 13m. Predominant funnelling occurred immediately above the position of the heading within the subsidence crater. The surface turf and semi-consolidated soils slumped as sheared blocks from the sides as the crater increased in size, indicating a predominance of lateral migration of the lower saturated deposits.

This incident underlined the importance of detailed geological information on surface deposits especially the presence of peat or any unconsolidated deposits within proximity of the mine. It also highlighted the dangers of working close to the surface and the need to properly prove the nature of the intervening ground before any working approaches within 46m of the surface.

Concluding remarks

Geological factors influence the stability, or instability, of a site even in the absence of any mining activities. Natural changes in the level and lateral movement of the ground surface are features which arise from seasonal and climatic changes.

The type of geological conditions encountered at the surface overlying longwall mining operations strongly influence the general character and magnitude of the resulting subsidence. The presence of faults and naturally fissured rocks can appreciably influence the nature of the subsidence and strain profiles. Strength and type of cover rock conditions can greatly influence the magnitude and limits of longwall mining subsidence.

CHAPTER 15

SUBSIDENCE ARISING FROM GROUND-WATER WITHDRAWAL, OIL AND GAS FIELD ACTIVITIES AND UNDERGROUND COAL GASIFICATION

Several underground processes are operated remotely as with ground-water withdrawal, oil and gas field operations and the underground gasification of coal. The fact that fluid and/or gas are removed from significant depths results in a measure of uncertainty arising regarding the full extent of the surface area affected by subsidence. Unless a distinct boundary exists in part or in full as can occur by virtue of geological structure, then the limits of fluid withdrawal may decrease gradually owing to the flow properties of the reservoir rocks/sands.

This chapter examines a number of aspects involved with the nature of surface subsidence as often occur with these forms of underground processes.

Subsidence arising from ground-water withdrawal

Poland (1972) has drawn attention to particular problems and their control in matters of land subsidence associated with ground-water withdrawal. An extension of Poland's work has been reported by Helm (1984) who examined field-based computational techniques with special reference to predicting subsidence due to fluid withdrawal. Helm suggested that the choice of predictive technique should be based on the availability of appropriate data from the field. In those cases where only estimates of the depth and thickness of compressible formations are possible, then simplified calculations for many situations have proved to be adequate.

Subsidence relates to vertical movement, and associated effects, of the land surface. Compaction in geological terminology refers to the decrease in thickness of sediments following the application of vertical stress. Consolidation in soil mechanics terminology relates to decrease in thickness of a laboratory sample subjected to compressive loading. The subsidence arising from withdrawal of ground-water is essentially a surface response to sediment compaction at depth, Helm (1984).

A small change in effective stress of an engineering soil at depth is accompanied by a small change in volume when considering a column of soil. The application of a sustained constant head of drawdown to a ground-water regime triggers a subsidence process which does not occur immediately. The response of the porous sediment forming the aquifer is to behave in accordance with time-consolidation theory which means that the subsidence rate will taper off gradually and can take many years. The magnitude of the drawdown head will influence the time of subsidence duration and also the limits of subsidence although the ground-flow properties also play a role. Helm (1984) suggests that empirical methods allow observed subsidence to be plotted against time so that extrapolation is possible for predicting future subsidence simply by selection of an appropriate curve fitting technique.

General surface behaviour to ground-water withdrawal. Two principal mechanisms have been advanced to explain ground behaviour following ground-water withdrawal:

1. The phenomenon of localised differential compaction.
2. The resulting horizontal contractions which arise owing to capillary effects in the zone above a lowering water table.

Holzer (1984) considers that based on US experience these particular points appear to be of significance to ground movement behaviour, although the second seems only of relevance in areas where surface fissures exhibit polygonal patterns. Earth fissures appear to mainly form in those zones where the near-surface aquifer system experiences thinning over ridges or even steps at the bedrock surface. Field evidence indicates that such fissures occur in those zones of maximum tension, ie curvature convex-upward. Fissure systems forming complex polygonal patterns appear to be essentially large contraction cracks.

Figure 201 illustrates the principal mechanisms of ground behaviour as based on US observations following the withdrawal of ground-water.

Nature of surface failure and subsidence resulting from ground-water withdrawal: US experiences

An interesting and comprehensive account of the nature of surface ground failures above unconsolidated sediments which have been subjected to ground-water withdrawal has been given by Holzer (1984). He reported that observed failures included long tension cracks or fissures at one end of the range through to surface faults (significant steps) at the other. These failures are a feature of land subsidence resulting from underlying unconsolidated sediment experiencing compaction during ground-water withdrawal. The fissures can range in length from tens of metres to kilometres, but generally open of the order of centimetres. Later erosion of such fissures commonly results in gullies of dimensions 1-2m width and 2-3m depth. Some fissures have been measured to depths of 5 to 10m using a weighted line lowered into the fissure. A fissure was logged as having a depth of 16·8m when logging terminated at the water table. The surface fault (step) features commonly exhibit scarps of 0·5m height with lengths of the order of a kilometre or so; some surface faults attain heights of 1m and a length of 16·7km has been observed. Scarp growth has been reported to be in the range of 4 to 60mm/year with most movement correlating with seasonally wet periods. Major step development has resulted in extensive surface damage at Houston - Galveston, Texas, metropolitan region.

Holzer (1984) has estimated that surface subsidence effects relating to ground-water withdrawal from underlying unconsolidated sediments has affected a total area of around 22 000km^2 in the United States. The main feature is that of loss of elevation namely subsidence, and has exceeded 1m in several areas whilst at the San Joaquin Valley, California the maximum subsidence has attained 8·5m. Holzer points out that fissures are generally first noticed after erosion along the line of the fissure particularly following rainstorms. The early stage of development would typically show collapse features along the line of the fissure, and such features would be generally connected by minor to hairline cracks. There is an indication of hydraulic connection between these surface features in most situations. A further important aspect drawn attention to by Holzer is that scarps formed by faults due to ground-water withdrawal usually appear similar to fault scarps of natural origin and that confusion can arise in differentiating between them. The following points are made in this respect:

1. ground-water fluctuations have been observed to relate to surface fault movement, although seasonal fault movements also occur,

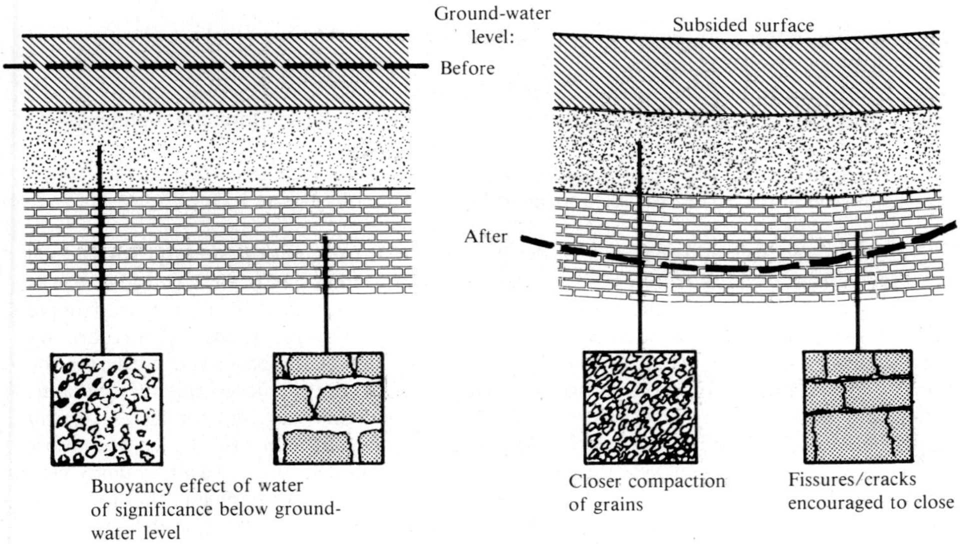

(a) Effect of ground-water withdrawal on localised differential compaction resulting in surface subsidence.

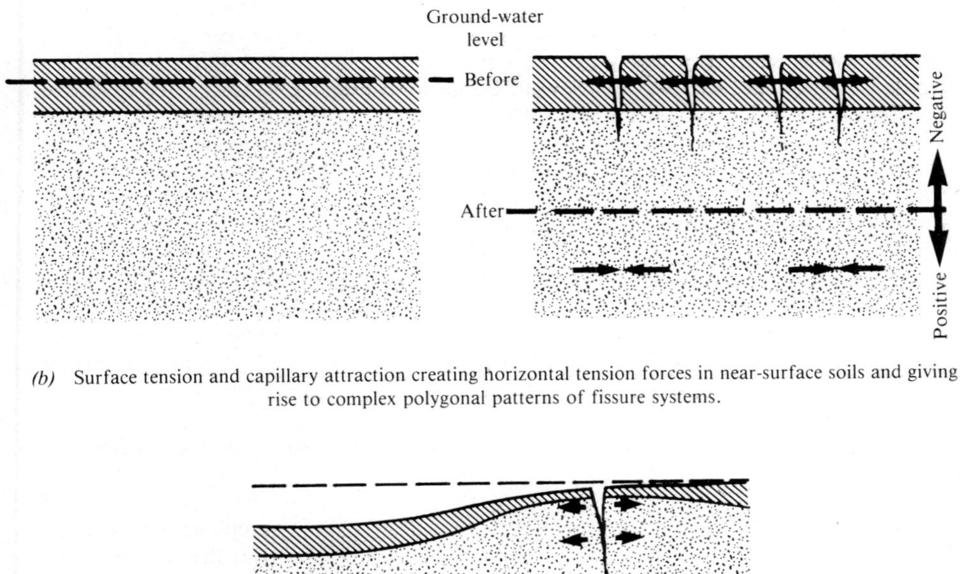

(b) Surface tension and capillary attraction creating horizontal tension forces in near-surface soils and giving rise to complex polygonal patterns of fissure systems.

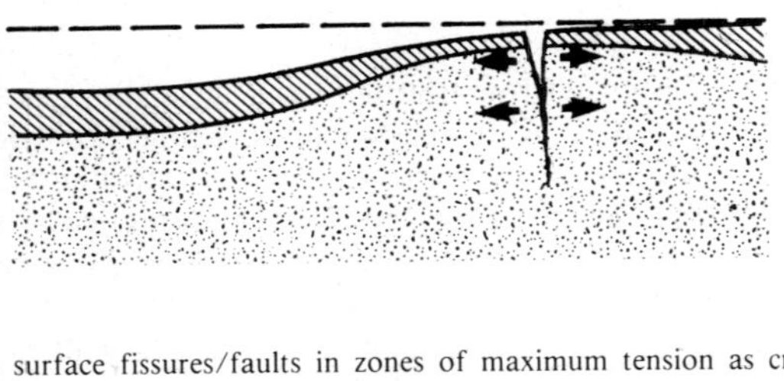

(c) Occurrence of large surface fissures/faults in zones of maximum tension as created by ground-water withdrawal.

Figure 201 Illustrating the principal mechanisms of ground behaviour following withdrawal of ground-water. Based on observations reported by Holzer (1984).

2. ground-water withdrawal appears linked with temporal and areal manifestation of localised fault occurrence,
3. earth fissures arising through water-withdrawal pose distinct hazards, namely *(a)* the displacements accompanying their formation, and *(b)* the resulting deep gullies which are modified by erosion,
4. most of the observed earth fissures occur above ridges in the bedrock surface which may be buried. The fissures appear to be controlled by bedrock conditions.

Holzer refers to land subsidence in the Picacho Basin, Arizona where water withdrawal has taken place over several years. Between the period 1963/64 to 1977 subsidence was observed to take place over the basin and spanning a distance of some 12km approximate diameter with maximum subsidence of 1·25m. Subsidence movements concentrated around the Picacho Fault which extends some 15km along the edge of the basin and bordering the Picacho Mountains. Fissuring was also observed to occur in the vicinity of this surface fault, which was first formed as a fissure in 1949 and had since developed relative vertical displacements (steps) of 0·2 to 0·6m. The fault scarp began to form in 1961. Creep rate across the scarp was around 60mm/year during the early stages but decreased to around 9mm/year by the period 1975-80. Holzer reports that the creep rate varies seasonally and that there is a correlation with water-level fluctuations. Field investigations indicate that the Picacho Fault is in the main associated with a pre-existing fault and this acts as a partial barrier to ground water flow in the alluvial aquifer.

Land subsidence at Houston - Galveston, Texas is also discussed in detail by Holzer. More than 160 surface faults of total length greater than 500km have been observed to be associated with subsidence attaining a maximum value of more than 2·7m with a subsidence affected area of some 90km or so approximate diameter. The land subsidence referred to here has been observed from 1906 to 1978. The surface faults tend to predominate in an approximate north-east direction, and their intensity is greatest around the central area of the subsidence basin. Although oil and gas production is fairly extensive in this region, it is considered that the land subsidence is due almost exclusively to ground-water withdrawal, with minor contributions from oil and gas production.

Detailed surveying observations made by the Nevada Department of Transport in the Las Vegas Valley are also referred to by Holzer, and these results indicate the formation of a localised subsidence depression in association with a fault. As no pumping wells were located in the area of the depression, this localised subsidence was attributed to sub-surface conditions rather than lowering of the water level.

Special problems arising from ground-water withdrawals above sink-hole prone carbonate bedrock conditions

Lowering of the water-table by ground-water withdrawal above carbonate rocks such as limestone and dolomite has given rise to particular sink-hole problems developing at the surface. Where such rocks occur at the surface, it is not uncommon to see pock-marked features of earlier erosion. The extent of such erosional features depends upon the sensitivity of such rocks to dissolution by surface- and ground-waters in addition to climatic conditions, vegetation, topography and the general character of structural weaknesses within the rock mass. Chapter 1 has discussed sink-hole development as a natural phenomenon in limestone country. Attention is drawn here to particular problems encountered when lowering the water-table over sink-hole prone carbonate bedrock conditions.

Effect of lowering the water-table. Figure 202 illustrates the main principles governing the effect of changed water-table conditions on the development of a sink-hole at the surface.

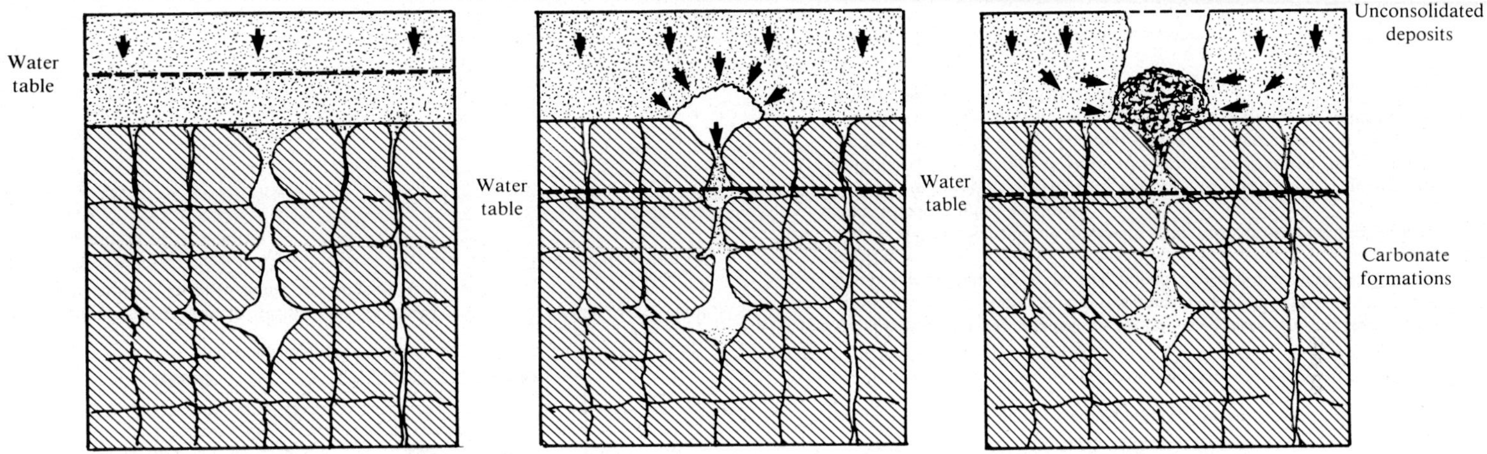

Stage 1

1. Water-table above carbonate bedrock subjected to steady state conditions and virtually in equilibrium.
2. Downward percolation of water into carbonate bedrock "controlled" by infilling of clays, silts, gravels etc within immediate bedrock fissures.

Stage 2

1. Water-table lowered to below carbonate bedrock.
2. Drainage of surface waters through to water-table line attempts to seek lines of least resistance. Large pre-existing erosion-widened fissures in bedrock form natural drainage attraction points.
3. Localised attraction of drainage above major fissured feature creates potential for surface deposit materials to wash down and trigger onset of a sink-hole development.
4. Lower connecting ground-water courses allow loose material to be washed down and transported underground.

Stage 3

1. Sink-hole develops to surface if material can be washed down into lower cavities — otherwise it could choke naturally before reaching surface if sufficient depth of cover exists.

Figure 202 Illustrating effect of ground-water withdrawal above sink-hole prone carbonate bedrock conditions.

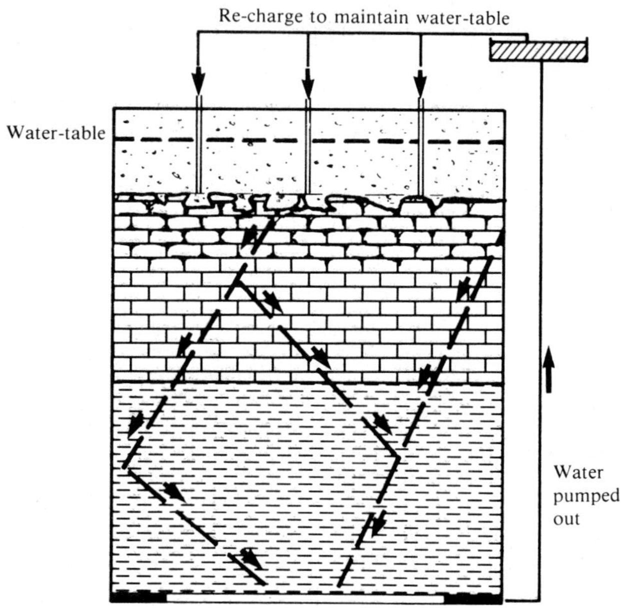

(i) Re-charging water-table during mining

1. Mining progressively results in increased inflow of water into mine, particularly via disturbance to connecting faults.
2. Water pumped out is used to re-charge the water-table and prevent, or control, formation of sink-holes.

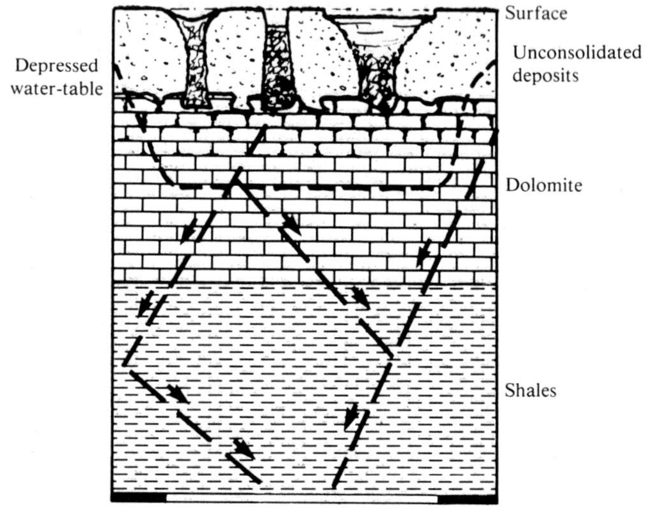

(ii) Effect of de-watered compartment on the surface

1. Water-table is lowered locally to create de-watered compartment in order to reduce inflow of water into mine.
2. Changed hydrogeological conditions at interface between unconsolidated materials and dolomite encourages development of sink-holes which can be of:
   (a) Local saucer-shaped, or funnel-shaped depression generally affecting small area.
   (b) Clearly defined subsidence hole with near vertical or even overhanging sides which is again localised.
   (c) Depression, usually circular, with stepped edges; the step can be of a few millimetres but generally of order of centimetres although steps of a few metres are not unknown. The diameter of the depression can be up to 100m or even greater, depending upon the rate at which the water-table is lowered and condition of the bedrock.

Figure 203 Illustrating effect of lowering the water-table below the dolomite bedrock horizon and subsequent effect on sink-hole development.

With the water-table above the carbonate bedrock level and located in the unconsolidated deposits, in effect is in a virtual state of equilibrium. The downward percolation of water into the carbonate bedrock is "controlled" by infilling of the immediate bedrock fissures with unconsolidated materials washed down from the surface deposits. If a flow-path for the waters draining into the carbonate, was suddenly enlarged say by breaking down of a clay plug, then the conditions could arise for the start of a natural sink-hole to develop. However, the latter condition could easily become choked by more material falling into the carbonate drainage cavity, and a state of equilibrium being restored.

Stage 2 of Figure 202 shows the effect of lowering the water-table below the carbonate bedrock horizon. Water draining through the unconsolidated materials will seek the line of least resistance and be attracted to major natural drainage features at the carbonate interface. The result can be to dislodge previously choked fissure systems and allow major drainage paths to come into operation. The water drainage pattern in the unconsolidated materials will encourage the formation of cavities at the contact horizon with the carbonate. As the unconsolidated materials fall into the carbonate fissure system, sufficient water flow now exists as to wash these loose sediments into lower cavities. This process can progress and eventually lead to the formation of a collapse-feature, often a sink-hole, appearing at the surface.

Sink-holes associated with ground-water withdrawals in carbonate bedrock conditions in the US. Newton (1984) has described sink-hole activity in carbonate terranes following ground-water withdrawals and remarked that the problem frequently arises in Alabama, Florida, Georgia, Maryland, Pennsylvania, South Carolina and Tennessee. He draws attention particularly to the problems created by sudden appearance of sink-holes resulting in collapses beneath highways, roads, railways, buildings, pipelines and other surface features. Newton refers to the conditions under which sink-holes develop and they are primarily dependent upon carbonate rock types such as limestone and dolomite being present; these rocks allow the storage and movement of water through interconnected openings created along joint and bedding planes, fractures and faults enlarged by the dissolution action of water. This author reports that thousands of natural sink-holes exist in areas of Alabama where carbonate bedrock is present, and they vary in size from a few metres to as much as 3km in diameter with depths of a few metres to more than 30m. The timing of occurrence can be of short duration after the introduction of man-made effects on the hydrogeological conditions, or in the case of natural sink-hole development may take periods of time involving up to thousands of years. The nature of the underlying bedrock plays a major role.

Sink-holes resulting from ground-water withdrawals in carbonate bedrock conditions in South Africa. The work of Jennings, Brink, Louw and Gowan (1965) has examined the development of sink-holes in the Transvaal dolomites of South Africa where pumpage of water has created cones of depression. These authors demonstrated that sink-hole and surface subsidence problems increased in those situations where the water-table was lowered.

Mining operations beneath the dolomite generally experience a gradual increase in the amount of water gaining access to the mine. Water commonly drains through disturbed fault planes to lower horizons. Water pumped out of the mine is commonly used to re-charge the water-table above the mining area and thus effect a control measure on the development of sink-holes at the surface. However, if the underground pumping operations become excessive, the overall economics of the situation may require other approaches to the problem. Attempts have been made to introduce grouting materials in the dolomite in order to try to control inflow into the mine below; such attempts have met with limited success especially in view of the magnitude of the problem and the general uncertainty as to where such remedial measures should be introduced. De-watering of a compartment above a mining area is generally only resorted to when other methods prove unsuccessful or inadequate. This allows the water entering the mine to be significantly decreased, as no re-charge of the water-table is carried out and consequently re-cycling the water is avoided. However, lowering the water-table results in the dolomite bedrock influencing the state of drainage, so that pre-existing drainage paths and cavities now allow scope for the unconsolidated materials to wash down into cavity systems deeper in the dolomite. Progressive washing down of the sediments and creation of cavities in the unconsolidated materials above the dolomite provides the conditions for sink-hole development at the surface.

The form of sink-hole and surface depression development depends upon the thickness and nature of the unconsolidated materials near the surface, as well as the size and geometry of the underlying cavity, and the rate at which such a cavity is formed by internal erosion and collapse. Clearly defined sink-holes may appear suddenly and be of significant depth and diameter, whilst other surface subsidence features may be that of a large diameter of up to hundreds of metres but with a relatively small step and a central area which has lowered of the order of centimetres. The nature of the surface subsidence feature can differ appreciably. There is a tendency for such surface subsidence features to cluster together in view of tending to reflect particular drainage patterns and cavity development at the dolomite bedrock horizon which promote internal erosion. Additionally areas of high water content within the unconsolidated materials appear to be those zones of importance in attracting near-surface drainage and are thus likely to experience cavity formation. Such saturated or significantly partially saturated zones will be in a weaker state and thus be sensitive to collapse processes.

Organic soil subsidence. Stephens, Allen and Chen (1984) have reported that organic soil subsidence is mainly a feature of drainage and development of peat. The reasons for subsidence are densification, particularly shrinkage and compaction, or from loss of mass through biological oxidation, burning, hydrolysis and leaching, erosion and mining. Densification results shortly after the implementation of drainage. Oxidation and erosion are generally slow with minor losses of mass. Losses due to mining activities depend upon direct removal of peat from the site and consequently has a more localised effect which will vary considerably from site to site.

The English Fens drainage activities began in 1652. These low-lying peat moors experienced alternate cycles of effective drainage which led to increased subsidence and corresponding changes in water tables. As the water-table was lowered by improved drainage, the peat surface subsided and thus created a need for further drainage operations to maintain land usage. Stephens *et al* report that those peat lands have been subjected to an annual subsidence rate in the range of 0·5 to 5cm/year with up to 3·48m of subsidence since drainage began in the 17th century. Several countries experience long-term subsidence effects due to organic soils, and particularly in The Netherlands, USA, USSR, Norway and Ireland.

Subsidence rates are influenced by the nature of the peat, the depth to the water-table, and temperature. The intensity and distribution of peat drainage operations are the main governing factors on the general development of surface subsidence. There appears to be an approximately linear relationship between the average depth to the water-table and the average subsidence according to Stephens *et al.* Subsidence of organic soils is faster in warmer regions than when compared to similar deposits in cooler climates.

Geothermal fluid withdrawal effects on subsidence. The work of Stilwell, Hall and Tawhai (1975) has demonstrated that fluid withdrawal from the geothermal field at Wairakei on the North Island of New Zealand has given rise to up to 4·5m subsidence with accompanying horizontal ground displacements of up to 0·5m. Mixtures of steam and predominantly water are yielded by the geothermal reservoir. Geothermal fluid production began in 1950 and it appears that subsidence affects an area in excess of $3km^2$ essentially over the main geothermal reservoir which has a thickness in the range 370 to 790m.

Narasimhan and Goyal (1984) give a review of subsidence aspects relating to geothermal fluid withdrawal and draw attention to experiences at Larderello in Italy, Cerro Prieto in Mexico, Wairakei in New Zealand, and The Geysers, California in the USA. These authors remark that land subsidence may accompany geothermal fluid withdrawal where favourable hydrogeological and exploitation conditions exist. The cause of subsidence is attributed to volume changes in the reservoir undergoing depletion of geothermal fluid storage although thermal contraction is also considered to play a role. Narasimhan and Goyal examined different bases of subsidence prediction in such conditions but concluded that the best course of action in establishing reliable data is that of comprehensive deformation monitoring of both surface and subsurface responses to fluid withdrawal so as to enable prediction models to be validated. Field evidence indicates that subsidence arising from geothermal fluid withdrawal tends to be in the form of a general depression which reflects the size and position of the underground reservoir, although major faults can have a significant limiting influence on subsidence development at the surface in some cases.

Subsidence over oil and gas fields

The oil and gas production activities at the Goose Creek oil field, Texas, gave rise to the first detailed reports on resulting surface subsidence during the period 1900-1920, Pratt and Johnson (1926). The Bolivar Coast oil fields, Venezuela, also experienced subsidence during the 1920's, van der Knaap and van der Vlis (1967), whilst Gilluly and Grant (1949) refer to subsidence above the Wilmington oil field, Long Beach, California during the 1930's. Subsidence has been reported to be associated with the Groningen gas field, The Netherlands, as discussed by Schoonbeck (1976), in addition to the Huntington Beach oil field of California, as referred to by Martin and Serdengecti (1984), and also oil and gas fields in the USSR, Ilijn (1977).

It is clear from the remarks of Martin and Serdengecti (1984) that subsidence over oil and gas fields has been widely reported and has occurred in several countries. The magnitude of the subsidence observed has been almost 9m at the Wilmington oil field. These authors note that surface subsidence probably occurs over all oil and gas reservoirs where a pressure decline is experienced, even though subsidence seems to have been detected at only a few of the many thousands of oil and gas fields which have been developed. The potential subsidence appears to be insignificant for most oil and gas fields.

Martin and Serdengecti (1984) suggest that where major subsidence occurs over oil and gas fields, then two *in situ* rock failure conditions appear to prevail. Firstly, there is the general weakness of some rock types regarding grain behaviour with accompanying reductions in porosity and the thickness of the reservoir. Secondly, the stress state may result in *in situ* failure and movement along fracture and fault planes. These authors state that the maximum subsidence (S) will be a function of:

(a) the associated one-dimensional compaction,
(b) the stress transfer factor and is related to stress transfer from the surrounding rocks to the reservoir rock,
(c) the subsidence spreading factor which relates the maximum reservoir compaction to that of the maximum surface subsidence.

The one-dimensional compaction is the product of the one-dimensional compaction coefficient, the pressure drop and the reservoir thickness. The stress transfer factor is the ratio of the maximum reservoir compaction to the one-dimensional compaction; this is an indication of the amount of stress transferred to the reservoir rock as a result of the pressure drop. The subsidence spreading factor is the ratio of maximum surface subsidence to the maximum reservoir compaction. This ratio is reported to be about 1·0 for shallow reservoirs of large lateral extent. This factor decreases with increasing depth below surface of the reservoir and for decreasing lateral extent, Geertsma (1973).

A general expression for maximum subsidence (S) has been given by Martin and Serdengecti (1984) as equation (67).

$$S = \begin{pmatrix}\text{one-dimensional}\\ \text{compaction}\end{pmatrix} \begin{pmatrix}\text{Stress transfer}\\ \text{factor}\end{pmatrix} \begin{pmatrix}\text{Subsidence}\\ \text{spreading factor}\end{pmatrix} \qquad (67a)$$

$$= \left(C_m \, \Delta p \, h\right) \left(\frac{\Delta h_m}{C_m \, \Delta p \, h}\right) \left(\frac{S}{\Delta h_m}\right) \qquad (67b)$$

where,

C_m = one-dimensional compaction coefficient
Δp = the pressure drop
h = reservoir thickness
Δh_m = maximum reservoir compaction

The general conclusions drawn by Martin and Serdengecti are:

1. The majority of oil and gas reservoirs give rise to only small amounts of reservoir compaction and associated surface subsidence.
2. Reservoir rock compaction and resulting surface subsidence exhibit inelastic behaviour of the reservoir rock and possibly that of the surrounding and overlying rocks.
3. The principal factors influencing oil-field subsidence appear to be: reservoir fluid pressure, depth, geometrical setting, and mechanical properties of reservoir rock and surrounding and overlying rocks.

These authors suggest that subsidence arising from reservoir compaction over oil and gas fields can be controlled by fluid injection in order to achieve pressure maintenance. In the case of strong water-drive reservoirs, the restriction of reservoir withdrawals in order to allow the water influx to maintain the reservoir pressure is entirely feasible as a subsidence control measure.

Subsidence over the Ekofisk oil field. The Ekofisk oil field consists of a fractured chalk reservoir located centrally in the North Sea, and subsidence was first recognised towards the end of 1984. Wiborg and Jewhurst (1986) have given subsidence details for the Ekofisk oil field. They reported that some 2·6m of subsidence was observed up to mid-1985 and subsidence rates of 0·4 to 0·46m/year with up to 0·7m/year centrally have occurred since 1979-80. The cause of the subsidence is explained as reservoir pressure depletion by fluid withdrawal as associated with production operations, and these authors argue that for the seabed movement to be arrested necessitates the maintenance of pressure by means of fluid injection (ie water, gas) back into the oil producing formation.

Ekofisk is a major gathering location for gas, oil and condensate produced from wells on the Norwegian shelf, with gas pipelines connecting with Teesside, England and Emden, West Germany.

The chalk reservoir rock at Ekofisk often has porosities greater than 40%, although the matrix permeability is low being of the order of 0·1md. Shales and clays comprise the bulk of the overburden with an estimated coverload pressure of 62 MPa. The original pressure of the reservoir was 48 MPa. Consequently the chalk reservoir rock was required to support the net pressure differential of 14 MPa. Oil production resulted in the reservoir pressure reducing to 24-28 MPa so that the net pressure for support by the chalk increased to 34-38 MPa. Some 65% to 85% of the reservoir compaction at Ekofisk was indicated by testing and measurement to have appeared as seabed subsidence. Surprisingly, effective permeabilities are virtually unchanged even after continuous production of 14 years.

Laboratory investigations by Wiborg and Jewhurst indicated that chalk with porosities less than 30-32% should not undergo significant compaction, even without the maintenance of reservoir pressure. These authors have related laboratory studies and field measurements to the Ekofisk subsidence problem. Table 26 gives five key parameters regarded of significance to evaluating the probability of subsidence.

Key Parameter	Reservoir rock formation	
	Danian Chalk	Cretaceous Chalk
High porosity	48%	35%
Thick reservoir	183m	122m
Large pressure reduction	22 MPa (1985)	24 MPa
Large areal extent	8km × 5km	
Shallow reservoir	3000m	3200m

Table 26 Key parameters of significance for evaluating the probability of appreciable subsidence at the Ekofisk oil field, North Sea. After Wiborg and Jewhurst (1986).

With reference to Table 26, these authors regard reservoir depth to be a parameter of significance at Ekofisk, and point out that at other localities reservoirs with substantial resulting surface subsidence have been producing from about 1500m or less.

At Ekofisk, a bathymetric survey from 1970 (±1m accuracy) and surveys performed during 1984/85 and later in 1985 (±0·7m accuracy) of the seabed provided data which allowed a subsidence depression of 2·6m maximum subsidence to be indicated. The shape of the subsidence depression resembles the size and general shape of the underlying reservoir some 3km below.

The Ekofisk subsidence depression as during 1985 was indicated to be about 6km long (north-south) and 4km wide (east-west).

The subsidence at Ekofisk is of special importance in view of platforms and other structures being located on the seabed at a depth of 70m. Submerging part of the structure further into the sea decreases the margin of safe elevation from wave and sea level action. Tilting of large seabed mounted structures can be highly significant in view of the overall height of such structures. Wiborg and Jewhurst refer to the need to make changes on existing platforms when the subsidence reaches 4m. The Ekofisk tank is located in the subsidence area; it has a storage capacity of one million bbl and is a concrete structure. This tank is planned to be modified so that it can accept up to 6m subsidence. They show results of subsidence predictions and essentially extrapolate the present subsidence data. Assuming no action is taken to control the Ekofisk subsidence, these authors show that for natural depletion of the reservoir subsidence is likely to level out to 6 to 7m beyond the year 2010. Conversely using 350 MMscfd injection into the reservoir should control subsidence immediately and ensure subsidence is kept to not more than about 4m by the year 2010 and thereafter level off to 4 to 5m. These predictions are based on a present subsidence rate of 0·45m/year.

Figure 204 shows a diagrammatic representation of the Ekofisk subsidence depression based on the interpretation of Maury, Sauzay and Fourmaintraux (1987). They draw attention to the problems of safety for seabed mounted structures in respect of wave height and tilting following subsidence. Additionally the seabed can experience changed foundation conditions owing to subsidence, especially regarding movements and behaviour under stress.

Surface subsidence behaviour above oil fields according to influence function prediction methods. Considering the general character of surface subsidence development above oil field operations, due account should be taken of the prediction methods used in conventional mining and in particular the influence function method described earlier in this book. The present authors have considered an oil field situation where the anticlinal structure of the reservoir results in the volume of the oil abstracted region having its maximum thickness in the centre and then tapers to zero at its boundaries. Under these conditions it has been assumed that the oil abstracted has created fissure spaces which can be regarded as equivalent to a mined-out zone and can thus be treated as a mining subsidence calculation.

Figure 205 shows the basic approach adopted for applying subsidence engineering principles to the oil field situation. The fact that there is tapering towards the edges means that subsidence trough development will be more confined than with conventional mining situations. Additionally there is likely to be a large spread of subsidence which is of minor to no significance in view of the edge effect. Under these oil field geometry conditions, the reservoir width to depth ratio will be of much less significance than with conventionally mined-out extractions. The limits of subsidence are likely to be very similar to conventional mining for similar cover rock conditions. The nature of the overlying rocks will be of major significance; strong and competent cover rocks will encourage bridging across oil abstracted areas. The depth of the reservoir coupled with effective width are of course singularly important in reducing subsidence effects at the surface.

Figure 206 presents subsidence predictions using influence function methods for an oil field situation.

Surface subsidence resulting from underground coal gasification

Underground coal gasification (UCG) involves burning coal underground by means of a selected mixture of gases which are injected into the coal seams through boreholes. The product gases of the burning process are extracted through a separate borehole. The underground burning process gradually creates a cavity which is commonly approximately circular within a seam of limited height. Thick coal seams allow roughly spherical or ovaloid shaped cavities to frequently form.

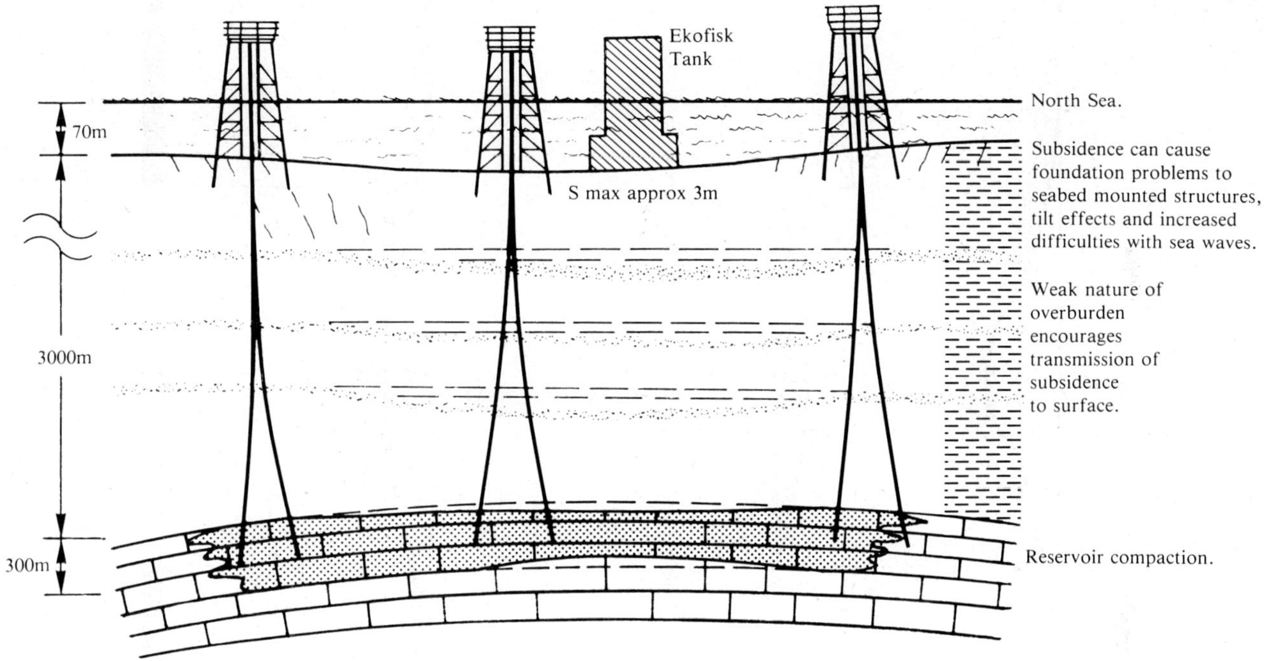

Figure 204 Subsidence of a North Sea oil field. After Maury, Sauzay and Fourmaintraux (1987).
(Diagrammatic representation only and not to scale)

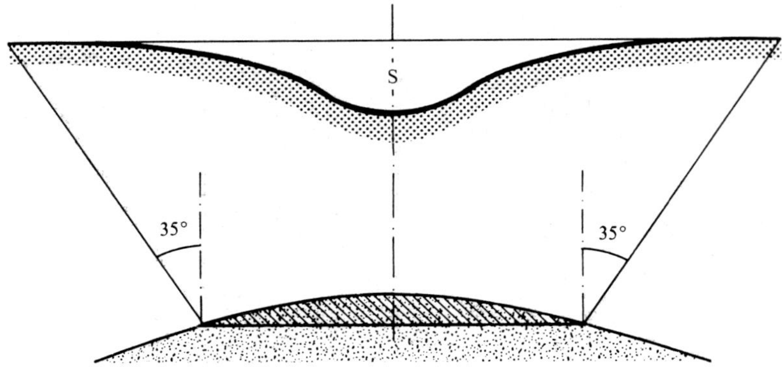

(a) Typical character of oil field subsidence profile

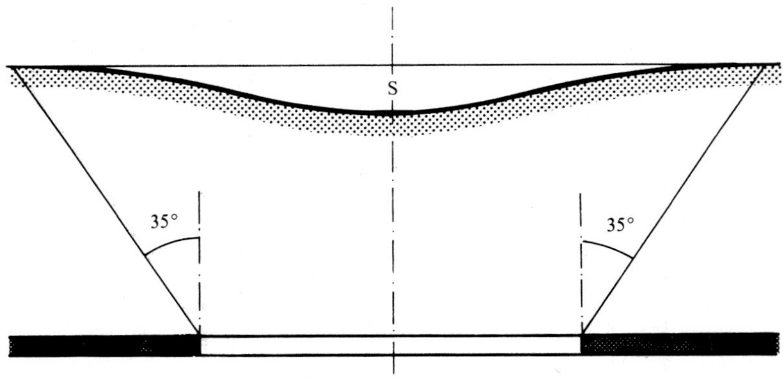

(b) Typical longwall extraction subsidence profile

Figure 205 Comparison of surface subsidence profiles for oil field and longwall mining situations.

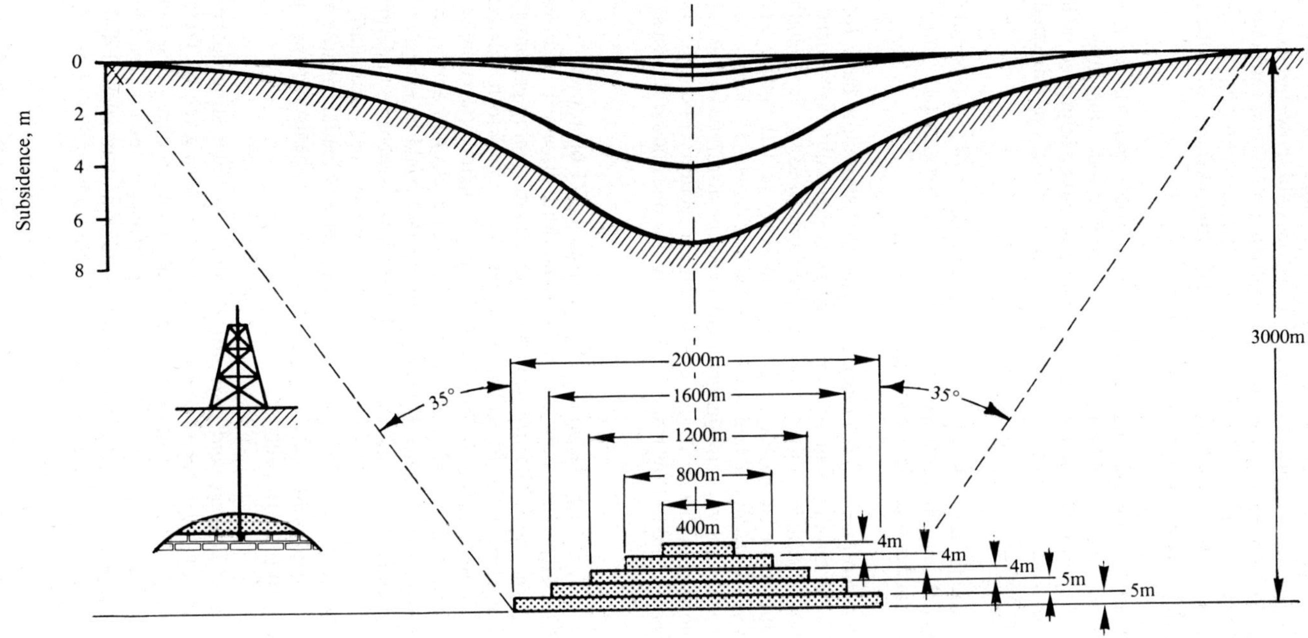

Note: Oil withdrawal assumed to create voids which can be regarded as equivalent to a mined-out area for the purposes of subsidence calculation.

Figure 206 Application of influence function method to calculating surface subsidence above oil field represented by several strips of decreasing length.

The UCG cavity experiences stress changes due to induced stress redistribution as the cavity increases in size and stress effects due to the thermal response of rocks surrounding the cavity. Experiments carried out by the authors on British Coal Measures rocks at the temperatures encountered in UCG cavities indicated that sandstones did not experience any significant changes in compressive strength. Shales and mudstones did, however, experience appreciable change especially in breaking down to form multiple thin layers. Carbonaceous material within such rocks readily promoted breakdown of the host rock during the burning process. Clay-based materials exhibited a baked characteristic after burning, with increased hardness properties and improved resistance to breakdown by water.

Experimental observations suggest that a coal seam roof comprising of thin carbonaceous shale/mudstone layers would experience detachment of individual layers owing to thermal effects. Their ability to bulk under such conditions would be improved owing to the increased occurrence of voids within the collapsed roof rocks. Rocks in this condition have improved drainage properties.

The collapse of UCG cavities can result in possible contamination of aquifers which are within proximity of such effects. However, these cavities can possess surface subsidence potential if the conditions favour their development.

Mode of roof collapse and subsidence development. Figure 207 shows the basic form of roof collapse and subsequent development of subsidence as is generally experienced with UCG cavities. The collapse potential of the roof is of course influenced by the strength and general competence of the immediate rocks overlying the seam and the nature of the overburden to the surface. Well-jointed and thinly-layered rocks will encourage upward collapse of the rocks above the cavity. Consequently aquifers can be disturbed by this form of collapse process. Sink-holes can occur if the depth and extraction conditions favour development. Weak clays may tend to flow towards the collapsed zone and result in a surface depression.

The nature of surface subsidence features resulting from UCG operations will have similarities with those associated with room and pillar operations discussed in an earlier chapter of this book. However, the cavities formed do not have significant inter-connection, and consequently have limited potential for collapsed material to flow into lower unfilled cavities as is the case with room and pillar layouts. The potential height of the collapse zone will be mainly governed by the bulking characteristics of the roof rocks and on the basis of earlier discussions on this aspect in Chapter 8, a height of 3-5 (M) could be expected for a single cavity with limited connections to adjacent boreholes. Where extensive underground cavity development has occurred in association with UCG operations and particularly with wide connections to neighbouring boreholes in the same seam, then a potential collapse height of 6-10 (M) may occur. The symbol M in this context would refer to the average extraction height of the cavity as formed prior to collapse.

Subsidence studies relating to underground coal gasification. An interesting account of subsidence modelling applied to underground coal gasification has been given by Trent and Langland (1983) who reported on finite-element and finite-difference studies. These authors suggested thermal effects to be important and that the finite-difference method appeared to allow more scope for studying such effects. Sutherland, Schuler and Benzley (1983) have reported experimental work using centrifuge simulations. These studies allowed progressive failure of the strata to be demonstrated.

Jegbefume and Thompson (1983) have examined the roles played by temperature and non-elastic behaviour on roof collapse and resulting subsidence development arising from

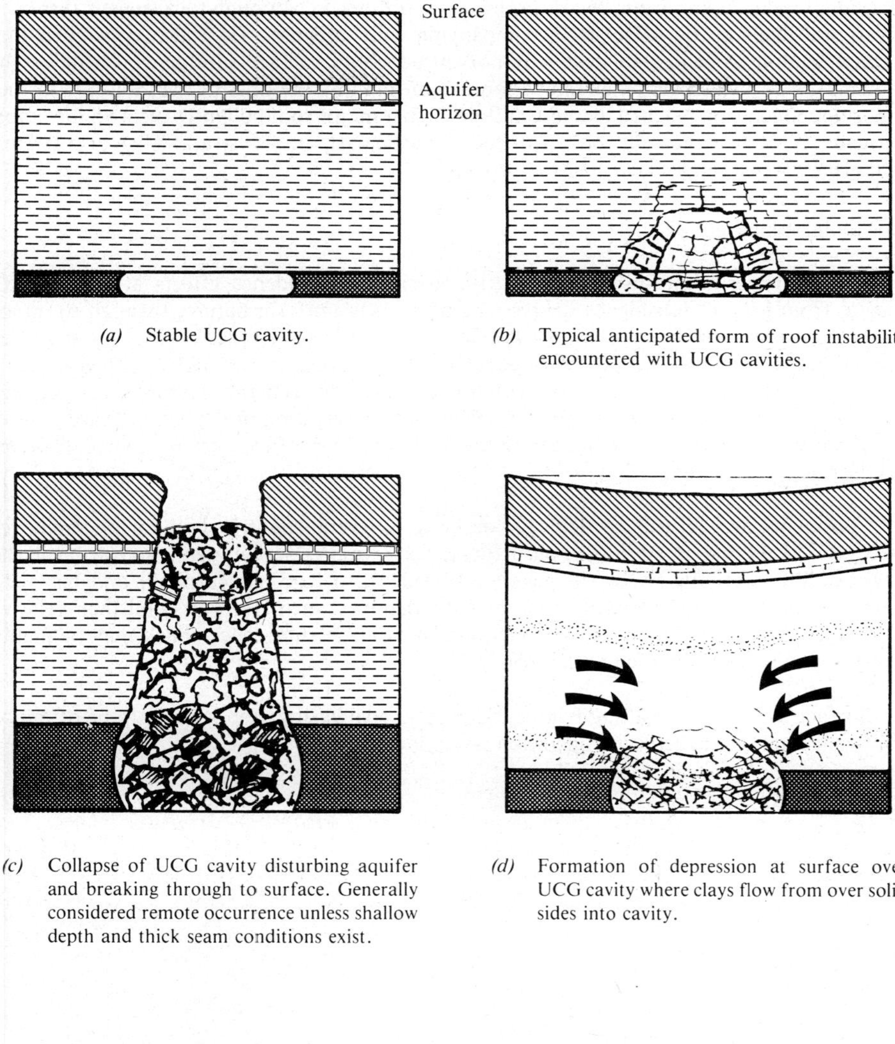

(a) Stable UCG cavity.

(b) Typical anticipated form of roof instability encountered with UCG cavities.

(c) Collapse of UCG cavity disturbing aquifer and breaking through to surface. Generally considered remote occurrence unless shallow depth and thick seam conditions exist.

(d) Formation of depression at surface over UCG cavity where clays flow from over solid sides into cavity.

Figure 207 Underground coal gasification (UCG) cavities: roof collapse and subsidence features.

underground coal gasification. The influence of drying of the surrounding rocks was considered to be an important factor in reducing subsidence, although they suggest that such effects are likely to be offset by accompanying roof collapse. They also concluded that thermal loading appeared to have a minor influence on surface subsidence owing to the limited extent to which very high temperatures transmit into the sides of the cavity. A further important conclusion reached by these authors is that major roof collapses seem an inevitable consequence of underground coal gasification, particularly in soft strata. They also suggest such effects would appear early in the process.

Concluding remarks

Ground-water withdrawal can give rise to special subsidence effects at the surface, ranging from general subsidence depressions of a fairly uniform nature, through to major fissure occurrence and even sink-hole development. The near-surface rock types and their general geomorphological and hydrogeological conditions substantially influence their behaviour following ground-water withdrawal and the general character of surface subsidence developments. Thorough assessment and investigation of the near-surface ground conditions can greatly assist in indicating the likely response of the surface to ground-water withdrawal.

Subsidence over oil and gas fields has occurred in several countries both on land and over water. The likelihood of subsidence occurring depends upon reservoir depth and size, and the nature of the overlying rocks. In general, subsidence occurrence from oil and gas field operations is not very significant in view of the depth of such reservoirs and their relative thicknesses. Undersea oil and gas production can result in subsidence which may be highly significant to seabed mounted structures.

Surface subsidence resulting from underground coal gasification is likely to be of minor significance unless the depth of operation is shallow.

CHAPTER 16

SUBSIDENCE ASPECTS OF ABANDONED MINE WORKINGS

Abandoned mine workings exist in many countries and they are not necessarily recorded and in some cases unknown. The most commonly encountered abandoned mine workings are those involved with the underground extraction of coal seams. For subsidence to be a problem above old mine workings there must of course be the potential for subsidence to occur. Those workings which have allowed extensively mined areas to remain partially or fully open, as with room and pillar operations and similar partial extraction layouts, generally possess significant potential for later roof and/or pillar failures to result in subsidence. Consequently, subsidence associated with abandoned mine workings relates to the occurrence of surface ground instability problems generally outside of the life of the operational mine.

The presence of old mine workings can pose problems to future developments. Uncollapsed ancient mine galleries and rooms may result in becoming unstable by disturbances created during foundation preparation work for new structures or even by later surcharging effects from later building/construction work at the surface. Additionally, ancient mine workings close to the surface which may have collapsed most probably represent areas with localised pockets of loosely compacted ground. As evidenced during open-cast mining operations these pockets of loose ground readily fall out when lateral support is removed. Of major significance, however, is the presence of old mine shafts some of which may be unknown and others whose accurate location is uncertain. Additionally, the abandoned condition of the shaft may not have been recorded. Abandoned shafts of ancient origin have given rise to sudden collapses without any warning. There is a long history of such incidences in the older mining areas of developed countries.

Abandoned mine workings pose a measure of uncertainty in many situations, especially regarding their potential to cause surface instability of consequence to present and future developments.

The magnitude of old mine workings in the UK has been highlighted by Littlejohn (1979) who reports that more than 70 000 have been recorded with not less than another 30 000 being unrecorded, and these workings occurred mainly during the three centuries up to the end of the 1800's. He also draws attention to some 80 000 old shafts remaining in the UK, and importantly some of these remain unchartered.

Ancient iron-mine working in the UK

Subsidence features have been reported at a third century mine in England. The Lydney mine, at Lydney Park, Gloucestershire, is reputed to be the first British iron-mine and was worked during the Roman occupation. It has similarities with the Roman iron-mines of Spain. This site was excavated and reported upon by Wheeler and Wheeler (1932). The site reveals evidence of ancient mining shafts. Hollows now exist of some 0·3m depth in the centre with diameters of 4 to 6m. All these old workings have largely choked with broken rock. An important feature, however, is that surface subsidence evidence of the existence of ancient mines dating back well in excess of 1500 years can still be identified.

General features of abandoned mine workings

Figure 208 illustrates some principal features of abandoned mine workings. The earliest form of coal mining operation applied on a wide scale in the UK appears to have been that of bell pits where a shallow shaft was sunk to a seam and coal worked out radially. The bell pit was gradually enlarged until the extraction was judged to be at its relatively safe limit for support by either natural or artificial means. Undoubtedly, the early miners learned from experience as to the approximate limits of bell pit size in view of the risk of roof collapse and also problems posed by water ingress and ventilation. Bell pits were frequently partly filled or simply collapsed after abandonment. The main problem presented by them today is the relatively loose nature of the infill or collapsed condition which signifies localised low bearing capacity. The depth of such bell pits has been given by up to 12m by Littlejohn (1979) with seam extraction diameters of 8-20m. Wardell and Wood (1966) report that they would not normally expect bell pit workings to be deeper than around 9m although they recognise that this would be influenced by the thickness and quality of the coal seam being worked.

The natural progression from the limited bell pit form of working was made possible by the invention of Newcomen's steam-driven pump in 1712. This allowed mining to extend deeper and further from the shaft as water could be coped with more effectively. In relatively water-free conditions, or situations where water could drain naturally from the mine as influenced by surface topography, room and pillar operations had already been established before Newcomen's invention. The role played by pillars in providing support to the roof was recognised during the early stages of large-scale underground mining in the UK. Such pillars were seen as providing support to an operational mine. Typical pillars were approximately square and of around 2m side length and on 5-6m centres. Some slight variations on these dimensions have been observed in England and Scotland by Littlejohn (1979).

Early forms of conventional room and pillar layout have been discussed in Chapter 9 and illustrated earlier in Figure 99.

Pillar effects. The forms of surface instability arising over room and pillar workings have also been discussed in Chapter 9, and it has been explained that both trough formation and sink-hole development can occur over such mine workings. Room and pillar mine workings of abandoned mines can still have open rooms some hundreds of years after abandonment, although with shallow coal mine workings this is very rare owing to the weak nature of the rocks and the effects of time and water on changing their strength properties further. Long-term pillar failures or failures of rock above pillars can result in the occurrence of subsidence up to some one hundred years or more after abandonment. Pillar configuration and dimensions are important factors as explained in Chapter 9. However, it is generally the upward progression of a collapsed mine junction that most commonly causes later subsidence effects at the surface. The nature of the immediate roof influences the likelihood of later subsidence. Where the roof beds are generally weak consisting of clays/shales then such rocks can exhibit flow behavioural properties with migration from over the pillars into the mine rooms and give rise to a widespread depression form of subsidence effect at the surface. The duration of such an effect would be influenced by the time-dependent response of the failed rock migrating into the mine rooms. The basic principle of this rock movement process in giving rise to long-term subsidence effects is illustrated in Figure 209 and has been observed by the authors in different countries in both coal and other mineral extraction operations employing room and pillar working. The shear strength of the roof rocks plays a major role in such subsidence developments. Additionally the ingress of water can affect the rate and magnitude of such subsidence effects.

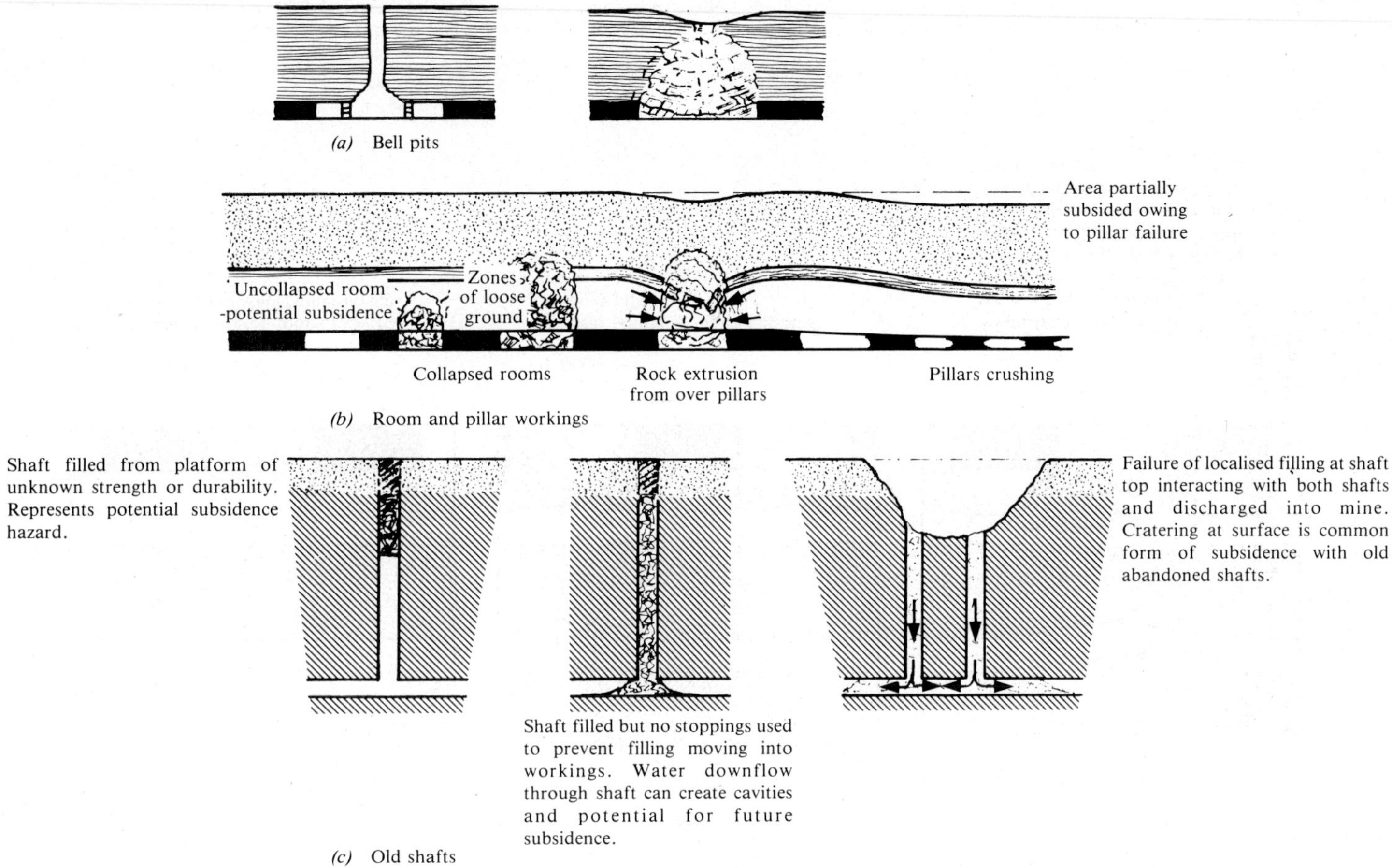

Figure 208 Some common subsidence features of abandoned mine workings and shafts.

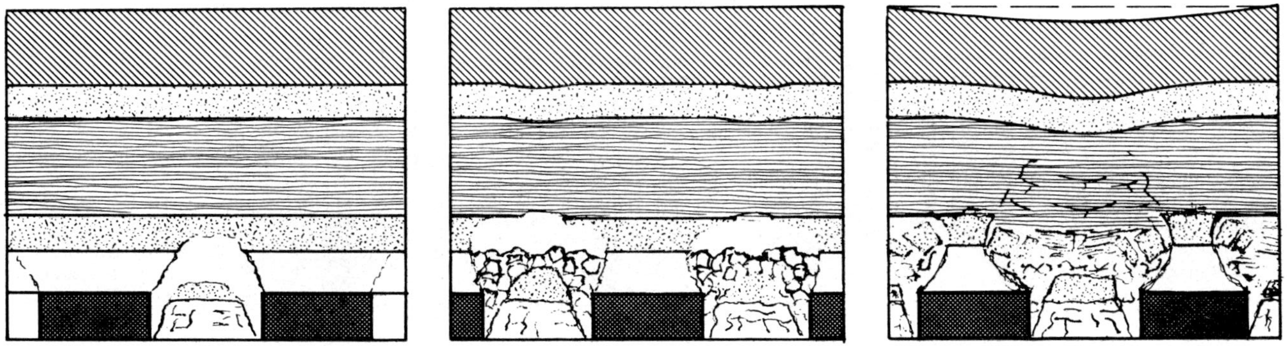

Stage 1

Initial failure of roof in mined-out rooms of stable pillar layout.

Stage 2

Shear failure of immediate strata above pillars.

Stage 3

Strata failure over pillars allows shear cones to form and tendency for broken rock to extrude/fall into mine rooms. Subsidence at surface is generally of fairly lengthy duration owing to time-dependent processes around the mining horizon.

Figure 209 Development of general surface subsidence depression over abandoned room and pillar mine workings.

Old shaft problems. The shafts of abandoned mines can present special long-term problems. The nature of the plug employed in shafts abandoned before detailed records were kept is frequently unknown. Some cappings have failed in time through deterioration in strength and allowed shafts of appreciable depth to suddenly become exposed. Conversely a shaft filled with material of the consistency of mud can flow into a mine if disturbed by say dewatering of the old mine workings from some later adjacent operations. Figure 208 shows some problems related to abandoned mine shafts. Old mine shafts can interact to cause a single crater as reported by Dean (1967), see Figure 210.

Character of old mine workings. An example of old mine workings which were holed during foundation excavation work in the North-Western Coalfield of the UK has been given by Wardell and Wood (1966). These authors considered the workings to be possibly much earlier than 1865. Consequently the re-discovery of these abandoned mine workings some 100 years later, enabled a detailed examination to be made of their condition and the findings are presented in Figure 211. The coal seam (Wigan Six Feet) had an immediate shale roof of 0·6 to 2m thickness above which was a coarse-grained sandstone (known as the Trencherbone Rock) and boulder clay at the surface. A detailed survey was made by the National Coal Board and revealed that the immediate roof shales had failed and falien down in virtually all of the old mine rooms and roadways. However, collapses of roof to the surface were only encountered at the western edge of the area examined where the depth of cover was 6 to 9m. The maximum working span across the mine rooms was around 3m. Wardell and Wood argue that the competent sandstone bed allowed a relatively large span of opening to depth below surface ratio to remain stable for over at least a century especially in the eastern part of the mine where this ratio was 1:3·5. These authors suggest that on the basis of the visual evidence there was no reason to suppose that the unsupported spans would not remain stable indefinitely. They also point out that there were no indications of pillar collapse or failure even though some very small pillars existed.

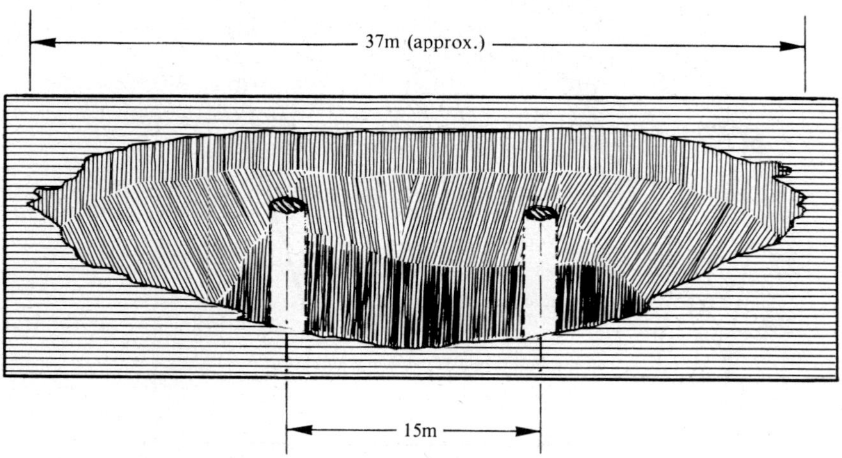

Figure 210 Crater caused by collapse in 1956 of two adjacent shafts, which had been filled previously, at Northwood, North Staffordshire, UK. Based on photographic material presented by Dean (1967).

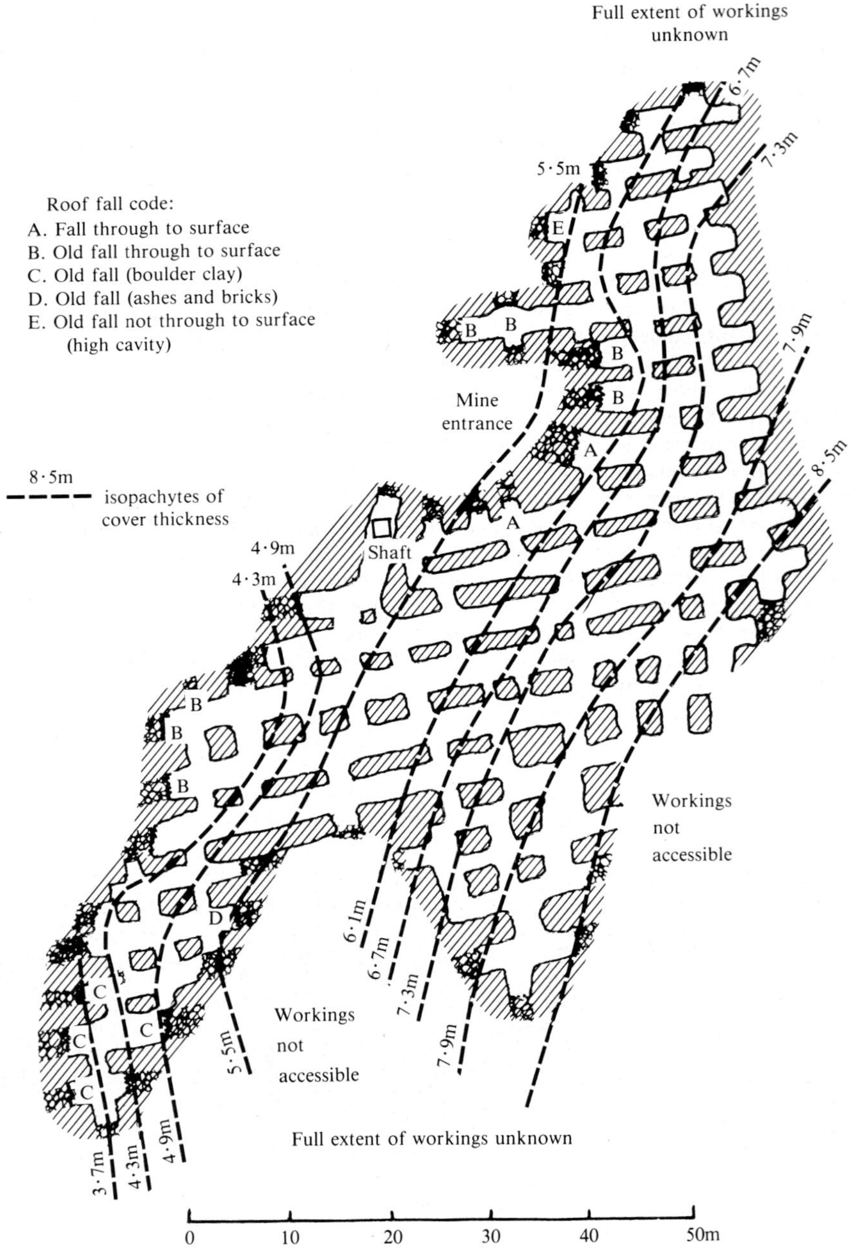

Figure 211 Example of condition of re-discovered old mine workings some 100 years or so after abandonment, after Wardell and Wood (1966).

An inference drawn by Wardell and Wood (1966) is that for shallow mine workings, the spans across rooms remain sensibly self supporting and rigid where they are less than 0·2 of the depth of cover. The inference is based on many years of field observations and applies essentially to normal Coal Measures strata.

Wardell and Wood point out that room widths in the form of room and pillar operations which prevailed during the 1800's and earlier were generally in the range 2 to 4·5m.

A paper by Piggott and Eynon (1977) also discusses ground movements associated with the presence of shallow abandoned mine workings with particular reference to coal mining. These authors suggest that on the basis of mining experience, the maximum height of collapse above a mine junction in room and pillar workings might reach 10 times the original roadway height although is more likely to be 3-5 times the roadway height. This assessment is in keeping with the discussion presented earlier in this book on such aspects (see page 196) where 4-10 (M) is indicated to be the caving height in these mining conditions.

Identification and investigation

ICE (1977) discusses general aspects of identification and investigation of potential problems arising from abandoned mine workings especially where such workings are unchartered and records may not exist. Attention is particularly drawn to seeking information from Geological Survey Maps, dating from 1857 (the first edition) with later editions from 1929. Mineral valuation offices possess records relating to land stability. British Coal (formerly the National Coal Board) has carried out collations of old records of mine workings, and these are lodged in regional offices and are generally available for consultation.

The *Catalogue of Abandoned Mines* covers minerals other than coal and oil shales and can be consulted at the Mining Records Office of the Health and Safety Executive, London.

ICE (1977) classifies methods of investigation under the following headings:

(a) visual inspection;
(b) aerial photography;
(c) drilling techniques;
(d) geophysical methods.

Visual inspection is particularly relevant where shallow workings exist as hollows, mounds and undulations relating to previous mining operations can be frequently indicated. Aerial photography is a natural extension of visual inspections, as depression and undulation patterns can be detected. Drilling is a positive means of undertaking detailed site investigation work both for identification of the strata and endeavouring to locate abandoned underground workings. Drilling into a cavity or zone of broken ground associated with underground mined-out rooms or other excavations can usually be detected by changes in the character of the drill's performance. Geophysical methods allow the opportunity for strata variations and geological disturbances such as faulting to be investigated. Additionally coal seams can be identified by geophysical probes lowered down boreholes.

Bell (1975) gives a comprehensive discussion on site investigation aspects with particular reference to mining subsidence areas.

Old mine shafts and adits: hazards and treatment

Old mine shafts can pose significant hazards and Piggott and Eynon (1977) draw particular attention to such features in that they regard surface collapses from the presence of the mine entries, ie shafts and adits as being the most severe forms of surface instability associated with abandoned mine workings.

Dean (1967) has discussed the problems of old mine shafts, pointing out the wide variations in depth and diameter, and in the nature of the filled condition whether in whole or in part. Many shafts are no longer easily detected owing to merging with the surrounding land. He remarks that it was not until 1872 that legislation was introduced enforcing compulsory keeping of working plans which also record the positions of shafts. The magnitude of the mine shaft problem is probably underlined by the fact that for some 200 years prior to 1872, shafts were being sunk and later abandoned without any records being kept showing their location or indicating depth or nature of the filling operations employed. Dean states that even after 1872 the records were rarely complete during the early days after the introduction of this legislation.

Dean (1967) presents a catalogue of old mine shaft collapses which attracted public attention in the Western Coalfield, UK. All the incidences indicate sudden manifestation, one of which occurred during April 1945 at Abram, Wigan when an old filled shaft suddenly collapsed and cratered in a railway sidings which immediately led to a whole train of 13 laden wagons and locomotive being engulfed and disappeared down the shaft. A further example of an old mine shaft collapse was linked to failure of the old shaft lining after being re-exposed following later construction work to re-utilise part of the shaft. Dean argues that this demonstrates the unreliability of old shaft linings on being re-exposed. A number of the incidents within this catalogue refer to collapses of filling in old shafts, and this seems to be the most common cause of failure. One old shaft at Hindley, near Wigan, and over 100 years old was covered by a brick cupola which suddenly collapsed and cratered. Dean explains that the water-table had been recently lowered and was the probable cause of the collapse of the shaft lining. He expresses the view that many other occurrences might be related to removing hydrostatic support previously provided by water-logged strata.

Causes of collapse. In the section on causes of collapse of old shafts Dean identifies haphazard filling of shafts as being of major significance and especially the lack of sufficient regard to future safety. One old practice involved erecting a scaffold, frequently of wood, close to the shaft-top and thereafter filling with loose materials. Dean refers to several such shafts where the scaffold was only around 2m below the surface, and remarks on his surprise that such wooden scaffolds last so long before collapse.

The most common cause of collapse of old shafts is attributed to filling with unsuitable material. Shafts are frequently connected to water-logged workings. Dean infers that during the filling of old shafts, erection of stoppings at the shaft-bottom or shaft insets probably was not carried out, or that any barricades used were too weak to resist the subsequent pressures exerted by the fill. As the fill material became saturated during flooding of the mine workings, the fill probably became transposed as behaving as a column of dense fluid with the ability to flow into the old workings should the barriers be non-existent or too weak to resist flow. Where a platform has been erected across an abandoned shaft with fill above it, subsequent flooding can increase the vulnerability of the platform to collapse and result in the entire column of fill falling to the shaft-bottom. The result is for the upper part of the shaft to suddenly daylight and quite frequently a crater forms as the sides collapse into the shaft void.

Dean considers two conditions relating to shaft filling:

1. where shafts are about to be abandoned and which are still accessible; and
2. old shafts which are inaccessible owing to hazardous conditions, namely water or gas presence or the shaft walling being in an unstable condition.

With the first situation he draws attention to the necessity to securely barricade all insets to the mine workings, in addition to securing scaffoldings and landing platforms before withdrawal. Solidification of the fill used results in some settlement and Dean advises on ensuring that either fencing off be employed or a reinforced concrete cap be placed over with an adequate overhang. An example of a raft used to cover an abandoned shaft is given in Figure 212.

Guidance on disused shafts

A detailed and comprehensive document has been published in connection with disused mine shafts and adits, NCB (1982). This handbook discusses sources of information about disused shafts, accuracy of information and records, geological and mine gases aspects, fill properties and behaviour, and clay seals. The design of shaft treatments is dealt with in detail, which is followed by a lengthy account on implementation of shaft treatments.

Procedures for locating abandoned mine shafts have been reviewed, DoE (1976). Attention is focussed on various methods available for locating such shafts ranging from examination of documents and maps through to field proving. Useful case histories are included which demonstrate quite clearly how various investigations and interpretations can be applied to field conditions. This publication is a valuable information source document on several aspects pertaining to the location of abandoned mine shafts.

Wardell and Wood (1966) comment that practically all areas where old shallow mine workings exist also have the problem of abandoned mine shafts. The most successful method for locating shafts no longer giving any visible signs at the surface, is considered by them to be that of stripping off the top-soil and vegetation in order to allow detailed inspections to be made of the freshly exposed surface especially for patches of filling material. They also recommend that potential old mine shaft sites should be drilled in order to prove the quality and extent of the filling employed. Even where an old abandoned shaft has been proved to be filled and the fill well-consolidated, these authors still consider it prudent to employ an amply large reinforced concrete capping before building over or near to it, and to take any residual settlement effects of the fill into account. They point out that drilling in the proximity of an abandoned shaft calls for special precautions as the drilling operation itself may trigger collapse if the shaft is not properly filled and that the shaft is in a weakened condition.

Ancient mine workings: further remarks on effects and treatment

In another publication Wardell (1966) examines the problems posed by past and future mineral workings in a development area. He discusses problems which arise from ancient shallow mine workings. He remarks upon the customarily used room and pillar method of extracting seams of mineral employed in such mines which included coal, clay, ironstone and limestone. Where these ancient mine workings are less than around 46m below the surface, Wardell comments that there can be a continuing risk of surface instability from underground collapse. Although the likelihood of such collapses occurring is influenced by mining depth and extraction height and width, sizes of mine pillars in addition to the strength properties of the surrounding rocks, Wardell points out that these collapses are not predictable in time or location. Piggott and Eynon (1977) concur with this conclusion, and add that unless the geometry of underground mine workings is fully established in relation to a proposed development, then some means of arresting or stabilising the process of collapse needs to be found and this they argue is usually done by filling the voids associated with the abandoned mine operations. The four types of filling operation described by them are:

1. hydraulic filling,
2. pneumatic or gravity stowing,

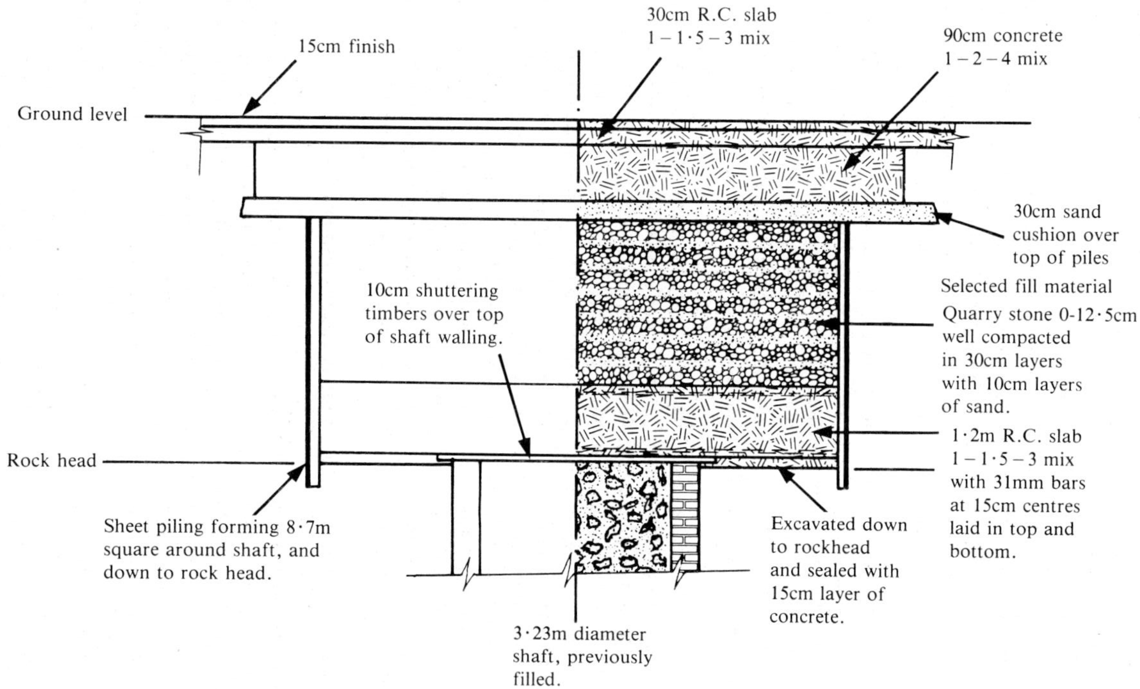

Figure 212 Details of rafting used to cover an old shaft located within a St. Helens industrial concern—completed on 5th August, 1965. After Dean (1967)

3. grouting, and
4. bulk filling with concrete or other materials, either from surface or underground sources.

These authors emphasise that when stowing underground voids, due account needs to be taken of the possible effects of any circulating groundwater and that consideration may need to be given to incorporation of appropriate barriers. They also point out that pressure grouting from the surface is the most commonly used method for filling underground voids of ancient mine workings.

Walton and Cobb (1984) have reported on void migration and overbreak features of old mine workings developing into surface subsidence manifestations. They comment on the general frequency of crown-holes exhibiting circular plan shapes whilst in section they develop with steep-sided cracks and conical depression characteristics. An important feature brought out by these authors is that the varying size of crown-holes commonly reflects bedrock variations and differing conditions of superficial deposits.

Ancient chalk workings at Bury St. Edmunds, UK, have given rise to sink-hole occurrences and significant surface instability to residential properties. The type of collapse is that of classical sink-hole forms as occur with room and pillar mine layouts.

Aspects of surface stability behaviour over old coal mine workings have been discussed by Price, Malkin and Knill (1969) who examined the problems of foundations of multi-storey blocks on the Coal Measures.

Effects of abandoned coal mines: US experiences

Subsidence above abandoned coal mines associated with the Pittsburgh Coal Bed in Pennsylvania has been discussed by Bruhn, Magnuson and Gray (1978). Their findings have been referred to in Chapter 9 (see page 199). It is added here that underground coal mining has taken place in the United States for at least 200 years, and vast areas exist where shallow abandoned coal mines with significant potential for subsidence are present. As the method of working which has most commonly prevailed in underground mines has been room and pillar operation, then subsidence occurrences are generally associated with the collapse of old workings and propagation of such effects to the surface in the form of sink-holes or widespread depressions. These authors report that subsurface conditions for abandoned mines are seldom known, and that time of occurrence, magnitude and extent of subsidence and accompanying strains cannot be predicted.

Some general conclusions drawn by Bruhn and co-authors are listed:

1. Unless total extraction has been carried out there is apparently no interval (depth) at which the surface above the abandoned mine is completely safe from subsidence.
2. Again, unless total extraction has been practised, the occurrence of subsidence may take place long after mining has ceased. Such subsidence may not necessarily be limited to single episodes.
3. The activities of man, in some cases, may initiate or hasten the onset of subsidence above abandoned coal mines.

Abandoned underground coal-mined lands and in particular surface subsidence related features are discussed by Johnson and Miller (1979). They remark that such abandoned lands due to underground mining offer scope for rehabilitation by way of filling operations. These authors argue that backfilling abandoned mine voids is the most suitable means of achieving surface stability. Abandoned underground coal mines can be susceptible to ignition particularly from spontaneous combustion. This is more commonly encountered along coal outcrops. Coal seam fires of this type can be a source of subsidence damage in addition to being hazardous to population centres, vegetation, livestock and wildlife etc.

Abandoned limestone workings: West Midlands, UK

DoE (1983) reports that in 1978 subsidence occurred due to old limestone workings of depth 150m at Wednesbury and caused significant damage to warehouses and industrial plant. It had been thought previously that underground collapses of such old workings at this depth would not be of significance to the surface, but this 1978 incident caused revision of this assumption and prompted courses of action to deal with the subsidence problem.

Appreciation of the subsidence problem. The magnitude of this old limestone mines problem is illustrated by the estimate that around 7 million cubic metres (19 million tons) of limestone were removed by quarrying and mining, mainly between 1750 and 1900. The concentration of the quarrying and mining of limestone in this part of the West Midlands was linked to the emergence of the Industrial Revolution. The limestone mines were abandoned without any remedial or filling works being carried out. However, the shafts were considered to have been filled with debris. Some of the mines became flooded after abandonment. The earliest known abandoned limestone mine dates back to 1828, whilst the last was in 1935. The mines consisted of gallery workings and essentially mined-out rooms of an irregular configuration with mineral support pillars of various sizes and shapes. A feature of the limestone rich strata is their high fabric strength but the joints between large blocks have occasional infilling with clay/mudstone which is subject to weathering and can lead to general weakening of the rock structure around the mine excavations. The report draws attention to rock failure above the pillars at depths of 100m and more, and this can lead to gradual weakening of the support pillars. Also at depths of less than 70m, to roof level, surface instability is likely to take the form of a crown hole rather than that of general subsidence. Efforts have been directed towards attempting to assess the potential for collapse by taking age of workings, character of the strata, depth below surface and mining layout into account. An important observation has been that up to 1983 some 100 collapses of limestone mines have resulted in surface disturbances of sufficient significance as to warrant identification since 1828; and these disturbances have continued to occur at a steady rate of around one every 1·5 years. Overburden strata type has been found to influence the likelihood of particular types of subsidence disturbance occurring at the surface. Over two-thirds of the 100 subsidence incidents have occurred where the depth below surface has been less than 30m. Of the subsidence events only one of general subsidence has been recorded above mines less than 70m deep. For mines with depths greater than 100m, all three events have been of the general subsidence depression category. This has influenced the general strategy for action that crown-holes are likely where the mining depth is less than 100m; in other situations general subsidence is anticipated.

The general approach to a strategy for action is put forward in the report as:

1. *Minimum action.* This entails preventing public access to accessible workings, and filling subsidence holes as they occur. In effect this translates to no action for workings deeper than 100m.
2. *Monitoring.* The aim of this action is to use instruments to detect movements of the strata generated by the presence of the mine so as to periodically assess the potential for collapse.
3. *Investigation.* This involves mine examination and that of the overlying strata to determine physical properties for the purpose of determining the potential for collapse.

4. *Mine treatment.* This action is considered desirable to preclude possible future surface disturbances where there is judged to be every likelihood of harmful damage arising to structures or services. Such action may involve collapsing, strengthening, excavating or mine filling operations.
5. *Treating structures and services.* This entails the action of treatment to allow structures and services to withstand general subsidence effects.

It is clear that the major programme of action being undertaken in the West Midlands limestone mines locality is leading to improved understanding of the overall picture of subsidence manifestations from these abandoned underground workings. This is leading to increased confidence in the general state of stability above such mines. The subsidence problem from these abandoned limestone mines is likely, however, to continue and consequently calls for continued efforts in identification of the relative risk categories of the surface and the desirability of various courses of action in implementation of monitoring schemes and/or remedial works.

O'Riordan, Cole and Henkel (1984) have discussed collapses of ground in abandoned limestone mines in the West Midlands and presented results to show that although such mining activities ceased between 50 and 150 years ago surface disturbances have continued to occur at a uniform rate since the time of abandonment. These authors confirm that surface instability generally takes the form of sink-hole developments with occasional occurrences of general subsidence depressions.

Coal Measures	Upper Coal Measures		Abandoned limestone mines
	Lower Coal Measures		
	Unconformity		
Silurian	Upper Ludlow Shales	max 12m	
	Aymestry Limestone	4·5m	None known
	Lower Ludlow Shales	approx 150m	
	Upper Wenlock Limestone	6-7m	Several
	Nodular Beds	30 to 36m	
	Lower Wenlock Limestone	9-10m	Several
	Wenlock Shales	approx 200m	
	Barr Limestone	10m	2 known

Table 27 Geological sequence relating to occurrence of abandoned limestone mines in the West Midlands, based on O'Riordan and co-authors (1984).

Geological and subsidence data. A summary of the relevant geological data relating to the abandoned limestone mines is presented in Table 27, whilst mining and subsidence data are given in Table 28.

Wednesbury abandoned limestone mine subsidence occurrence.

O'Riordan and co-authors (1984) have given an account of the occurrence of a significant subsidence depression at Wednesbury which happened during 1978 over a mine in the Upper Wenlock Limestone lying at a depth of around 140m below the surface. The mine workings consisted of rooms of 12-15m width and square pillars of 8-12m side length. Figure 213 presents a layout of the mine workings as recorded during an 1879 survey. The mined-out

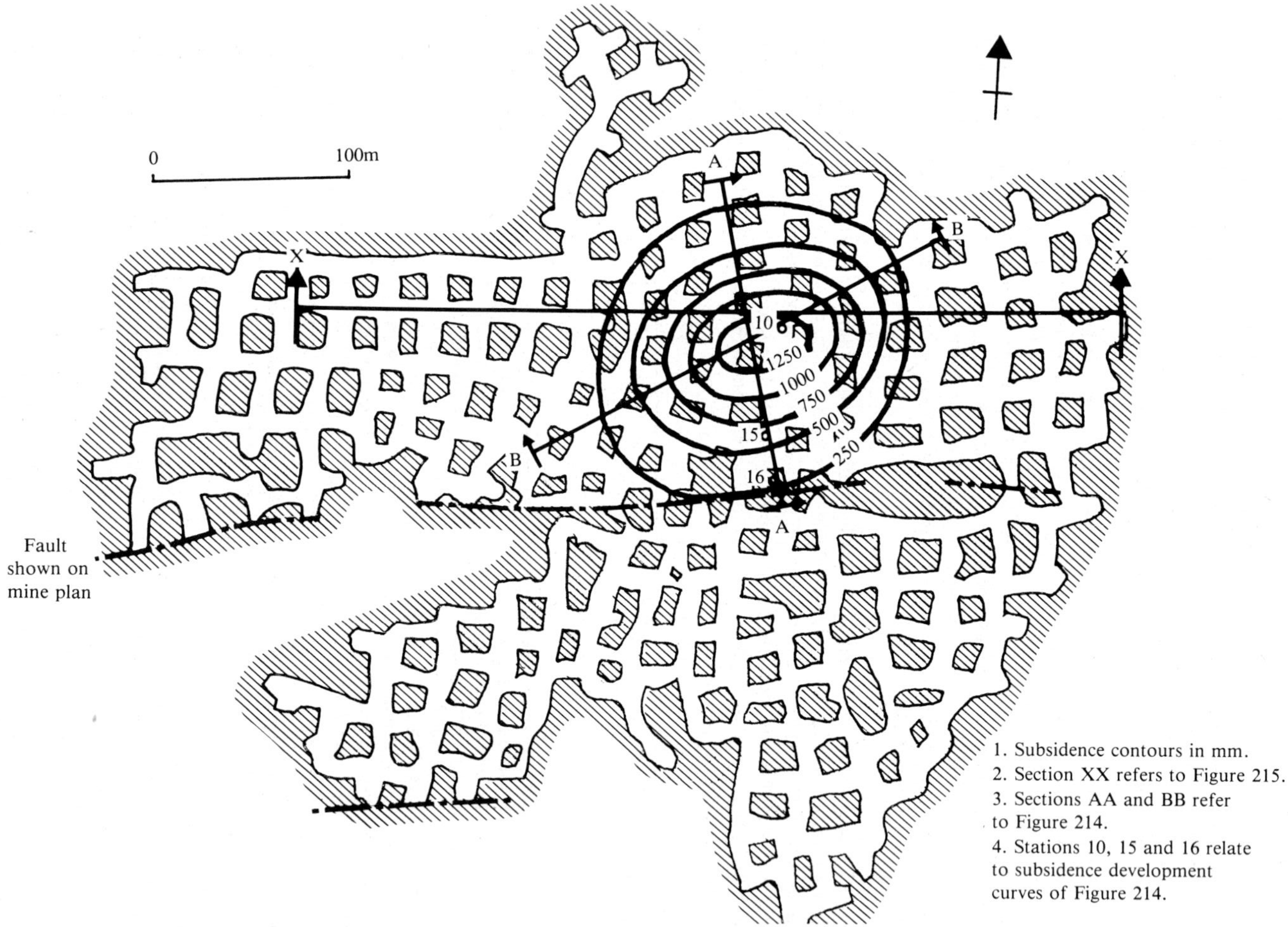

Figure 213 Surface subsidence contours and outline of mine workings, Cow Pasture mine, Wednesbury, UK, following mine workings collapse in 1978. After Braithwaite and Cole (1986).

Range of dimensions of rooms and pillars	generally 7-20m
Percentage extraction in pillared areas	70-90%
Mining extracted heights of rooms:	Upper Wenlock around 4m
	Lower Wenlock around 8m
Deepest recorded mine workings (Sandwell)	over 300m
Area affected by mining subsidence: relative to known mined area	about 5%
Percentage of abandoned mines (total 31) having given surface instability	58%
Limestone workings resulting in surface stability problems	Upper & Lower Wenlock
Period of most intense collapses	Within 30 years after abandonment

Table 28 General extraction and subsidence data relating to abandoned limestone mines, West Midlands, based on O'Riordan and co-authors (1984).

rooms are considered to have a height in the range 4·5-5m, and are flooded. Subsidence observations made during and following this mine collapse event are also included in Figure 213 and have been plotted in relation to the mine workings. Drill holes have indicated that the limestone workings still contain significant voids within the rooms with heights of up to about 4m in the main area of the subsidence depression. The maximum subsidence was of the order of 1·35m.

An important observation reported by O'Riordan, Cole and Henkel (1984) was the relatively highly stressed nature of pillars under the subsidence-affected area. These authors attribute the main cause of the collapsed ground associated with this abandoned limestone mine as being due to strength deterioration of the rocks forming the roof of the mine; the pillars supporting the roof were not regarded as having lost their strength or being directly responsible for the subsidence occurrence. The present authors would add here that there is a strong possibility of rock failure in the weaker strata above the pillars, and this probably encouraged the pillars to punch into the weaker roof beds; the subsidence occurrence would be prolonged in view of time-dependent processes being involved.

Figure 214 shows observed subsidence data plotted on a time basis and this clearly illustrates the time-dependent character of this subsidence occurrence.

Figure 215 presents a section of the mine showing ground conditions across the area of the collapse of 1978 according to Braithwaite and Cole (1986). These authors have contributed significantly to improved understanding of the subsidence mechanisms involved in these abandoned limestone mining areas of the West Midlands, particularly by way of thorough field investigations involving various remote sensing probes, detailed subsidence measurements, in addition to geophysical and drilling examinations. A natural extension would appear to be further understanding of the role played by mining configuration especially the size and distribution of pillars in relation to rooms, the role played by rock strengths and subsequent deterioration, followed by examination of the failure process at the mining horizon and assessment of its relevance to development of subsidence taking depth of mining into account. There would appear to be scope for increased appreciation of areas at greater risk in respect of subsidence so that phasing of remedial action can be correspondingly effectively executed.

Further discussion on subsidence aspects of abandoned limestone mines has been given by Brook and Cole (1984) who provide a valuable overview of the problems and clarify the objectives of the programme of investigation and related courses of action being taken.

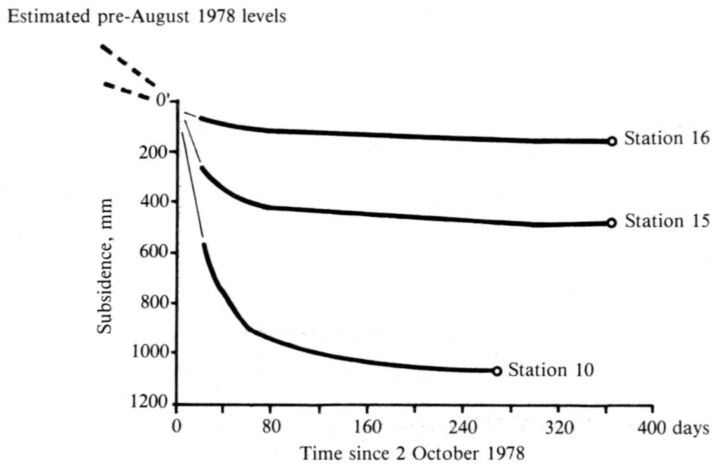

(a) Subsidence development with time for levelling stations nos 10, 15 and 16.

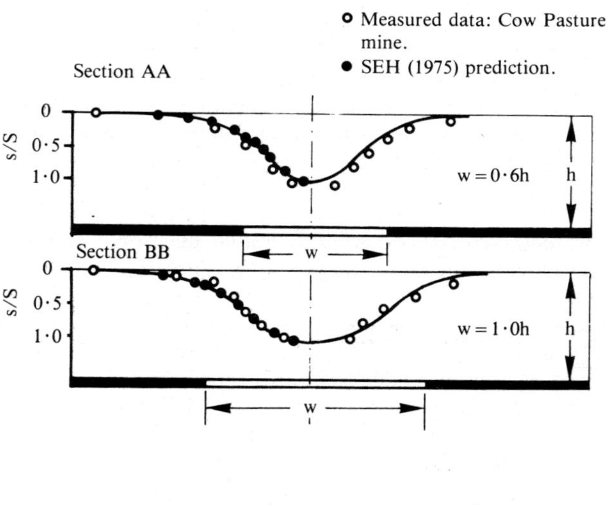

(b) Transverse subsidence profiles

Figure 214 Subsidence characteristics above Cow Pasture mine relating to the 1978 collapse of mine workings. After Braithwaite and Cole (1986).

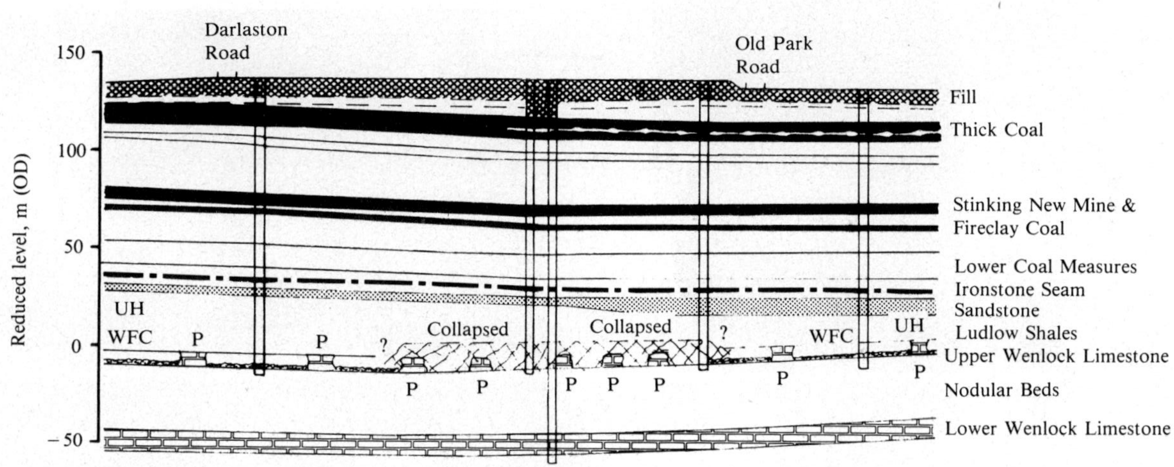

Symbols: UH = Unconformity horizon; P = Pillar; WFC = Water filled cavity

Figure 215 Vertical section XX showing ground conditions across area of 1978 collapse referred to in Figure 213. After Braithwaite and Cole (1986).

Figure 216 Subsidence features associated with mineral vein abandoned mines. Top: sink-holes along line of outcrop of vein. Bottom: subsidence detail of crown pillar failure, Carrock Fell, Cumbria, UK.

Mineral vein abandoned mines: subsidence occurrences

A large number of mineral vein abandoned mines exist in many countries. Such veins frequently pitch steeply and outcrop at the surface. It is common practice to leave a crown pillar at the surface outcrop in order to promote stability and prevent surface rocks collapsing into the working mine in addition to controlling water inflow into the mine. After abandonment of such mines, crown pillar failures can occur in time especially where the pillar dimensions were relatively small and the pillar material strength properties have deteriorated in time.

Figure 216 shows two examples of the nature of subsidence occurrences at such abandoned mines at Carrock Fell, Cumbria, UK. The upper photograph shows the type of sink-hole occurrences which develop along the line of the outcrop of the mineral vein. The lower photograph indicates the nature of crown pillar failure giving rise to the old mine workings suddenly daylighting. Several subsidence holes exist in this locality, and the majority appear to be quite stable, although joint weakening is apparent close to the line of outcrop and can result in localised wall failure. Surface soils readily slump into the mine workings as shown by Figure 216.

Abandoned open stope workings are particularly vulnerable to this form of subsidence feature, and can occur many years (100 years or more) afterwards. Such subsidence occurrences appear to be confined to shallow situations particularly near to the mineral vein surface outcrop.

Concluding remarks

Abandoned mine workings give rise to concern in view of the potential for later subsidence of the surface. In some cases there is uncertainty as to the existence of underground mine workings, or even as to the presence of significant voids within such workings. The magnitude, extent and timing of subsidence occurring above abandoned mine workings are all influenced by the nature and configuration of the workings in addition to depth and thickness of the extraction. The type of strata in close proximity to the mining horizon also makes a contribution to long-term behaviour of ground movement and the possibility of later subsidence of the surface. Those mine workings which leave appreciable underground voids as with room and pillar mines can represent potential subsidence after abandonment if the geotechnical and mining factors favour its development. Such potential subsidence can still exist some 100 years or more after abandonment. Longwall mining with total caving eliminates this form of long-term potential subsidence.

Current studies and investigations at abandoned mine sites are demonstrating the value of co-ordinated programmes of observation, evaluation and subsequent remedial actions aimed at controlling, and in some cases eliminating the long-term potential subsidence problem. This is particularly important for those areas planned for future development.

Old mine shafts still represent a special hazard of abandoned mine workings and they call for particular vigilance and effective courses of action.

CHAPTER 17

EFFECTS OF MINING SUBSIDENCE ON SURFACE STRUCTURES, DESIGN CONSIDERATIONS AND PRECAUTIONARY MEASURES

Introductory remarks

Orchard (1956-57) drew attention to the main causes of surface damage arising from mining subsidence as being subsidence, tilt, tensile strain and compressive strain. He remarked that the first of these concerned the loss of height in areas liable to flooding and consequently the only practicable precautionary measures consisted of:
(a) limiting the amount of extraction, and
(b) performing surface works such as raising the banks of rivers or canals, etc.

An important aspect identified by Orchard was the role which can be played by layout and phasing of underground workings especially in reducing the potential for formation of ponds or depressions liable to flooding. An example described by him is that of firstly extracting mineral from below the seawards end of a river or the outfall end of a sewer; such phasing of mine working thus ensures continuation of the flow. In such situations gradient of flow is the predominant feature, and the influence of mining subsidence introducing a risk to impairment of the efficiency of a water sewer or water course can be evaluated where the gradient is known.

Tilt is of special importance to tall structures such as chimneys, etc, and excessive leaning of the stack can become dangerous although this rarely occurs. Where excessive leaning has occurred it has generally arisen due to a number of seams being worked only on one side.

Shaft pillars of inadequate design can result in shafts losing their vertical alignment. A common problem, however, is that of the shaft lining and fittings becoming damaged owing to the upper part of the shaft subsiding whilst the lower part remains unaffected by adjacent mining extraction operations. The lining in the upper part of the shaft can experience damage due to vertical compression.

Mining data:	Durham case	Yorkshire case
Extraction height, M (m)	0·91	1·68
Maximum subsidence, S (% M)	6	8·7
Face widths, w (m)	74	74
Depth of seam, h (m)	556	648
Pillar widths, p (m)	74	111
Percentage extraction (%)	Less than 50	40
Lateral strains	No measurements taken	Hardly measurable
w/h ratio	0·133	0·114
Subsidence calculated for one panel according to partial subsidence relationship (mm)	37 (4% M)	50 (3% M)

Table 29 Longwall mining partial extraction data illustrating effect on reducing the amplitude of subsidence. After Orchard (1956-57).

Orchard suggests that where slight or moderate tilting is unaccompanied by other damage, it can be often tolerable or even unnoticeable. However, he recognises that with certain structures the provision for jacking is sometimes provided in view of the influence of small changes in tilt to certain industrial processes.

Reducing the amplitude of subsidence by resorting to partial extraction is regarded by Orchard as a major control measure, and Table 29 serves to demonstrate this point.

With reference to the data presented in Table 29 an important feature observed was that in addition to the magnitude of subsidence being quite small in relation to the extracted seam height, the subsidence basin was very flat and well spread out which had the effect of producing negligible lateral strains. Orchard argues that although the method is wasteful of coal, it is particularly effective in those areas where land lowering cannot exceed more than a few centimetres or insignificant damage is to be caused.

ICE (1977) discusses structures in relation to mining subsidence and provides a useful account of preventative measures, communications, land drainage and services aspects.

Response of surface structures to undermining

King, Whittaker and Shadbolt (1974) examine various aspects of the effects of mining subsidence on surface structures and comment that although surface behaviour can be determined, it is difficult to quantify in specific terms how surface structures will respond to the effects of undermining. Taking account of the substantial differences in the nature of various surface structures, these authors express the view that assessments can be made, with reservations on experience and understanding of particular structures of the likely reaction when affected by movement. They list the following factors to be taken into account when making and assessment of the likely behaviour of a structure subjected to mining subsidence effects:

1. The size, shape and orientation of the structure in relation to the underground mine workings.
2. The foundation design and type of superstructure.
3. The methods of construction and quality of materials used.
4. The age and standard of maintenance and repair.

The last factor is highlighted by an example in comparing an old, dilapidated building, whose state is suffering from the effects of poor maintenance and repair, and consequently will be more sensitive to, and behave more violently than a similar building which is in a good state of maintenance subsequent to earlier movements.

Classification of subsidence damage

In their paper King, Whittaker and Shadbolt (1974) note that many years of experience in studying case histories involving damage to surface structures due to mining subsidence in the UK has enabled a classification to be established. This allows assessments to be made of likely mining effects of underground workings on a particular structure. It is clear that individual elements of ground movement have different effects and result in differing degrees of importance depending upon the type of structure being considered, as for example vertical subsidence is a major aspect in low lying areas prone to flooding and where drainage is a problem. Consequently, a few centimetres of subsidence can incur heavy subsidence damage costs. Ground tilt is of special significance to drainage schemes and sewerage networks;

highways and rail tracks are sensitive to tilt effects resulting from mining subsidence; tall structures including water towers are also sensitive to tilt; some types of industrial plant cannot tolerate any appreciable dislevelment.

Buildings and conventional structures mainly experience damage through mining subsidence as a result of differential horizontal movements and flexing of the ground around the subsidence depression. Consequently, states of compression/extension are produced within surface structures affected by subsidence. King, Whittaker and Shadbolt (1974) suggest that discernible mining subsidence damage appears in conventional structures when the level of effective strain exceeds 0·5-1·0mm/m.

Extension and compression

Damage arising from ground compression or extension is similar according to Orchard (1956-57) and lists the principal factors as follows:

1. The intensity of the strain in the ground.
2. The dimensions of the affected structure.
3. The coefficients of friction between the foundations of the structure and the ground.
4. The susceptibility of the superstructure to damage.

Orchard remarked that the first factor can be calculated in advance to within reasonable limits of accuracy whilst the second factor is immediately ascertainable. On the third factor he comments that in a great many cases the coefficient of friction is 1·0, and means that the same sub-soil changes in length are also suffered by the building. This is the case with traditional buildings where the foundations are more or less located in deep trenches within the sub-soil and correspondingly firmly anchored thereto. Orchard contrasts this foundation practice with the technique of placing the building on a lightly reinforced concrete raft which itself rests on a layer of granular material such as sand, thereby reducing the coefficient of friction to possibly less than 0·5.

With regard to the fourth factor concerning the susceptibility of the superstructure to damage, Orchard suggests that this is almost infinitely variable. He draws attention to the fact that strain is not always transmitted to the superstructure; there is sometimes sliding along a damp-proof course which he suggests can be a useful but generally accidental layer along which sliding can take place. Further, the more flexible types of building material can accommodate a certain amount of extension and compression; the effects can be almost unnoticeable if spread out over numerous voids such as doors and windows. Conversely, some buildings, such as long continuous walls and especially if plastered, can show the effects of slight movement.

Orchard argues that extension and compression effects resulting from mining subsidence can be taken into account by way of introducing certain modifications to the design of new structures and these aim to reduce or to absorb strain effects. The most obvious approach is suggested as keeping the unit small. However, flexibility within the structural design is considered to be also a very important factor.

Relating ground movement to surface damage

SEH (1965,1975) gives a summary of UK coalfields subsidence observations with special comments on the relationship of ground movements to surface damage. The following discussion is based essentially on these two publications.

Figure 217 Switchback effect on a motorway caused by settlement of fill material over a previous opencast coal site.

Damage due to subsidence

It is the vertical component of ground movement that is responsible for changes in gradient in the ground. This can adversely affect such features as surface drainage, tall structures and factory plant and machinery. Attention is drawn to the difference in tilt response of a small building by comparison to a sewer or drainage pipeline. Whilst the small building may undergo tilting by the same amount, the sewer or pipeline may be severely tilted locally. Consequently, the overall effect of tilting needs to be taken into account since localised tilts may frequently be tolerable without undue influence on the overall functioning of the service. Sewer lengths are generally considered between manholes, and calculations are performed to determine gradients from the invert levels at the manholes.

Of importance is the fact that any adverse condition may be only of a temporary nature and as mining progresses the original gradient may be eventually restored. Backfalls over distances may be tolerable temporarily; deep manholes can act in assisting to even out the overall gradient.

Open water courses may require consideration in respect of raising the height of banks. Also, overflowing of filter beds at sewage disposal works, and of water reservoirs, etc, may need consideration regarding the amount of free-board tolerable for effective operation.

Subsidence troughs in addition to giving rise to potential ponding and flooding, can cause a switchback effect which cannot be tolerated on railways or motorways. Abrupt troughs can arise with incomplete extraction of shallow workings, especially where coal pillars are left between longwall extractions.

Figure 217 shows a switchback effect on a motorway caused by settlement of fill material over a previous opencast coal site.

Where the tilting of factory machines and plant results in such equipment being inoperative and cannot be rectified by built-in methods, then this can be classed as damage. Such machinery commonly have screw feet or other re-levelling facilities, in order to allow adjustments to be made to maintaining the equipment in a level condition.

Careful consideration needs to be given to the effects of undermining tall buildings and chimney stacks. Tilt changes in tall buildings can affect the working of lifts.

Damage due to horizontal strains

Subsidence damage most commonly arises as a result of horizontal extension and compression. The pulling open of joints in brickwork or fracture occurrence in masonry are common characteristics of tension, whilst squeezing-in of such voids as doors and windows and the occurrence of horizontal movement along well-defined lines of thrust are characteristics of compression.

The nature of the structure (construction materials, shape, age and design) greatly influences the degree of damage and only approximate rules can be applied for the purpose of predicting the severity of damage. The intensity of strain and size of structure to which the strain is applied are regarded as the main factors.

Experience gained over many years and from photographic records of various types of damage arising from mining subsidence has enabled a damage scale classification to be devised based on observations in UK coalfields. It was devised by King and Orchard in the early 1960's and is used by the National Coal Board for classification of subsidence damage and is presented here as Table 30. Five classes of damage are considered, and the factors of strain and building length employed in compiling the classification are reported as giving only general guidance with respect to predicting damage intensity.

Figure 218 shows the relationship of damage to length of structure and horizontal ground strain with special reference to the National Coal Board classification of subsidence damage and should be used in conjunction with Table 30.

A study carried out by the National Coal Board over a 10-year period in the Midlands coalfields of the UK examined the various categories of subsidence damage to dwelling houses within the region of 53,824 hectares affected by mining subsidence. During this period of time some 455 million tonnes were mined and there were approximately 200,000 dwelling houses within the area influenced by mining, of which the numbers damaged corresponding to the categories given in Table 30, were found to be as follows:

1.	Very slight or negligible	20%
2.	Slight	10%
3.	Appreciable	1%
4.	Severe or very severe	0·125%

The number of dwelling houses which suffered damage was 62,250 or about 31%. These results indicate the spread of subsidence damage categories and show that the 'severe' or 'very severe' categories are comparatively rare, NCB (1982), Shadbolt (1972).

Change of Length of Structure	Class of Damage	Description of Typical Damage
Up to 0·03m	1. Very Slight or Negligible	Hair cracks in plaster. Perhaps isolated slight fracture in the building, not visible on outside.
0·03m - 0·06m	2. Slight	Several slight fractures showing inside the building. Doors and windows may stick slightly. Repairs to decoration probably necessary.
0·06m - 0·12m	3. Appreciable	Slight fracture showing on outside of building (or one main fracture). Doors and windows sticking; service pipes may fracture.
0·12m - 0·18m	4. Severe	Service pipes disrupted. Open fractures requiring rebonding and allowing weather into the structure. Window and door frames distorted; floors sloping noticeably; walls leaning or bulging noticeably. Some loss of bearing in beams. If compressive damage, overlapping of roof joints and lifting of brickwork with open horizontal fractures.
More than 0·18m	5. Very Severe	As above, but worse, and requiring partial or complete rebuilding. Roof and floor beams lose bearing and need shoring up. Windows broken with distortion. Severe slopes on floors. If compressive damage, severe buckling and bulging of the roof and walls.

Example: A structure say 25 metres in length (which would include most normal residential structures) if subjected to a strain of 1 mm/m would only suffer Very Slight damage. The table describes this class of damage as:

> 'Hair cracks in plaster. Perhaps isolated slight fracture in the building, not visible on the outside'.

The same building would have to receive a strain three times as large as this, i.e. 3 mm/m before it suffered Appreciable damage.

Table 30 National Coal Board (now British Coal) Classification of Subsidence Damage. After SEH (1965, 1975)

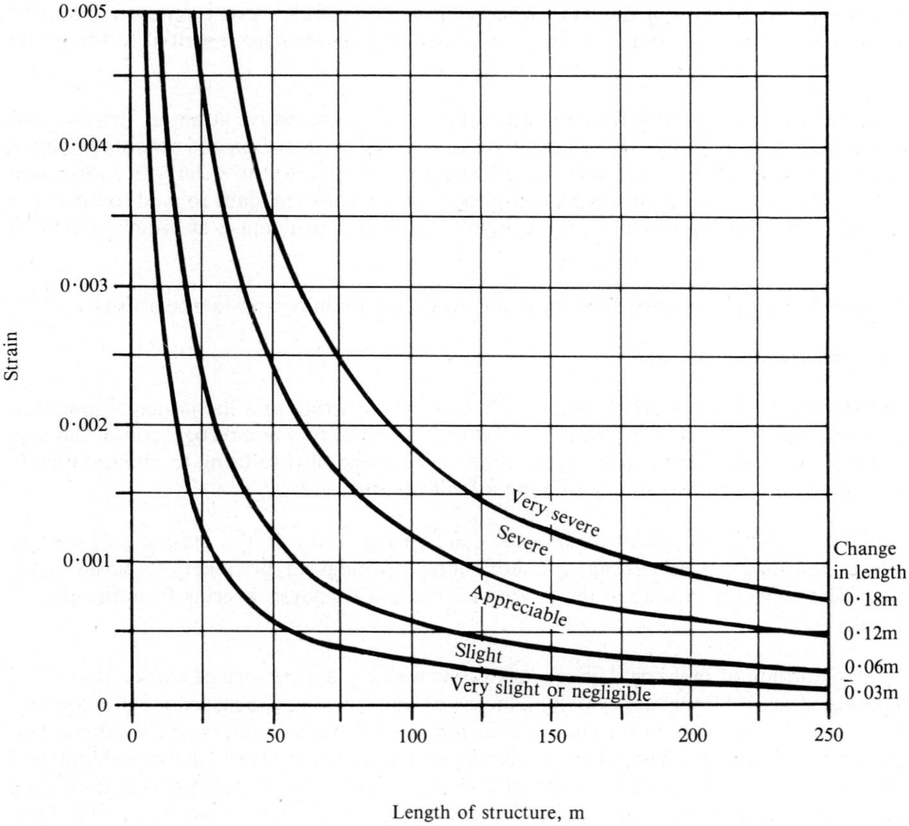

Figure 218 Relationship of damage to length of structures and horizontal ground strain. After SEH (1965, 1975).

Figures 219 to 224 inclusive illustrate features of damage to surface structures arising from mining subsidence.

When considering the possible degree of damage to long structures, it is essential to take into account those parts of the structure within the particular strain zone and not simply multiply the whole length by the maximum strain. It is possible that some buildings may be partly in tension and partly in compression. Additionally, the nature of foundation anchorage of a structure within the sub-soil, and whether it is secured throughout its length or not will also influence the severity of damage. When considering a bridge structure, each abutment must be regarded as a separate structure when assessing the damage potential to the whole bridge.

An airfield runway, now disused, was subjected to compressive strain of the order of 2 mm/m from deep workings. The runway was of wartime construction and consisted of mass concrete of around 230mm thickness and laid in panels of about 6·1m × 3m with a macadam running surface. The effect of mining subsidence was to cause the slabs to tend to lift at the pour joints. The runway surface proved highly vulnerable to damage at around 2 mm/m compression.

Figure 225 illustrates airfield runway damage due to mining subsidence effects.

The characteristics of damage

SEH (1975) considers briefly some main types of structure and the principal modes in which each may be affected by ground movement, in order to assist recognition of damage due to mining subsidence (as opposed to other causes) and also to bring to attention likely behaviour of structures when undermining a built-up area.

Buildings. Compressive strain arises with concave curvature of the ground surface and results in crushing, overthrusting and horizontal openings in brickwork. Tensile strain accompanies convex curvature of the ground resulting in fractures tapering from the ground upwards.

Early warning of building damage may occur with the appearance of cracks, doors and windows binding and floor tiles exhibiting loosening and heaving. Cracks tend to develop first at ground level with crack occurrence in evidence at voids such as doors and windows with both tension and compression zones. A feature of tensile strain effects is the possibility of opposite walls tending to move bodily apart with floor and roof joists decreasing their bearing surface. Conversely, severe compression can cause the feet of opposite walls to move closer together; the wall tops are likely to continue being strutted apart by roof members.

Recognition of whether the structure is in a zone of principal tensile strain is important regarding differentiating with non-mining damage. An example quoted by SEH (1975) is that opposite walls of a building which both lean outwards in a zone of tension cannot be due to mining but most probably suffering from roof-spread, namely the roof timbers thrusting the wall tops outwards owing to the timbers being ineffectively cross-tied thus encouraging settling with consequential outward movement.

Wall movement along the damp-proof course is indicative of compression. Void closure in the wall below the damp-proof course occurs to some extent under compression; however, owing to the lower coefficient of friction incorporated within the damp-proof course, as compared to the brickwork above, it tends to retain its original length. Door openings commonly have the effect of encouraging greater wall compression at ground floor level with the result of door voids experiencing closure and windows distort rhomboidally.

Figure 219 Damage due to tilt

Figure 220 Severe damage resulting in fractures in outside walls

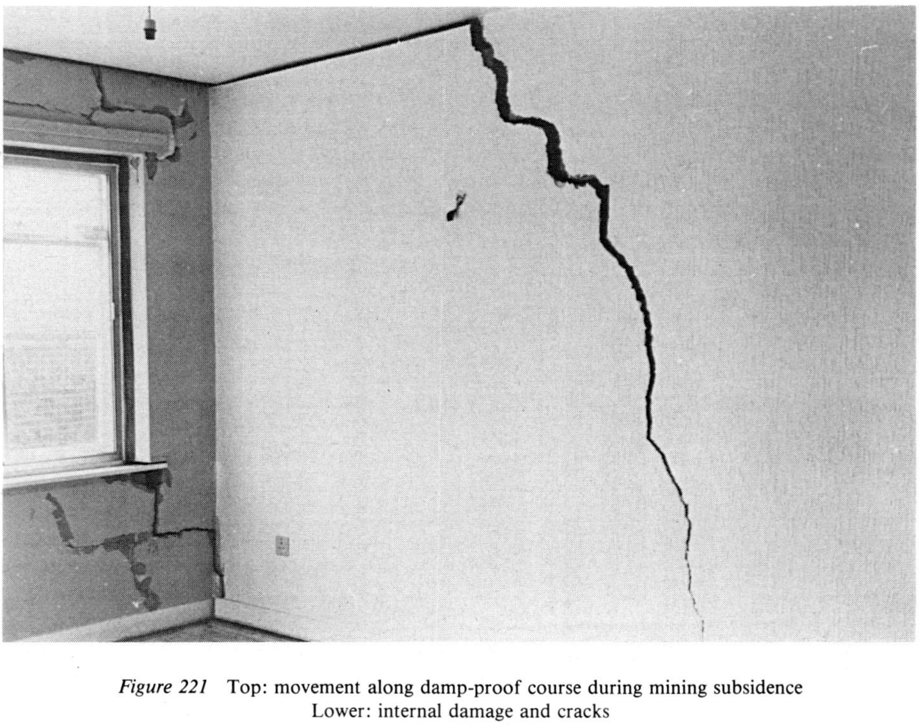

Figure 221 Top: movement along damp-proof course during mining subsidence
Lower: internal damage and cracks

Figure 222 Top: effects of compression on railings
Lower: compression effects on wall and kerbing accentuated by the presence of a fault

Figure 223 Distortion of mineral railway line due to mining subsidence

Ground concavity, although this can give exceptional effects with very shallow workings, it depends upon depth and in general does not result in noticeable concavity within structures; localised floor depressions are in general due to other causes as is the case with sagging roof lines. Attention is drawn to unexpected aspects of damage such as twisting behaviour or uneven foundations which present difficulties in recognising unless probable ground movement has been firstly calculated.

Bridges. Bridge piers, or supports, can experience movement towards or from each other depending upon the nature of the ground strains. Tilt or twist can also be present and often gives rise to serious effects to decking or arches. Long bridges may be greater than the length of the subsidence wave. Subsidence damage appearance on bridge arches is often characterised by a complex of cracks. The manifestation of compressive damage takes the form of brick facings and horizontal joint openings exhibiting crushing and spalling. Extension can result in bridge supports moving sufficiently outwards as to cause arch flattening effects to emerge. Comment is made on many bridge arches which are unaffected by subsidence but exhibit open joints and arch flattening by virtue of loss of keystone tightness probably because of perished mortar.

Roads. SEH (1975) points out that a highway is deemed to consist of carriageway, footpaths, verges, kerbs, storm-water drainage, fences, cycletracks and ancillary services namely sewers, gas mains, water mains, electricity cables and telephone cables, in so far as they affect the highway.

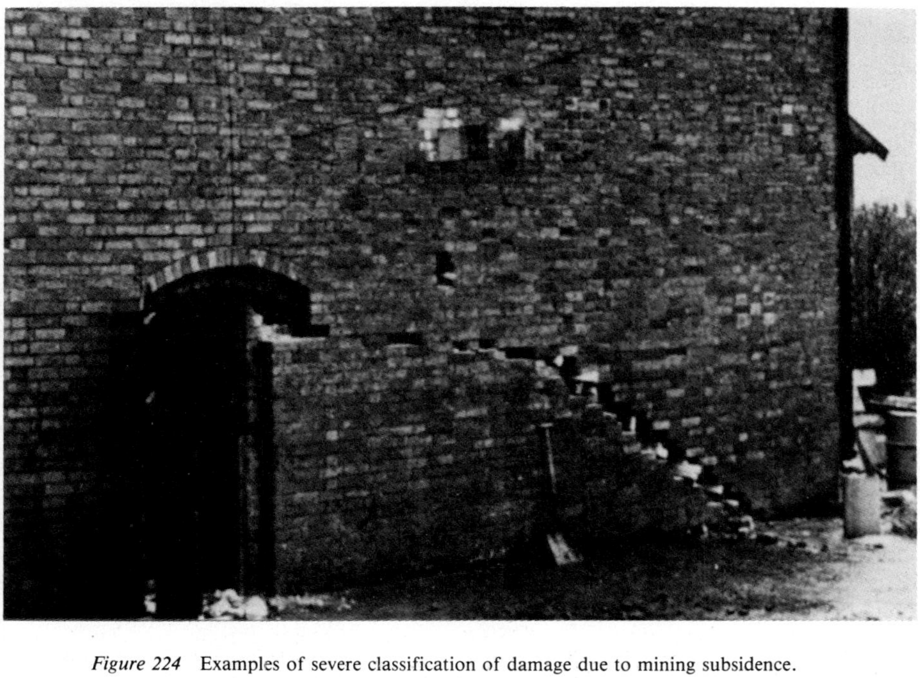

Figure 224 Examples of severe classification of damage due to mining subsidence.

Figure 225 Mining subsidence damage to airfield runway. Top: general behaviour of runway to subsidence; bottom: localised effect of geological fault intercepting surface under part of runway.

Mining subsidence damage to such structures consists of mainly:

1. Distortion of horizontal and vertical alignment.
2. Fracturing, leading to deterioration of foundations.
3. Corrugations on the running surface.
4. Damage and displacement of kerbs, channels, flagging and fences.
5. Disruption of drainage.
6. Consequential damage, an example being water action from fractured mains.

Attention is drawn to motorways and the source of danger to high-speed cars resulting from severe local changes in gradient, whilst disruption to drainage can allow water accumulation on a carriageway thus introducing an additional hazard.

Sewerage. The effects of subsidence on sewers regarding changes in vertical movement have been remarked upon earlier. Cracked and broken pipes and damage to other elements including the disposal works can introduce malfunctioning of the system.

Water supply. Water mains breakages commonly occur at intersection points where service pipes tap into the mains. Tilting has an appreciable effect on a reservoir, but occurrence of cracks may lead to direct loss of water.

Canals affected by subsidence require systematic raising of banks and water-sealing with clay. Some bridge raising may be required. Locks, and culverts associated with locks, may need consideration regarding the effects of subsidence.

Gas and electricity supply. Gas mains experience similar effects to pipe mains referred to earlier. Gas works plant and buildings can suffer damage due to horizontal strain. Gas-holder tilting can be detrimental to capacity and operational functioning.

Underground electricity supply cable systems when subjected to extension usually experience cable joint type of damage. Moisture may gain access through sheathing which has suffered cracking. Overhead lines with long spans of 240 to 300m employ pylons which can become subjected to tilt or experience vertical lowering with resulting changes in cable tension and clearances. Pylon distortion can result from uneven subsidence arising from shallow workings.

Pseudo mining damage

Subsidence damage can arise from factors not attributable to mining. Several phenomena can cause damage to surface structures. This section examines the more important pseudo mining damage causes as referred to in SEH (1975), and more information is given in the original source documents, namely Building Research Station (UK) digests.

Shrinkable clay. The natural response of surface deposits of clays to climatic and seasonal changes has been shown to be significant. This has been discussed in Chapter 14, and see Figure 178. Shrinkable clay is a common source of movement of surface structures particularly buildings on shallow foundations. Uneven lowering of structures can occur, in addition to extensions to older structures tending to exhibit differential movement on wet clay.

Sulphate attack. Interaction between Portland cement and soluble sulphates in moist conditions gives rise to the phenomenon of sulphate attack. Brickwork containing soluble salts react slowly with Portland cement in the presence of water. Damage is most common where brickwork is in contact with wet conditions over significant periods of time. The reaction causes expansion of the mortar with horizontal cracks emerging in the joints. Mortar facing may flake or fall off. Continued expansion gives rise to brickwork deformation. The later stages of the reaction witness mortar deterioration and possibly conversion to a wet powder. Household chimney stacks have developed significant leaning owing to sulphate attack.

Differential settlement. This occurs where variations in the strength and deformation properties of the sub-soils give rise to uneven settlement behaviour of structural foundations. Wet clays and made-up ground covering part of the site are common causes. Defective drainage from surface structures can also result in differential settlement.

Rust damage. Iron and steel pins and other components embedded in masonry and concrete, when subjected to corrosion can produce cracking by the action of chemical reaction causing expansion. Glass panes in steel window frames can crack by similar action.

Thermal effects. The loosening of clay floor tiles on a concrete sub-floor owing to shrinkage and thermal movement at the tile-concrete interface can result in tile lifting in the form of an arch, but on breaking of the arch the tiles fall loose.

Roof spread. Faulty construction can cause roofs to spread. Also timber decay in old buildings can give rise to cracking in external walls by the same roof spreading action of pushing the walls outwards at the top.

In concluding this discussion on pseudo mining damage due account needs to be taken that innumerable defects commonly occur in building. Frost action, a collapsed drain, sagging roof timbers are just a few of the more commonly encountered causes of such damage. Pseudo mining damage is frequently characterised by uneven lowering over short lengths of the structure.

Structural precautions: new structures

King, Whittaker and Shadbolt (1974) comment that the incorporation of flexibility into the design of new surface structures is the commonest precaution against the effects of subsequent mining subsidence and various techniques are used.

SEH (1975) discusses various aspects of structural precautions in relation to new structures and the following review summarises the main points.

Location. Where a fault is known to intercept the surface, it is recommended to avoid erecting any structure within several metres of its known position. The point is made, however, that a fault (or fissure) situated near a building can be of beneficial effect in that strains can tend to concentrate at the fault plane.

Foundations. It is recommended that in subsidence areas, buildings erected should be designed, where possible, to be either rigid or flexible. Foundations firmly anchored in sub-soil are bound to move in taking up some curvature in response to ground movement. Reinforced foundations are likely to behave similarly unless thay are adequately deep to act as a beam or cantilever. Change in length of buildings during subsidence movement does not necessarily coincide with that of the immediate ground owing to the latter behaving plastically and allowing some compression and even localised uplift to occur. This is especially the case with shallow foundations. Old foundations which are loose and compressible may sometimes fail.

With reference to new buildings which need to resist horizontal ground movement, the foundation should permit the superstructure to slide over it. Flexible superstructures can accommodate curvature of the ground induced by subsidence, whilst rigid structures can bridge across curved ground and, if designed for, be corrected by jacking action. Tall building blocks can be designed sufficiently rigid as to tilt *en bloc* and be correspondingly jacked level if desired.

Foundation slabs should preferably have a smooth underside and be placed on polythene sheeting, and itself be laid on a friable layer of around 150mm or more of sand. The slab requires to be laid at ground level to avoid any significant side thrusts being transmitted to it during the ground movement phase.

Should strip foundations be employed in a subsidence affected region, these should be reinforced and can be laid on sand. For strip foundations laid in trenches, the leaving of a space at the ends which is then filled with compressible material assists in reducing compressive strain effects being transmitted to the structure. The same principle applies to basement walls.

Superstructures. Incorporating flexibility in all parts of the superstructure is highly desirable where traditional building without any special foundation design is employed. Lime mortar is to be preferred in brickwork. Damp-proof courses, should be of the bitumastic type. Window frames should be of wood. Ceilings should be of fibre-board rather than plaster. Any arches of brickwork or stone should be also avoided. Macadam or other flexible material should be used for drives and service roads.

Drainage. Pipeline flexibility incorporating telescopic joints should be features of the drainage system employed. Pitch-fibre pipes can be used for service connections. Drains should possess ample falls to cater for gradient changes.

Provision of gaps. The provision of gaps within new structures can assist in protecting them from the effects of ground strains, although weather covering of the gaps needs to be incorporated.

Flexible structures. Buildings erected by C.L.A.S.P.* led the way regarding structural design based on scientific study of ground movement and subsequently many schools, offices, etc, have been built employing this principle of flexible structure both in the UK and on the Continent. A concrete raft foundation carries a lightweight steel frame superstructure which is pin-jointed for flexibility and includes spring-loaded diagonal braces to act against wind pressure. Cladding and internal finishes are suspended to allow movement to occur without distortion.

Rigid structures. In order to isolate foundations from the sub-soil, strongly reinforced concrete rafts or beams with the capability of supporting the superstructure over ground subjected to movements should be employed. Cellular rafts have proved a successful form of foundation design for tall structures requiring rigidity. Blocks of flats have been constructed employing such foundation principles.

Structural precautions: existing structures

Existing structures can be modified before ground movement occurs, if the anticipated damage justifies the expense, or that it may be deemed desirable to minimise the damage

*Consortium of Local Authorities Special Programme.

without undue regard to cost. Examples of the latter category include schools and hospitals where suitable alternative accommodation may not be available should repairs necessitate building evacuation. Churches, ancient monuments, buildings of historic or architectural importance, several different types of industrial complexes and factories, etc, also come under this category of structures which can gain significant protection against subsidence damage by way of modifications implemented prior to being affected by ground movements. Delicate buildings by virtue of construction or age may require temporary supports and/or strengthening before being affected by mining. Tie rods are only occasionally used where ground extension may significantly displace roof trusses off their bearings. External shoring of house walls is a common form of temporary support. Compression effects on structures can be reduced by removing a section from rigid paving, or in the case of superstructures by cutting out slots in floors, paved areas or through corridors or even through buildings.

Service pipelines can be modified by incorporating flexible and telescopic joints.

Trenching around buildings is an effective method of reducing damage from horizontal strains, and in particular that of compression. Trenches are excavated close to the structure and down to just below the foundations. The trenches are filled with compressible material for support of the sides.

King, Whittaker and Shadbolt (1974) point out that although there is a powerful case for incorporation of flexibility into new buildings in mining areas, most buildings have been built without regard to the effects of mining subsidence. The provision of deliberate lines of weakness can effectively reduce subsidence effects by encouraging localisation of strains. Preventive works aim at promoting greater flexibility within surface structures. The cutting of interconnecting corridors can allow buildings to be made into smaller individual units; waterproofing such cuts with flexible coverings can be easily achieved. Trenches around buildings are particularly advantageous where significant compressive strains are anticipated. These authors report that in general terms, trenching has had the effect of reducing compressive strain damage by up to 50%.

Figure 226 shows the main principles of protecting surface structures against the effects of mining subsidence.

Application of preventive measures to a large hall

An extensive programme of preventive measures for protecting a large hall against mining subsidence has been described by King, Whittaker and Shadbolt (1974) where an extraction of almost critical dimensions, namely $w = 260$m, $h = 260$m and $M = 1{\cdot}68$m, was advanced underneath an important structure listed as a building of architectural or historical value. The early nineteenth-century building had two storeys with a partial basement, stucco walls, slate roof and four-column open Ionic porch at the main entrance. Regency plaster ceilings existed with many other special interior architectural features.

The effect of undermining was assessed and subsidence predictions indicated $S = 1{\cdot}5$m, $+E = 2{\cdot}0$ mm/m, $-E = 3{\cdot}5$ mm/m, and $G = 15{\cdot}8$ mm/m, where S, E and G relate to the maximum predicted values of subsidence, strain and tilt respectively. In view of the size of the surface buildings it was anticipated that severe damage would occur unless preventive measures were adopted. The location of buildings together with wall cuts and extent of trenching implemented are shown in Figure 227. Flexibility was incorporated into service pipes where they passed through walls and floors. Window panes were taped, particularly where metal frames existed, as a measure against the risk of flying glass when subjected to compression effects. Mesh was suspended under ornamental ceilings.

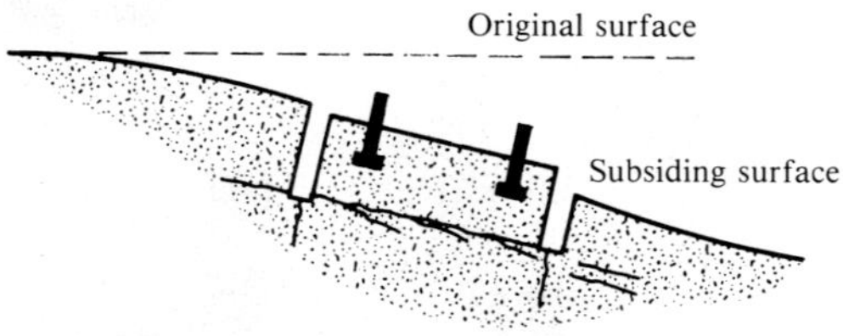

(a) Subsidence

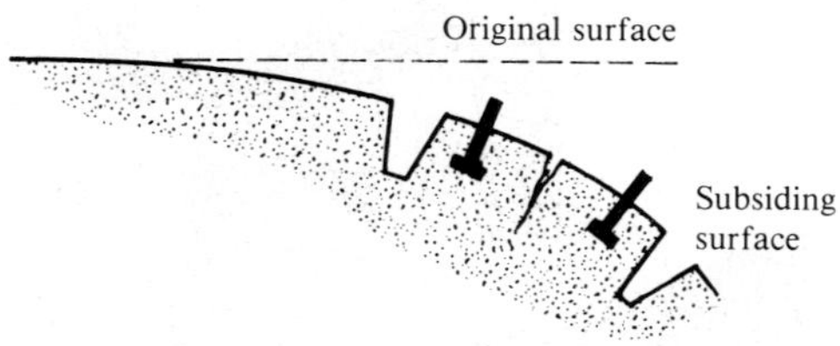

(b) Curvature (convexity)

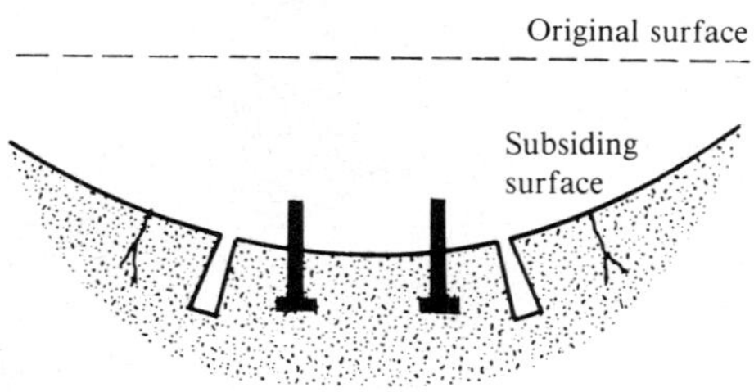

(c) Curvature (concavity)

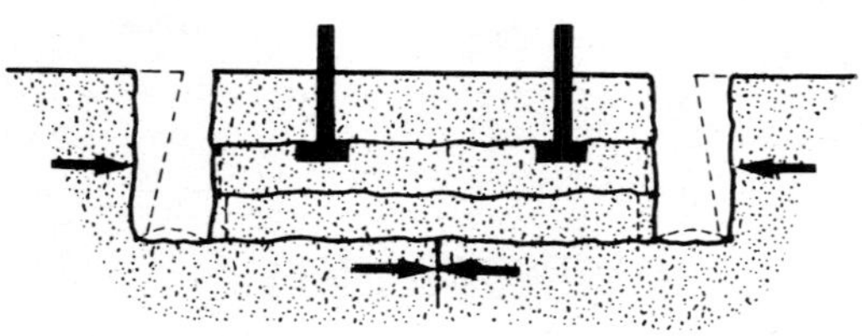

(d) Differential displacement

Figure 226 Illustrating the main principles involved in protecting surface structures against the effects of mining subsidence by trenching around the foundations. After King, Whittaker and Shadbolt (1974).

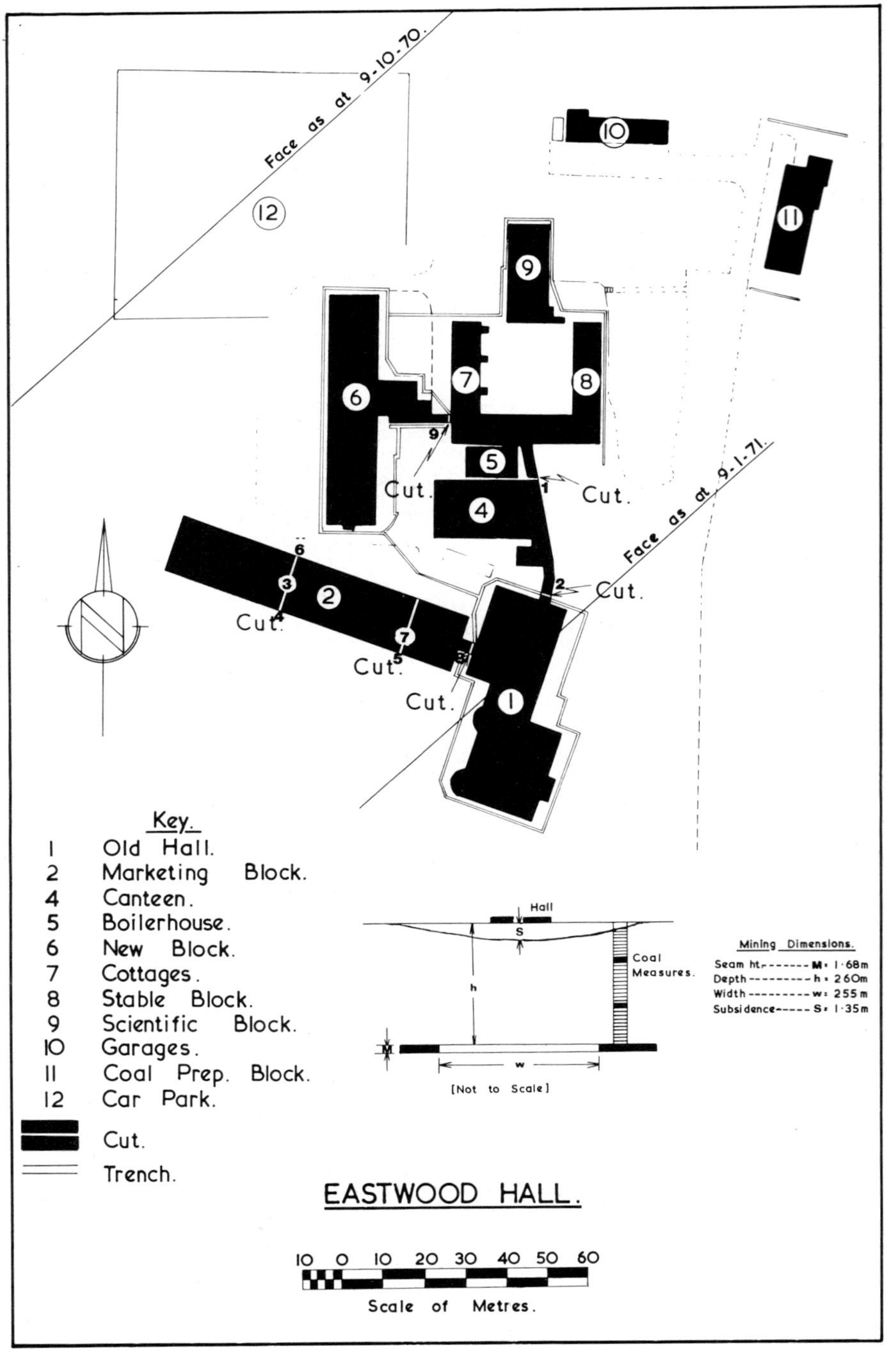

Figure 227 Plan of Eastwood Hall showing location of wall cuts and trenches used to protect surface buildings against mining subsidence. After King, Whittaker and Shadbolt (1974).

Figure 228 shows wall cut displacement behaviour in relation to longwall face position. These results show no appreciable opening of the cuts before or after undermining. Conversely, the large degree of cut closure indicates that compression was the main contributory factor to damage caused by strain in this particular case. The same reasoning applies to the foundations of the structure also. The measures adopted to protect the hall proved highly successful in preventing excessive damage or vacation of any of the offices.

These authors list the following conclusions on trenching applications and practice regarding reducing subsidence damage to existing structures:

1. Damage extent to a structure depends mainly on size and intensity of ground strain although other factors are also involved.
2. A strain intensity of 1 mm/m can be absorbed fairly evenly within standard brickwork of the average dwelling house, of length less than 15m. This order of strain can be significant to a terrace dwelling structure of 60m length.
3. Compressive strain effects cause most damage to surface structures; trenches around buildings and cutting buildings into smaller units can give protection against ground compression effects.
4. Trench depth should exceed the foundation depth for effective protection against ground compression.
5. Terraced dwelling structures only need trenches at the ends of the major axis. These structures show marked effects at subsidence strains as low as 1·5 mm/m unless protected.
6. For 'slight' categories of predicted subsidence damage, a single trench sited to give maximum protection against compression is generally considered sufficient, although there are a few exceptions.
7. Where 'very severe' categories of subsidence damage are anticipated, trenching around the entire structure should be considered. Some special types of surface structure may require trenching on all sides at lower categories of subsidence damage.
8. The preventive works carried out at Eastwood Hall reduced a predicted 'very severe' class of damage to an apparent class of 'slight' to 'appreciable' damage.

Detailed observations of subsidence magnitude and limit angle have been discussed for the Eastwood Hall site by Whittaker and Pasamehmetoglu (1981) and their results are presented in Table 31. The instrumentation observation stations referred to were specially designed for the investigation and were spread over the entire site. Table 31 compares predicted and actual values of subsidence and limit angle. The predicted values compare very favourably with the observed ones; the accuracy is substantially better than 10%. The subsidence limit angles were generally less than the accepted 35° value for UK coal mining situations. The spread of limit angle values ranged from 25½° to 34°, even though the ground conditions of this site were fairly uniform.

Underground mining designs for reduction of subsidence effects

Surface protection against damage from mining subsidence has been successfully achieved on a large scale by adoption of the 'partial extraction system'. The method involves working relatively short longwall faces (panels) and leaving rib pillars of around 0·25 to 0·2 (depth) and panels 0·25 to 0·33 (depth). Extensive areas of coal reserves have been worked using this method, and especially where previously such reserves had been considered commercially unworkable. It is common to achieve around 50% extraction with the partial extraction system, although with some layouts this has been as high as 70% even under industrialised and heavily built up areas.

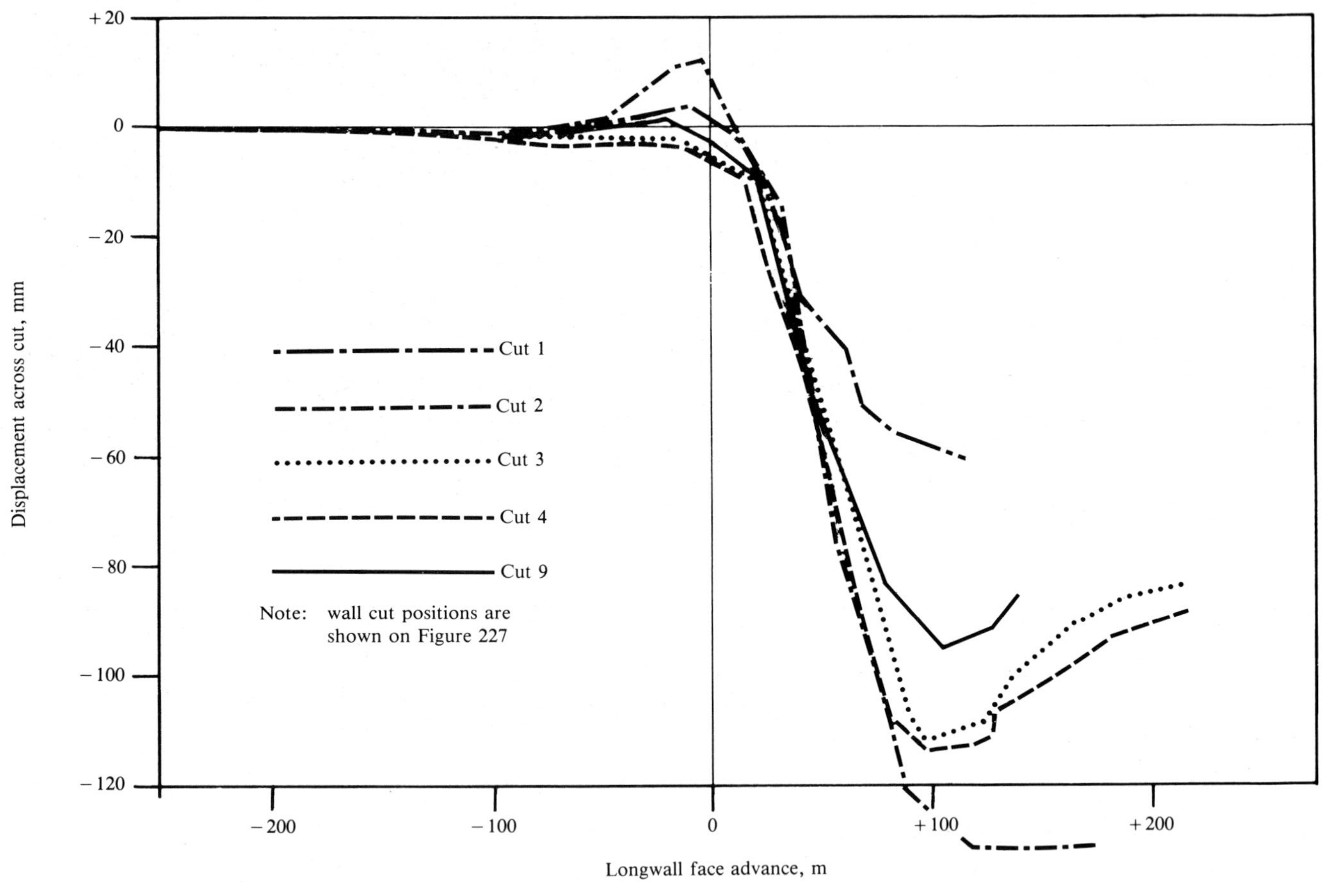

Figure 228 Observed behaviour of wall cuts during the mining subsidence phase, Eastwood Hall. After King, Whittaker and Shadbolt (1974).

	Maximum subsidence (S)		Limit of mining subsidence		
Station	Predicted (mm)	Actual (mm)	Act.S -Pred.S / Actual S (%)	Distance ahead of face (h)	Angle of draw (°)
T1	930	972	4·3	0·51	27
T2	910	872	−4·3	0·51	27
T3	910	909	−0·1	0·50	26½
T4	800	815	1·9	0·50	26½
T5	650	632	−2·8	0·55	29
T6	1080	1151	6·2	0·59	30½
T7	1200	1202	0·2	0·50	26½
T8	1250	1213	−3·1	0·50	26½
T9	915	942	2·9	0·59	30½
T10	950	1021	7·0	0·58	30
T11	1030	1025	0·5	0·60	31
T12	1080	1062	1·7	0·60	31
T13	1160	1138	1·9	0·48	25½
T14	1220	1220	0	0·68	34
T15	1250	1229	1·7	0·68	34
T16	1120	1172	4·4	0·68	34
T17	1200	1205	0·4	0·56	29
T18	1120	1135	1·3	0·56	29
T19	1000	1015	1·5	0·64	32½
T20	1050	1050	0	0·64	32½
Mean			|2·3 ± 2·1|		29·6 ± 2·8°

Table 31 Comparison of predicted and actual subsidence together with limit of mining subsidence ahead of longwall. After Whittaker and Pasamehmetoglu (1981).

Partial extraction data

Table 32 presents data observed at several sites in the UK where partial extraction layouts have been employed, and this serves to demonstrate the effectiveness of such mining designs in reducing subsidence at the surface. Examples of successful partial extraction layouts are also given in NCB (1961). The advantages of systematic partial extraction layouts based on longwall extraction with substantial intervening pillars in reducing surface subsidence have been drawn to attention by SEH (1975). This publication remarks that the small magnitude of subsidence and gentle curvature of the resulting profile over commonly adopted panel widths of h/4 is in essence its overall result. The interaction of individual surface subsidence profiles with partial extraction layouts produces a smoothing out of the overall subsidence profile and a tendency for some of the strains to cancel each other out. SEH (1975) presents design charts to assist in determining subsidence for different combinations of panel and pillar widths in relation to depth.

Stowing

The use of stowing mined-out areas behind longwall faces has been referred to earlier in this book. This form of controlling subsidence is rarely practised at the present-day owing to the cost of the operation. Stowing by pneumatic means has been demonstrated to reduce surface subsidence by up to 50%. It is generally only of significance, however, where the width/depth ratio of the extraction is 0·6 or greater. The effect of reducing subsidence by this method when w/h is substantially less than 0·6 is not significant. Partial extraction layouts

Case	Seam depth (m)	Pillar width (m)	Panel width (m)	Width overall (m)	Extracted height (m)	Max subsidence (mm)	Subsidence (%)
Handsworth	110	32 46 37	27	166	1·37	44	3·2
Hucknall	177	44 44	44 44 44	220	1·24	63	5·1
Wollaton	220	46 46	46 46 46	229	1·37	89	6·5
Cadeby	686	110 110 110 110	73 73 73 73 73	805	1·68	235	14
Bradford	823	92 92 92 92	64 64 64 64 64	686	1·78	279	15·7
Wearmouth	549	76 76	73 73 73	372	0·89	79	9
Snibston	253	50	59 53	163	1·83	201	11
Glenochil	203	58	38 38	134	0·86	40	4·6
Easthouses	616	168	152 198	518	1·80	287	15·9
Barbara	170	40	50	230	1·2	60	5

Table 32 Partial extraction data for longwall mining layouts designed to reduce surface subsidence. After Orchard (1964)

have proved to be more successful where sub-critical extractions have required a measure of surface subsidence control. Where very thick extractions have involved several lifts being taken under heavily built-up areas as in Upper Silesia, Poland, then solid stowing has found favour in allowing more coal to be mined than would be otherwise possible. Solid stowing of the goaf region behind longwall extractions in European coalfields can be rarely justified on economic grounds as a subsidence control measure. However, stowing of the goaf regions is practised in some countries such as in India, where the requirement is mainly governed by the necessity to achieve improved safety in the underground operations, rather than solely as a means of controlling surface subsidence.

Harmonic extraction

The basic aim of harmonic extraction is the carefully phased removal of all the coal from the critical area relating to the surface structure so that the latter is lowered uniformly and in a manner allowing it to be more or less free from horizontal strain, NCB (1963). The method has been used when extracting coal under churches, schools, hospitals, factories and reservoirs, etc. An important aim is to leave the structure in a strain-free position, although it will be lower due to subsidence. The method involves undermining a structure by successive narrow width longwalls each of which only produces a minor subsidence effect on the surface. By appropriate staggering of such multiple extractions, then a measure of cancelling out of some of the strains becomes possible, and the development of subsidence is more spread out. Orientation of the surface structure needs to be taken into account when designing a harmonic extraction layout. This form of layout has not proved popular for controlling surface subsidence effects owing mainly to the rigidity of the system regarding underground extraction sequence control.

Surface support pillars

The use of support pillars to give protection to certain surface structures has been applied extensively in the past. The most common size of pillar left to give full support to the surface has been 0·33 (depth) in width from the edge of the mined-out area to a vertical line projected from the surface to meet the seam (in flat seam conditions). This margin of support has proved adequate in many situations. Most structures can accommodate the very small effects of subsidence resulting from the edge effect at 0·33 (depth). The sensitivity of certain surface structures, such as some very old churches, etc, has resulted in leaving a support pillar of 0·5 (depth) width between the edge of the workings and the structure. General experience in the UK coal mining industry has been that the 0·33 to 0·5 (depth) width of the support zone up to the structure from the mined-out area, has proved adequate for the majority of special structures requiring total support. Subsidence protection zones wider than these margins are rarely used, although some in the range 0·7 to 1·0 (depth) have not been unknown. In general, the site conditions and type of structure requiring protection, eg power station, high-speed railway line, etc, need to be taken into account in relation to the mining situation in order to arrive at the exact size of pillar for total support.

Pillars for total support of a surface structure can result in sterilising vast quantities of mineral, so that their adoption needs to be very carefully considered in relation to the other options available for undertaking precautions to mitigate subsidence and its associated effects.

Further discussion on surface support pillars is given in a later section of this chapter when aspects of shaft pillar design are examined.

Effect of curvature

Figure 229 shows a graphical representation of the influence of increasing depth on the induced elongation and superelevation of a surface structure due to ground curvature according to King, Whittaker and Shadbolt (1974). The surface structure considered in this theoretical examination, was 20m length and situated directly above the position of maximum tensile strain. The graph demonstrates the major role played by depth in accentuating the elongation and superelevation behaviour of surface structures, especially at depths less than around 200m below the surface in supercritical longwall extractions. According to this analysis, the behaviour of the structure was inversely proportional to the square of the depth. These authors pointed out, however, that other factors relating to structural design and

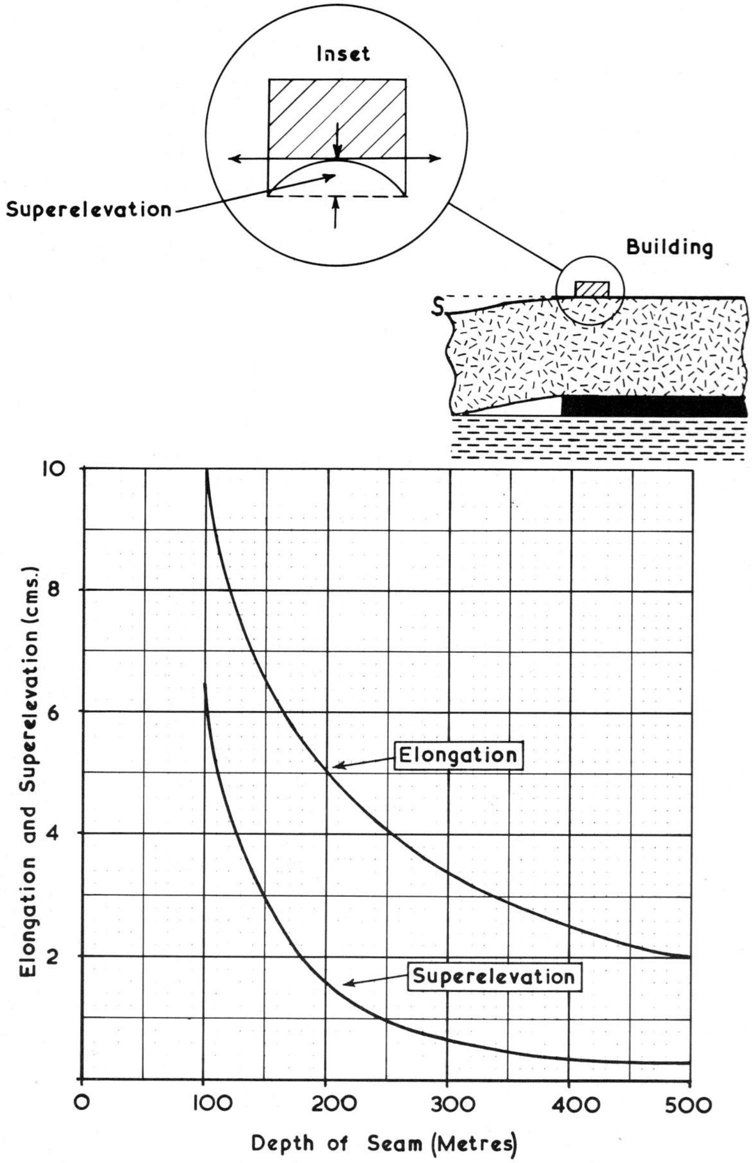

Figure 229 Effect of depth of seam on elongation and superelevation of a surface building affected by mining subsidence. After King, Whittaker and Shadbolt (1974).

Notes:
1. Supercritical longwall extraction with total caving.
2. Building is 20m long.
3. Maximum subsidence (S) is 0·67m.

behaviour in addition to site characteristics may prove of greater significance to that of ground curvature.

Ground movement and structures

The highly successful conferences on ground movements and structures held during 1977, 1980 and 1984 at the University of Wales Institute of Science and Technology, Cardiff, discussed several aspects relating to surface subsidence, and particularly prediction, measurement and behaviour of ground movements, Geddes (1977, 1980, 1984). Abandoned mine workings and associated problems are presented mainly from structural engineering points of view.

Surface damage from subsidence has been reviewed by Voight and Pariseau (1970). They remarked that most pipeline failures are due to compression, telescoping, and not from tension. Also, they comment upon proposals for strain limit values in different countries; for example, Japan: concrete foundations 0·5 mm/m, wooden buildings 1 mm/m. These authors concur with SEH (1965) on design precautions in areas of mining subsidence. Their assessment of the state of predictive art in subsidence engineering revealed that subsidence can be engineered, and that arguments regarding complexity of rock masses over Coal Measures being a limiting factor are no longer tenable.

Effects of mining subsidence on colliery spoil heaps

Several field investigations were carried out in the UK coalfields in order to monitor the general character of surface deformational behaviour of spoil heaps subjected to the effects of mining subsidence. The purpose of the study was to compare the response of spoil heaps to undermining and relate these to anticipated movements as predicted by standard subsidence calculation procedures, Forrester and Whittaker (1976). Standard subsidence measuring methods were applied to collate data on the deformational behaviour of the spoil heaps.

Table 33 presents a summary of the predicted and observed subsidence results for the spoil heaps investigated.

Figures 230 and 231 show comparisons of predicted and observed subsidence profiles for two of the spoil heaps studied, whilst Figure 232 shows the observed character of spoil heap displacements. Figure 233 shows the observed pattern of cracks which occurred during the subsidence of one of the spoil heaps studied. After the crack pattern had formed it was observed to travel across the spoil heap surface at the rate of progress of the longwall working.

The following general conclusions were drawn by Forrester and Whittaker (1976) on undermining spoil heaps:

1. Spoil heap subsidence was observed to be greater than predicted in both magnitude and extent; maximum subsidence was observed to be as much as 67% greater than that for natural ground. Predicted magnitudes were generally low estimates.
2. Spoil heaps responded earlier to the effects of subsidence than for the natural ground.
3. Observed maximum surface strains were greater for spoil heap structures than for natural ground.
4. The incidence of surface cracking was not covered by the standard prediction procedures; cracks on the surface of the tip were observed to occur when the observed tensile ground strain exceeded 3 mm/m. Compression humps became apparent on the spoil heap at compressive strain magnitudes of the order of 10 mm/m.

Case study	Position of maximum subsidence (indicated by ↓)	Spoil heap height at Sp position H	Depth to seam h	Width/depth ratio w/h	Extraction thickness M	Predicted maximum subsidence Sp	Observed maximum subsidence on spoil heap St	Observed maximum subsidence on the natural ground Sg
		(m)	(m)		(m)	(m)	(m)	(m)
A	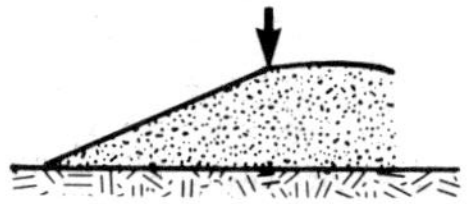	27	308	0·70	1·68	1·30	1·69	1·41
B	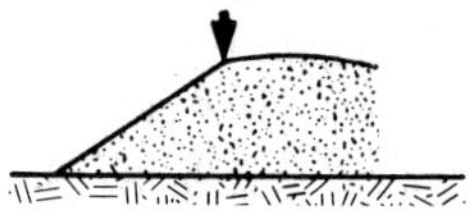	60	258	0·94	1·22	1·01	1·41	-
C	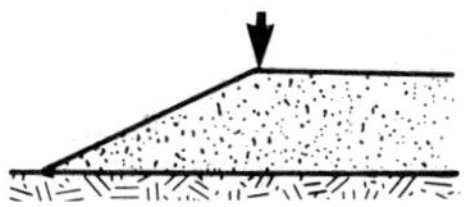	27	700	0·35	1·22	0·37	0·54	0·38
D	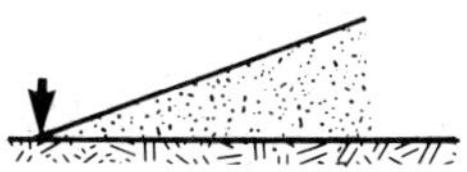	0	93	1·46	1·12	1·00	-	0·84
E	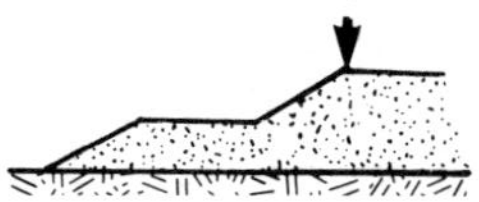	23	108	1·33	1·12	1·00	1·25	0·75
F	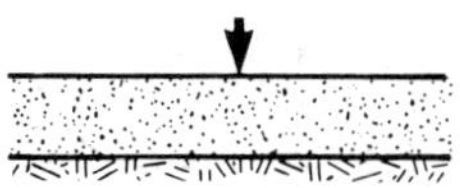	23	120	1·40	1·12	1·00	1·15	0·78*

*Extrapolated from D and E

Table 33 Summary of predicted and observed subsidence results. After Forrester and Whittaker (1976).

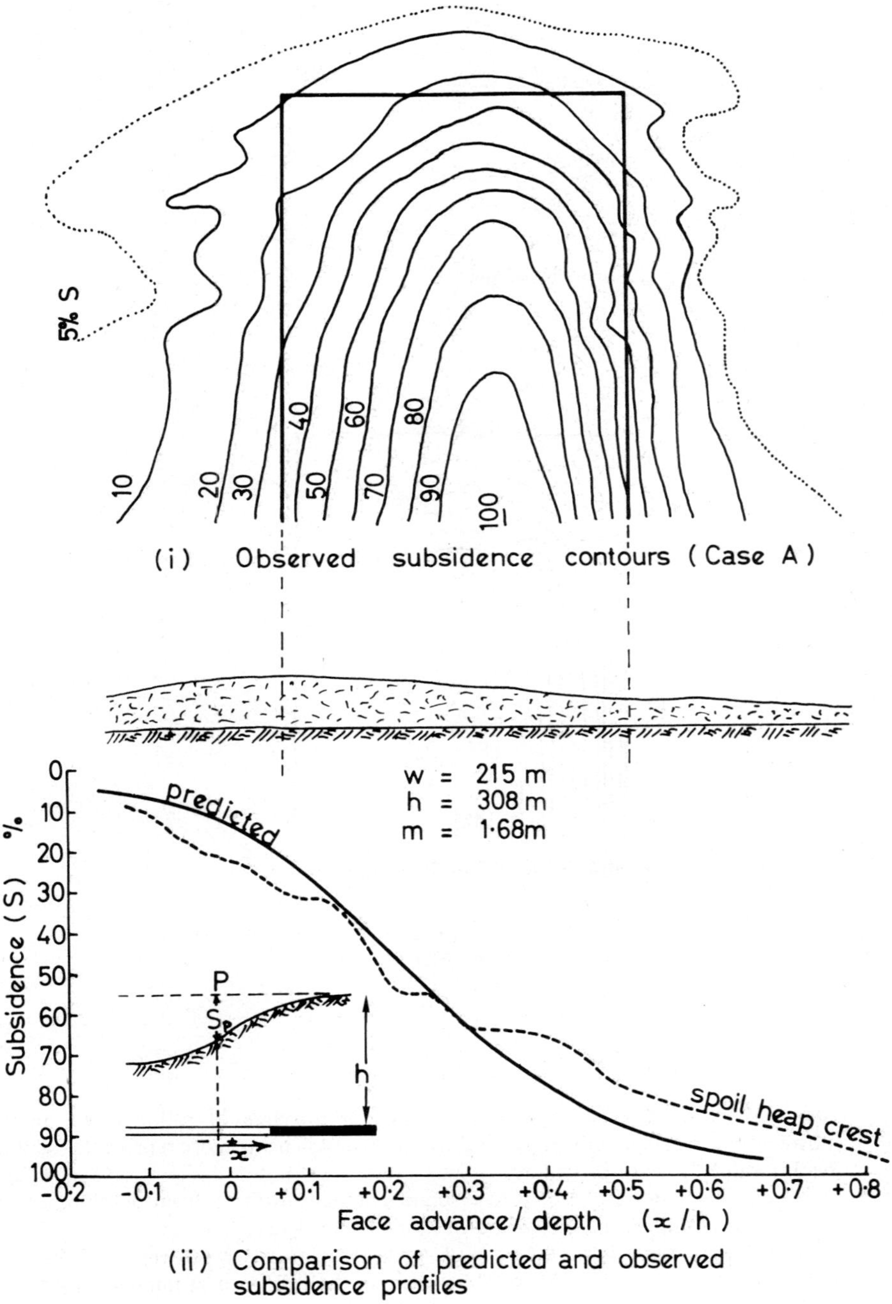

Figure 230 Predicted and observed subsidence for spoil heap A. After Forrester and Whittaker (1976)

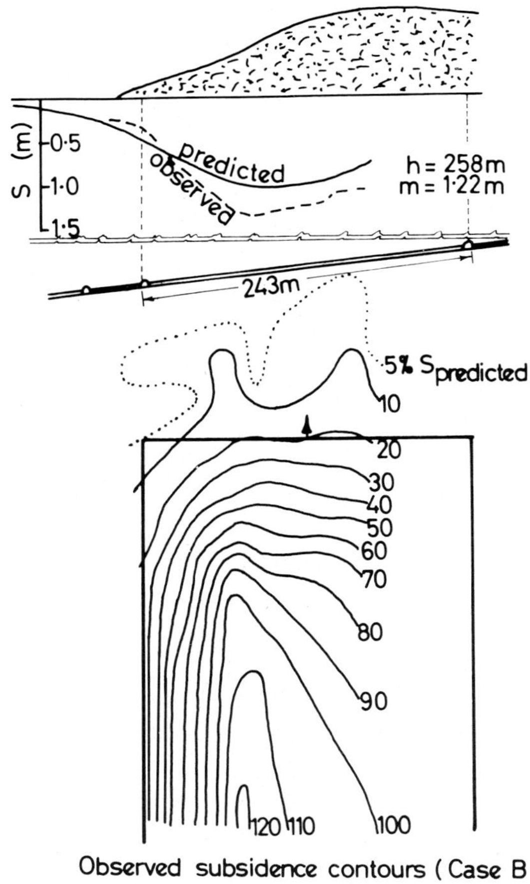

Figure 231 Predicted and observed subsidence for spoil heap B. After Forrester and Whittaker (1976)

5. The presence of a mine spoil heap resulted in the area of influence of mining subsidence increasing, with angles of draw of up to 45° being more representative than the 35° value for natural ground.
6. Mine spoil heaps were observed to respond gradually with uniform development of ground movement during the period of mining subsidence.
7. Typical displacement features observed, were those of excess settlement at the crest and heave at the toe of a mine spoil heap under the influence of mining subsidence.

A tentative linear relationship is indicated by these results and this is shown in Figure 234. The results suggest that there is a relationship between excess settlement per unit spoil heap height and the S/h ratio.

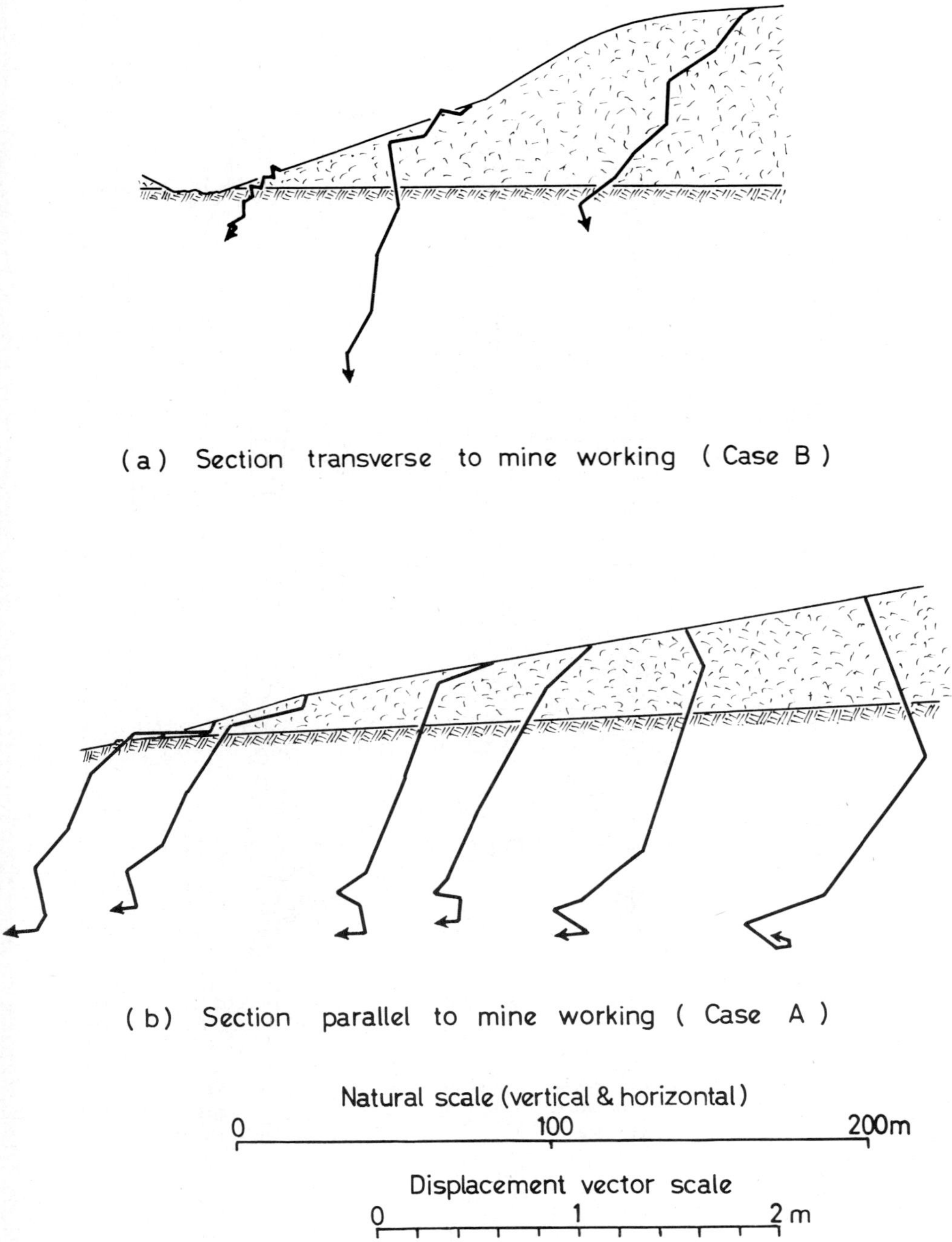

Figure 232 Spoil heap displacement characteristics due to mining subsidence. After Forrester and Whittaker (1976).

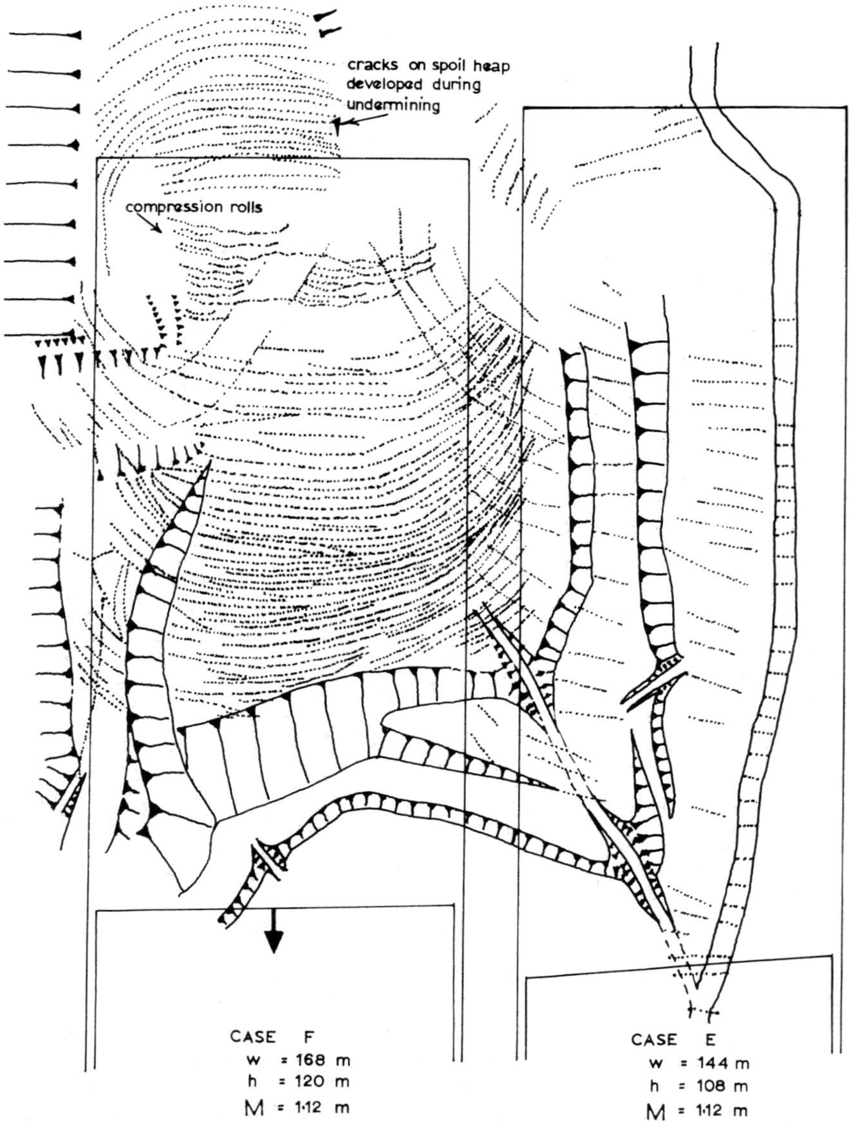

Figure 233 Pattern of cracks on spoil heap which developed during undermining. After Forrester and Whittaker (1976).

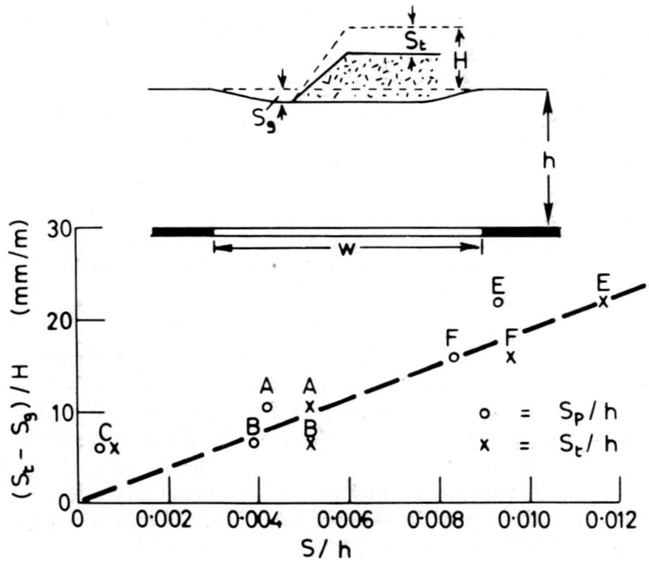

Figure 234 Relationship between excess settlement of mine spoil heap and the S/h ratio. After Forrester and Whittaker (1976).

Working under the sea

The extraction of coal seams underneath the sea is governed by specific instruction in the UK. The Production Department Instruction, PI/1968/8, forms the basis for safe design of undersea coal mining operations, NCB (1968). The Instruction considers undersea workings to be defined as those lying beneath the sea and tidal water, with the demarkation line being displaced landwards from the high water mark for a distance equal to half the depth of the workings immediately below the high water mark. Two methods of mining are considered by the Instruction:

Longwall extraction. The minimum depth of cover between the top of the seam and the seabed is stipulated as 105m, of which at least 60m of Carboniferous strata must exist above the coal seam. Additionally, the extraction thickness and width of working are required to be designed so that the tensile strain at the seabed does not exceed 10 mm/m. This applies to single and multiple seam working. Table 34 gives guidance on thickness of extraction for total extraction working at different depths as included in the Instruction.

Thickness of cover between extraction and seabed (m)	Thickness which may be extracted (m)
105	1·7
110	1·8
120	2·0
140	2·3
160	2·7
180	3·0
200	3·3
220	3·7
240	4·0

Table 34 Maximum thickness of coal which may be totally extracted by longwall operation for different depths of cover according to PI/1968/8, NCB (1968).

Room and pillar partial extraction. The minimum thickness of cover between the seam and the seabed is 60m of which at least 45m is required to be Carboniferous strata immediately above the seam. The minimum dimension of the pillars is required to be 0·1 (depth) and if the extraction thickness exceeds 2m this assessment of pillar width should have the seam thickness added to it. The room width is limited to not more than 6m. The 0·1 (depth) quantity needs to also include an addition of 0·5 (depth of sea water) in the "depth" value of the former. The Instruction requires to be taken into account those mine floors susceptible to becoming plastic when wet, and in such conditions where the pillars may be affected the minimum pillar width is one-sixth of the cover thickness (including the half-depth of sea water allowance).

If pillar extraction is to be carried out then the Instruction requires the design basis to conform with that for longwall extraction.

Due account is required by the Instruction to be given to information on the nature of any potentially dangerous aquifers, as well as subsidence effects on permeability. Also, account is to be taken of seabed levels, nature and thickness of seabed drift deposits, and the position of buried river channels both known or suspected. Geological faults are required to be taken into account, again whether known or suspected, and particularly the indication of their positions. The significance of such faults in reducing the Carboniferous strata cover, vertically and laterally, or outcropping into the seabed and/or seabed drift deposits, or intersecting known aquifers need also to be taken into account. This information is to be shown on the development plan.

The effect of solid stowing undersea workings if adopted, should be regarded as equivalent to halving the extraction thickness for the purpose of deducing subsidence effects on the seabed.

General comments on undersea working criteria

The criteria for undersea working are based on many years of experience in mining coal under the seas around Britain. The Instruction referred to above is judged to possess a significant margin of safety against inundation from the sea due to breaking through cover rocks disturbed by mining operations. The thickness of cover is regarded as the major factor. The tensile strain value of 10 mm/m, essentially acts as a limit on the amount of underground longwall extraction and consequently acts as a control on the distance to which lines of fractures extend from the mining horizon towards the sea, and also seabed subsidence fractures extending towards the extraction. This aspect is further discussed in the next chapter where the results of model tests are examined in relation to fracture propagation from longwall extraction operations.

The undersea working criteria for room and pillar mining operation should be viewed alongside the discussions given earlier in this book on the development of sink-holes and subsidence generally in connection with such workings. Special account needs to be given to the room height and the potential height of a collapsed mine room junction, and particularly if there is a possibility of a collapsed chimney being washed out into adjacent rooms should an overlying aquifer be tapped. The nature of the immediate cover rocks needs to be considered in relation to permeability and their caving potential in respect of encouraging upward development of collapsed mine junctions with the risk of forming a sink-hole.

In conclusion, when designing mining layouts for undersea conditions, special regard should be given to the thickness of cover to the seabed and to what extent it may be reduced by hidden erosional features such as old river channels. Faults within the cover thickness to the seabed need identification and their significance evaluating in view of their potential to transmit water between different horizons. Whittaker and Aston (1985) have reported on undersea longwall mining subsidence with special reference to geological and water occurrence criteria in the North-East of England Coalfield. Particular attention is drawn to the importance of the proximity and nature of major faulting in such layouts.

The nature and deformational character of the cover rocks together with permeability and the strains to which they are subjected also play important roles in prompting safe designs for undersea working.

Shaft pillar considerations

Purpose of shaft pillar. The purpose of the shaft pillar is essentially to protect the shaft from any mining damage. This may include loss of vertical alignment due to tilting caused by ground movement; shaft lining or fittings damage resulting from ground strains; and damage to the overall construction including insets and connecting tunnels arising from the effects of excessively high rock stress which can occur if the shaft pillar is too small. Additionally there is the important role played by the shaft pillar in protecting surface structures at the shaft top against the effects of mining subsidence; winding towers and other major structures and associated plant, machinery and equipment need considerations in respect of effects resulting from mining subsidence.

UK practice. General UK practice in the coal mining industry is to employ a shaft pillar whose diameter at the mining horizon is approximately equal to the depth below surface, for flat seam and level surface conditions. In effect this allows for mining extractions, involving total extraction by longwall working, to work up to a position of about half-depth from the shaft. This size of shaft pillar affords adequate protection for the surface structures immediately around the shaft.

Where the coal seams are inclined or the ground surface is sloping, the shaft pillar will need to be designed to take such conditions into account. The effect of steeply pitching seams on influencing the position of the subsidence trough and on changing the magnitude of the surface ground strains is discussed in Chapter 10. Conversely, the effect of sloping ground surfaces on influencing the character of surface subsidence has been discussed in Chapter 11.

The discussion here is mainly concerned with total extraction around the shaft pillar. Those mining layouts involving room and pillar operation without pillar extraction, do not generally impose strict limitations on the design of the mine pillar. This is providing the pillars have an ample factor of safety against failure and that there is a minimal risk of any roof falls giving rise to potential surface instability problems.

With multi-seam situations, progressive working of lower seams usually requires correspondingly larger pillars, and due account to be taken of the magnitude of the cumulative tensile strain at the surface.

Development of stress within the shaft pillar

The development of vertical stress within the shaft pillar can be considered as shown in Figure 235. This shows the shaft pillar to be loaded by an inverted frustum of a solid cone of rock whose sides have been considered as sheared due to total extraction around the pillar. The principles of this method of analysis of considering load transfer to the pillar from mined-out areas and incorporating a shear angle due to longwall working has been discussed by King and Whittaker (1971). The treatment of the problem involves determining the average vertical stress ($\bar{\sigma}_p$) in the shaft pillar for a flat coal seam. Subsidence theory forms the basis of the analysis, and it is assumed that the roof shears over the extracted region at an angle ϕ. Projection of this angle to the surface defines the size of the frustum acting on this pillar. The shear line is considered to meet the surface when the extraction width adjacent to the shaft pillar exceeds 1·2 h. The shaft pillar will be carrying its maximum load when the extraction width exceeds 1·2 h, namely under supercritical extraction conditions.

Considering a depth (h) below surface condition as represented in Figure 235 and a shaft pillar whose diameter is p and a shear angle (ϕ), then using a value of $\phi = 31°$ (corresponding to an extraction width of 1·2 h) as is known to apply to Coal Measures rocks, this allows the average vertical stress to be deduced. The main steps in the analysis are as follows:

$$\text{Average vertical stress } (\bar{\sigma}_p) = \frac{9{\cdot}81\gamma}{1000(\pi p^2/4)}\left[\begin{matrix}\text{Volume resting}\\ \text{on pillar (m}^3)\end{matrix}\right] \text{kN/m}^2 \quad \ldots(67)$$

$$\text{Volume} = 1/3\,\pi\left[(1/2p + h\tan\phi)^2(1/2p\cot\phi + h) - (p^3/8)\cot\phi\right] \quad \ldots(68)$$

$$\bar{\sigma}_p = \frac{9{\cdot}81\gamma 4}{1000\ 3p^2}\left[(1/2p + h\tan\phi)^2(1/2p\cot\phi + h) - (p^3/8)\cot\phi\right] \quad \ldots(69)$$

Using a value of $\phi = 31°$ in equation (69) allows it to be rearranged as equation (70).

$$\bar{\sigma}_p = \frac{9{\cdot}81\gamma}{1000}\left\{\frac{4}{3p^2}\left[(1/2p + 0{\cdot}6h)^2\left(\frac{p}{1{\cdot}2} + h\right) - \frac{p^3}{4{\cdot}8}\right]\right\} \quad \ldots(70)$$

Equation (70) can be expressed so that the average stress is a function of p/h and this is given in equation (71).

$$\bar{\sigma}_p = \frac{9{\cdot}81\gamma h}{1000}\left\{\frac{4}{3}\left(\frac{h}{p}\right)^2\left[\left(\frac{p}{2h} + 0{\cdot}6\right)^2\left(\frac{p}{1{\cdot}2h} + 1\right) - \left(\frac{p}{h}\right)^3\frac{1}{4{\cdot}8}\right]\right\} \quad \ldots(71)$$

where γ = average density of strata overlying the pillar
p = diameter of shaft pillar
h = depth of pillar below surface
ϕ = angle of shear of roof strata due to longwall extraction around shaft pillar.

The amount of extraction around the shaft pillar is assumed to be supercritical, namely the extraction width exceeds 1·2 h.

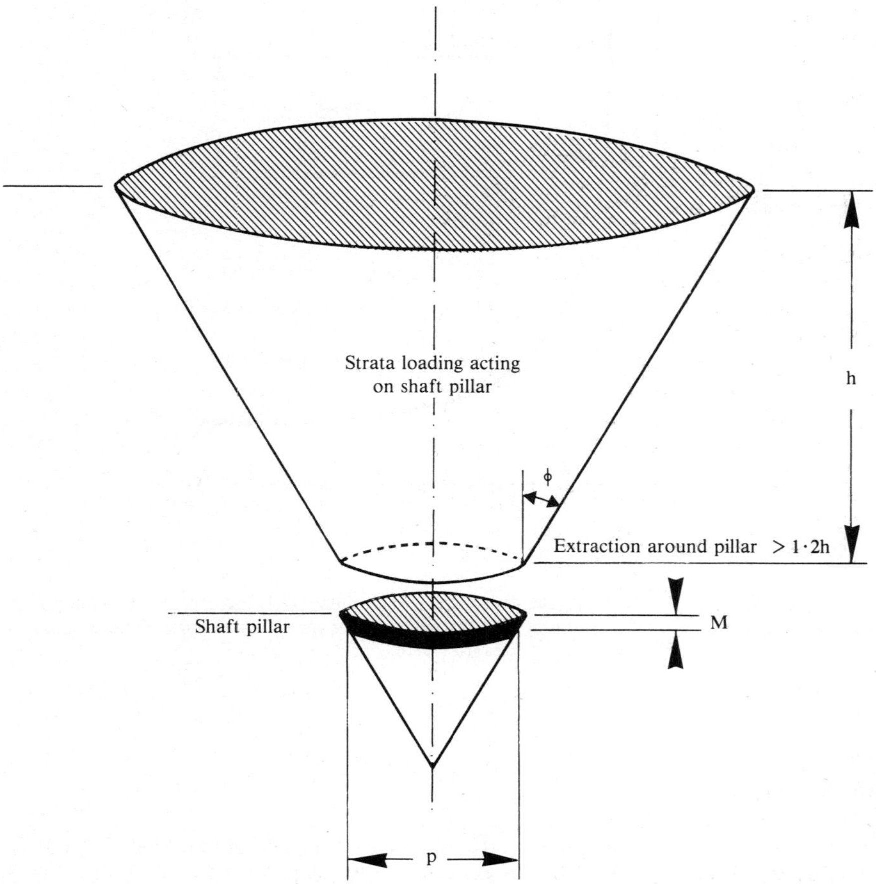

Figure 235 Illustrating basic assumptions in determining strata loading acting on shaft pillar in a flat seam.

Equation (71) has been used to examine the influence of the shaft pillar diameter to depth ratio (p/h) on the average vertical stress acting on the pillar. Figure 236 shows this relationship graphically. This shows that for the UK coal mining industry practice of employing shaft pillars of around p/h = 1, this gives an average vertical stress acting on the pillar of 2 - 3 $\bar{\sigma}_p$. This is the state of stress after formation of the shaft pillar and indicates that roadways and other excavations within the pillar could experience appreciable increased loading during the later stages of shaft pillar development. This would be of greater significance in deep mining situations. Figure 236 shows that for small shaft pillars of around p/h = 0·5 and less, there is a major increase in $\bar{\sigma}_p$.

Further discussion on mine pillar design and shaft pillar stability is given by Whittaker and Singh (1979), and Whittaker, Singh and Zadeh (1980).

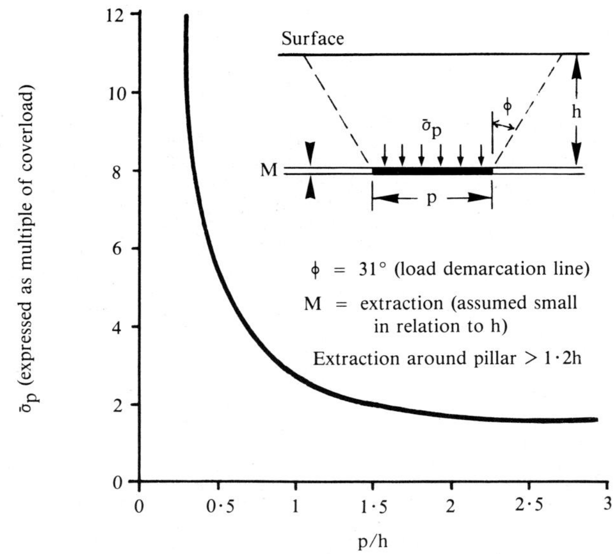

Figure 236 Average vertical stress ($\bar{\sigma}_p$) acting on a circular shaft pillar of diameter (p) and at depth (h) below the surface, assuming an extraction width around the pillar greater than 1·2h. After Whittaker, Singh and Zadeh (1980).

Concluding remarks

Damage to surface structures arising from mining subsidence and associated effects can be conveniently classified and related to ground movement parameters. There are various forms which mining subsidence damage can take, and these are generally distinguishable from pseudo mining damage particularly that arising due to the presence of shrinkable clays.

Several different structural precautions can be taken ranging from new structures to existing ones. Incorporation of flexibility into a structure significantly assists in reducing its damage potential from subsidence effects.

There are a number of different underground mining designs which can be adopted to reduce surface subsidence effects. Partial extraction layouts employing narrow longwalls with substantial intervening pillars is a highly effective method of reducing subsidence effects at the surface; it also offers considerable opportunity to design layouts with virtual precision accuracy in specifying low levels of subsidence. Partial extraction designs have been used to mine coal seams under heavily built-up and industrialised areas where previously mining would not have been contemplated. Longwall mining with stowing of the extracted area is rarely carried out at the present day owing to the high cost of such operations and also the fact that it can significantly slow down the rate of face advance. Stowing operations give a measure of subsidence control which can be achieved using a partial extraction layout. However, in very thick seams, stowing has found favour in Poland where it has allowed almost total extraction under heavily built-up areas. There is always the problem of providing the quantities of stowing material required for thick extractions, and in some situations such materials may not be available. Consequently, other designs need to be considered and this is where partial extraction layouts offer a major advantage in not being governed by the availability of stowing materials. Additionally, stowing in mining situations where the width/depth (w/h) ratio is appreciably less than 0·5 - 0·6 achieves only marginal benefits regarding reducing subsidence at the surface. Stowing is beneficial in mainly shallow mining conditions. Harmonic mining has practical difficulties in employing to achieve its maximum benefits; it has limited advantages when compared to partial extraction designs.

Surface support pillars are employed only occasionally, owing mainly to the significant sterilisation of mineral reserves. Where they are used the 0·33 - 0·5 (depth) margin between the extraction edge and line of the structure (level seam conditions) has been proved by experience to provide ample protection against subsidence effects. However, the mining and site conditions in addition to the nature of the surface structure involved need to be properly assessed in terms of the size of the support pillar required. Other methods of reducing subsidence and associated effects should be considered also.

Colliery spoil heaps respond to mining subsidence similarly to natural ground except that the crest tends to experience increased settlement, and there is more displacement down the slope of the spoil heap. The angle of draw tends to be increased locally by the presence of the spoil heap.

Mining subsidence effects in respect of undersea workings need to be carefully evaluated and the UK coal industry's Instruction provides the basis for safe extraction designs.

CHAPTER 18

GROUND MOVEMENT BEHAVIOUR AS INDICATED BY SUBSIDENCE MODELS

The use of models to indicate the likely behaviour of ground when subjected to mining extraction processes and resulting caving and general movement, has been widely recognised in many countries as a powerful tool in the prediction and interpretation of subsidence phenomena. The most commonly used form of subsidence model has been that of physical representation of the mining extraction and overlying rocks using a sand/plaster/water mix or similar combination of materials, suitably scaled in strength and size so as to allow accurate simulation of ground movement following extraction. Many other forms of physical modelling technique have been used ranging from gelatine subsiding under its own weight, to plaster materials subjected to forces generated in a centrifuge. Later techniques have seen the emergence of powerful computed-based models particularly those of finite difference and finite element methods which have been adapted to allow simulation of ground movement behaviour. The choice of subsidence modelling technique depends mainly upon the situation being simulated and on the information sought. Two particular techniques are discussed here and results are presented to demonstrate the form of output achieved and general interpretations which can be made.

Physical models based on sand and plaster and which subside under their own weight

A large subsidence model testing facility has been used at the University of Nottingham, UK, for a number of years and this has enabled studies to be made of fracture development in ground above longwall mining extractions. This model employs gravity loading to generate caving, fracturing and subsidence of the ground above the mining excavation. A major advantage of the technique is the opportunity to observe fracture development and caving propagation as the longwall face proceeds from its starting position. Rock strength can be readily adjusted by selecting the appropriate sand to plaster ratio in the mixing stages of model construction. A particular advantage of this modelling technique is the fact that the effects are visible during all stages of development. Additionally, variations in stratigraphy can be incorporated as well as allowing the mining extraction process to take the form of several lifts if required.

The test rig has internal dimensions of 3m length, 1·5m height and 0·15m thickness and has been referred to by Whittaker, Reddish and Fitzpatrick (1985). Extraction dimensions are pre-determined and built into the physical model so that the mining phase can be easily accomplished by simply progressive removal of blocks from near the base of the model. The physical model is constructed in layers, and with the results referred to here the layers were cast 1·25cm thick using a sand/plaster mix with a thin interface of sawdust to facilitate bed separations to form. Dyed layers were used at intervals to ease photographic comparisons. These authors reported that the entire model casting operation took two days and that a drying period of 24 hours was allowed at 20°C after removal of the front casting shuttering. A convenient grid, commonly 10cm square, was drawn over the face of the model prior to the beginning of the test.

Dimensional analysis. The subsidence model was designed with due regard to the laws of dimensional analysis.

The variables of the problem are expressed in terms of their fundamental dimensions, usually mass, length and time (however, force, length and time are occasionally used). The dependent variables can be expressed, if required, in terms of all the independent variables. The subsidence model considered here is judged to depend on the following parameters:

1. The longwall extraction geometry, G (m)
2. The rock's tensile strength, T (N/m^2)
3. The rock's compressive strength, C (N/m^2)
4. The rock's Young's modulus, E (N/m^2)
5. The rock's Poisson's ratio, υ (dimensionless)
6. The rock's unit weight, W (N/m^3)

There are other parameters of less significance but in order to simplify the analysis, only those listed above have been used as the modelling control parameters for the purpose of demonstrating the validity of this experimental technique. Subsidence can be expressed in terms of these parameters as follows:

$$S = f\,(G, T, C, E, \upsilon, W)$$

Buckingham's theorem allows simplification of this expression since "a complete equation can be reduced to a functional relationship between a complete set of independent dimensionless products". A relationship needs to inter-relate the model with reality, and this is achieved by the various scale factors involved. Table 34 gives a summary of the dimensional analysis details used by Whittaker, Reddish and Fitzpatrick (1985) in their report on ground fractures due to longwall mining subsidence.

Results from physical models of mining subsidence

Figure 237 illustrates the general character of subsidence above a caved longwall extraction. The situation modelled corresponds to a shallow mining setting (h = 84m, w = 118m, M = 4m), with strong overburden, UCS = 75 – 125 MPa. Well-defined fracture lines occur at the edges of the extraction. Bed separations are prominent features exhibited by this model test, and indicate the potential for further closure. The central region of the collapsed roof strata shows the beds to have lowered in an almost intact condition near to the extraction horizon, although significant separations exist at higher horizons. The fracture lines above the edges of the longwall extraction show that such regions can represent flow paths for water and gas from upper horizons gaining access to the working horizon. The major separation above the longwall could act as a temporary reservoir for water percolating from adjacent or overlying strata, and result in later release through interconnecting fractures to the working longwall.

A further feature shown by Figure 237 is the ability of strong overburden at the surface to cause bridging across the lower caved beds overlying the longwall extraction. The capacity for bridging of the strata in this manner is dependent upon the strength and general competence of the rocks near to the surface, in addition to the width of the extracted region. Figure 237 shows tensile fractures at the surface and over the ribsides.

Figure 238 shows the progressive development of roof collapse as the longwall extraction advances from left to right. Large bed separations appear to be a significant feature during the early stages of longwall face extraction. Additionally, the roof caved in large blocks up to a distance of around 100m from the start position (about equal to the depth in this shallow situation) and thereafter caved more regularly in the form of controlled cantilever failures.

Figure 237 Physical model of caved strata above longwall extraction with strong overburden. Mining data: h = 84m; w = 118m; M = 4m.

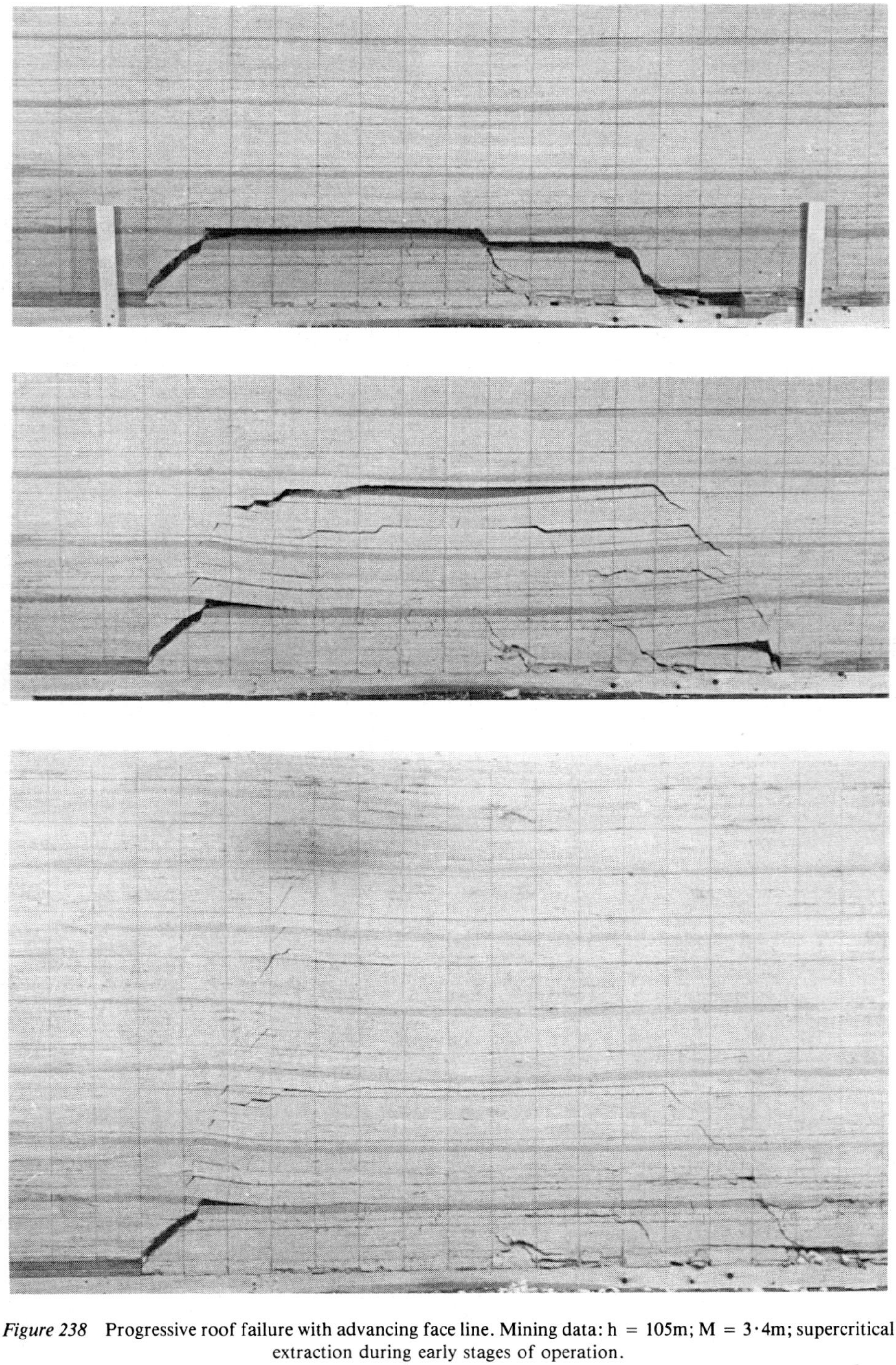

Figure 238 Progressive roof failure with advancing face line. Mining data: h = 105m; M = 3·4m; supercritical extraction during early stages of operation.

Variable	Dimensions	
S (subsidence)	L	
G (geometry)	L	M = mass
T (tensile strength)	$ML^{-1}T^{-2}$	L = length
C (compressive strength)	$ML^{-1}T^{-2}$	T = time
E (Young's modulus)	$ML^{-1}T^{-2}$	
υ (Poisson's ratio)	dimensionless	
W (unit weight)	$ML^{-2}T^{-2}$	

From Buckingham's theorem, the dimensional relationship can be expressed:

$$\frac{S}{G} = f\,(T/C,\ E/C,\ \upsilon,\ WG/C)$$

This needs to be valid for both the model and reality so that

$$\frac{Sm}{Gm} = \frac{Sr}{Gr}$$

where the subscripts m and r refer to the model and reality respectively.

It follows that:

$$\frac{Tm}{Cm} = \frac{Tr}{Cr}$$

$$\frac{Em}{Cm} = \frac{Er}{Cr}$$

$$\upsilon m = \upsilon r$$

$$\frac{Wm\ Gm}{Cm} = \frac{Wr\ Gr}{Cr}$$

Rearrange, to obtain:

$$\frac{Sr}{Sm} = \frac{Gr}{Gm}$$ } The *geometric scale factor*

and,

$$\frac{Tr}{Tm} = \frac{Er}{Em} = \frac{Cr}{Cm} = \frac{Wr\ Gr}{Wm\ Gm}$$

As Poisson's ratio is considered to have little effect in respect of this subsidence simulation, then it is ignored in this treatment.

It will be noted that Gr/Gm is readily determined, and in the analysis involved here

$$Gr/Gm \text{ for the model} = 105m/1{\cdot}0795m = 92{\cdot}27$$

A further ratio which is known is Wr/Wm as this corresponds to the ratio of the respective densities. Using an average value for Coal Measures rocks of 2350 kg/m^3, gives

$$\frac{Wr}{Wm} = \frac{2350\ kg/m^3}{1711\ kg/m^3} = 1{\cdot}37$$

(continued)

It is also necessary to determine the *strength scale factor,* particularly for Young's modulus and the model strata strengths (as they have the same units, they also have the same scale factors), and this is given by:

$$\frac{\text{Wr Gr}}{\text{Wm Gm}} = 133{\cdot}26 \Big\} \text{ The } \textit{strength scale factor}$$

Strength data of model material:

Sample	Young's modulus ($\times 10^6$ N/m^2)	Compressive strength ($\times 10^6$ N/m^2)
1	7·384	0·0821
2	8·636	0·0706
3	6·78	0·0856
4	7·04	0·0867
Average	7·384	0·0821

Rock properties simulated:

Uniaxial compressive strength	10·94 MPa
Young's modulus	0·984 GPa
Density	2350 kg/m^3

These strength data seem a little low, but it is suggested that when relating laboratory measurements of strength to *in situ* values a factor of as much as 5-10 times the latter needs to be usually applied. Taking this last point into account, indicates that the strength values determined for the model appear reasonable.

Table 34 Dimensional analysis aspects of physical modelling of mining subsidence due to a longwall extraction in Coal Measures strata. After Whittaker, Reddish and Fitzpatrick (1985).

Figure 239 shows the fully developed situation of the sequence given in Figure 238. A supercritical subsidence condition has been achieved, and the general nature of the fractures regarding distribution and opening is illustrated. The fractures occur over the mined-out region, and their main concentration is associated with the longwall extraction edge effect. The fractures gradually dissipate from the extraction edge although they are still visible at the surface. There is a tendency for a general line of fracturing to develop, although such a line is continuous near to the extraction, it changes to an intermittent one towards the surface. The fractures have tended to close in the region over the extraction and significantly away from the edge effect. In considering the significance for such lines of fractures to allow the flow of water from upper horizons to the mining horizon, due regard needs to be given to the continuity and degree of inter-connection of such fractures. As illustrated in Figure 239, these extraction edge fractures become less frequent and of decreased magnitude in the upper horizons. The type of overburden will greatly influence the development and occurrence of such fractures and strong beds such as sandstones are likely to behave in this way. Pliable mudstones are more likely to deform during the subsidence phase with a decreased tendency for significant fracture development especially at higher horizons.

Figure 239 Illustrating development of fractures above edge of longwall extraction and general fracture opening and distribution between the mining horizon and the surface. Mining data: h = 105m; M = 3·4m; supercritical extraction.

The pattern of fracture development as indicated by Figure 239 is for an initial continuous line grading into an intermittent line of fractures which gradually dissipate towards the surface. It has been previously reported that the overburden exhibits three distinct zones of behaviour, a lower third of significant fracture development, a middle third almost free of fracture effects, and an upper third (in contact with the surface) which exhibits fracture development. The results shown in Figure 239 do not support the middle third concept of general freedom from fracture effects. These results suggest that a line of fractures could develop through to the surface if the extraction height was of sufficient magnitude and the depth of cover sufficiently shallow. This modelling technique was used to examine this aspect further and the principal findings are reported here.

The strength and composition of the overburden influences the magnitude of the resulting surface subsidence. Strong overburden can greatly reduce the maximum subsidence particularly in shallow depth conditions. An example of the influence of strength of overburden is illustrated in Figure 240 *(a)* and *(b)*. The strong overburden resulted in low magnitudes of maximum subsidence of the order of 10-20% (of M) for models M1 and M2 using a longwall width w = 136·5m; the condition of caving and subsidence for model M1 at this face width is shown in the lower photograph of Figure 238. Also shown in Figure 240 *(b)* are the longitudinal subsidence development graphs for these models, after the face had advanced more than 2h from its starting position. In the latter condition the overburden was unable to continue bridging and ultimately collapsed to give the subsidence profiles shown. The weaker overburden gave almost 100% (M) maximum subsidence, which was around 10% greater than that predicted by SEH (1975).

The subsidence results obtained using these physical models justified their use to examine the nature of fracture development in the overburden above longwall extractions.

Further research findings have been reported by Gaskell, Reddish and Whittaker (1988) on comparing physical model subsidence results with SEH (1975) predictions. Figure 240 *(c)* and *(d)* show results of subsidence and principal strains as measured from the model. Reasonable consistency with SEH (1975) is shown for the surface subsidence profile as observed and compared with that predicted. The physical model rock strength (UCS) was 0·06 MPa. The principal strains shown in Figure 240 *(d)* were calculated from measured displacements directly from the face of the model. A semi-circular concentration of principal strains is in evidence, as is the tendency for bridging to take place across the extraction.

Test results relating subsidence to fracture development

The physical model employed was constructed to specifically investigate the role played by extraction height on fracture development. The seam height modelled was equivalent to 10·8m which was extracted in six lifts. This resulted in maximum tensile ground strains at the surface of between 7·6 and 68·8 mm/m.

Figure 241 shows the resulting surface subsidence profiles compared with SEH (1975) for the first five lifts; there is generally good agreement between the results compared for the successive development of each profile. The half profile subsidence results for increasing extraction thickness are presented also in Figure 241, and these show a consistent trend.

A further aspect needing to be taken into account is that of the subsidence data embodied in SEH (1975) having been collated from field studies whose extractions were in the range 1-3m generally with virtually no data for extractions greater than around 3m. Consequently, in the investigation described here, SEH (1975) subsidence data when extrapolated to an extraction height of 8 to 9 m is well beyond the usual limits of applying such data for

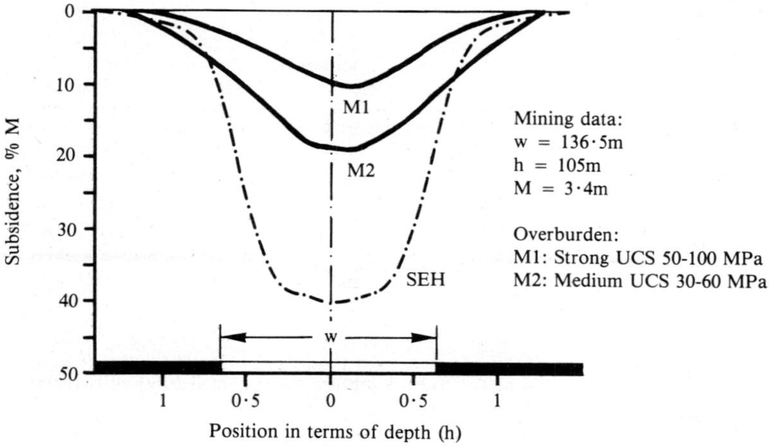

(a) Transverse subsidence profiles

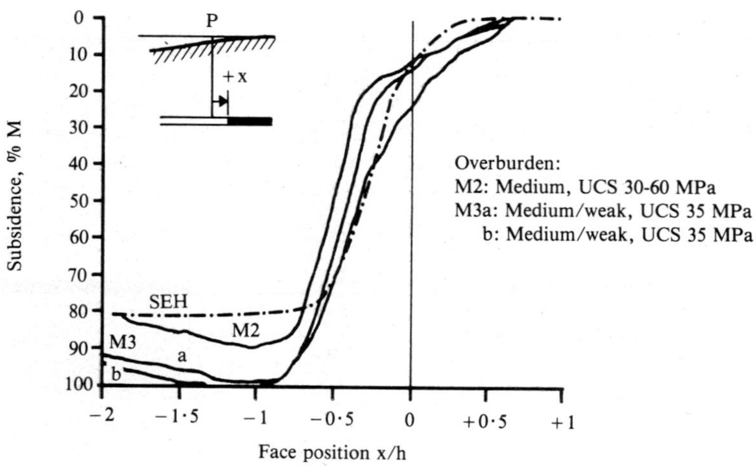

(b) Longitudinal subsidence profiles

Figure 240 Subsidence profile results for physical models compared with SEH (1975) prediction. After Whittaker, Reddish and Fitzpatrick (1985).

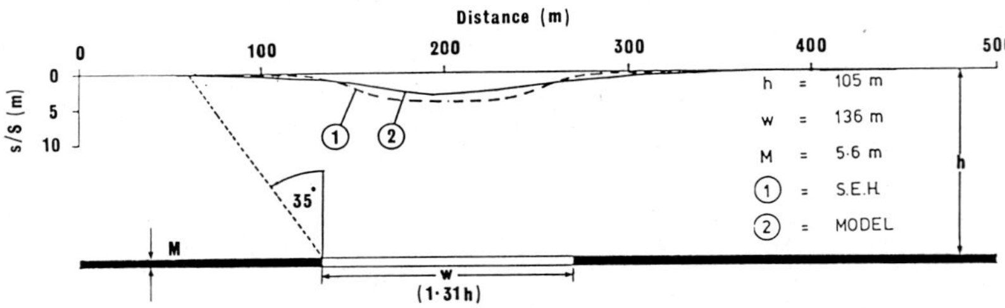

(c) Comparison of subsidence profiles: physical model with SEH (1975)

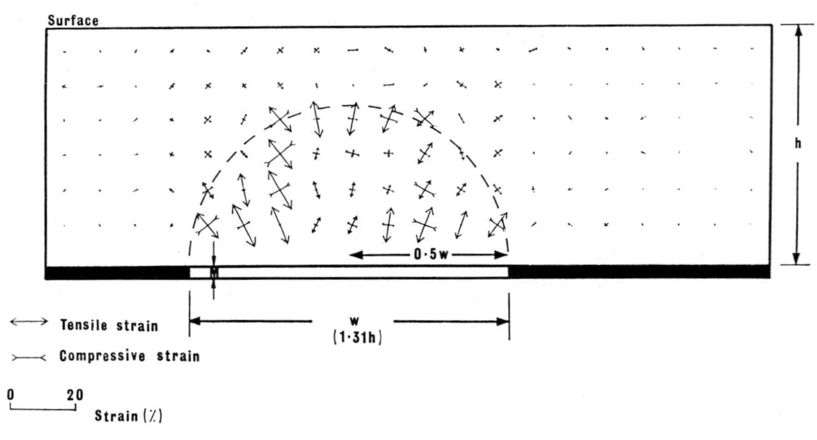

(d) Principal strains calculated from measured data obtained from *(c)* above.

Figure 240 (continued) Subsidence profile results for physical model compared with SEH (1975) prediction. After Gaskell, Reddish and Whittaker (1988).

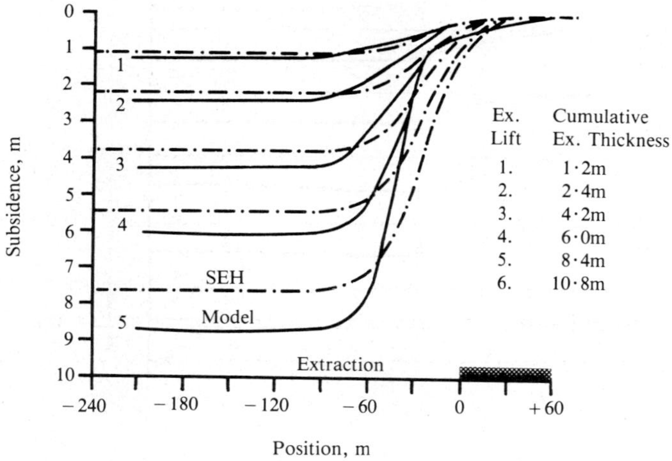

(a) Comparison between SEH predicted and physical model subsidence profiles

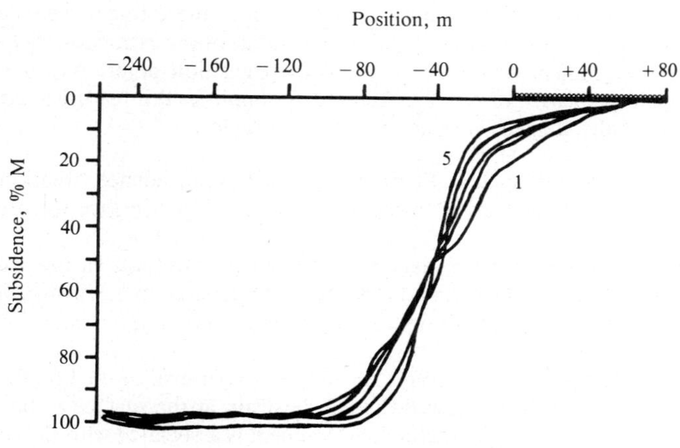

(b) Half-profile subsidence results for increasing extraction thickness

Figure 241 Surface subsidence profiles for increasing extraction thickness, model 4. After Whittaker, Reddish and Fitzpatrick (1985).

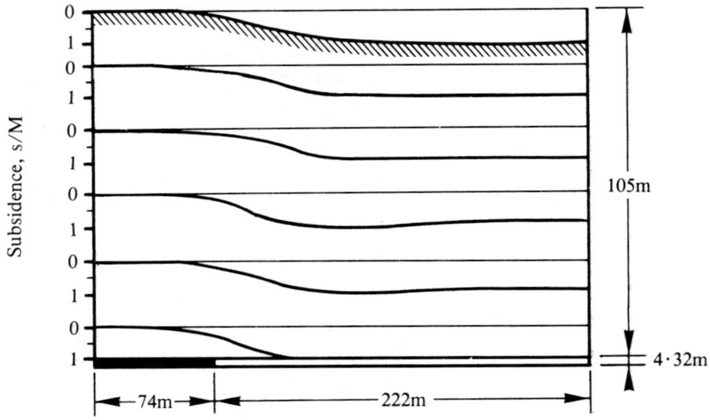

Figure 242 Surface and subsurface subsidence profiles (model 4). After Whittaker, Reddish and Fitzpatrick (1985).

prediction purposes. Taking this latter point into account it is interesting to note that Figure 241 *(b)* which shows subsidence curves plotted as a percentage of the extraction, demonstrates a definite trend in the gradient of the curve about the transition point. As the extraction increases, the gradient steepens. Conversely, SEH (1975) simplifies this feature to considering the curves to be identically shaped, varying only in magnitude.

In view of the satisfactory surface fit of these modelled subsidence situations, it was judged that this allowed valid deductions to be made as regards subsurface features.

Figure 242 presents the subsidence data for both the surface and various subsurface positions in respect of the longwall fracture development model used in the investigation. The subsidence curves steepen in close proximity to the extraction horizon.

The progressive development of fractures for successive removal of the lifts forming the extraction of this thick seam which allowed the tensile strain at the surface to be varied, is summarised in Figure 243. The major fracture development is associated with the edge of the extraction. Consequently, if mining dimensions permit, the face-line also becomes an origin of major fracture development. A principal feature exhibited is the regular progression of the fractures which appear related to the predicted tensile strain (+E) at the surface; this is particularly the case in the range 0 to +30 mm/m. The fractures demonstrate similarity in nature and magnitude. For the situations above +30 mm/m maximum tensile strain, a second line of fractures started significant development, and these are seen to be approximately parallel to the first line whilst displaced over the extracted region by a distance of the order of h/3 - h/3·5. As the lifts are extracted, fractures widen in addition to resulting in continuing to connect up.

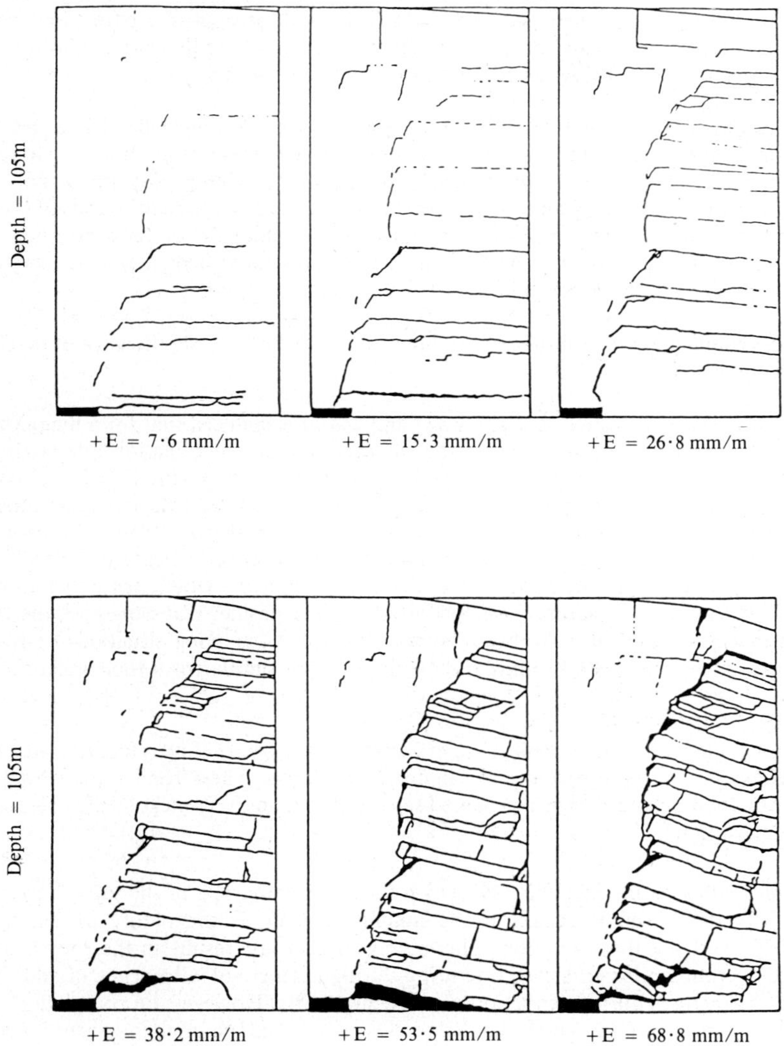

Figure 243 Fracture pattern development over ribside of longwall for progressive magnitudes of extraction height. After Whittaker, Reddish and Fitzpatrick (1985).

The fractures shown in Figure 243 have been examined in terms of fracture propagation in relation to the value of maximum tensile strain (+E) at the surface. Figure 244 shows an interpretation of these fracture developments. Two specific aspects have been addressed in the analysis, firstly the maximum height extended by the fractures which are judged to definitely interconnect with the extraction horizon, and secondly the extent of appreciable fracture development above the mining horizon but which do not necessarily interconnect. The first category of fracture development is interpreted as being indicative of where free flow from an overlying aquifer would readily occur, whilst the second could be indicative of where there might be a risk of water inflow seeping from an overlying aquifer.

The results indicate that fracture development from a longwall ribside gradually decreases in significance; there is no abrupt change in development of the fractures in respect of their upward occurrence. Local geology and depth of mining play important roles, especially in influencing the magnitude and extent of fracture development. Bands of clay and aquicludes generally located in the overburden can act as major factors in controlling water seeping from overlying horizons even though stronger fractured beds may exist above and below such pliable and impervious bands.

Fracture development in overburden above longwall extraction and significance of maximum tensile ground strain

From the results presented in Figures 243 and 244, it is deduced that for a magnitude of +10 mm/m maximum tensile strain at the surface, fracture propagation approaches the surface to a height of around h/3 for fracture interconnection and h/2 for maximum influence of fracture development. These deductions are made in respect of the model to investigate the effects of fracture development for a depth (h) of 105m. The geological situation modelled is one which readily promotes fracture development. In actuality, the existence of pliable mudstone beds within the strata sequence would tend to inhibit the magnitude and extent of fracture development above the ribside. Consequently, the results presented in Figures 243 and 244 most probably represent the worst situations of fracture development in that the beds forming the overburden are simulated in their most sensitive condition.

It is concluded from these results that an acceptable factor of safety is incorporated in the UK guidelines for longwall operation in undersea workings. These results do not apply to geological faults which need to be considered separately and in the context of their setting and physical description.

An interesting feature of Figure 245 which shows a photograph of the subsidence model after six lifts have been extracted at the mining horizon, is the nature of the ground immediately overlying the extraction. The mining of each lift results in the roof beds and immediate overburden experiencing fracturing during caving and subsidence in addition to reopening of fractures during extraction of subsequent lifts. However, on completion of the extraction of each lift, major cracks tended to close and caved material compact to a certain extent in the extracted region unaffected by the ribside. The general appearance of the caved ground is one of general integrity of the beds over the central region of the extraction. It is the ribside that exhibits major fracture development following multi-lift or single-lift thick extractions. The central region shows marked restoration of its ability to restrain water inflow from any overlying aquifers. Any inflow in such conditions would be mainly associated with the longwall edge/ribside fracture zone.

Major fracture development occurred at the surface after +30 mm/m (+E) and the maximum depth to which such fractures penetrated downwards was of the order of 7·5m.

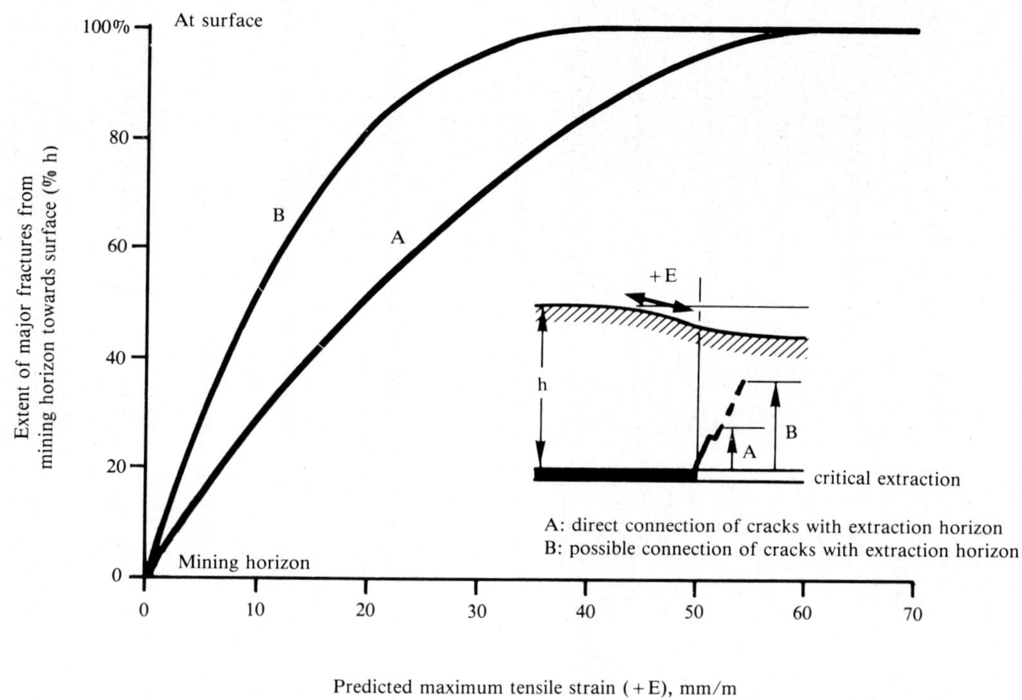

Figure 244 Extent of fractures above longwall extraction in relation to predicted maximum tensile strain (+E)

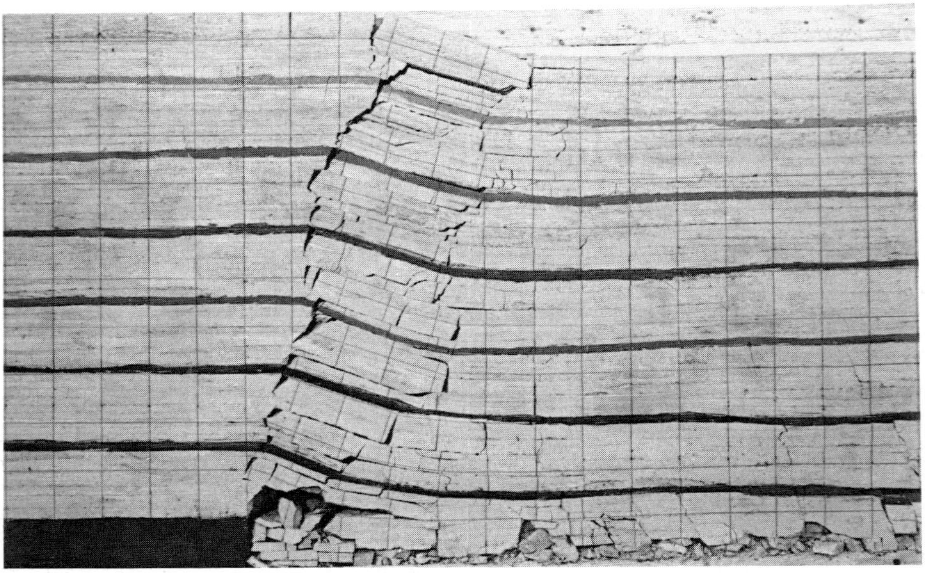

Figure 245 Demonstrating characteristic behaviour of caved goaf and closing of fractures after six successive longwall extractions.

This suggests that such surface cracks are essentially local effects with relatively minor deep-seated significance in normal longwall mining operations.

This particular example serves to indicate the useful role of physical models in mining subsidence investigations particularly in seeking guidance on likely fracture development within the subsurface strata.

Numerical modelling with special reference to finite element analysis

The *finite element method* (FEM) and its application to engineering problems has been comprehensively discussed by Zienkiewicz (1967, 1971 and 1977) and by Smith (1971) and is essentially a powerful numerical modelling procedure. The *method of finite differences* solves boundary-value problems in partial differential equations, with the differential equation being replaced by an approximating difference equation thereby permitting reduction of the problem to solving a system of algebraic equations. Solution of such systems usually requires resorting to some iterative technique, Forsythe and Wasow (1960), and Sokolnikoff and Redheffer (1966). The latter authors point out that a disadvantage of such numerical techniques is the generation of numerical values for unknown functions. These values are given for a set of discrete points. Conversely, the finite element method simplifies the partial differential equations governing the situation under consideration, and then proceeds to provide exact solutions.

The finite element method simplifies the dependent variable, which is displacement in the discussion here, over a small region of the area under analysis. The situation is subdivided into small elements considered to be connected at nodes and hence the derivation of the name. Displacement variation throughout the body is simplified and the nodal values in an element related to the forces applied at the nodes. The principles of virtual work and minimisation of potential energy are employed.

A full discussion of the principles of the finite element process is given by Zienkiewicz (1977) and the reader is recommended to consult this text for detailed appreciation of the application of the technique to wide ranging engineering problems. Zienkiewicz defines the finite element process as that of a method of approximating continuum problems involving:

1. division of the continuum into elements of finite number; the elements' behaviour is defined by parameters also of finite number, and
2. solution of the assemblage of elements as a complete system is governed by the same mathematical rules which apply to standard discrete problems.

He points out that the finite element method is particularly advantageous for dealing with plane problems in view of the high accuracy, low cost and the versatility it offers, in addition to the ease with which transverse isotropy characterising stratification, and body force problems can be dealt.

Zienkiewicz summarises the basic concept of finite element analysis of an elastic continuum using the displacement approach as:

1. establishment of the finite element structure of the continuum using imaginary lines or surfaces;
2. assumed interconnection of the elements at nodal points located on the boundaries of the elements;
3. nodal point displacements form the problem's basic unknown parameters;
4. the state of displacement within each element as related to its nodal displacements, requires to be defined by a set of functions;
5. the displacement functions specify the state of strain within each element as a function of the nodal displacements;
6. having established the strains, and with knowledge of any initial strains and the material's constitutive properties the state of stress both within and at the boundaries of each element can be defined.

Recognition of a number of approximations needs to be appreciated. Displacement continuity between elements may not always be completely satisfied by the displacement functions employed, and frequently leads to minor violation of the compatibility condition. Furthermore by use of equivalent forces concentrated at the nodes, only overall satisfaction of the equilibrium conditions is achieved, and this usually introduces localised violation of the equilibrium conditions within and at the boundaries of each element. It should be noted that the element shape chosen, the form of displacement function employed, and the degree of approximation adopted, according to Zienkiewicz are important factors to be taken into account in such analyses although there is considerable scope for ingenuity and skill on the part of the user. Finite element analysis of elastic solids using the displacement approach is the most widely adopted method especially in view of the ease with which it can be interpreted and understood. Both linear elastic and non-linear analyses are dealt with conveniently.

Application of the finite element method to mining subsidence and related effects

Hazine (1977) considered the application of the finite element method to a longwall mining subsidence situation and made the following general comments:

1. By virtue of the symmetrical configuration of the longwall transverse subsidence trough, only a half profile needed analysis with suitable restraints.

2. Restraint of the model should be achieved horizontally on both sides in addition to the vertical direction along the model base; elsewhere was required to be free to move.
3. A satisfactory subsidence profile was achieved using a value for Young's modulus of 0·55 GPa.
4. Poisson's ratio was found to have a generally insignificant effect on the results.
5. Elastic, isotropic analysis produced subsidence profiles which were much shallower and broader than observed in practice.
6. A subsidence model based on transverse isotropy employing a modular ratio in the order of Ev/Eh = 50 to 100 achieved a closer match of subsidence profile with SEH (1975).
7. Surface ground strain profiles were best obtained using curvature considerations.

Reddish (1984) critically reviewed Hazine's general findings and made the following remarks:

(a) Modular ratios of the order of 50-100 were considered unrealistic and produced unsatisfactory displacement profiles.
(b) Non-linear modelling was considered to be a much improved approach over that of excessive variation of the modular ratio.

Fitzpatrick (1987) examined aspects of modelling of mining subsidence mechanisms and the prediction of ground movements. In particular he discussed the application of finite element analysis to mining subsidence, and the results presented here have been obtained using his particular computer program.

Fitzpatrick extended the work on finite element modelling of mining subsidence and developed a program which incorporated the Mohr-Coulomb failure criterion.

The basic computational procedure consists of:

1. Input mesh data, element properties and loading/restraint requirements.
2. Calculation of the element stiffness matrices and assembling into the global stiffness matrix.
3. Application of back-substitution to the set of equations linking the system of applied forces (gravity loading) with the global stiffness matrix and the resultant displacement vector, in order to determine the latter.
4. Calculation of the elemental stresses and strains.
5. Application of the failure criterion and updating of the element properties as considered necessary.
6. Should the load stepping be complete, then finish, if not return to (2).
7. Output the nodal displacements and the element stresses and strains for post-processing.

The program is run twice for each situation so that the gravity loading of intact strata results can be subtracted from those for the mined-out case, so that the result of mining can be determined. Consequently the results for the intact and mined-out cases are used as inputs to a post-processing program which performs the necessary subtraction operation.

There are a number of graphical capabilities: mesh plots of the undeformed, intact deformed and mined-out states; vector displacement plot; principal stress and strain plots; profile plots of subsidence, displacement and strain.

The mined-out element was considered by Hazine (1977) and Reddish (1984) for earlier elastic analyses to be modelled by a traditional element of very low magnitude of Young's modulus. This could introduce difficulties where the element inverted or incomplete closure of the extraction occurred. Adjustment of this value of Young's modulus usually required several computer runs. Fitzpatrick's approach to the mined-out element was to use a very low value for Young's modulus up to the strain (vertical) approaching unity and to use a very high value for Young's modulus thereafter. This was considered to be a fairly realistic representation of the build-up of loading in the mined-out area behind a longwall extraction. The term 'failure' is in effect only an analogy in this situation as it is a theoretical procedure.

The following subsidence situations have been examined using the finite element model as developed by Fitzpatrick (1987). The basic data used in the finite element models are summarised in Table 35.

Four particular mining situations are considered, namely a single longwall of different w/h ratios; shaft pillar dimensional effects; interaction between workings in two seams; and effect of surface ground slope.

Model strata property	x-direction		y-direction
Young's modulus, intact condition (MPa)	600		600
Young's modulus, failed condition (MPa)	60		60
Uniaxial compressive strength, intact condition (MPa)	2		2
Uniaxial tensile strength, failed condition (MPa)	0·2		0·2
Poisson's ratio	0·2		0·2
Rock density (kg/m^3)		2350	
Bulk modulus (MPa)		100	

Table 35 Model strata properties used in finite element analysis of mining subsidence

Effect of different longwall width/depth ratios on subsidence and subsurface behaviour, stress and strain configurations

In order to gain an appreciation of the influence of different longwall width/depth ratios on subsidence behaviour, a finite element model was devised having a depth (h) of 100m, extraction height (M) of 2m and longwall widths (w) of 50, 100 and 200m respectively.

Figure 246 shows a comparison of the vector displacement diagrams for the three face widths considered and the following comments are made:

1. There is a major increase in the magnitude and extent of ground displacement as the w/h ratio increases beyond a value of around 1, and is most probably associated with the breakdown of the bridging properties of the strata across the longwall extraction at higher w/h ratios.
2. The surface line of Figure 246 *(c)* clearly illustrates areas of increasing lateral movement towards and over the ribsides.
3. Incomplete closure is taking place at the edges of the extraction.
4. This model indicates insignificant floor lift, due to stress removal following mining, particularly around the central area of the longwall extraction.
5. The greatest lateral displacements associated with the longwall mining process, are indicated to occur at the surface boundary.

The surface subsidence profile for each of these w/h ratios has been plotted and is compared to the SEH (1975) predicted result in Figure 246 *(d)*. It is concluded from the comparison of these finite element and SEH (1975) predicted subsidence profiles, that the former appears to represent fairly competent overburden in this particular situation. The finite element predicted result has a significant measure of agreement with the empirically predicted profile. Due regard needs to be given to the type of overburden strata which the finite element model represents, and there is of course scope for changing the properties in

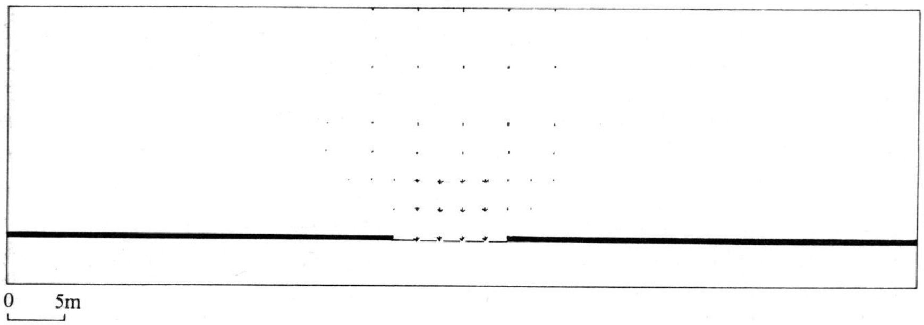

(a) Displacement pattern: w = 50m, h = 100m, M = 2m, L > 2h

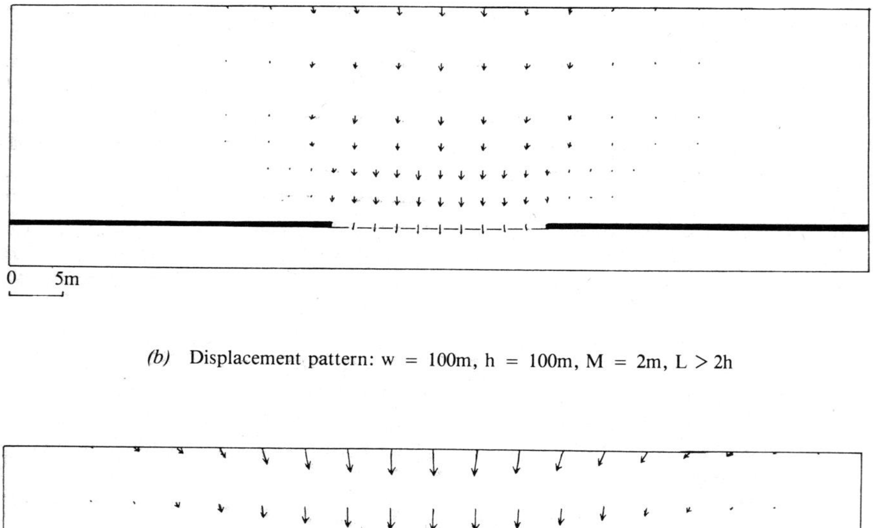

(b) Displacement pattern: w = 100m, h = 100m, M = 2m, L > 2h

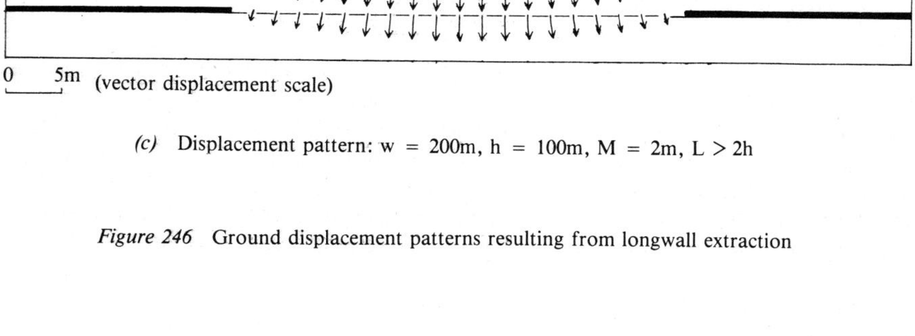

(c) Displacement pattern: w = 200m, h = 100m, M = 2m, L > 2h

Figure 246 Ground displacement patterns resulting from longwall extraction

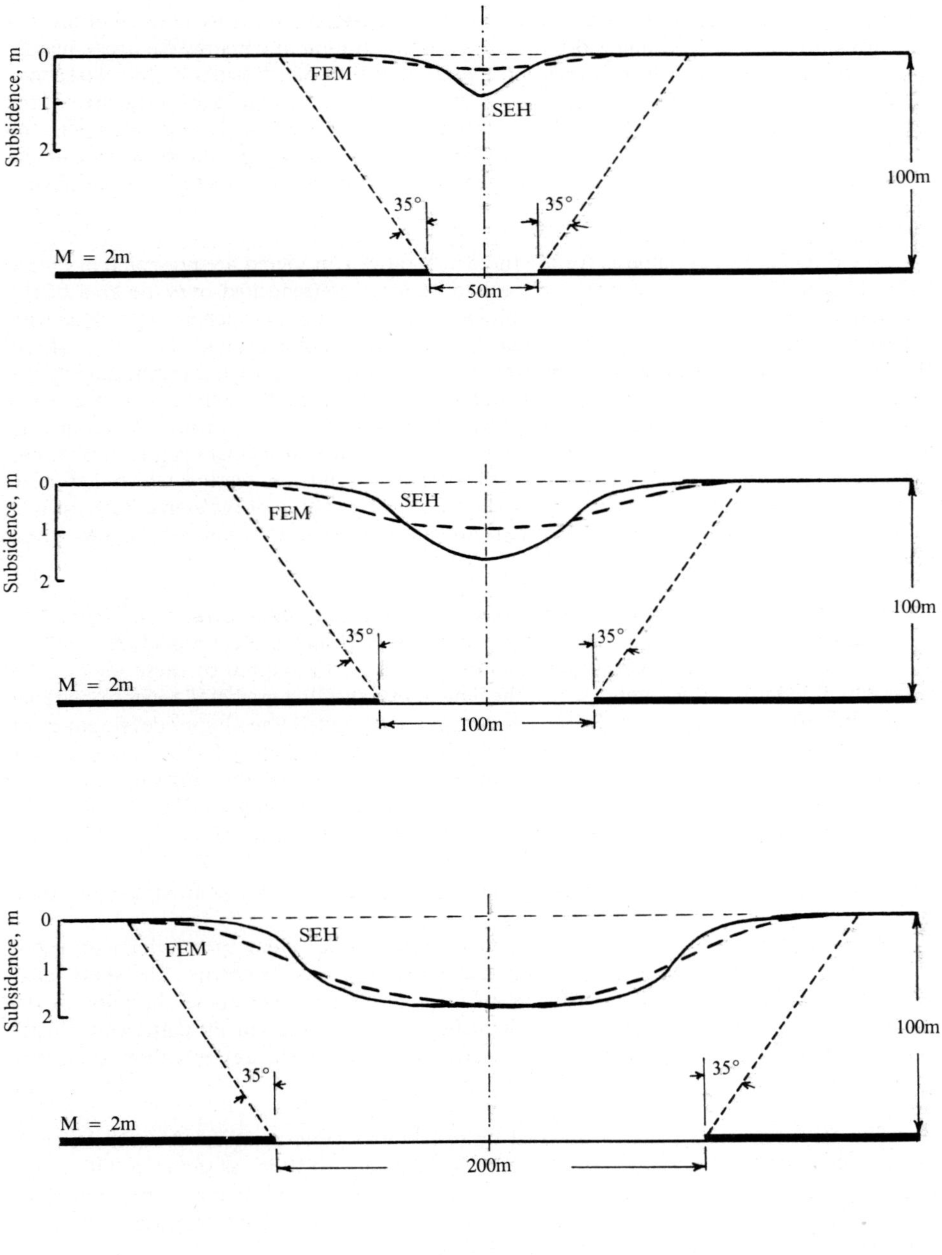

Figure 246 (d) Comparison of finite element method and empirical (SEH) predictions of surface subsidence above a longwall mining extraction.

this form of model in order to produce particular results especially regarding the magnitude of subsidence.

Figure 247 clearly shows that the extent and magnitude of stress redistribution is a function of the longwall mining width. There is a concentration of compressive stress, mainly in the vertical direction, which is concentrated around the solid ribsides of the extraction. Appreciable tensile stress occurs close to the extraction horizon around the central area of the longwall for the w/h ratios of 0·5 and 1 particularly. This is probably associated with the model showing incomplete closure at the face centre-line. Figure 247 *(c)* shows that the model experienced significant closure of the mined-out zone after the supercritical subsidence condition was exceeded.

The principal strain patterns for the three w/h ratios considered are presented in Figure 248. The significant principal strains appear to be more concentrated over the area of the excavation, than by comparison to the significant principal stresses which are associated with the solid ground. There is a general indication that at a point around the centre of the longwall extraction, the principal strains appear to illustrate a particular trend, namely the principal tensile strains are approximately radial whilst the principal compressive strains are approximately tangential. The strain magnitude decreases progressively towards the surface, and in general over most of the area of the model the induced principal compressive strains are larger than the induced principal tensile strains. A further feature which has been indicated by physical models, is that of the central mined-out area of the extraction shown in Figure 248 *(c),* demonstrating the occurrence of an approximately semi-circular low strain region.

An appreciation of the non-linear nature of the model is demonstrated in Figure 249, which shows zones of material failure as predicted by the finite element model. As material failure has progressed, corresponding updating of the elastic moduli of these regions has taken place. Figures 249 *(a)* and *(b)* show the limited extent of the predicted zones of material failure, but after the w/h ratio becomes supercritical, this model shows major development of the failed zones. Considering the surface boundary condition, Figure 249 *(b)* shows tensile failure over the solid ribsides before any compressive failure over the longwall extraction, whilst the more extensively mined condition of Figure 249 *(c)* shows surface compressive failure immediately above the extraction flanked by zones of tensile failure.

Effect of shaft pillar size on the resulting patterns of ground displacement, stress and strain

Figure 250 shows the effect of increasing the size of shaft pillar from a diameter (p) of 50m (0·5h) to 200m (2h) on the resulting ground displacement patterns. The shaft pillar diameter of 0·5h is indicated to result in appreciable displacement above the pillar at the surface. The effect of increasing the diameter on the displacement pattern is clearly demonstrated by Figure 250 *(b)* and *(c),* especially on reducing surface effects above the centre of the pillars.

The principal stress patterns induced by totally extracting the seam around the shaft pillar is illustrated in Figure 251. For a shaft pillar diameter of 0·5h, as shown in Figure 251 *(a),* the pillar gives rise to a high concentration of stress throughout its entire width around the mining horizon. As the pillar diameter is increased there is a significant decrease in the stress magnitudes around the central area of the shaft pillar.

Figure 252 shows the principal strain patterns induced in the ground between the mining horizon and the surface due to total extraction around the shaft pillar. With small diameter shaft pillars, the most significantly strained part of the shaft is indicated to be vertical

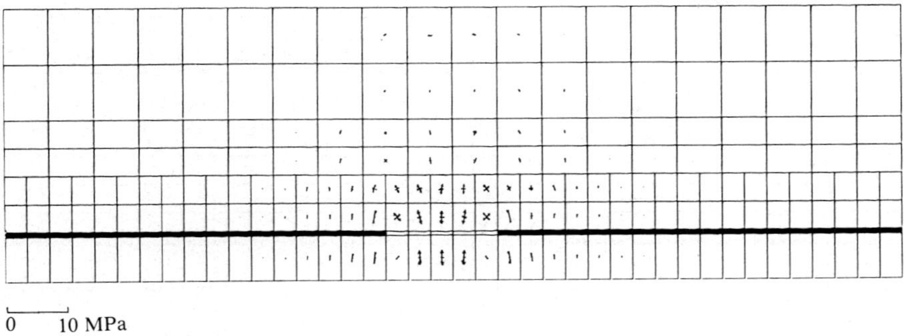

(a) Principal stress pattern: w = 50m, h = 100m, M = 2m, L > 2h

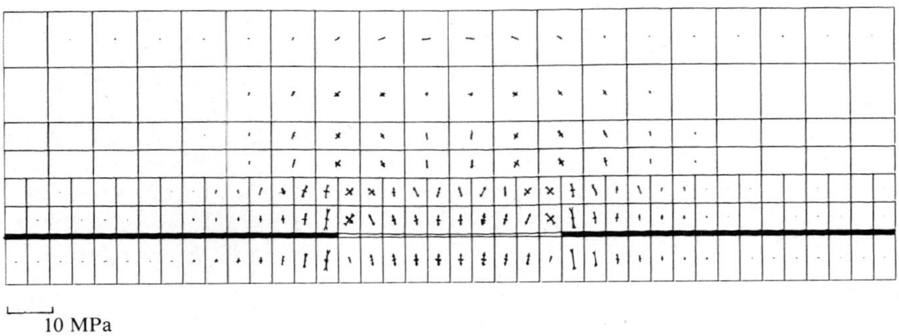

(b) Principal stress pattern: w = 100m, h = 100m, M = 2m, L > 2h

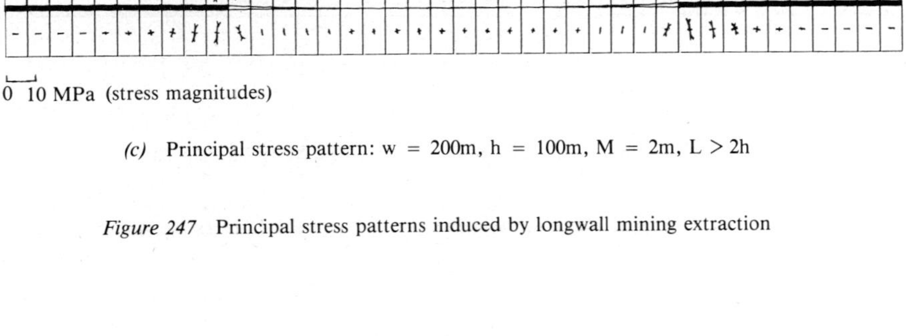

(c) Principal stress pattern: w = 200m, h = 100m, M = 2m, L > 2h

Figure 247 Principal stress patterns induced by longwall mining extraction

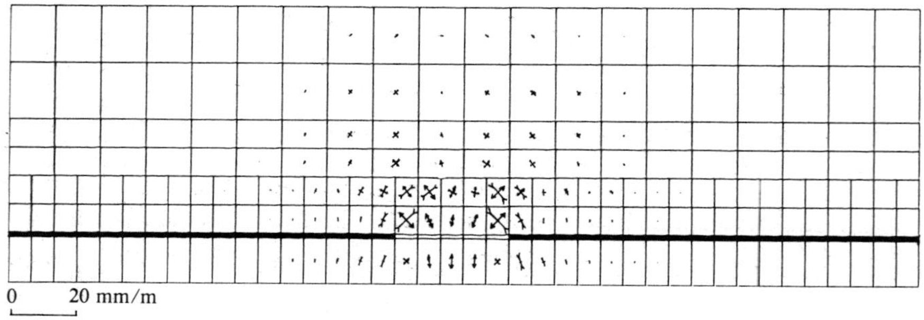

(a) Principal strain pattern: w = 50m, h = 100m, M = 2m, L > 2h

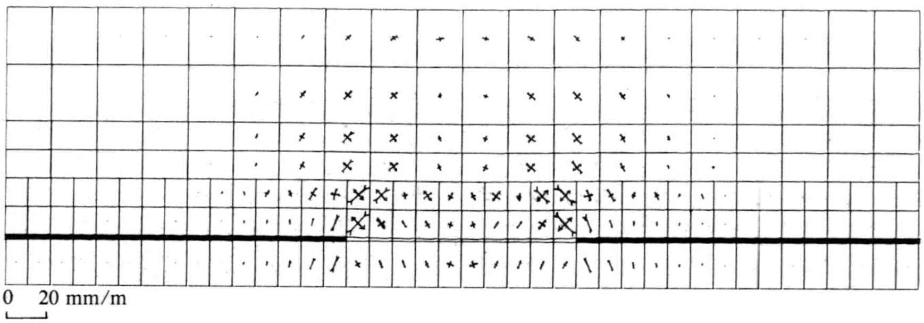

(b) Principal strain pattern: w = 100m, h = 100m, M = 2m, L > 2h

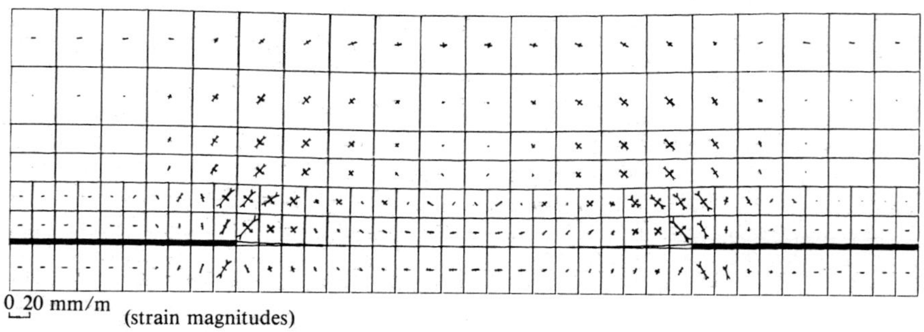

(c) Principal strain pattern: w = 200m, h = 100m, M = 2m, L > 2h

Figure 248 Principal strain patterns induced by longwall mining extraction

(a) Zones of material failure: w = 50m, h = 100m, M = 2m, L > 2h

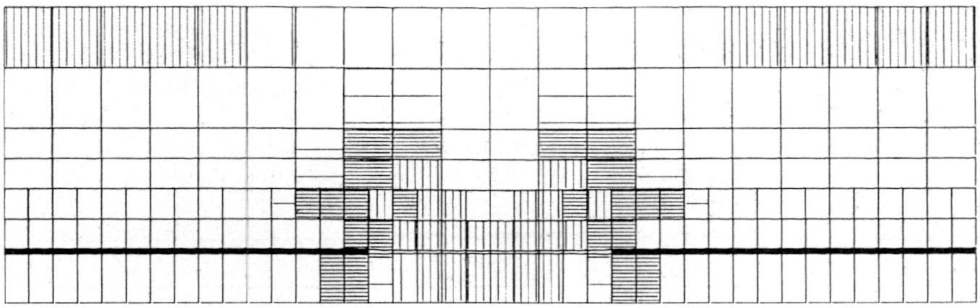

(b) Zones of material failure: w = 100m, h = 100m, M = 2m, L > 2h

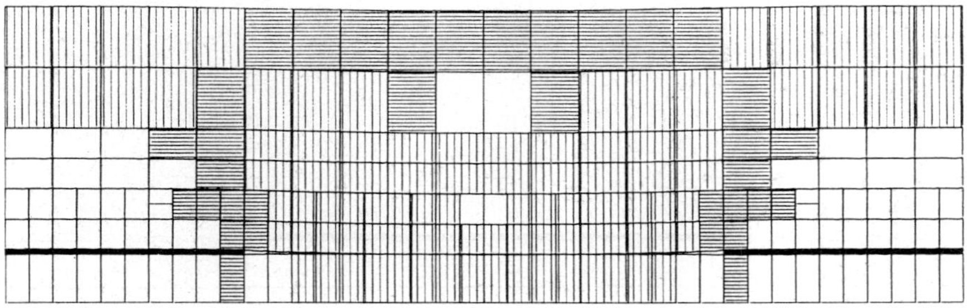

Vertical hatching corresponds to tensile failure.
Horizontal hatching corresponds to compressive failure.
Density of hatching indicates order of failure.

(c) Zones of material failure: w = 200m, h = 100m, M = 2m, L > 2h

Figure 249 Zones of material failure induced by longwall extraction

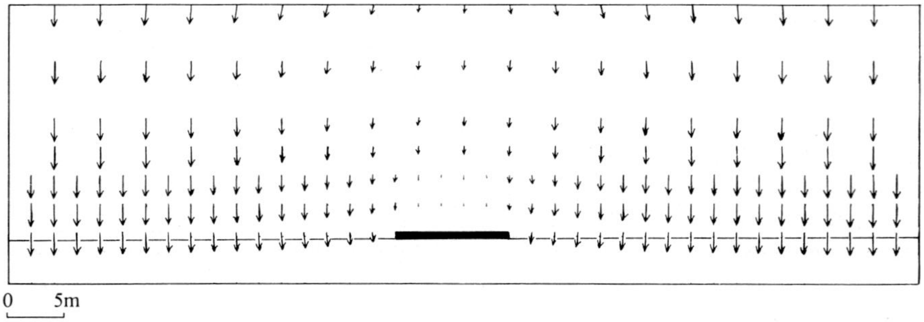

(a) Displacement pattern: p = 50m, h = 100m, M = 2m

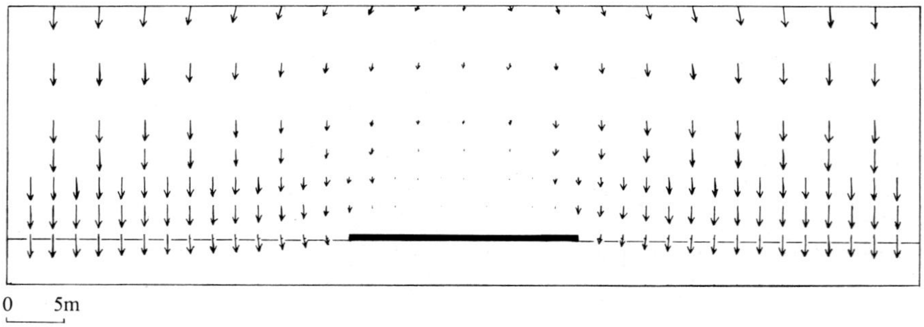

(b) Displacement pattern: p = 100m, h = 100m, M = 2m

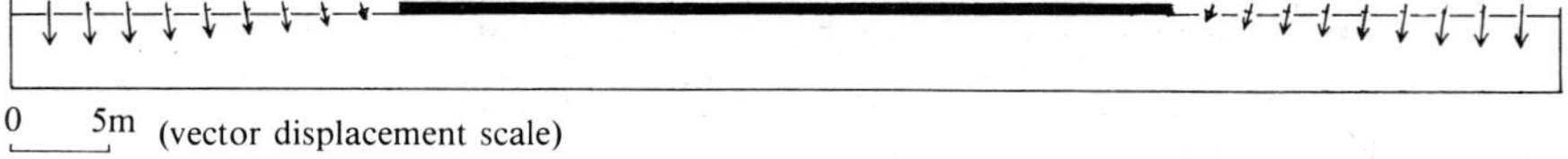

(c) Displacement pattern: p = 200m, h = 100m, M = 2m

Figure 250 Displacement pattern around shaft pillar due to total extraction on all sides

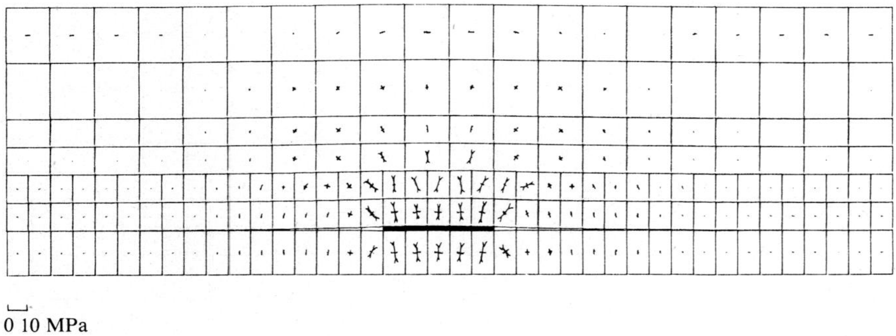

(a) Principal stress pattern: p = 50m, h = 100m, M = 2m

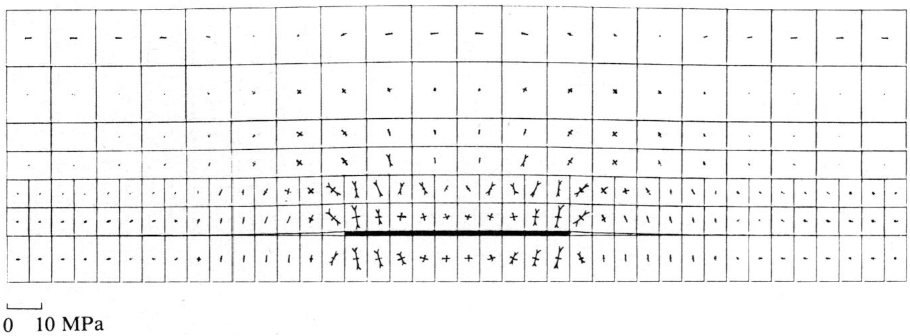

(b) Principal stress pattern: p = 100m, h = 100m, M = 2m

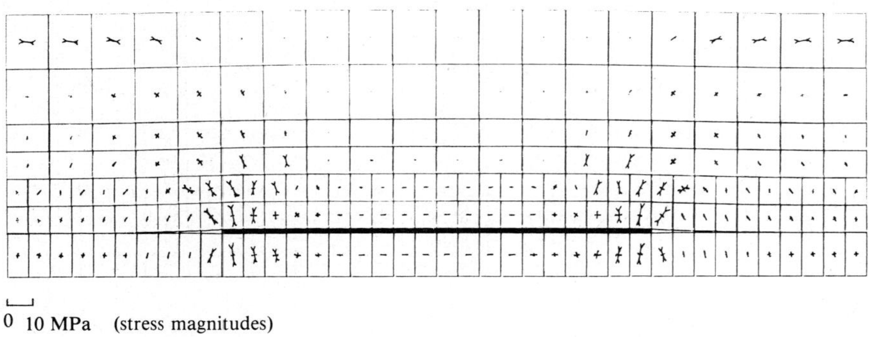

(c) Principal stress pattern: p = 200m, h = 100m, M = 2m

Figure 251 Principal stress patterns induced by total extraction around shaft pillar

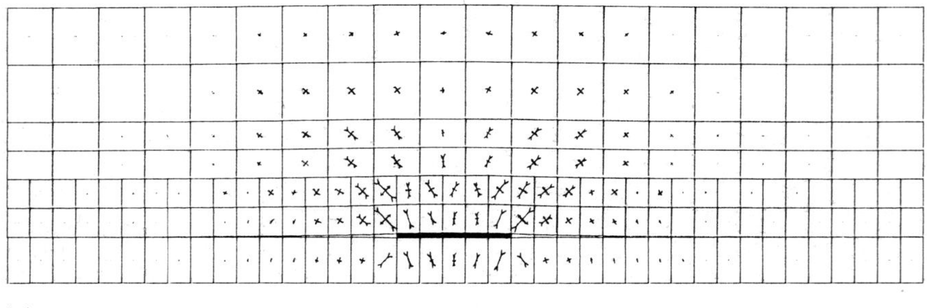

0 20 mm/m

(a) Principal strain pattern: p = 50m, h = 100m, M = 2m

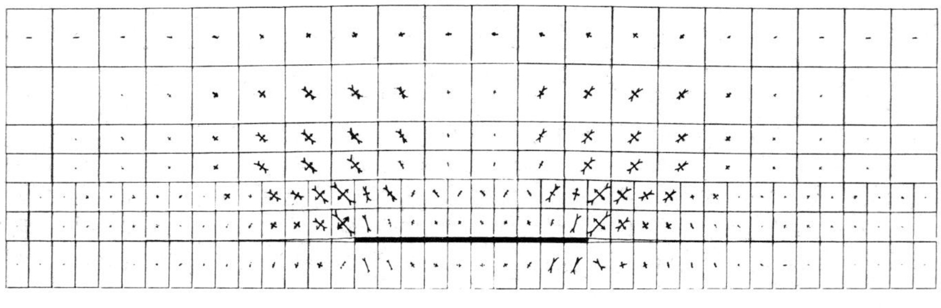

0 20 mm/m

(b) Principal strain pattern: p = 100m, h = 100m, M = 2m

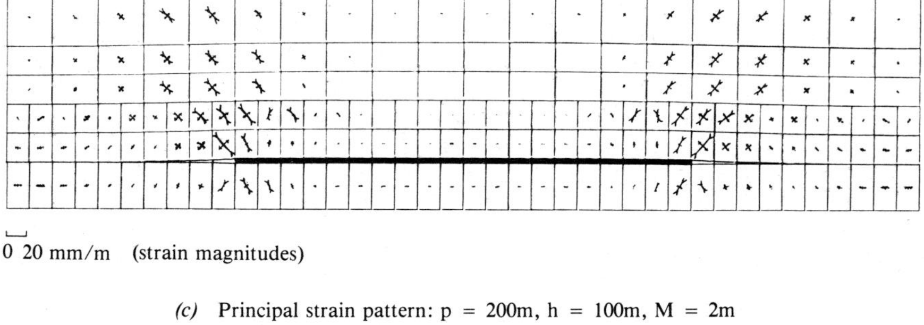

0 20 mm/m (strain magnitudes)

(c) Principal strain pattern: p = 200m, h = 100m, M = 2m

Figure 252 Principal strain patterns induced by total extraction around shaft pillar compression near the pillar horizon. Tensile strain at the surface boundary is indicated to be the most predominant source of damage to surface structures. It would appear that an undersized shaft pillar can result in two main effects, firstly that of excessive vertical compressive strain giving rise to potential damage at the base of the shaft, and secondly horizontal tensile strain in and around the top of the shaft.

Interaction effects between longwall mining extractions in two seams

Interaction between longwall workings in different seams calls for special consideration to be given to layouts which minimise such effects on the stability of roadways and faces.

Figure 253 shows two common forms of longwall layout and the resulting ground displacements caused by the extractions employed. The effect of locating the rib pillar over the extraction in the lower seam has been to produce increased subsidence as indicated by the displacement vectors.

Considering the principal stresses induced by interaction between the longwall extractions in the two seams, it is seen from Figure 254 that the in-line layout gives rise to marked concentrations of vertical compressive stress in the proximity of the pillars and rib edges. Conversely the staggered layout promotes stress dissipation in the location of the pillar. These results demonstrate the transmission of vertical compression stress effects from an upper seam rib edge to a significant depth below it, see Figure 254 *(a)*.

Figure 255 shows the principal strains induced by interaction of these longwall mining layouts. The two ground strain patterns make interesting comparison, as for instance there is markedly increased strain between the seams within the mined-out area with the staggered layout, see Figure 255 *(a)*.

Figure 256 shows a comparison of the zones of material failure associated with the interaction of the longwall extractions in these two layouts. There is appreciable compressive failure indicated with the in-line pillar arrangement of Figure 256 *(b)*, and similarly with the rib edges.

Effect of sloping ground surface on displacement, stress and strain patterns associated with a longwall mining extraction

Two different values of slope have been considered, namely 15° and 30° dip above a flat seam where w = 100m, mean depth = 100m and M = 2m.

Figure 257 shows the displacement patterns to be asymmetric and similar to the surface effect predicted earlier in Chapter 11.

The induced stress patterns illustrated in Figure 258 show a marked effect on the down-slope ribside, again with asymmetrical patterns resulting from the effect of the sloping ground.

Figure 259 shows the principal strain patterns for the two values of surface ground slope. Increased strain magnitudes occur in the ground between the mining horizon and the surface over the ribside on the down-slope side of the longwall and thus indicates greater likelihood of ground fracturing on this side of the extraction.

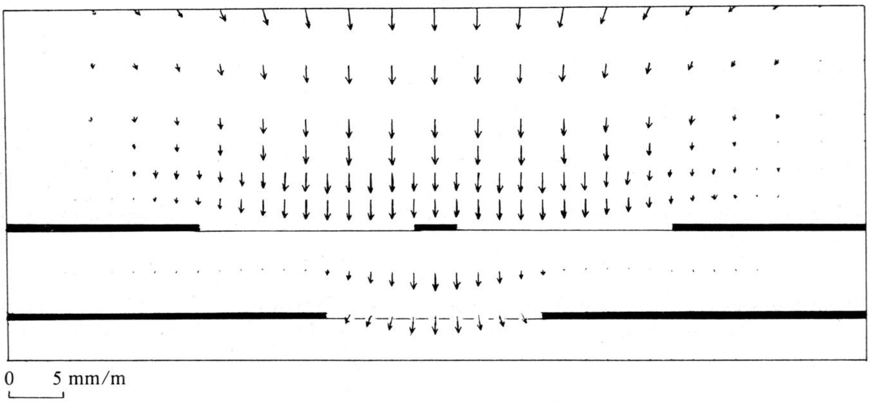

(a) Staggered layout: w = 100m, p = 20m, h = 100m and 140m, and M = 2m (both seams)

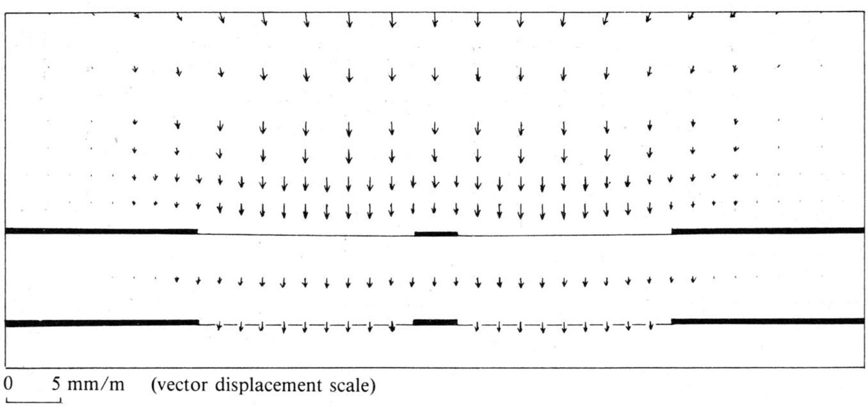

(b) Pillars in-line layout: w = 100m, p = 20m, h = 100m and 140m, and M = 2m (both seams)

Figure 253 Vector displacements associated with interaction between longwall extractions in two seams

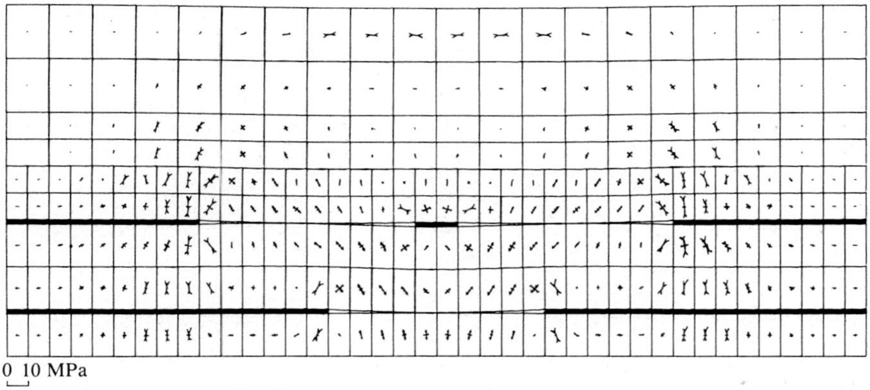

(a) Staggered layout: w = 100m, p = 20m, h = 100m and 140m, and M = 2m (both seams)

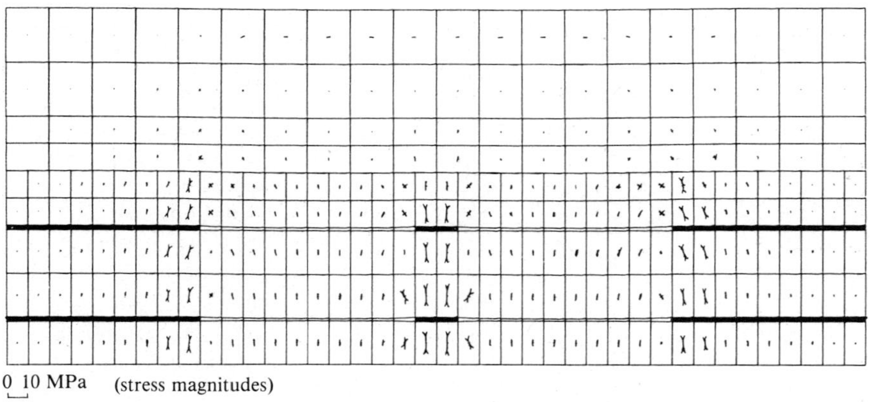

(b) Pillars in-line layout: w = 100m, p = 20m, h = 100m and 140m, and M = 2m (both seams)

Figure 254 Principal stresses induced by interaction between longwall extractions in two seams

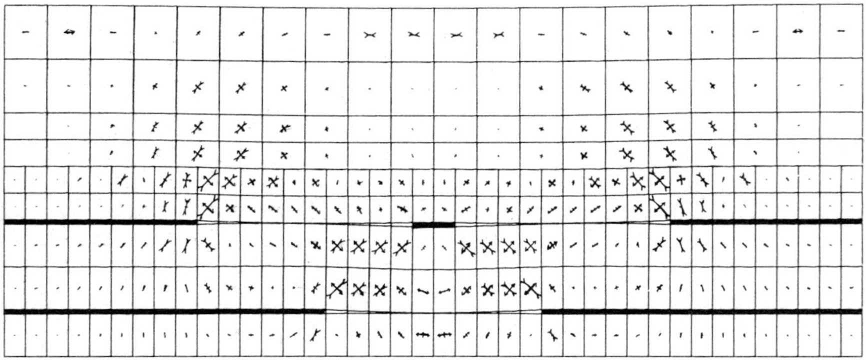

0 20 mm/m

(a) Staggered layout: w = 100m, p = 20m, h = 100m and 140m, and M = 2m (both seams)

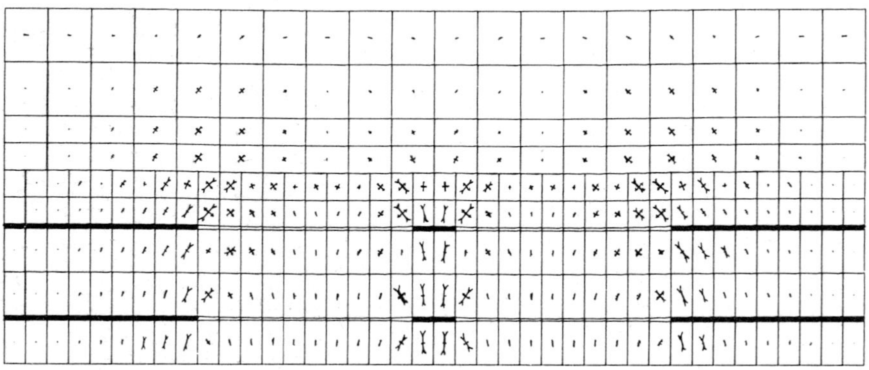

0 20 mm/m (strain magnitudes)

(b) Pillars in-line layout: w = 100m, p = 20m, h = 100m and 140m, and M = 2m (both seams)

Figure 255 Principal strains induced by interaction between longwall extractions in two seams

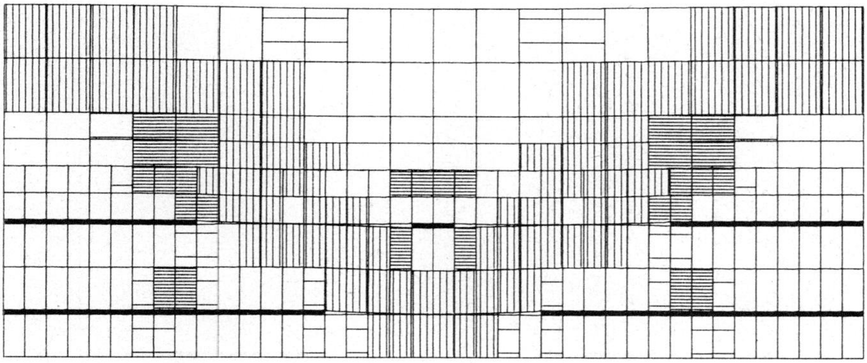

(a) Staggered layout: w = 100m, p = 20m, h = 100m and 140m, and M = 2m (both seams)

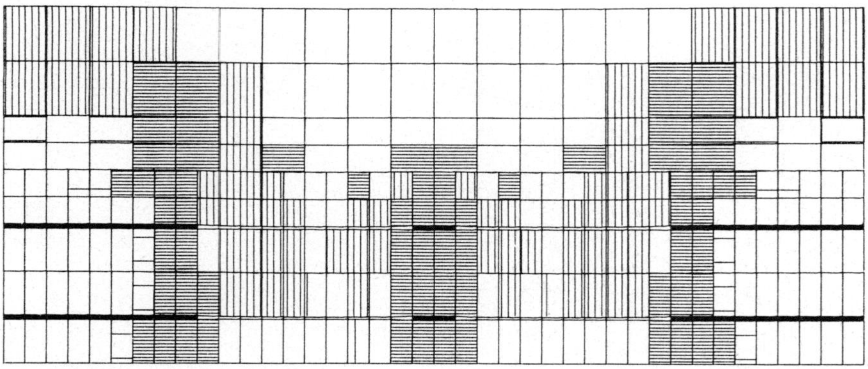

Vertical hatching corresponds to tensile failure; horizontal hatching corresponds to compressive failure. Density of hatching indicates order of failure

(b) Pillars in-line layout: w = 100m, p = 20m, h = 100m and 140m, and M = 2m (both seams)

Figure 256 Zones of material failure induced by longwall extractions in two seams and producing interaction effects

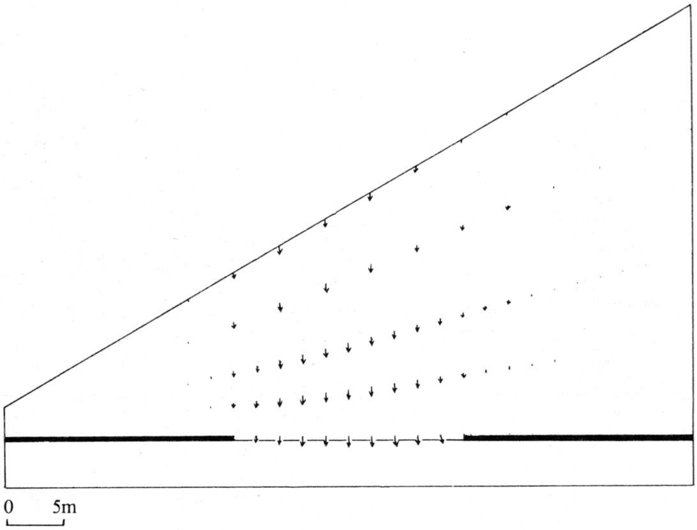

(a) Ground slope = 30°; h_{mean} = 100m, w = 100m, M = 2m

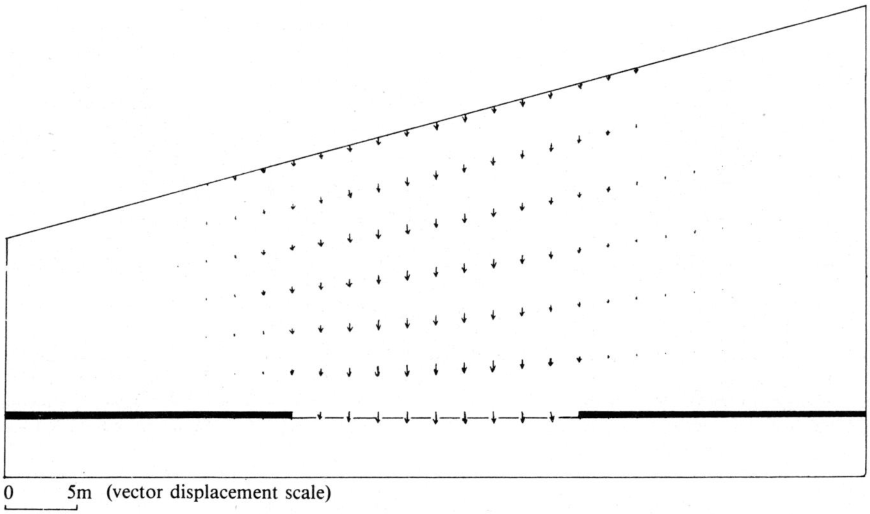

(b) Ground slope = 15°; h_{mean} = 100m, w = 100m, M = 2m

Figure 257 Displacement patterns arising from mining under sloping ground

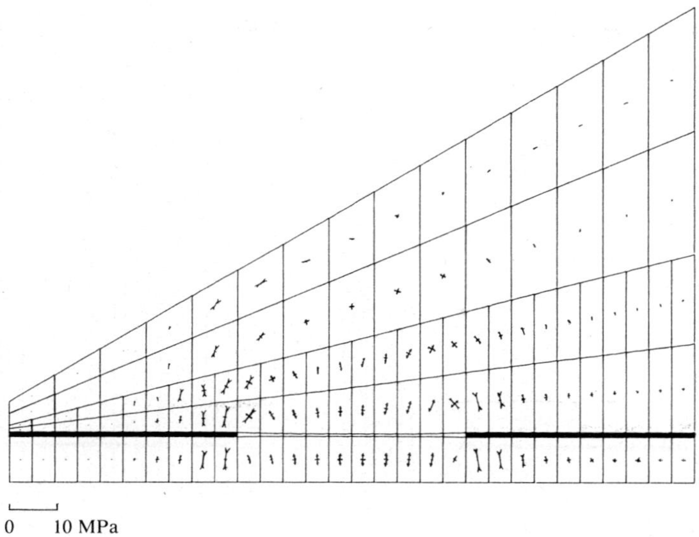

(a) Ground slope = 30°; h_{mean} = 100m, w = 100m, M = 2m

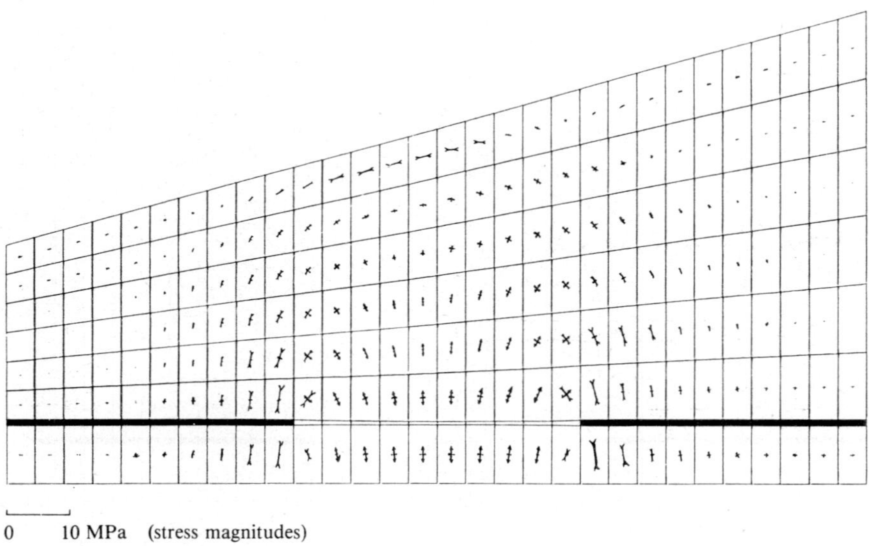

(b) Ground slope = 15°; h_{mean} = 100m, w = 100m, M = 2m

Figure 258 Principal stresses induced by longwall mining with sloping ground surface

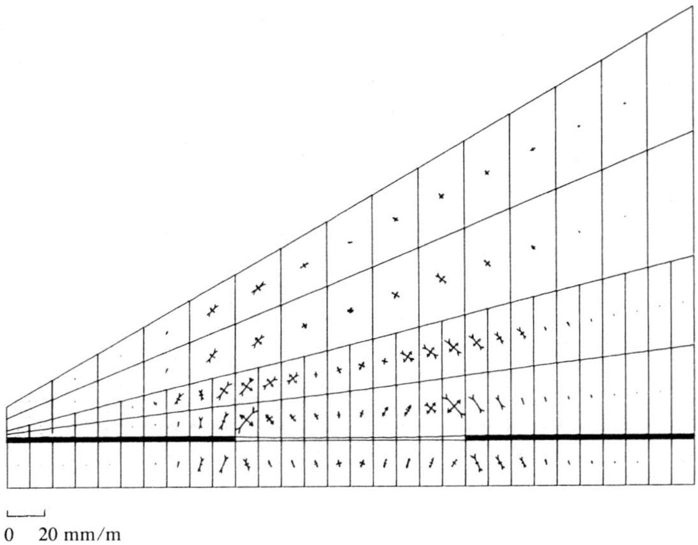

(a) Ground slope = 30°; h_{mean} = 100m, w = 100m, M = 2m

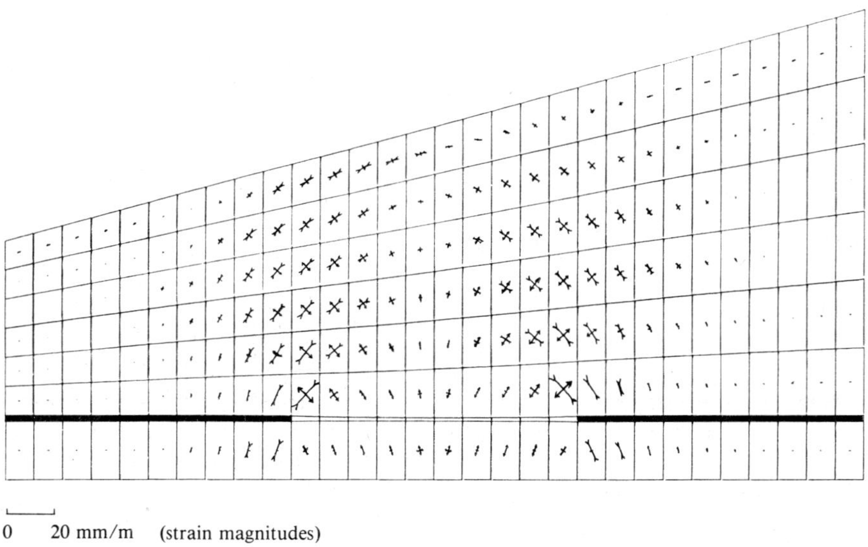

(b) Ground slope = 15°; h_{mean} = 100m, w = 100m, M = 2m

Figure 259 Principal strains induced by longwall mining with sloping ground surface

Concluding remarks

The application of physical and numerical modelling to various problems related to subsidence is well-established as an important investigative procedure. These techniques are employed in several countries and provide a useful tool for assisting in the design and planning of future mining operations in addition to elucidating current ground control problems.

Physical models allow scope for indicating the location, orientation and extent of ground fractures propagating from underground excavations. Additionally, the significance of such ground fractures can be identified in relation to the particular geological features within the overburden. For example, special rock formations and major faulting can be built into the physical model and thereafter undermined in order to observe the response of such geological features to the processes of mining.

The finite element method is a most useful tool for undertaking subsidence and ground response investigations due to undermining. The type of subsidence model used to perform such investigations needs to take the nature of the overburden into account fairly accurately. Finite element modelling can allow detailed insight to be made of the nature of surface subsidence. Also the accompanying patterns of displacement, induced principal stresses and strains can be ascertained for the ground between the mining horizon and the surface affected by longwall extraction.

APPENDIX 1

MINING SUBSIDENCE AND DISPLACEMENT PREDICTION USING INFLUENCE FUNCTION METHODS

This appendix is based on the publication of Ren, Reddish and Whittaker (1987) and discusses a flexible influence function/zone area approach to subsidence and horizontal displacement prediction.

Basic influence function theory and zone area method

Subsidence calculation. The degree to which an underground extraction which induces subsidence occurrence and affects the surface, can be conveniently described using an influence function. On extraction of an infinitesimal element dA at depth (h), this creates the occurrence of an elementary trough at the surface, with the maximum depth of the trough located vertically over the extraction element dA; the trough decreases towards the limit line. The magnitude of subsidence at any point along the trough varies according to position from the centre-line of the trough, so that subsidence can be expressed as a function of the position of point P in terms of radial distance from the trough central point, or the zone angle ξ.

Figure A1-1 shows the terminology relating to the basic influence function definition.

It follows that:

$$K'z = f(r)$$

$$\text{or} \quad K'z = f(\xi)$$

Figure A1-2 shows the surface point P above an extraction area, and from this it is possible to deduce the amount of influence caused by the extraction. Dividing the extraction into a number of extraction elements enables the amount of influence each element makes on the amount of subsidence experienced at point P to be deduced; each of the elements has a different influence on point P by virtue of their differing positions.

The element dA_1 can be clearly seen to have a greater influence than element dA_2. Conversely extraction element dA_3 is outside of the influence area and has no effect on the point P. The corresponding amount of influence caused by an extraction element is determined by the influence function. Taking reciprocity into account, the influence function profile should coincide with the function of an elementary trough profile. Consequently, the influence function reflects the effect of an extraction element on the surface.

There are several different types of influence function and are either derived empirically or are based on theoretical assumptions. Influence functions all have the common aim of accurately reflecting the subsidence behaviour of the surface, but as some differences in ground profile occur owing mainly to geological setting variations as occur from country to country, or in some cases, even from coalfield to coalfield, then such functions have been based predominantly on knowledge of subsidence characteristics in a particular locality or

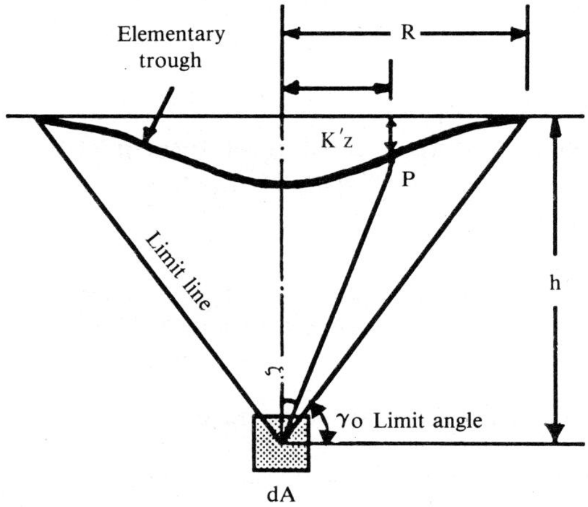

Figure A1-1 Influence function definition: extraction based

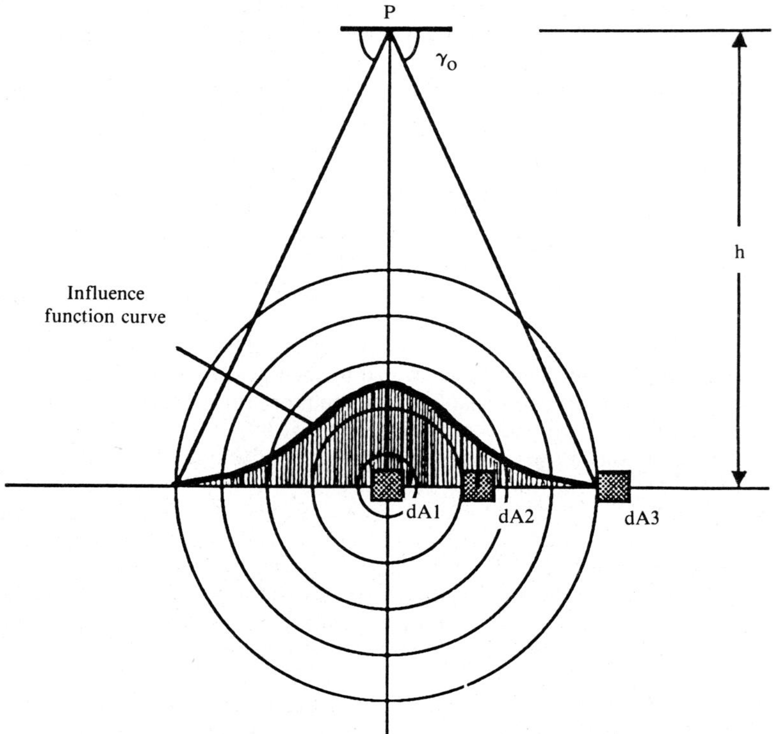

Figure A1-2 Influence function definition: surface point based

country. Selection of a particular influence function needs therefore to take account of the prevailing geological conditions. Four examples of influence functions are listed below:

1. Stochastic influence function:

$$Kz = \frac{1}{R^2} e^{-\pi r^2/R^2} \qquad \ldots(A1\text{-}1)$$

where

R = the radius of the influence circle
r = the radial distance from point P

2. Bals' influence function:

$$Kz = \cos^2 \zeta \qquad \ldots(A1\text{-}2)$$

where

ζ = zone angle, which is in the range $0 \leqslant \zeta \leqslant \gamma$ (angle of draw)

3. Sanns' influence function:

$$Kz = 2{\cdot}256 \frac{e^{-4r^2}}{r} \qquad \ldots(A1\text{-}3)$$

4. Ehrhardt and Sauer influence function:

$$Kz = 0{\cdot}1392\, e^{-0{\cdot}5r^2} \qquad \ldots(A1\text{-}4)$$

On extraction of an infinitesimal element from within the influence area results in causing subsidence of magnitude dS at the surface point P, such that

$$dS = So\, Kz\, dA$$

where So is the maximum possible subsidence.

It follows that an extraction of area A will result in subsidence S as expressed by equation A1-5 which is of generalised form.

$$S = So \int\int_A Kz\, dA \qquad \ldots(A1\text{-}5)$$

Evaluation of this double integral is required in order to determine the magnitude of the influence function over the limits of area A in terms of polar coordinates (r and θ).

Subsidence can be expressed in terms of maximum subsidence:

$$S(i) = S/So = \int\int_A Kz\, dA$$

The stochastic function in this form gives:

$$S(i) = \int\int_A \frac{1}{R^2} e^{-\pi r^2/R^2}\, dA$$

With reference to Figure A1-3, if A represents the area of the annular zones r_1 to r_2 (integration changed into the polar system), it follows that:

$$S(i) = \int\int_A \frac{1}{R^2} e^{-\pi r^2/R^2}\, r\, d\theta\, dr$$

$$= \int_o^{2\pi} d\theta \int_{ri-1}^{ri} r \frac{1}{R^2} e^{-\pi r^2/R^2}\, dr$$

$$= e^{-\pi\left(\frac{ri-1}{R}\right)^2} - e^{-\pi\left(\frac{ri}{R}\right)^2} \qquad \ldots (A1\text{-}6)$$

The quantity S(i) represents the relative subsidence at point P after extraction of the annular ring (ri, ri – 1). Additionally, S(i) is called the influence factor or weighting factor. Equation A1-6 allows determination of the weighting factors of each ring to be readily achieved, where the stochastic function has been selected. Subdivision of each ring into a series of equal sectors, as shown in Figure A1-3, is an important controlling factor on refinement of calculation. Each equal sector within the same ring has the same weighting factor, that is S(i)/(number of dividing sectors). Consequently, the weighting factors of each sector can be calculated.

Displacement calculation using the focal point method

The calculation of displacement can be readily achieved with the aid of the focal point theory. The basic assumption here is that each extraction element, for example dA, exerts an influence on a surface point P such that it attracts point P to move towards it by an amount dV. As shown in Figure A1-4, this vector dV can be represented by two orthogonal components dVxy and dVz. It is shown in Figure A1-4 that the quantity dVz represents subsidence whilst that of dVxy represents horizontal radial displacement resulting from the extraction of element dA.

On the basis of this theory, it is possible to calculate the displacement in conjunction with the subsidence. The focal point theory provides the directions of surface point movement, whilst the influence function gives the magnitude of subsidence. After determining these two factors, the amount of horizontal displacement requires to be calculated and this can be achieved using equation A1-7.

$$dVxy = \tan \zeta\, dVz \qquad \ldots (A1\text{-}7)$$

where ζ = the angle between the vertical and the line joining the surface point P with the extraction element dA as illustrated in Figure A1-4.

Programming method

The program calculates subsidence and displacement at required surface points and the general procedure involves the following principal steps. The geometrical dimensions, depth and extraction thickness are required for each mining panel, in addition to the coordinates of the required surface point. A series of annular rings are constructed around this point and within its area of influence. The rings are divided into sectors as illustrated by Figure A1-5. Each small zone, dA for example, can be considered as potential extraction elements with their own weighting factors indicating the degree of influence on the surface point P.

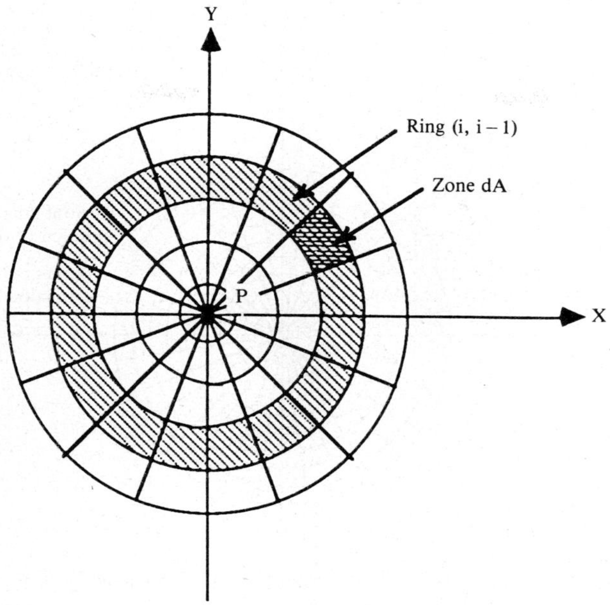

Figure A1-3 Treatment of influence circle as annular zones of influence

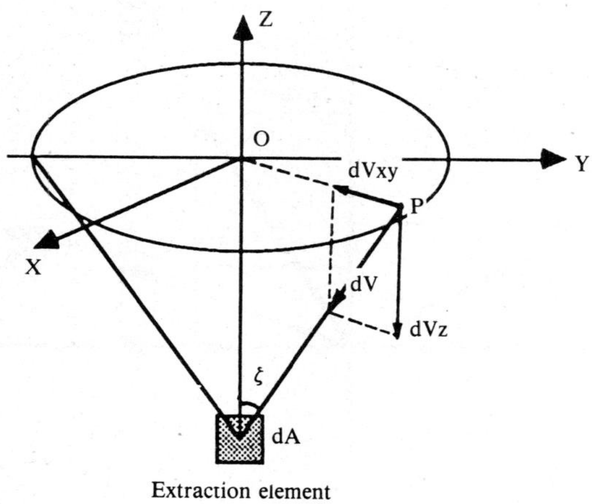

Figure A1-4 Schematic representation of focal point theory

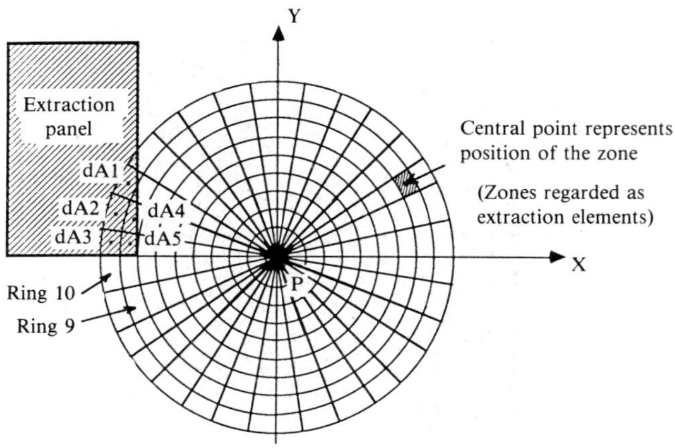

Figure A1-5 Illustrating subsidence calculation method

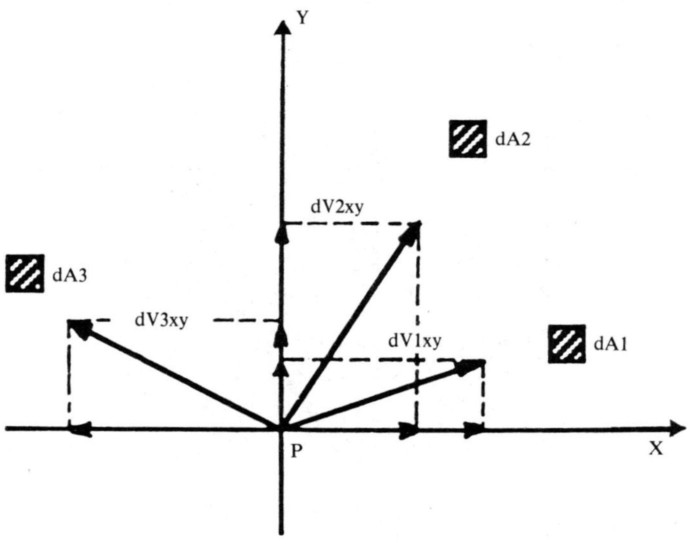

Figure A1-6 Effects of extraction elements on surface horizontal displacement at point P

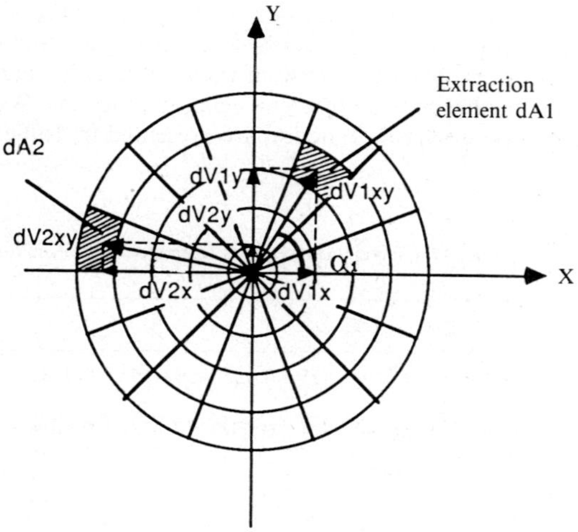

Figure A1-7 Example of displacement calculation method

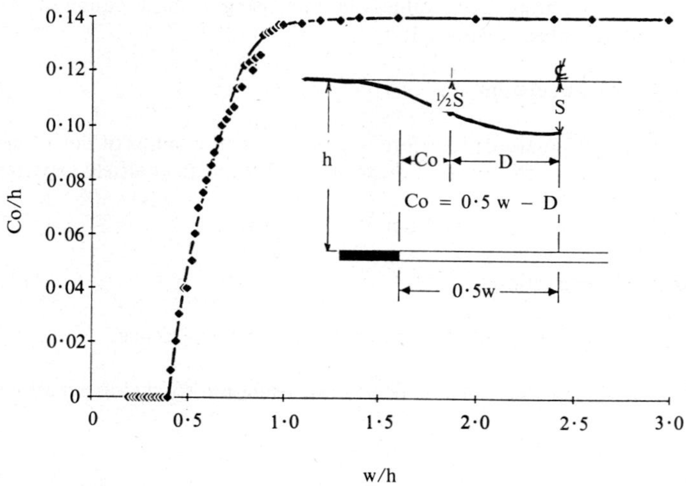

Figure A1-8 Face edge correction factor derived from SEH (1975)

Equation A1-6 permits determination of each weighting factor for the stochastic influence function procedure referred to here. Equation A1-6 gives the total weighting factor of a ring, and then it is divided by the number of elements within that ring to provide the extraction element weighting factor.

Increasing the number of rings and sectors naturally increases the accuracy of the calculation. The calculation procedure described here involves using 640 elements, with one element accounting for approximately 1/640 of the total (this is only approximate owing to the elements of different rings having different weighting factors). Weighting factors of each extraction element for the stochastic function are presented in Table A1-1.

Table A1-1 Weighting factors from stochastic function

Ring	1	2	3	4	5	6	7	8	9	10
S(i)	0·0323	0·0911	0·134	0·1555	0·1557	0·1392	0·1131	0·0842	0·0579	0·0369
For element *S(i)*/64	0·0005	0·0014	0·0021	0·0024	0·0024	0·0022	0·0018	0·0013	0·0009	0·0006

The following mathematical condition needs to be satisfied in the calculation of weighting factors:

$$\iint_{Ao} Kz \, dA = 1$$

where Ao represents the whole area of the influence circle.

Adjustment of the weighting factor values obtained using the basic equations A1-1 to 5 is required so that the total integral equals 1.

Program for subsidence calculation

The first step in the calculating procedure is to determine the radius of the influence circle R. Using a standard value for the angle of draw of 35° (European coalfield experience) gives

$$R = h \tan 35° = 0{\cdot}7h$$

where h is the depth of extraction.

The angle of draw can be selected on the basis of local experience.

The coordinates of each element are determined using the following equations:

$$X = Xp + r \cos\alpha$$

where α is increased in steps of 360°/64, and

$$Y = Yp + r \sin\alpha$$

where r is increased in steps of R/10. Xp, Yp are the coordinates of the surface point P whilst α and r are in the ranges:

$$0 < \alpha \leqslant 360^\circ$$

$$0 < r \leqslant R$$

The coordinates X, Y represent the middle point of each element. These are used to represent the element position within the coordinate system as indicated by Figure A1-5.

Of the two ways of calculating the subsidence at point P, the usual one involves firstly determining the proportion extracted of each annular ring, and thence multiplying by the weighting factor of that ring followed by summing all the rings to give the value of subsidence. This approach, however, has proved difficult to program conveniently. A sieve method is used to overcome this problem; this basically checks whether each element is extracted or not, within the panel. Where the element is inside the panel, its weighting factor is added to a variable. Summing the weighting factors of all the elements inside the seam panel gives the relative subsidence at point P. This is given by:

$$\frac{S}{So} = \Sigma S(i)$$

where So is the maximum possible subsidence and is expressed in terms of the extraction thickness.

The value of So is normally influenced by the following general relationship:

$$So = aM$$

where M is the extracted seam height. For European longwall extractions the value of $a = 0 \cdot 9$ is commonly observed where full caving is carried out.

Figure A1-5 illustrates in general terms the calculation procedure involved. A series of annular elements are constructed around the surface point P and each element subsequently tested to determine whether it lies within the extracted area. Elements A1, A2, A3, A4 and A5 are inside the panel, and are referred to as being extracted. Consequently, the subsidence at point P should be:

$$Sp/So = 3S\ (10)/64 + 2S\ (9)/64$$

As the three elements on ring (10) namely A1, A2 and A3 and the two elements on ring (9) given as A4 and A5, are extracted, then S (9) and S (10) are the weighting factors of ring (9) and ring (10) respectively. The program contains a subroutine which has been written to test every element around the surface point P, and add all the weighting factors of the elements inside the panel into a variable Sp. It follows that the final value of Sp determines the subsidence.

Program for displacement calculation

Equation A1-7 in conjunction with the determination of subsidence, can be used for calculation of surface displacement and the program achieves this as follows. The horizontal orthogonal component dVxy is determined from the spatial movement vector dV, and this has its direction towards the extraction element. Figure A1-4 illustrates this feature. The displacement of point P is the summation of the total influence of all the extraction elements. Each extraction element causes the point P to move towards it by an amount determined by the subsidence resulting from that element (dVz) and the angle (ζ) it subtends to the point P. From the known parameters, dVxy can be determined for that element. Consequently the total movement of point P is given by summing all the extracted elements' horizontal movements (ΣdVxy). A general representation is given in Figure A1-6. The horizontal displacement is further divided into X and Y directional components, such that:

$$dVx = dVxy \cos\alpha$$
$$dVy = dVxy \sin\alpha$$

where dVx is the orthogonal component of vector dVxy on the X-axis, and dVy is the orthogonal component of vector dVxy on the Y-axis.

Summing the influence of every extraction element gives the displacement at point P and is given as follows:

$$\left.\begin{aligned} Vx &= \Sigma dVx \\ Vy &= \Sigma dVy \end{aligned}\right\} \qquad \ldots (A1\text{-}8)$$

Equation A1-8 shows that if all of the elements are extracted, then there will be no horizontal displacement at point P (panel centre point, for example). Displacement calculation is carried out in the program at the same time as calculating subsidence and is performed in a similar manner.

Figure A1-7 illustrates the extraction of elements A1 and A2 which will produce surface displacement at point P of magnitude:

$$Vpx = dV_1xy \cos\alpha_1 + dV_2xy \cos\alpha_2$$
$$Vpy = dV_1xy \sin\alpha_1 + dV_2xy \sin\alpha_2$$

Profile correction using panel size adjustment and effect of width/depth ratio

Direct application of the zone area method gives subsidence over the ribside which is half of the maximum value of subsidence produced by that longwall working. Field observations in European coalfields indicate that the half maximum subsidence position is normally located at a given distance inside of the ribside edge. This feature is believed to be due to the edge effect. An adjustment needs to be made in the program to take this edge effect into account. The adjustment incorporated within the program is based on subsidence data reported in SEH (1975), and is regarded as a reliable data source for calibration of the zone area method.

Figure A1-8 has been deduced from SEH (1975) in order to demonstrate the relationship between the half maximum subsidence position Co and the width/depth (w/h) ratio. This illustration shows that for a supercritical extraction, the value of Co remains constant and is expressed by:

$$Co = 0{\cdot}14\ h$$

Having determined the w/h ratio for the panel under consideration, Co can then be determined and the zone area method can then be adjusted to give the half maximum subsidence value in the correct position, and in this program in accordance with SEH (1975). The adjustment is simply achieved by making the panel of different size to that given in the

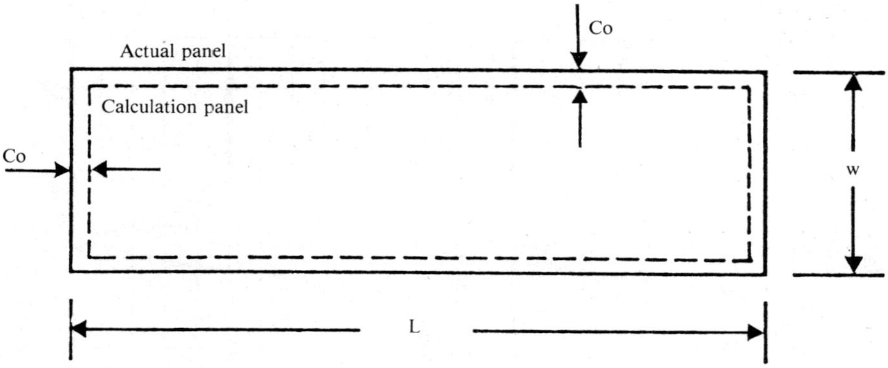

Figure A1-9 Face edge correction

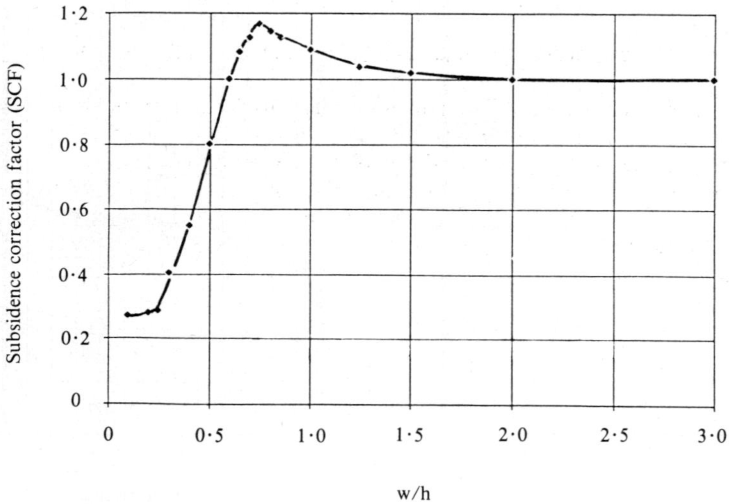

Figure A1-10 Relationship between the subsidence correction factor (SCF) and the panel width/depth (w/h) ratio; data derived by comparison to SEH (1975)

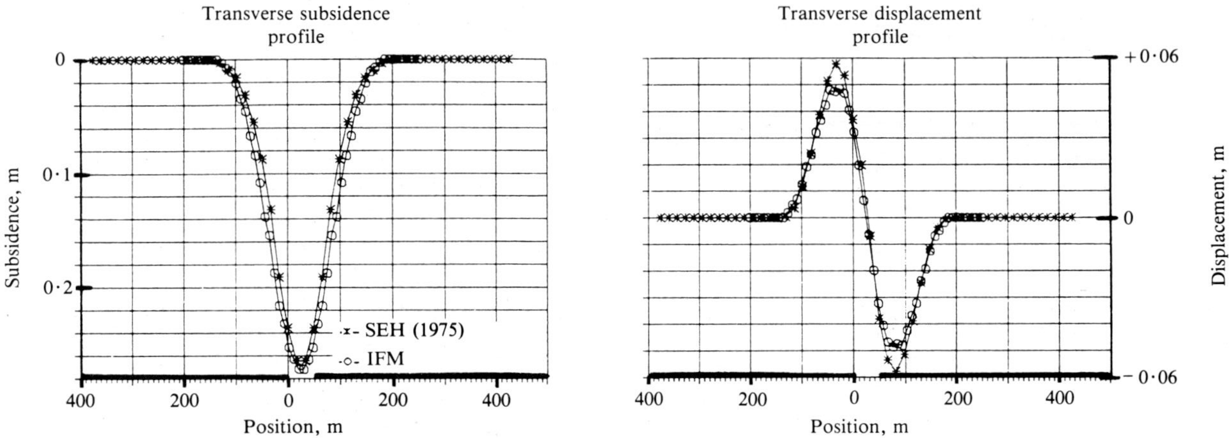

mining data: w = 50m; M = 3m; h = 200m; L = 1000m

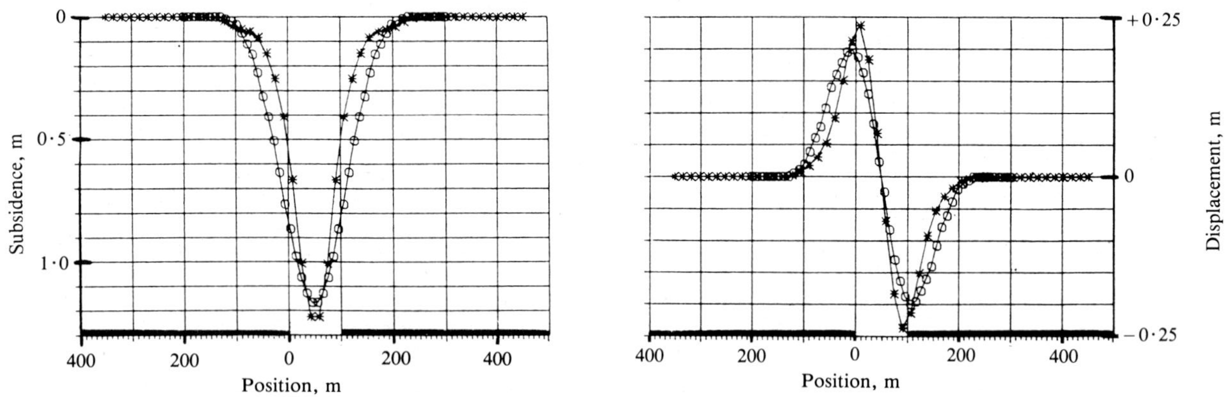

mining data: w = 100m; M = 3m; h = 200m; L = 1000m

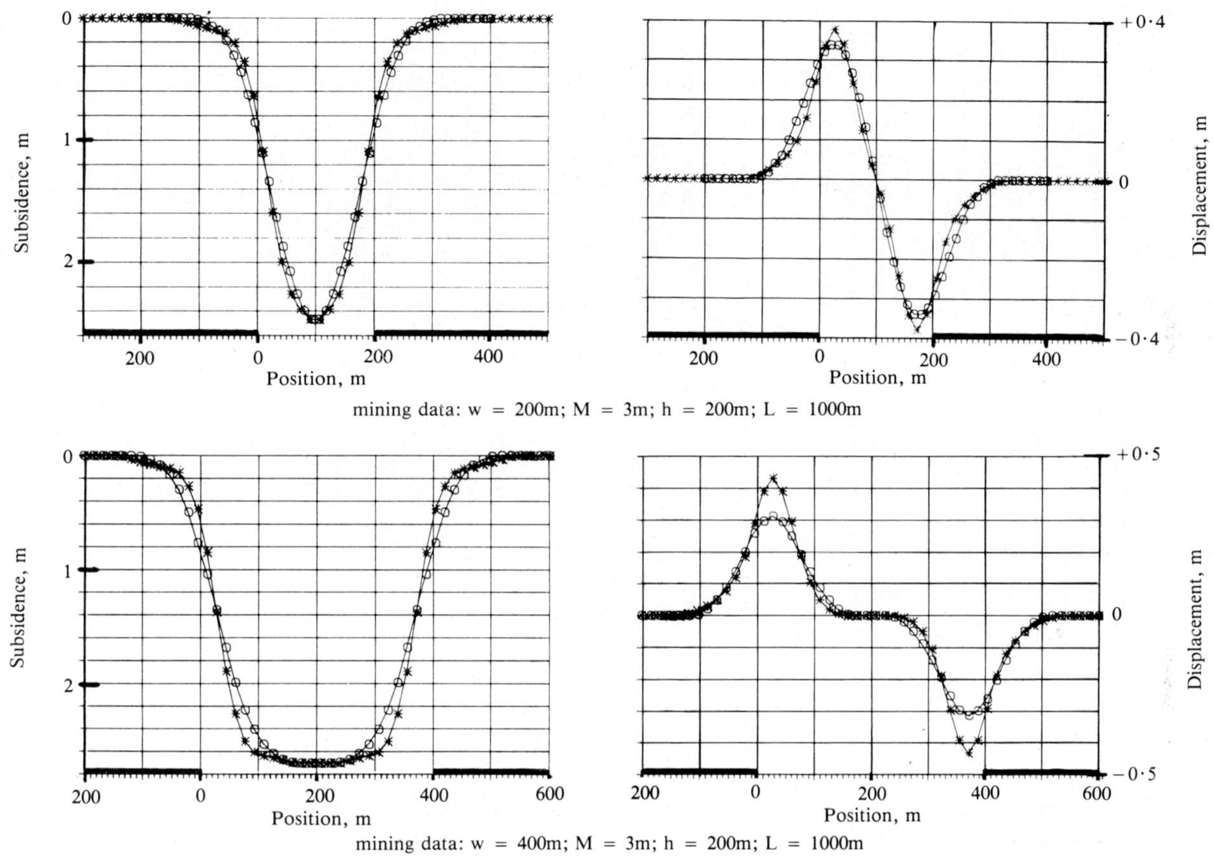

Figure A1-11 Comparison of influence function method (IFM) and computerised SEH empirical prediction subsidence results original input data, and this is illustrated by Figure A1-9. For panel w/h ratios below 0·4, the adjustment for Co is taken as zero (Figure A1-8) as experience with the calculation procedure has indicated that negative values of Co (as with small w/h ratios) does not help the adjustment of the half maximum subsidence value position.

Alteration of the panel size naturally influences the magnitude of the subsidence, and consequently a further adjustment to subsidence is made using Figure A1-10. This graphical relationship has been established empirically by comparing results from the program to those predicted by SEH (1975) for a wide range of w/h ratios. The factor calculated compensates for the effect of reducing the panel size and for the difference between the zone area method and the SEH (1975) prediction. This allows improvement to be made to the predicted subsidence curve shape. Figure A1-10 could also have been established from locally measured subsidence data, so as to achieve an accurate assessment of surface movement prediction in particular geological conditions.

In summary, therefore, Figure A1-8 serves to adjust the zone area method so as to locate the half maximum subsidence point in the correct position. Figure A1-10 adjusts the magnitude of subsidence to the corrected value. Displacement is determined directly from subsidence and is hence also adjusted.

Program output validation comparisons with SEH (1975)

The accuracy of prediction of subsidence using this program has been ascertained by comparison with SEH (1975) standard subsidence data. The main test case employed for comparison of results involved using a longwall panel extraction, totally caved in a level seam situation and having a flat surface topography. The depth below surface was 200m, the extracted seam height was 3m, the longwall panel run was 1000m whilst the face widths employed in the calculations were 50m, 100m, 200m and 400m respectively.

Graphical output of subsidence and horizontal displacements for transverse profiles are presented in Figure A1-11. The subsidence profiles show favourable comparison with SEH (1975) and the program gave an accuracy of generally better than 5% when compared to the standard subsidence data. A similar order of accuracy is evident when comparing the horizontal displacement program results with those predicted by SEH (1975). These results demonstrate the validity of the influence function model based on the stochastic theory, for subsidence and surface development predictions relating to European coalfield conditions employing the longwall mining method of working.

REFERENCES

AIMM (1973) *Subsidence in Mines.* Ed. A. J. Hargraves, Proceedings of Symposium of Australasian Inst. Min. and Metallurgy, Wollongong.

Abel, J. F. and Gentry, D. W. (1978) A longwall subsidence prediction model. ASCE Spring Convention Coal Mine Subsidence, Pittsburgh, Pennsylvania, April, 56-76.

Adamek, V. and Jeran, P. W. (1981) Evaluation of existing predictive methods for mine subsidence in the US. Proc. 1st Annual Conf. Ground Control in Mining, Morgantown, West Virginia, 209-219.

Akimov, A. G. (1958) On methods of precalculating ground surface movements. Ugol, 2, 20-23.

Allgaier, F. K. (1982) Surface subsidence over longwall panels in the Western United States. Proc. Conf. on State-of-the-Art of Ground Control in Longwall Mining and Mining Subsidence, Hawaii, USA.

Arcamone, J. (1980) *Methodologie d'étude des affaissements miniers en exploitation totale et partielle.* Ph.D. Thesis, Institut National Polytechnique de Lorraine, France, 100pp.

Arcamone, J., Schroeter, P. and Dejean, M. J. P. (1986) State of the art of mining subsidence in France. Proc. 88th Annual General Meeting of C.I.M., Montreal, May, Paper No. 84, 17pp.

Astin, J. (1968) A visco-elastic analysis of ground movement due to an advancing coal face. Journal of Eng. Maths, **2,** 9-22.

Attewell, P. B. and Farmer, I. W. (1974) Ground disturbance caused by shield tunnelling in a stiff, overconsolidated clay. Engng. Geol. **8,** 361-381.

Attewell, P. B., Yeates, J. and Selby, A. R. (1986) *Soil movements induced by tunnelling and their effects on pipelines and structures.* Blackie, Glasgow.

Aynsley, W. J. and Hewitt, G. (1961) Subsidence observations over shallow workings, including pneumatic stowing and rapidly advancing faces. Trans. I. Min. E., **120,** 552-569.

Bals, R. (1931-32) Beitrag zur Frage der Vorausberechnung bergbaulicher Senkungen. Mitt. Markscheidew., **42-43,** 98-111.

Baron, H. (1951) *Schachtbeanspruchungen beim Abbau eines Schachtsicherheitspfeilers unter machtigem Deckgebirge.* Dissertation. Technical University, Aachen.

Bauer, R. A. and Hunt, S. R. (1981) Profile, strain and time characteristics of subsidence from coal mining in Illinois. Proc. Workshop on surface subsidence due to underground mining, Morgantown, West Virginia, 207-219.

Bawden, W. F. and Mottahed, P. (1986) Comparison of three subsidence prediction techniques applied to Saskatchewan potash mining. 88th Annual General Meeting of C.I.M., Montreal, Paper No. 89, 34pp.

Beevers, C. and Wardell, K. (1954-55) Recent research in mining subsidence. Trans. I. Min. E., **114,** 223-244.

Bell, F. G. (1975) *Site investigations in areas of mining subsidence.* Newnes-Butterworth, London.

Berry, D. S. (1963) Ground movement considered as an elastic phenomenon. The Mining Engineer, **37,** 28-39.

Berry, D. S. (1964) Reply to M. D. G. Salamon in Discussion on 'Ground movement considered as an elastic phenomenon' by D. S. Berry. The Mining Engineer, **42,** 304-305.

Berry, D. S. (1964*a*) A discussion of the stochastic theory of ground movement. Rock Mechanics and Engineering Geology, **2,** 213-227.

Berry, D. S. (1964*b*) The ground considered as a transversely isotropic material. Int. J. Rock Mech. Mining Sci., **1,** 159-167.

Berry, D. S. (1977) Progress in the analysis of ground movements due to mining. Proc. of Conf. Large Ground Movements and Structures, Cardiff, Ed. J. D. Geddes, Pentech Press, 781-810.

Berry, D. S. and Sales, T. W. (1960) An elastic treatment of ground movement due to mining. Part 1: J. Mech. Phys. Solids, **8,** 280-292.

Berry, D. S. and Sales, T. W. (1961) An elastic treatment of ground movement due to mining. Part 2: J. Mech. Phys. Solids, **9,** 52-62.

Berry, D. S. and Sales, T. W. (1962) An elastic treatment of ground movement due to mining. Part 3: J. Mech. Phys. Solids, **10,** 73-83.

Beyer, F. (1945) *On predicting ground deformation due to mining flat seams.* Habilitation Thesis, Tech. Univ., Berlin.

Blès, J. L. and Feuga, B. (1986) *The fracture of rocks.* North Oxford Academic Publishers Ltd., 131pp.

Bott, M. H. P. (1976) Mechanisms of basin subsidence — an introductory review. Tectonophysics, **36,** 1-4.

Braithwaite, P. A. and Cole, K. W. (1986) Sub-surface investigations of abandoned limestone workings in the West Midlands of England by use of remote sensors. I.M.M., Section A, Mining Industry, A181-A190.

Bräuner, G. (1973*a*) Calculation of ground movement in European coalfields. Proc. Symp. Subsidence in Mines, Australasian Inst. Min. Met., 10/1-10/8.

Bräuner, G. (1973*b*) *Subsidence due to underground mining. Part 1 Theory and practices in predicting surface deformation.* US Department of the Interior, Bureau of Mines, 56pp.

Breckels, I. M., McCaul, C. and Whittaker, B. N. (1980) Comparison between measured and simulated displacements that result from tunnel and shaft interaction in Lower Chalk. Trans. IMM, Sec. A, **90**, 13p.

Briggs, H. (1929) *Mining Subsidence.* Edward Arnold & Co., 215pp.

Bruhn, R. W., Magnuson, M. O. and Gray, R. E. (1978) Subsidence over the mined-out Pittsburgh Coal. A.S.C.E. Convention 'Coal Mine Subsidence' session, Pittsburgh, Pa, USA, April, 26-55.

Bryan, A. (1951) Accident at Knockshinnock Castle Colliery, Ayrshire. Ministry of Fuel and Power, H.M.S.O., London, Cmd. 8180, 48pp.

Bulman, H. F. (1905-06) Contribution to Discussion on Halbaum (1905). Trans. I. Min. E., **31**, 6-9.

CANMET (1985) *Report on Seminar on Canadian Potash Mining and Ground Control.* Energy, Mines and Resources Canada, Report MRP/MRL 85-1, 75pp.

Callon, J. (1874) *Cours d'Exploitation des Mines.* Vol. 2, 334pp.

Calvert, A. F. (1915) *Salt in Cheshire.* E. & F. N. Spon.

Cameron-Clarke, I. S. (1986) The distribution and nature of surface features related to shallow undermining on the East Rand. SANGORM, 1986 Symposium, 33-38.

Cherny, G. I. (1966) Determination of surface subsidence and deformation by rock movements in form of rheological flow. Izvestiya VUZ, Gorny Zhurnal, **9**, 3-9.

Conroy, P. J. and Gyarmaty, J. H. (1982) Planning subsidence monitoring programs over longwall panels. Proc. Conf. on 'State-of-the-Art of Ground Control in Longwall Mining and Mining Subsidence', Eds. Y. P. Chugh and M. Karmis, Soc. Min. Eng. Am. Inst. Min. Met. and Pet. Eng., New York, USA. 225-234.

Cooper, A. H. (1986) Subsidence and foundering of strata caused by the dissolution of Permian gypsum in the Ripon and Bedale areas, North Yorkshire. Published in *The English Zechstein and Related Topics.* Eds. Harwood, G. M. and Smith, D. B., Geological Society Special Publication, **22**, 127-139.

Corden, C. H. H. (1964) *The recording of boundary distortion in mining subsidence.* Ph.D. Thesis, University of Leeds, UK, 146pp.

Corden, C. H. H. and King, H. J. (1965) A field study of the development of surface subsidence. Int. J. Rock Mech. Min Sci., **2**, No. 1, 43-55.

Dahl, H. D. (1972) Two- and three-dimensional elastic-elasto-plastic analyses of mine subsidence. Fifth Int. Strata Control Conf., London, 5pp.

Dean, J. N. (1967) Old mine shafts and their hazards. The Mining Engineer, **126**, 368-376.

Degirmenci, N., Reddish, D. J. and Whittaker, B. N. (1988) A study of surface subsidence behaviour arising from longwall mining of steeply pitching coal seams. 6th Coal Congress, Zonguldak, Turkey, May, 27p.

Dejean, M. J. P. and Martin, F. (1973) Amplitude of subsidence of underground openings subject to the influence of mining adjacent and below. Proc. Symp. *Subsidence in Mines,* Australasian Inst. Min. Met., 4/1-4/5.

Dickinson, J. (1898) Subsidence due to colliery workings. Proc. Man. Geol. Soc., **25**, p600.

Dixon, J. S. (1885) Some notes on subsidence and draw. Trans. Min. Inst. Scotland, **7**, p224.

DoE (1976) *Reclamation of derelict land: Procedure for locating abandoned mine shafts.* Planning, Regional and Minerals Directorate, Department of the Environment, London, 88pp.

DoE (1983) *Limestone mines in the West Midlands: the legacy of mines long abandoned.* Department of the Environment, London, 24pp.

Donovan Jacobs, J. (1975) Some tunnel failures and what they have taught. In: *Hazards in tunnelling and on falsework.* ICE, London, 37-46.

Dortmund Board of Mines (1897) On the influence of coal mines under marl capping upon the Earth's surface in the Dortmund District. Zeit für das Berg Hutten und Salinen Wesen, **45**, p372.

Drent, S. (1957) Some considerations on the connection between time-curves and the thickness of the non-Carboniferous overburden in the South Limburg Coalfield. Proc. European Conference on Ground Movement, 49-56.

Dumont, G. (1874) *Des Affaisements du Sol Attributés a l'Exploitation Houillère.* Luttich.

Dunrud, C. R. (1976) *Some engineering geologic factors controlling coal mine subsidence in Utah and Colorado.* US Geological Survey Professional Paper 969, 39pp.

Dunrud, C. R. and Osterwald, F. W. (1978*a*) *Effects of coal mine subsidence in the western Powder River Basin, Wyoming.* US Geological Survey Open-File Report 78-473, 71pp.

Dunrud, C. R. and Osterwald, F. W. (1978*b*) Coal mine subsidence near Sheridan, Wyoming. Bulletin of the Association of Engineering Geologists, XV(2), 175-190.

Ege, J. R. (1984) Mechanisms of surface subsidence resulting from solution extraction of salt. In 'Man-induced land subsidence', Ed. Holzer, T. L., Reviews in Engineering Geology Vol. VI; The Geological Soc. of America, 203-221.

Ehrhardt, W. and Sauer, A. (1961) Die vorausberechnung von Senkung, Schieflage und Krummung uber dem Abbau in flacher Lagerung. Bergbauwissenschaften, **8**, 415-428.

Evans, W. H. and Jones, T. J. (1946) An investigation of load on packs at moderate depths, Pt. II. Trans. I. Min. E., **105**, Pt. 6.

Fayol, H. (1885) Sur les mouvements de terrain provoqués par l'exploitation des mines. Bul. de la societé de l'industrie mineral, IIe série, Vol. 14, p818. See also translation by Galloway, W. (1897) Subsidences caused by the workings in mines. Proc. South Wales Inst. of Engrs., Vol. 20, p311.

Fisekci, M. Y. and Chrzanowski, A., Das, D. M. and Larocque, G. (1981) Subsidence studies in thick and steep coal seam mining. Proceedings of the Conference on Ground Control in Mining, Morgantown, West Virginia, USA.

Fitzpatrick, D. J. (1987) *Modelling of mining subsidence mechanisms and prediction of ground movements.* Ph.D. Thesis, University of Nottingham, UK.

Fläschenträger, H. (1938) Die Kostenverteilung bei gemeinsam verursachten Bergschäden im Ruhrgebiet. Mitt. Markscheidew., **49**, 95-137.

Fläschenträger, H. (1957) Considerations on ground movement phenomena based on observations made in the left bank Lower Rhine region. Proc. European Congress on Ground Movement, 58-73.

Fletcher, H.. (1866-67) *Royal Commission on Coal; Minutes of Evidence,* H.M.S.O.

Forrester, D. J. and Whittaker, B. N. (1976) Effects of mining subsidence on colliery spoil heaps. Int. J. Rock Mech. and Mining Sc. and Geomech. Abstracts, Vol. 13, 113-120 and 121-133.

Forster, J. (1967) *Stability investigations applied to the mining of evaporites.* Ph.D. Thesis, University of Newcastle upon Tyne, UK, 201pp.

Forsythe, G. E. and Wasow, W. (1960) *Finite difference methods for partial differential equations.* John Wiley & Sons, Inc., New York.

Frankham, B. S. and Mould, G. R. (1980) Mining subsidence in New South Wales. Proc. Australasian Inst. Min. Metall., **275**, 167-179.

Franks, C. A. M. and Geddes, J. D. (1984) A comparative study by numerical modelling of movements on sloping ground due to longwall mining. Proc. 3rd Int. Conf. on Large Ground Movements and Structures, Cardiff, Pentech Press, 377-396.

GIMS (1958) *The movement of the rock masses and of the surface in the main coalfields of the USSR.* General Institute of Mine Surveying (G.I.M.S.), Ugletekhizdat, Moscow, 250pp.

Galvin, J. M. (1982) *Total extraction of coal seams: the significance and behaviour of massive dolerite sills.* Chamber of Mines of South Africa, Research Report 19/82, 80pp.

Geddes, J. D. (1977) *Large ground movements and structures.* Pentech Press, London: Plymouth, 1064pp.

Geddes, J. D. (1980) *Ground movements and structures*. Pentech Press, London: Plymouth, 964pp.

Geddes, J. D. (1984) *Ground movements and structures*. Pentech Press, London: Plymouth, 876pp.

Geertsma, J. (1973) Land subsidence above compacting oil and gas reservoirs. Journal of Petroleum Technology, 734-744.

Geng, D. Y. and Peng, S. S. (1983) Longwall mining techniques for minimizing surface structural damage. S.M.E.-A.I.M.E. Annual Meeting, Atlanta, Georgia, Preprint No. 83-99.

Gentry, D. W. and Abel, J. F. (1978) Rock mass response to mining longwall Panel 4N, York Canyon Mine. Mining Engineering, **30(3)**, 273-280.

Gentry, D. W. and Abel, J. F. (1978*a*) Surface response to longwall coal mining in mountainous terrain. Bulletin of the Association of Engineering Geologists, XV(2), 191-220.

Gentry, D. W., Stewart, C. L. and King, R. P. (1981) Rock mechanics instrumentation program — York Canyon Mine, Raton, New Mexico. Final report, Vol. 1, US Dept. of Energy, Contract No. De-AC01-74ET 12530, 93-170.

Gentry, D. W., Stewart, C. L. and King, R. P. (1981) Rock mechanics instrumentation program for Kaiser Steel Corporation's demonstration of shield-type longwall supports at York Canyon Mine, Raton, New Mexico. Final Report, D.O.E., DE-AC01-74ET12530.

Gilluly, J. and Grant, U. S. (1949) Subsidence in the Long Beach Harbour Area, California. Geological Soc. of America, Bulletin, **60**, 461-530.

Goldreich, A. H. (1913) *Die Theorie der Bodensenkungen in Kohlengebieten,* Berlin.

Gonot, J. *Theory of the Normal.* Published in Belgium 1858 and referred to in Young and Stoek (1916).

Grand Saline Sun (1976) Salt well cave-in attracts wide interest. Grand Saline Sun, Texas, May 6.

Grard, C. (1969) Mining subsidence and the means of permitting limitation of the effects on the surface. Revue de l'Industrie Minérale, **51**, 35-70.

Grond, G. J. A. (1951) The precise topographical measurements in coal mine underground works. Proc. Int. Conf. Rock Pressure and Support in Workings, Liège, 52-66.

Grond, G. J. A. (1953) *A critical analysis of early and modern theories of mining subsidence and ground control.* Published by The Department of Mining, University of Leeds, 57pp.

Grond, G. J. A. (1957) Ground movements due to mining with different types of strata and at different depths. Proc. European Congress on Ground Movement, 115-125.

Groothoff (1922) Bodembegegingen tengevolge van de mijnbouw. Verhandelingen van het Geologisch Mijnbouwkundig Genootschap voor Nederland en Kolonien.

Gurtunca, R. G. and Schumann, E. H. R. (1986) Computer simulation of surface subsidence using a displacement discontinuity method. SANGORM 1986 Symposium, 81-87.

Hackett, P. (1958-59) An elastic analysis of rock movements caused by mining. Trans. I. Min. E., **118**, 421-433.

Halbaum, H. W. G. (1903) The action, influence and control of the roof in longwall workings. Trans. I. Min. E., **27**, 211pp.

Halbaum H. W. G. (1905) The great planes of strain in the absolute roof of mines. Trans. I. Min. E., **30**, 175pp.

Hall, M. and Orchard, R. J. (1963) Subsidence profile characteristics. Chart. Surv., **98** (8), 422-428.

Hamilton, J. (1923) *Royal Commission on Mining Subsidence; Minutes of Evidence,* H.M.S.O.

Hansmire, W. H. and Cording, E. J. (1972) Performance of soft ground tunnel on the Washington Metro. R.E.T.C. Chicago, Proc. **1**, 371-389.

Hausse, R. (1885) *Beitrag zur Bruchttheorie Erfahrungen über Bodensenkungen und Gebergsdruckwirkungen.* Jahrbuch für das Berg und Hutten Wesen im Königreich Sachsen.

Hausse, R. (1907) Von der Niedergehen des Gebirges beim Kohlenbergbau und den damit zusammenhangenden Boden und Gebaudesenkungen. Zeit für das Berg Hutten und Salinen Wesen, 324-446.

Hazine, H. I. (1977) *A study of the development of surface strains produced by mining.* M.Phil. Thesis, University of Nottingham, UK.

Hedley, A. M. (1923) *Royal Commission on Mining Subsidence; Minutes of Evidence,* H.M.S.O.

Hedley, D. G. F. (1972) An evaluation of roof stability at a Canadian salt mine. Fifth Int. Strata Control Conf., London, Paper No. 30, 6pp.

Hellewell, E. G. (1988) The influence of faulting on ground movement due to coal mining: the UK and European experience. The Mining Engineer, January, **316**, 334-337.

Helm, D. C. (1984) Field-based computational techniques for predicting subsidence due to fluid withdrawal. In Holzer, T. L., Ed. Man-induced land subsidence. Rev. Eng. Geology VI, The Geological Soc. of America, 1-22.

Herak, M. and Stringfield, V. T. (1972) *Karst.* Elsevier Publishing Company, Amsterdam, 551pp.

Hiramatsu, Y. and Oka, Y. (1968) Precalculation of ground movements caused by mining. Int. J. Rock Mech. Min. Sci., **5**, 399-414.

Hiramatsu, Y., Okamura, H.. and Sugawara, K. (1979) Surface subsidence and horizontal displacement caused by mining inclined coal seams. Proc. 4th Congress International Society for Rock Mechanics, 665-670. Montreux, 2-8 Sept 1979, Vol 1, Publ. Rotterdam: A. A. Balkema.

Hoffman, H. (1964) The effects of direction of working and rate of advance on the scale-deformation of a self-loaded stratified model of a large body of ground. Proc. Int. Conf. Strata Control, New York, 397-411.

Holla, L. (1985*a*) Empirical prediction of subsidence movements in the Southern Coalfields of New South Wales. Proc. Sixth Int. Congress of Int. Soc. of Mine Surveying, Harrogate, 557-567.

Holla, L. (1985*b*) The minimisation of surface subsidence by design of mine workings. Proc. Australasian Inst. Min. Metall., **240**, 53-59.

Holla, L. (1986) Evaluation of surface subsidence characteristics in the Newcastle Coalfield of New South Wales. The Coal Journal (Australia), **11**, 12pp.

Holmes, A. (1965) *Principles of physical geology*. Thomas Nelson, London, 1288pp, Second Edition.

Holzer, T. L. (1984) Ground failure induced by ground-water withdrawal from unconsolidated sediment. In Holzer, T. L. Ed. *Man-induced land subsidence*. Rev. Eng. Geology VI, The Geological Soc. of America, 67-105.

Hood, M., Ewy, R. T. and Riddle, L. R. (1983) Empirical methods of subsidence prediction — a case study from Illinois. Int. J. Rock Mech. Min Sci. and Geomech., Abstr., **20**(4), 153-170.

Howell, R. J. (1979) Private communication: reference (40) of O'Rourke and Turner (1979).

Hull, E. (1881) Proceedings of the Royal Dublin Society; referred to in Calvert (1915), 342pp.

Hunt, S. R. (1980) *Surface subsidence due to coal mining in Illinois*. Ph.D. Thesis, University of Illinois at Urbana-Champaign, 129pp.

I.C.E. (1977) *Ground subsidence*. The Institution of Civil Engineers, London, 99pp.

Ilijn, A. S. (1977) Earth surface subsidence at the areas of gas and oil pumping out. Proc. 2nd Int. Symp. on Land Subsidence, Anaheim, California. Int. Assoc. Hydrological Sciences Publications, **121**, 665pp.

Jegbefume, E. U. and Thompson, T. W. (1983) The influence of temperature and non-elastic behaviour on roof collapse and subsidence resulting from underground coal gasification. In Situ, **7(2)**, 143-174.

Jennings, J. E., Brink, A. B. A., Louw, A., and Gowan, G. D. (1965) Sink-holes and subsidences in the Transvaal dolomites of South Africa. In Proc. Int. Conf. on Soil Mech. and Found. Eng., 6th, Montreal, 51-54.

Jeremic, M. L. (1964) Exploitation methods for rock salt deposits in Poland. Mining and Metallurgical Bulletin, Belgrade, No. 2, 264-274.

Jeremic, M. L. (1975) Subsidence problems caused by solution mining of rock-salt deposits. Proc. 10th Canadian Rock Mechanics Symp., Kingston, 2-4 Sept, VI, p203-223.

Jicinsky, W. (1884) *The influence of coal mining upon the surface.* Monograph of the Ostrau-Karwin Coal District. Reviewed by Briggs (1929) in *Mining Subsidence,* p45-47.

Johnson, W. and Miller, G. C. (1979) *Abandoned coal-mined lands.* US Department of the Interior, Bureau of Mines, 29pp.

Jones, T. J. (1945) Strata movements induced by mining. Trans. Inst, Min. Surveyors, **25,** Pt. 1.

Kapp, W. A. (1973*a*) Mine subsidence. Proc. Symp. *Subsidence in Mines,* Australasian Inst. Min. Met., 1/1-1/9.

Kapp, W. A. (1973*b*) Subsidence: Kemira Colliery, New South Wales. Proc. Symp. *Subsidence in Mines,* Australasian Inst. Min. Met., 7/1-7/9.

Kapp, W. A. (1977) Subsidence investigations in the Northern Coalfield, New South Wales, and their application to the design of mine layouts in residential areas. 11th Commonwealth Min. & Metall. Congr., Hong Kong; IMM, London, Paper 32, 11pp.

Kapp, W. A. (1982) Subsidence from deep longwall mining of coal overlain by massive sandstones. Proc. Australasian Inst. Min. Metall., **281,** 23-36.

Kapp, W. A. (1985) Mine subsidence in the Newcastle District, New South Wales. Trans. Inst. of Engrs., Australia, CE27, **4,** 331-340.

Karmis, M., Haycocks, C., Webb, B. and Triplett, T. (1981) The potential of the zone area method for mining subsidence prediction in the Appalachian Coalfield. Workshop on Surface Subsidence due to Underground Mining, Morgantown, West Virginia, 12pp.

Karmis, M., Haycocks, C. and Triplett, T. (1984) Ground settlement and deformation characteristics above undermined areas: experiences from the Eastern US coalfields. Fourth Australia-New Zealand Conf. on Geomechanics, Perth, May, 647-653.

Karmis, M., Jarosz, A. and Schilizzi, P. (1987) Monitoring and prediction of ground movements above underground mines in the eastern US. Proceedings of 6th International Conf. Ground Control in Mining, Ed. S. S. Peng, West Virginia University, Morgantown, June, 184-194.

Karmis, M., Triplett, T., Haycocks, C. and Goodman, G. (1983) Mining subsidence and its prediction in the Appalachian Coalfield. Proc. 24th US Symp. on Rock Mechanics, June, 665-675.

Kay, S. R. (1898-99) The effect of subsidence due to coal workings upon bridges and other structures. Proc. Inst. Civil Engrs, London, **135**, 115-123.

Keinhorst, H. (1928) Bei Bodensenkungen auftretende Bodenverschiebungen und Bodenspannungen. Gluckauf, **64**, 1141-45.

Keinhorst, H. (1934) Considerations on the problem of mining damage. Gluckauf, **70**, 149-155.

Kerridge, F. B. (1952) Contribution to Discussion on North (1952). South Wales Institute of Engineers, **67**, 155pp.

Khair, A. W., Chaffins, R. D. and Quinn, M. K. (1987) Study of ground movements over a longwall mine. Proceedings of 6th International Conf. Ground Control in Mining, Ed. S. S. Peng, West Virginia University, Morgantown, June, 141-155.

King, H. J. (1964) Discussion on 'Ground movement considered as an elastic phenomenon' by D. S. Berry. The Mining Engineer, **40**, 187-188.

King, H. J. and Smith, H. G. (1954) Surface movements due to mining. Colliery Engineering, August.

King, H. J. and Whetton, J. T. (1957) Mechanics of mine subsidence. Proc. European Congress on Ground Movement, 27-36.

King, H. J., Whittaker, B. N. and Shadbolt, C. H. (1974) Effects of mining subsidence on surface structures. *Mining and the Environment*. IMM, London, 617-642.

Knothe, S. (1953) Rate of advance and ground deformation. Bergakademie, **5**(12), 513-518.

Knothe, S. (1957) Observations of surface movements under influence of mining and their theoretical interpretation. Proc. European Congress on Ground Movement, 210-218.

Knox, G. (1913) Mining subsidence. Proc. Int. Geol. Congress, Vol. 12, p798.

Kochmanski, T. (1959) Integral-Theorie der Bodenbewegungen unter Einwirkung des Abbaus, Freiberger Forschungshefte, **A118**, 36-58.

Kowalczyk, A. (1968) Die Einwirkung des Bergbaus auf die Erdoberflache in industrialisierten und besiedelten Gebieten, Bergakad, **7**, 389-394.

Kratzsch, H. (1967) Uber den Absenkungsfaktor in geneigter bis steiler Lagerung. Bergb.-Weiss., **14**, 413-423.

Kratzsch, H. (1975) Uber die Anwendung des Verfahrens des finiten Elemente in der Bergschadenkunde. Mitt Markscheidew., **82**, 49-65.

Kratzsch, H. (1983) *Mining Subsidence Engineering*. Springer Verlag, Berlin, 543pp.

Kumar, R., Saxena, N. C. and Singh, B. (1983) A new hypothesis for subsidence prediction. Journal of Mines, Metals and Fuels, p459.

Kumar, R. and Singh, B. (1967) Subsidence investigations in a coal mine. Research Report, No. 34, Central Mining Research Station, Dhanbad, India, May.

Kumar, R., Singh, B. and Sinha, K. N. (1973) Subsidence investigations in Indian mines. Proc. Symp. Subsidence in Mines, Australasian Inst. Min. Met., 6/1-6/7.

Landes, K. K. and Piper, T. B. (1972) *Effect upon the environment of brine cavity subsidence at Grosse Ile, Michigan — 1972.* Solution Mining Research Institute, 812 Muriel Street, Woodstock, Ill. 60098, 52pp.

Lane, W. T. and Roberts, J. H. (1929) *The principles of subsidence and the law of support.* Alfred A. Knopf, 319pp.

Lee, A. J. (1965) *The effect of faulting on mining subsidence.* M.Sc. Thesis, University of Nottingham, UK, 136pp.

Lee, A. J. (1966) The effect of faulting on mining subsidence. The Mining Engineer, **71,** 735-743.

Leeds University (1957) Proceedings of the European Congress on Ground Movement, 218pp.

Legget, R. F. (1962) *Geology and Engineering,* McGraw-Hill Book Company, New York, 884pp.

Lehmann, K. (1919) Bewegungsvorgange bei der Bildung von Pingen und Trogen. Gluckauf, **55,** 933-42.

Lenge, A. (1957) Inspection and correction of damage to railways caused by mining in the Saar. Proc. European Congress on Ground Movement, 106-114.

Littlejohn, G. S. (1979) Surface stability in areas underlain by old coal workings. Ground Engineering, March, **12,** No. 2, 22-30.

Litwiniszyn, J. (1957) The theories and model research of movements of ground masses. Proc. European Congress on Ground Movement, 202-209.

Litwiniszyn, J. (1964) On certain linear and nonlinear strata theoretical models. Proc. Int. Conf. Strata Control, New York, 84-394.

Litwiniszyn, J. (1972) Polish Academy of Science, Krakow. Personal communication.

Lloyd, W. D. (1918-19) The effect of coal mining on the overlying rocks and on the surface. Trans. Inst. Mining Engrs., **57,** 74-93.

Louis, H. (1922) A contribution to the theory of subsidence. Trans. I. Min. E., **64,** 257pp.

MacCourt, L., Madden, B. J. and Schumann, E. H. R. (1986) Case studies of surface subsidence over collapsed bord and pillar workings in South Africa. SANGORM, 1986 Symposium, 25-32.

Mahar, J. W. and Marino, G. G. (1982) Building response and mitigation measures for building damage in Illinois. Proceedings of Workshop on Surface Subsidence due to Underground Mining, Morgantown, V.W., USA. Ed. S. S. Peng and M. Harthill, Department of Mining Engineering, West Virginia University, 238-252.

Maize, E. R., Thomas, E. and Greenwald, H. F. (1941) *Studies of roof movement in coal mines.* Study of subsidence of a highway caused by mining coal beneath. U.S.B.M. Report 3562, 11pp.

Marino, G. G. and Mahar, J. W. (1985) *Subsidence damaged homes over room and pillar mines in Illinois.* Report for Illinois Abandoned Mined Lands Reclamation Council, Springfield, Illinois, 450pp.

Marr, J. E. (1961) Subsidence observations in the South Lancashire Coalfield. Sheffield University Mining Magazine, 24-35.

Marr, J. E. (1975) The application of the zone area system to the prediction of mining subsidence. The Mining Engineer, **176,** 53-60.

Marshall, G. J. and Berry, D. S. (1966) Calculation of the stress around an advancing longwall face in visco-elastic ground. Proc. First Congress Int. Soc. Rock Mech., Lisbon.

Martin, J. C. and Serdengecti, S. (1984) Subsidence over oil and gas fields. In Holzer, T. L. Ed. *Man-induced land subsidence.* Rev. Eng. Geology VI, The Geological Soc. of America, 23-34.

Martos, F. (1958) Concerning an approximate equation of the subsidence trough and its time factor. Proc. Int. Strata Control Congress, Leipzig, 191-205.

Martos, F. (1958) Uber eine Naherungsheichung der Senkungsmulde und deren Zeitfaktor. Proc. Int. Bureau on Rock Mechanics, Berlin: Akademie-Verlag, 102-109.

Maury, V., Sauzay, J-M., and Fourmaintraux, D. (1987) Geomechanical approach to oil and gas production — main problems, first results. Proceedings ISRM 6th Int. Congr. Rock Mech., Vol 1, 181-189.

McCaul, C., Morgan, J. M. and Boden, J. B. (1976) *Measurement of ground movement due to excavation of a shallow tunnel in Lower Chalk.* T.R.R.L. Supplementary Report 199 UC, Department of the Environment.

McClain, W. C. (1963) *A study of the strata movement in the partial extraction of potash.* Ph.D. Thesis, University of Newcastle upon Tyne.

Miller, H. D. S. and Stoakes, M. A. (1985) Subsidence associated with Louisiana salt domes. C.I.M. Conf. on Subsidence in Soft Rock Mining, Saskatoon, Saskatchewan, 20pp.

Moebs, N. N. and Barton, T. M. (1985) Short-term effects of longwall mining on shallow water sources. Mine Subsidence Control, Proceedings of Bureau of Mines Technology Transfer Seminar, Pittsburgh, Pa, September, Bureau of Mines IC 9042, 13-24.

Montz, H. W. (1930) Subsidence from anthracite mining. Trans. A.I.M.M.E., Coal Division, 98-134.

Morrison, R. G. K. (1976) *A philosophy of ground control.* Department of Mining and Metallurgical Engineering, McGill University, Canada, 182pp.

Mraz, D. Z. (1985) Proposed subsidence model for Saskatchewan potash mines. Paper presented to Seminar of R.M.S.C. Sub-committee on Soft Rocks, November, 9pp.

NCB (1965, 1975) *Subsidence Engineer's Handbook.* National Coal Board, London, 111pp.

NCB (1950-51) Memorandum on design of mine workings to secure effective strata control. Divisional Strata Control Research Committee, Durham and Northern (N and C) Divisions, National Coal Board. Trans. I. Min. E., **110**, 251-265.

NCB (1961) *Partial extraction as a means of reducing subsidence damage.* National Coal Board, Production Department, Information Bulletin 61/231, 16pp.

NCB (1963) *Principles of subsidence engineering.* National Coal Board, Production Department, Information Bulletin 61/240, 27pp.

NCB (1968) *Working under the sea.* National Coal Board, London, Production Department Instruction, PI/1968/8, 4pp.

NCB (1982) *Mining subsidence report.* National Coal Board, North East Leicestershire Prospect, Bestwood, Nottingham, UK, 37pp.

NCB (1982) *The treatment of disused mine shafts and adits.* NCB Mining Department, London, 88pp.

Narasimham, T. N. and Goyal, K. P. (1984) Subsidence due to geothermal fluid withdrawal. In Holzer, T. L. Ed. *Man-induced land subsidence.* Rev. Eng. Geology VI, The Geological Soc. of America, 35-66.

Newton, J. G. (1984) Sink-holes resulting from ground-water withdrawals in carbonate terranes — an overview. In Holzer, T. L. Ed. *Man-induced land subsidence.* Rev. Eng. Geology VI, The Geological Soc. of America, 195-202.

Niederhofer, G. (1962) Neuere Verfahren für die Vorausberechnungen von Bodensenkungen, vornehmlich über Abbauen in steiler Lagerung, Mitt. Markscheidew., **69**, 33-75.

Niemczyk, O. (1949) *Bergschadenkunde.* Gluckauf, Essen.

Nieto, A. S. (1979) Evaluation of damage potential to earth dam by sub-surface coal mining at Rend Lake, Illinois. Proc. 10th Ohio River Valley Soils Seminar on Geotechnics in Mining, Lexington, Kentucky, 9-18.

North, F. J. (1952) Some geological aspects of subsidence not due to mining. South Wales Institute of Engineers, **67**, 127-153.

O'Donahue, T. A. (1907) *Mining Formulae.* Published by Science and Art of Mining, Wigan.

Ogden, H. and Orchard, R. J. (1959-60) Ground movements in North Staffordshire. Trans. I. Min. E., **119**, 259-272.

O'Reilly, M. P. and New, B. M. (1982) Settlements above tunnels in the United Kingdom, their magnitude and prediction. Proc. Tunnelling '82, Brighton, 173-181, Discussion Trans. I.M.M. 1983, A35-A48.

O'Riordan, N. J., Cole, K. W. and Henkel, D. J. (1984) Collapses of abandoned limestone mines in the West Midlands of England. In: *Design and performance of underground excavations.* E. T. Brown and J. A. Hudson (Eds.), ISRM, Cambridge, 401-408.

O'Rourke, T. D. and Turner, S. M. (1981) Empirical methods for estimating subsidence in US coalfields. Proc. 22nd US Symp. Rock Mech., Massachusetts, Institute of Technology, Cambridge, 322-327.

Oravecz, K. I. (1986) Improved prediction of surface subsidence using the influence function approach. SANGORM 1986 Symposium, 73-80.

Orchard, R. J. (1954) Recent developments in predicting the amplitude of mining subsidence. Trans. R.I.C.S., **33,** 864pp.

Orchard, R. J. (1956-57) Surface effects of mining — the main factors. Trans. I. Min. E., **116,** Part I, 942-944.

Orchard, R. J. (1956-57) Surface effects of mining — the main factors. Trans. I. Min. E., **116,** Part 11, 941-955.

Orchard, R. J. (1957) Surface effects of mining — the main factors. Trans. I. Min. E., **116,** Pt. 11, 955-958, Discussion.

Orchard, R. J. (1957*a*) Prediction of the magnitude of surface movements. Colliery Engng., **34,** 455pp.

Orchard, R. J. (1957*b*) Prediction of the magnitude of surface movements. Proc. European Congress on Ground Movement, 39-46.

Orchard, R. J. (1964) Partial extraction and subsidence. The Mining Engineer, April, **43,** 417-427.

Orchard, R. J. and Allen, W. S. (1965) Ground curvature due to coal mining. Chart. Surv., May, **97** (11), 622-631.

Orchard, R. J. and Allen, W. S. (1974) Time-dependence in mining subsidence. Minerals and the Environment. I.M.M., London, 643-659.

Oyanguren, P. R. (1972*a*) Simultaneous extraction of two potash beds in close proximity. Fifth Int. Strata Control Conf., London, Paper No. 32, 5pp.

Oyanguren, P. R. (1972*b*) Discussion on 'Simultaneous extraction of two potash beds in close proximity'. Fifth Int. Strata Control Conf., London, Paper No. 32, 5pp.

Pasamehmetoglu, A. G. (1972) *An investigation into time-dependent aspects of mining subsidence.* Ph.D. Thesis, University of Nottingham, UK, 190pp.

Peck, R. B. (1969) Deep excavation and tunnelling in soft ground. Seventh Int. Conf. Soil Mechanics and Foundation Engineering, Mexico City.

Peng, S. S. and Geng, D. Y. (1982) Surface subsidence, overburden behaviour and structural damages due to longwall mining — two case studies. Proceedings 2nd International Conf. Stability in Underground Mining, University of Kentucky, 497-523.

Perz, F. (1948) Der Einfluß der Zeit auf die Bodenbewegung über Abbauen. Mitt. Markscheidew, **55**, 92-117.

Perz, W. (1957) Observations of subsidence over underground workings in the Austrian Tertiary Coal Basins. Proc. European Congress on Ground Movement, 92-98.

Phillips, D. W. and Henshaw, H. (1942) Underground and surface strata movements. Trans. Inst. Min. Surveyors, **21**, Pt. 1.

Piggott, R. J. and Eynon, P. (1977) Ground movements arising from the presence of shallow abandoned mine workings. In *Large Ground Movements and Structures,* Ed. J. D. Geddes, Pentech, 749-780.

Poland, J. F. (1972) Subsidence and its control. In Cook, T. D., Ed. *Underground waste management and environmental implications.* Memoir No. 18, Houston, Texas, Am. Assoc. Pet. Geologists, 50-71.

Pöttgens, J. J. E. (1978) Ground movements by coal mining in the Netherlands. Proc. Conf. Evaluation and prediction of subsidence. A.S.C.E., Pensacola Beach, Florida, 267-282.

Potts, E. L. J. (1964) The influence of time-dependent effects on the design of mine pillars. Int. Bureau of Rock Mechanics, 6th International Meeting, November.

Potts, E. L. J., Potts, W. H. and Szeki, A. (1972) Development of ground control techniques and mining design parameters in rock-salt mining. Fifth Int. Strata Control Conf., London, Paper No. 26, 8pp.

Pratt, W. E. and Johnson, D. W. (1926) Local subsidence of the Goose Creek Field. Journal of Geology, **34** (1), 577-590.

Price, D. G., Malkin, A. B., and Knill, J. L. (1969) Foundations of multi-storey blocks on the Coal Measures with special reference to old mine workings. Q. Jl. Engng. Geol., Vol 1, 271-322.

Rankin, W. J. (1987) Ground movements resulting from urban tunnelling predictions and effects. Proc. Conf. Eng. Group Geological Soc. Engineering Geology of Underground Movements, Nottingham University, September, Eds. F. G. Bell, M. G. Culshaw and J. C. Cripps, 13-33.

Reddish, D. J. (1984) *Study of ground strain in relation to mining subsidence.* Ph.D. Thesis, University of Nottingham, UK.

Rellensmann, O. and Wagner, E. (1957) The effect on railways of the ground movements due to mining. Proc. European Congress on Ground Movement, 74-82.

Ren, G., Reddish, D. J. and Whittaker, B. N. (1987) Mining subsidence and displacement prediction using influence function methods. Mining Science and Technology, **5,** 89-104.

Rom, H. (1964) A limit angle system. Mitt. Markscheidew., **71,** 197-199.

Rziha, F. (1882) Subsidence due to mining operations. Zeit für das Berg Hutten und Salinen Wesen, Oest, Vol. 30.

SANGORM (1986) *The effect of underground mining on surface.* I.S.R.M. Symposium, South African National Group on Rock Mechanics (SANGORM), Sandton, South Africa, 155pp.

Salamon, M. D. G. (1963-64) Elastic analysis of displacements and stresses induced by the mining of seam or reef deposits. J. South Afr. Inst. Min. Met., **64,** 128-149, 468-500.

Salamon, M. D. G. (1964) Discussion on 'Ground movement considered as an elastic phenomenon' by D. S. Berry. The Mining Engineer, **42,** 303-304.

Salamon, M. D. G. (1967) A method of designing bord and pillar workings. J. S. Afr. Inst. Min. Metall., **68,** 68-78.

Salamon, M. D. G. (1974) Rock mechanics of underground excavations. Proc. Third Congress of International Society for Rock Mechanics, Denver, Advances in Rock Mechanics, National Academy of Sciences, Washington, **1**(B), 951-1099.

Salamon, M. D. G. (1977) The role of linear models in the estimation of surface ground movements induced by mining tabular deposits. Proc. of Conf. Large Ground Movements and Structures, Cardiff, Ed. J. D. Geddes, Pentech Press, 187-208.

Salas, J. A. J. (1979) Two subsidence cases in Spain. I.S.R.M. 4th Int. Congress on Rock Mechanics, Montreux, Vol. 3, 369-373.

Sann, B. (1949) Betrachtungen zur Vorausberechnung von Bodensenkungen infolge Kohlenabbaus, Bergbau-Rundschau, **1,** 163-168.

Schleier, O. (1937) Zur Frage Senkungsvorausberechnung beim Abbau von Steinkohlenflozen in geneigter Lagerung. Mitt. Markscheidew., **49,** 16-27.

Schmidt, B. (1969) *Settlements and ground movements associated with tunnelling in soil.* Ph.D. Thesis, University of Illinois.

Schmidt, B. (1974) *Prediction of settlements due to tunnelling in soil: three case histories.* R.E.T.C., Proc. **2,** 1179-1199.

Schoonbeck, J. B. (1976) Land subsidence as a result of natural gas extraction in the Province of Groningen. Soc. of Petroleum Engineers of A.I.M.E., SPE. Paper 5751 presented to SPE — European Spring Meeting, Amsterdam, The Netherlands.

Schulte, H. F. (1957) The effects of subsidence on the strata immediately above a working with different types of packing in level measures. Proc. European Congress on Ground Movement, 188-198.

Schulz, A. (1867) Investigations on the dimensions of the safety pillars for the Saarbrücken Coal Mining Industry and on the angle of fracture at which the strata settle into mined rooms. Zeit für das Berg Hutten und Salinen Wesen.

Schumann, E. H. R. (1983) Controlled subsidence of Blackhill-Greenside railway line: Part 1. Chamber of Mines of South Africa, Research Report, 8/83.

Schumann, E. H. R. (1984) Surface subsidence due to the extraction of moderately thick coal seams at shallow depth. Proc. 3rd Conf. on *Ground Movements and Structures,* Cardiff, Ed. J. D. Geddes, Pentech Press, London: Plymouth, 248-263.

Schumann, E. H. R. (1986) The monitoring, computation and data analysis of surface subsidence. SANGORM 1986 Symposium, 13-23.

Schwartz, D. and Dubois, R. (1957) Effects of the treatment of the goaf (strip packing or caving) on the rock in the immediate vicinity. Proc. European Congress on Ground Movement, 152-158.

Selby, A. R. (1987) Surface movements caused by tunnelling in two-layer soil. Proc. Conf. Eng. Group Geological Soc. Engineering Geology of Underground Movements, Nottingham University, September, Eds. F. G. Bell, M. G. Culshaw and J. C. Cripps, 1-12.

Shadbolt, C. H. (1972) Subsidence engineering. Univ. Nottingham Min. Dept. Mag. **24,** 80-89.

Shadbolt, C. H. (1986) Personal Communication.

Shadbolt, C. H. and Mabe, W. J. (1968) Ground movement in the East Midlands Coalfield. Paper presented to R.I.C.S. (Minerals Section), Basford, Nottingham, UK, 15 May, 22pp.

Shadbolt, C. H., Whittaker, B. N. and Forrester, D. J. (1973) Recent developments in mining subsidence engineering. Mining Department Magazine, University of Nottingham, **35,** 59-74.

Shadbolt, C. H., Whittaker, B. N. and Forrester, D. J. (1973) Recent developments in mining subsidence engineering. Paper presented to R.I.C.S. Midland Counties Minerals Division, Basford, Nottingham, UK, 19 October, 15pp.

Sinclair, J. (1950-51) Mining subsidence in the South Yorkshire Coalfield. Trans. I. Min. E., **110,** 365pp.

Singh, B. (1983) Central Mining Research Station, Dhanbad, India. Personal communication.

Smith, G. N. (1971) *An introduction to matrix and finite element methods in civil engineering*. Applied Science Publishers, London, 222p.

Sokolnikoff, I. S. and Redheffer, R. M. (1966) *Mathematics of physics and modern engineering*. McGraw-Hill Book Company, New York. 752p.

Sonntag, G. (1958) Einfluß der Anisotropie auf die Beanspruchung des Gebirges in der Umgebung von Stollen. Bauingenieur, **33**, 287 and 344.

Sonntag, G. (1960) Die Beanspruchung des Hangenden in Abbaubetrieben mit Vollversatz auf Steinkohlengruben. Gluckauf, **96**, 434pp.

Sovinc, I., Hass, N. and Ribicic, M. (1982) Prediction and evaluation of subsidence above a thick coal seam. Proc. Int. Symp. Numerical Models in Geomechanics, Zurich, September, 814-819.

Stephens, J. C., Allen, L. H. and Chen, E. (1984) Organic soil subsidence. In Holzer, T. L. Ed. *Man-induced land subsidence*. Rev. Eng. Geology VI, The Geological Soc. of America, 107-122.

Stilwell, W. B., Hall, W. K. and Tawhai, J. (1975) Ground movements in New Zealand geothermal fields. In Proc. Second U.N. Symp. on Development and Use of Geothermal Resources, San Francisco. Washington, D.C., US Government Printing Office, Vol. 2, 1427-1434.

Stojanovic, A. and Pataric, M. (1979) Uber die stochastische Theorie der Gebirgsbewegung als Abbaufolge., Int. Symp. Mine Surveying (I.S.M.), Aachen, **4**, 169-180.

Stout, K. S. (1980) *Mining methods and equipment*. McGraw Hill Inc., New York, 218pp.

Sutherland, H. J., Schuler, K. W. and Benzley, S. E. (1983) Numerical and physical simulations of strata movement above idealised mine structures. In Situ, **7(1)**, 87-113.

Terzaghi, R. D. (1969) Brine field subsidence at Windsor, Ontario. Proceedings of Third Symposium on Salt, Cleveland 1969. Edits. J. L. Rau and L. F. Dellwig; Vol. 2, 298-307.

Thomas, T. M. (1952) Contribution to Discussion on North (1952). South Wales Institute of Engineers, **67**, 154-155.

Trent, B. C. and Langland, R. T. (1983) Subsidence modelling for underground coal gasification. In Situ, **7(1)**, 53-85.

Trevits, M. A. and King, R. L. (1986) Overview of the Bureau of Mines subsidence research program. 88th Annual General Meeting of C.I.M., Montreal, May, 14pp.

Trompeter, W. H. (1899) *Die Expansivkraft im Gestein als Hauptursache der Bewegung des den Bergbau umgebenden Gebirges*. Baedeker, Essen.

Tute, J. S. (1868) On certain natural pits in the neighbourhood of Ripon. Geol. Mag. **5**, 178-9.

Tute, J. S. (1870) On certain natural pits in the neighbourhood of Ripon. Proc. Yorks. Geol. Polytech. Soc. **5**, 2-7.

van der Knapp, W. and van der Vlis, A. C. (1967) On the cause of subsidence in oil-producing areas. 7th World Petroleum Congress, Mexico City, **3**, 85-105.

van der Merwe, J. N. (1986*a*) Design methods to arrive at the optimal placing and mining of inter panel pillars to alleviate their effects on surface. SANGORM 1986 Symposium, 133-144.

van der Merwe, J. N. (1986*b*) Analysis of surface subsidence over a longwall panel at 50m below surface. SANGORM 1986 Symposium, 145-150.

Voight, B. and Pariseau, W. (1970) State of predictive art in subsidence engineering. J. Soil Mech. and Foundation Div., Proc. A.S.C.E., **96**, March, 721-750.

Von Dechen, H. (1869) Opinion on the surface subsidence in and about the city of Essen. German Manuscript.

Wade, L. V. and Conroy, P. J. (1980) Rock mechanics study of a longwall panel. Mining Engineering, **32** (12), December, 1728-1734.

Wade, L. V. and Conroy, P. J. (1977) Rock mechanics study of a longwall panel. Soc. Min. Engrs. Fall Meeting, St. Louis, Mo, Pre-print No. 77-I-391.

Walters, R. F. (1977) *Land subsidence in Central Kansas related to salt dissolution.* Kansas Geological Survey Bulletin No. 214, 82pp.

Waltham, T. (1978) *Catastrophe — the violent earth.* Macmillan, London, 170pp.

Walton, G. and Cobb, A. E. (1984) Mining subsidence in ground movements and their effects on structures. Eds. P. B. Attewell and R. K. Taylor, Surrey University Press, Chapman and Hall, 216-241.

Wardell, K. (1950) A comparison between British and German experiences of mining subsidence. Trans. Inst. Min. Surv., **30**, 51pp.

Wardell, K. (1952) The surveying observations required for the determination of ground movements caused by mining. Trans. Inst, Min. Surv. **32**, p12.

Wardell, K. (1953-54) Some observations on the relationship between time and mining subsidence. Trans. I. Min. E., **113**, p471.

Wardell, K. (1954) Mining subsidence. Trans. R.I.C.S., **86**, p53.

Wardell, K. (1957) The minimisation of surface damage by special arrangement of underground workings. Proc. European Congress on Ground Movement, 13-19.

Wardell, K. (1966) Problems posed by past and future mineral workings in a development area. The Chartered Surveyor, Mineral workings, October, 184-188.

Wardell, K. (1969) Ground subsidence and control. Mining Congress Journal, **55(1)**, 36-42.

Wardell, K. and Webster, N. E. (1957) Some surface observations and their relationship to movement underground. Proc. European Congress on Ground Movement, 141-148.

Wardell, K. and Wood, J. C. (1966) Ground instability problems arising from the presence of old, shallow mine workings. Proc. Midlands Soil Mech. and Foundation Eng. Soc., **7,** Paper No. 36, 5-30.

Webster, N. E. (1951-52) Strata control and the influence on underground and surface damage. Trans. I. Min. E., **111,** 445-462.

Weissner, J. (1932) Gebirgsbewegungen beim Abbau flachgelagerter Steinkohlenfloze. Gluckauf, **68,** 945pp.

Wheeler, R. E. M. and Wheeler, T. V. (1932) Excavation of the Prehistoric, Roman and Post-Roman site in Lydney Park, Gloucester. Report of the Research Committee of the Society of Antiquaries of London, 68pp.

Whetton, J. T. (1957) A general survey of the ground movement problem. Proc. European Congress on Ground Movement, Leeds University, 9-12.

Whitelock, G. C. H. (1920) Subsidence due to coal mining. Trans. Inst. Mine Surveyors.

Whittaker, B. N. (1974) An appraisal of strata control practice. The Mining Engineer, **166,** 9-24.

Whittaker, B. N. and Aston, T. R. C. (1985) Undersea longwall mining subsidence with special reference to geological and water occurrence criteria in the North-East of England Coalfield. Mining Science and Technology, **2,** 105-130.

Whittaker, B. N. and Breeds, C. D. (1977) The influence of surface geology on the character of mining subsidence. *The geotechnics of structurally complex formations.* Assoc. Geotechnica Ital., Milan, v.**1,** 459-468.

Whittaker, B. N. and Pasamehmetoglu, A. G. (1981) Ground tilt in relation to subsidence in longwall mining. Int. Journal of Rock Mech. and Min. Sciences and Geomech. Abstr. **18,** 321-329.

Whittaker, B. N. and Pye, J. H. (1977) Design and layout aspects of longwall methods of coal mining. Proc. US Sixteenth Symp. Rock Mechanics, Minneapolis, Eds. C. Fairhurst and S. L. Crouch, Pub. American Soc. Civil Engineers, 303-314.

Whittaker, B. N. and Pye, J. H. (1977) Ground movements associated with the near surface construction operations of a mine drift in Coal Measures strata. Int. J. Rock Mech. Min. Sci. and Geomech. Abstr., **14,** 67-75.

Whittaker, B. N. and Reddish, D. J. (1984) Mining subsidence in longwall mining with special reference to the prediction of surface strains. In: A. B. Szwilski and C. O. Brawner (Eds.), Proc. 2nd Int. Conf. on Stability in Underground Mining. Publ. Soc. Min. Eng., Am. Inst. Min. Metall. Pet. Eng., New York, 576-588.

Whittaker, B. N. and Singh, R. N. (1979) Design and stability of pillars in longwall mining. The Mining Engineer, July, **214,** 59-70.

Whittaker, B. N. and Smith, S. F. (1987) Stability and operational aspects of room and pillar mining in UK sedimentary ironore deposits. *Underground Mining Methods and Technology,* Ed. A. B. Szwilski and M. J. Richards, Elsevier, 393-402.

Whittaker, B. N., Reddish, D. J. and Fitzpatrick, D. J. (1985) Calculation by computer program of mining subsidence ground strain patterns due to multiple longwall extractions. Mining Science and Technology, **3,** 21-33.

Whittaker, B. N., Reddish, D. J. and Fitzpatrick, D. J. (1985) Ground fractures due to longwall mining subsidence. Proc. I.M.W.A. Second Int. Congress on Mine Water, Granada, Spain, 1057-1072.

Whittaker, B. N., Singh, R. N. and Zadeh, A. M. H-. (1980) Underground excavation structural stability with special reference to construction work within shaft pillars. Int. Symp. *Safety of underground works.* Assoc. Belge des tech. et de l'urbanisme souterrains, Brussels, May, 19p.

Wiborg, R. and Jewhurst, J. (1986) Ekofisk subsidence detailed and solutions assessed. Oil and Gas Journal (Technology), Feb. 17, 47-51.

Williams, E. L. (1897) Proceedings of the Manchester Geological Society; referred to in Calvert (1915), 330pp.

Winstanley, A. (1938) Strata movements. Trans. Inst. Min. Surveyors, **19,** Pt. 1.

Young, L. E. and Stoek, H. H. (1916) *Subsidence resulting from mining.* University of Illinois Engineering Experiment Station, Bulletin No. 91, 204pp.

Zenc, M. (1969) Comparison of Bals' and Knothe's methods of calculating surface movements due to underground mining. Int. J. Rock Mech. Min. Sci., **6,** 159-190.

Zienkiewicz, O. C. (1977) *The finite element method.* McGraw-Hill Book Company, New York. First edition 1967; Second revised edition 1971; Third revised edition 1977. 787p.

Zilavy, B. (1968) *Vplyvy Poddolovania.* Nakladatelstvo Bratislava, 267pp.

BIBLIOGRAPHY

Brook, D. and Cole, K. W. (1984) Subsidence of abandoned limestone mines of West Midlands of England. Proc. 3rd Symp. Land Subsidence, Venice.

Gaskell, P., Reddish, D. J. and Whittaker, B. N. (1988) Subsurface ground movements associated with longwall mining. Proc. 7th Int. Conf. *Ground Control in Mining,* S. S. Peng (Ed.), Morgantown, WV, 195-204.

Kapp, W. A. (1986) Mine subsidence in New South Wales: its effects on surface features and structures. SANGORM Symposium, 18p.

King, H. J. and Whittaker, B. N. (1971) A review of current knowledge on roadway behaviour, especially the problems on which further information is required. Proc. Symp. *Roadway Strata Control;* I. Min. E., London, 73-87.

O'Rourke, T. D. and Turner, S. M. (1979) *Longwall subsidence patterns: a review of observed movements, controlling parameters, and empirical relationships.* Cornell University, Ithaca, New York: Report to USBM, Minneapolis, Geotechnical Engineering Report 79-6, 82p.

Ren, G., Reddish, D. J. and Whittaker, B. N. (1988) Mining subsidence and displacement prediction using influence function methods for steep seams. Mining Science and Technology.

Shadbolt, C. H. (1975) Mining subsidence and protective measures for surface structures. Chartered Surveyor: Land, Hydrographic and Minerals Quarterly, **3** (2), 29-32.

Shadbolt, C. H. (1977) Mining subsidence — historical review and state-of-the-art. Proc. Conf. *Large Ground Movements.* UWIST, Cardiff. Ed. J. D. Geddes, Pentech Press, London: Plymouth, 705-748.

Shadbolt, C. H. (1987) *A study of the effects of geology on mining subsidence in the East Pennine Coalfield.* Ph.D. Thesis, University of Nottingham, UK.

Siriwardane, H. J. and Amanat, J. (1984) Prediction of subsidence in hilly ground terrain using finite element method. In: A. B. Szwilski and C. O. Brawner (Eds) 2nd Int. Conf. on *Stability in Underground Mining,* SME and AIMMPE, New York, 554-575.

Trent, B. A., Bauer, R. A. and Du Montelle, P. B. (1988) *Bibliography of subsidence-related literature.* Illinois Mine Research Program; Illinois State Geological Survey, Natural Resources Building, Champaign, Illinois, 190p.

Wagner, H. and Schumann, E. H. R. (1985) The effects of total coal seam extraction on the surface and surface structures. Chamber of Mines of South Africa, Research Report 20/85, 1985; also Colloquim *Recent mining and metallurgical developments in the Eastern Transvaal,* S.A. Inst. Min. Met., Witbank.

SUBJECT INDEX

Subject Index

PLACE INDEX

AUTHOR INDEX